MATHEMATICAL METHODS *for* DIGITAL COMPUTERS

Edited by

ANTHONY RALSTON, Ph.D.
Bell Telephone Laboratories

HERBERT S. WILF, Ph.D.
Assistant Professor of Mathematics
The University of Illinois

NEW YORK · LONDON, John Wiley & Sons, Inc.

MATHEMATICAL METHODS

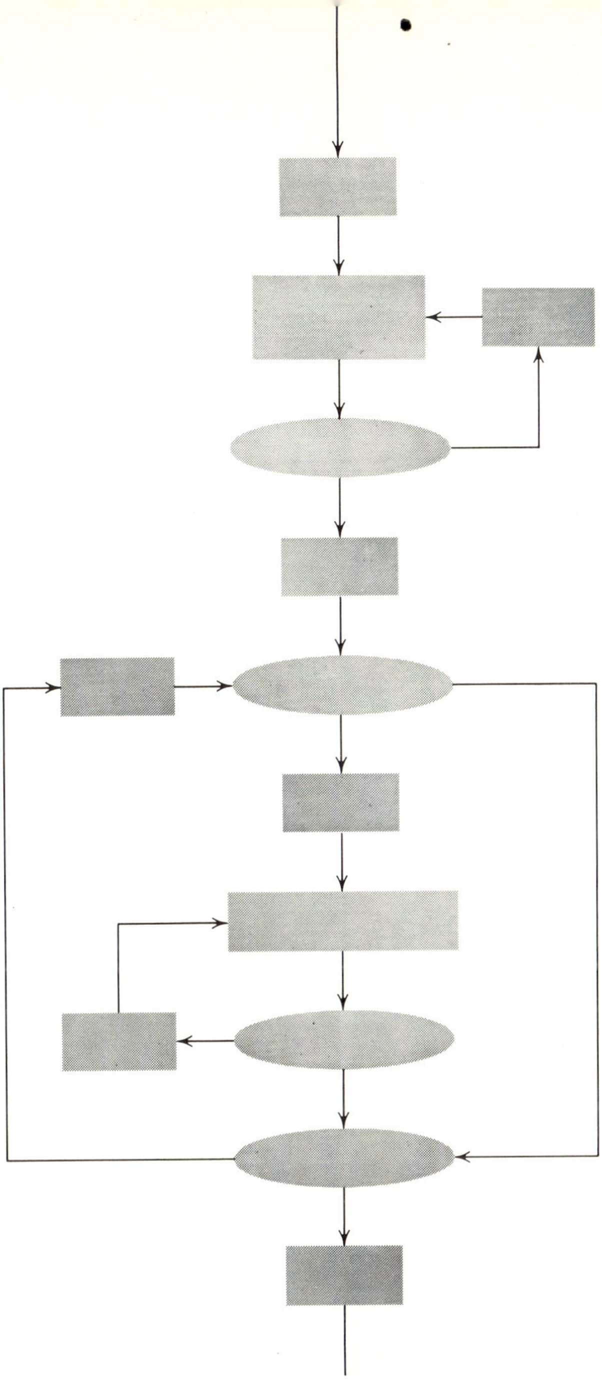

for

DIGITAL COMPUTERS

LIBRARY OF CONGRESS CATALOG CARD NUMBER: 60-6509
PRINTED IN THE UNITED STATES OF AMERICA

to Jayne and Ruth

PREFACE

Today an ever-increasing number of workers active in applied mathematics, the physical sciences, and engineering are finding use for a large-scale digital computer in the solution of problems. Many of these workers as a part of their professional competence have acquired an understanding of the mathematics underlying the solution of problems on a digital computer, whereas others find the mathematics involved to be a new or relatively unexplored area. For the first group this book will satisfy the rapidly growing need for a reference text for many of the more commonly used mathematical methods for digital computers. The second group will find this book helpful both as a reference work and as an introduction to the field of numerical analysis, although the book is in no sense a text on numerical analysis.

With the exception of the first chapter, which treats the methods by which elementary functions can be generated on digital computers, the chapters follow a standard format. This consists of a mathematical discussion of the method under consideration followed by a sequence of topics (calculation procedure, flow chart, sample problem, etc.) of particular interest to users of digital computers. Moreover, each chapter has been written by someone in close contact with the latest developments in his respective area. Taken together, the twenty-six chapters include not only chapters on many of the more commonly used mathematical methods for digital computers but also a few chapters on some of the promising newly developed methods and a few which illustrate the novel tactics (e.g., Monte Carlo techniques) that may be employed at times to solve problems on computers.

Although a basic mathematical maturity is required of the reader in order that he get full value out of each chapter, we believe that no more mathematical background is required than that which is generally possessed by people wishing to do work in the area of numerical analysis. However, even those who are lacking the mathematical competence to appreciate fully each chapter may still profit from the use of this book as a compendium of calculation procedures, computer flow charts, and references. To understand those parts of each chapter oriented toward computers does not require personal computing experience on the part of the reader but only a general understanding of fundamental computer concepts.

The second named editor wishes to express his gratitude to his former employers, the Nuclear Development Corporation of America, White Plains, New York, for their assistance and encouragement in the preparation of this book.

ANTHONY RALSTON
HERBERT S. WILF

September 21, 1959

CONTENTS

INTRODUCTION 1

PART I GENERATION OF ELEMENTARY FUNCTIONS 5

1. Generation of Elementary Functions 7
E. G. Kogbetliantz, Rockefeller Institute for Medical Research
International Business Machines Corporation

PART II MATRICES AND LINEAR EQUATIONS 37

2. Matrix Inversion and Related Topics by Direct Methods 39
Alex Orden, University of Chicago

3. The Solution of Linear Equations by the Gauss-Seidel Method 56
R. Van Norton, Institute of Mathematical Sciences, New York University

4. The Solution of Linear Equations by the Conjugate Gradient Method 62
F. S. Beckman, International Business Machines Corporation

5. Matrix Inversion by the Method of Rank Annihilation 73
Herbert S. Wilf, The University of Illinois

6. Matrix Inversion by Monte Carlo Methods 78
Florence Jeanne Oswald, Nuclear Development Corporation of America

7. The Determination of the Characteristic Roots of a Matrix by the Jacobi Method 84
John Greenstadt, International Business Machines Corporation

PART III ORDINARY DIFFERENTIAL EQUATIONS **93**

8. Numerical Integration Methods for the Solution of Ordinary Differential Equations 95
Anthony Ralston, Bell Telephone Laboratories

9. Runge-Kutta Methods for the Solution of Ordinary Differential Equations 110
Michael J. Romanelli, Ballistic Research Laboratories

10. The Numerical Solution of Boundary Value Problems 121
Eugene L. Wachspress, Knolls Atomic Power Laboratory

11. The Solution of Ordinary Differential Equations With Large Time Constants 128
J. Certaine, Nuclear Development Corporation of America

PART IV PARTIAL DIFFERENTIAL EQUATIONS **133**

12. The Numerical Solution of Parabolic Partial Differential Equations 135
Herbert B. Keller, Institute of Mathematical Sciences, New York University

13. Iterative Methods for the Solution of Elliptic Partial Differential Equations 144
J. W. Sheldon, Computer Usage Company, Inc.

14. A Monte Carlo Method for the Solution of Elliptic Partial Differential Equations 157
Carl N. Klahr, Technical Research Group

15. The Numerical Solution of Hyperbolic Partial Differential Equations By the Method of Characteristics 165
Mary Lister, The Pennsylvania State University

16. The Solution of Hyperbolic Partial Differential Equations by Difference Methods 180
P. Fox, Massachusetts Institute of Technology

PART V STATISTICS **189**

17. Multiple Regression Analysis 191
M. A. Efroymson, Esso Research and Engineering Company

18. Factor Analysis 204
Harry H. Harman, System Development Corporation

19. Autocorrelation and Spectral Analysis 213
Raymond W. Southworth, Yale University

20. Analysis of Variance 221
H. O. Hartley, Iowa State College

PART VI MISCELLANEOUS METHODS **231**

21. The Numerical Solution of Polynomial Equations 233
Herbert S. Wilf, The University of Illinois

22. Methods for Numerical Quadrature 242
Anthony Ralston, Bell Telephone Laboratories

23. Multiple Quadrature by Monte Carlo Methods 249
Herman Kahn, The RAND Corporation

24. Fourier Analysis 258
G. Goertzel, Nuclear Development Corporation of America

25. The Solution of Linear Programming Problems 263
Dean N. Arden, Massachusetts Institute of Technology

26. Network Analysis 280
T. R. Bashkow, Columbia University

Index 291

Introduction

I. PURPOSE

The recent development of automatic digital computers has already had drastic repercussions on numerical analysis. In the first place, much new methodology has been evolved to solve problems whose solution has been made feasible only by the speed of modern computing equipment. Secondly, the classical methods of numerical analysis have been subjected to a rigid scrutiny. Because of this, flaws in many of these classical methods have been discovered which were not apparent when they were applied to problems of lesser computational magnitude. The major purpose of this book is to present many—but by no means all—of the more commonly used tools of the modern numerical analyst along with some of the more promising newly developed methods. The motivation behind this presentation is not only to gather together in one place a partial survey of modern numerical methods but also, in each case, to acquaint the reader with the interplay between computer capabilities and the processes of analysis. The aim of each chapter is to describe the process which takes a problem from its formulation up to, but not including, the actual coding for a digital computer. In this way it is hoped that problem originators will be better able to understand the mysterious process by which a mathematics group in a modern industrial or university computing laboratory prepares its problems for coding. It should be emphasized that nowhere in this book is any particular computer hardware assumed.

Of the twenty-six chapters which follow, not all fall into the categories of "more commonly used" or "more promising" of modern numerical methods. A few have been chosen because they are apt illustrations of the novelty of methods which can be and sometimes are used on digital computers. Into this category fall the particular applications of Monte Carlo methods which are presented here. These stochastic methods have found their greatest utility in fields which are too specialized (e.g., neutron transport, queuing, etc.) to be discussed in this volume. In one case—Chapter 26—the chapter deals not so much with numerical methods as such, but rather with how a large class of practical problems, those of network analysis, can be reduced by numerical analysis to a form convenient for coding. In another case—Chapter 11—a quite specialized method applicable to a fairly limited class of problems is presented here to illustrate the usefulness of special methods in place of the application of more general methods in particular cases.

Most of the chapters, however, do deal with those methods which are today the standard tools of the numerical analyst or bid fair to become so. In order to keep the book down to a reasonable size, the editors were constrained to reject as chapters some topics which others might have included. Since the purpose of this book is quite definitely not to introduce the subject of numerical analysis, many areas of the computational art have been omitted. For example, much less space than is necessary in a text on numerical analysis has been devoted to the interrelationships and common formulations of many methods which are presented here as independent concepts. There is, however, every expectation that one who is thoroughly grounded in numerical analysis will find some fresh points of view here.

In looking for the best method to solve any given problem, the reader is warned that he will not necessarily find what he wants. The variety of problems encountered under each category discussed in this book makes it generally impossible to recommend a "best" method for a given class of problems (e.g., matrix inversion). Nevertheless, the contributors have generally been careful to point out the types of problems in which the method under discussion has its major applicability and those cases, if any, where it is probably best to avoid the method. A careful reading of the chapters dealing with a particular class of problems will, we hope, give some of the insight needed in the quest for "best" methods.

We believe that a careful study of the chapters in this book will profit the reader beyond what he learns about the particular methods discussed. For example, the reader is invited to study the flow charts in the chapters that follow, to trace the motivations for some seemingly outlandish steps, and to find the flaws which must surely still exist in many of them even though they have been, for the most part, extensively tested. By doing so he should learn a great deal about numerical analysis and the preparation of problems for a digital computer.

2. FORMAT OF THE CHAPTERS

Because of the diversity of material contained herein it has been neither possible nor desirable to maintain an absolutely rigid format for each chapter. In particular, because of the nature of its subject matter, Chapter 1 has a unique format. However, with few exceptions the succeeding chapters contain the following information:

Function

What is the function of the method which is about to be described? The contributor has endeavored to give a concise and accurate formulation of the particular problem which is considered in his chapter.

Mathematical Discussion

Insofar as the scope of the chapter extends, the contributor has tried to give a complete mathematical description of the problem which is to be solved and the method or methods by which he proposes to solve it. Here one may find relevant mathematical theorems with proofs or references to proofs, an error analysis if applicable and available including a discussion of the circumstances where the proposed technique may be expected to perform well or poorly, comparisons with other available techniques, and citations of the relevant literature.

Summary of the Calculation Procedure

All too often when a mathematical discussion is interlaced with the derivation of a method the reader may finish quite unaware of what precise sequence of steps is required to carry out the solution of the problem. In this section, therefore, the reader will find the method previously derived stated in "recipe" form; i.e., first do this, then this, etc.

Flow Chart

A flow chart is a computer code written in a universal language instead of the language of a particular computer. A flow chart can be either macroscopic (gross), microscopic (detailed), or somewhere in between. A coder, however, can work only from a microscopic flow chart. The flow charts contained herein are as microscopic as the generality of the material covered allows. In order to keep the discussion as general as possible, the flow charts for this book have been drawn for a hypothetical computer with an infinite rapid access memory, no input mechanisms, and an output facility which, when instructed to print the desired answer, prints it. In this way the logic has been separated as much as possible from the requirements of specific hardware. The present trend in hardware is toward precisely such a machine as described above.

The boxes on the flow chart have been restricted to what was considered a bare minimum of different types. All flow charts begin with a box labeled START and end with a box labeled STOP. The other types of boxes used are:

(*a*) The assertion box

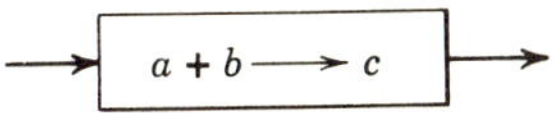

This box asserts that the operations contained within it are executed at this time

(*b*) The test box

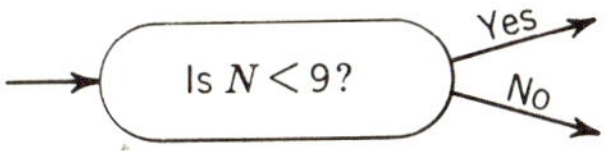

This box has one input line and two output lines. Either output is selected, depending on whether the answer to the question contained in the box is affirmative or negative.

(*c*) The remote connector 

This circle indicates that logical control is transferred to the point where another such circle appears with the same number in it. In the case of a variable remote connector

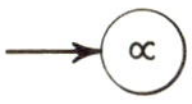

the exit point is set by the program itself, as indicated in the flow chart, and therefore a separate entry circle will be found elsewhere on the chart for each possible setting of the variable connection.

Description of the Flow Chart

A box by box description of the flow chart is given in order to aid the reader in following it.

Subroutines

Any standard subroutines required by the program are listed.

Sample Problem

A representative sample problem is worked through step by step in order to give the reader a feeling for the behavior of the process in practice.

Memory Requirements

The number of memory locations that are required for the program itself and working storage, data storage, etc., is given in terms of the parameters of the problem. It is assumed that a computer word holds one piece of numerical data which may be either an instruction or a calculated quantity.

Estimation of the Running Time

The approximate running time of a problem is given in terms of the parameters of the problem and the addition and multiplication times of whatever computer is used.

References

The references cited in the text (plus some others of particular interest) are listed. References are indicated in text by numbers enclosed in square brackets.

PART I GENERATION OF ELEMENTARY FUNCTIONS

Generation of elementary functions

I

E. G. Kogbetliantz
Rockefeller Institute for Medical Research
International Business Machines Corporation

INTRODUCTION

Electronic computing equipment is very often used by a physicist, a chemist, or an engineer to explore by mathematical methods all the possible results of a new chemical or physical situation, the outcome of which depends on many variable parameters α_k, $1 \leq k \leq n$.

Suppose—to fix our ideas—that the final result, which is naturally a number, is expressed by a definite integral. The integrand of this integral will involve some elementary functions of the variable of integration which moreover will depend on the values of the parameters α_k, considered as fixed constants during the integration.

To characterize all possible physical situations, the integral is to be computed many times for many combinations of various values of the parameters α_k. If their ranges are large the average number m of values of a parameter to be used may also be large. In all, the integral will be computed m^n times. During one such numerical integration the interval of integration (a, b) is subdivided into $p = (b - a)/h$ subintervals, where h is the length of a subinterval. Therefore, each elementary function involved in the integrand is to be computed p times during one integration. In all, the number of times an elementary function will be computed is expressed by a product $m^n p$ and very often its value is of the order of 10^5. Moreover, there are problems which necessitate the computation of many such integrals.

In other words, electronic computers are very often used to tabulate functions of many variables and, to achieve this result, the computations of some elementary functions may be repeated many thousands of times. Electronic computers are so rapid that a computation of an elementary function is performed in some milliseconds (the exact duration of computation depends on the accuracy required, on the method of computation used, on the particular elementary function considered, and on the type of computer). But such a computation may be repeated during a year tens or maybe hundreds of millions of times and it becomes clear that an economy of only 1 millisecond in the generation of an elementary function, achieved by using a more economical and faster subroutine, can save during one year 20–30 hours of machine time which is equivalent to \$10,000 to \$20,000.

In many cases an economy of 2–3 milliseconds is possible.

The aim of this chapter is to formulate the most economical mathematical procedures for the generation of elementary functions. To economize machine time it is necessary to reduce as much as possible the number of multiplications and divisions without increasing too much the number of instructions, i.e., without complicating the logical part of the subroutine. The additions and subtractions, as well as shifts, are so fast that their number may be disregarded.

In constructing a more economical mathematical procedure it is important to make it flexible: it must be sufficiently general to allow approximations of any prescribed accuracy. In other words, for one and the same elementary function such a procedure should define a family of subroutines of increasing accuracy, based on the same general method of computation, but differing in the number M of multiplications and/or divisions as well as in the number PC of precomputed and stored constants. These two important parameters M and PC depend on the kind of approximation (polynomial or rational), on the range of the argument N in which the approximation is applied, and on the accuracy prescribed, i.e., on the number Dg of correct digits expected in the result in the whole range of the argument.

Here we have to distinguish between fixed point and floating point computations. For a floating point subroutine the accuracy is defined by the number Dg of first correct significant digits, so that the *relative* error (i.e., the ratio of the absolute error to the computed value) is involved. But in a fixed point computation in most cases it is the *absolute* error which is considered as the measure of accuracy.

Therefore, in a subroutine programmed for a fixed point computation we have to consider and try to minimize the upper bound of the absolute error. Thus, if this bound does not exceed $5 \times 10^{-(n+1)}$, we say that the subroutine yields $Dg = n$ correct digits after the decimal dot even in the case when the resulting value of an elementary function has a string of m zeros preceding $n - m$ significant digits.

Naturally, in a subroutine programmed for a floating point computation it is the upper bound of the relative error which characterizes the accuracy of the subroutine. Thus, if this bound is at most equal to $5 \times 10^{-(n+1)}$, we say that this subroutine yields $Dg = n$ first correct *significant* digits.

It is also important to point out that for some electronic computers with a very rapid operation of multiplication, the division is so slow an operation that subroutines for them should be based on polynomial approximations only, eliminating completely the use of division.

Some years ago only, almost all subroutines for elementary functions were based on polynomial approximations, but now the superiority of rational approximations seems to be a generally recognized fact. In this chapter we will study both kinds of approximations because of the importance of polynomial approximations for computers with a slow division.

For the same accuracy (value of Dg) subroutines with a different number PC of precomputed and stored constants can be constructed.

In general, increasing PC enables one to use less operations, i.e., to decrease M. This is related to the *range* of the argument N in which the bound of the error does not exceed the value $5 \times 10^{-(n+1)}$ prescribed by the desired accuracy $Dg = n$. The range $0 \leqq N \leqq N_0$ of N is another important parameter since the error of an approximation depends on the range and in general increases very rapidly with N_0. This is why in practical applications any approximation is used in a reduced range $(0, N_0)$ which is only a fraction of the general range of the argument.

To take an example, a subroutine for the exponential function e^N should allow the computation of e^N for any value of the exponent in the infinite range $-\infty < N < +\infty$. No formula can be constructed which could approximate e^N in the whole range $(-\infty, +\infty)$ with a prescribed accuracy. Therefore, the first step to take in devising a subroutine for e^N consists in the reduction of the infinite range $0 < |N| < \infty$ to a finite range $0 < |N| < N_0$ and the same holds for any other elementary function.

In general, the error of any approximation

may be made as small as is necessary to achieve the prescribed accuracy by choosing a sufficiently small reduced range for the argument of the function. Therefore, M can be made small if the reduced range is small. On the other hand, reduction to a smaller range necessitates more stored constants as well as more instructions, which means loss of machine time and memory space.

The existence of two different types of electronic computers—binary and decimal—entails the necessity of adapting a subroutine to the type of computer. Therefore, for each elementary function we need subroutines of two kinds: binary subroutine and decimal subroutine. The reduction of the argument's range, in particular, is very different in a binary and in a decimal subroutine.

Finally, one more point is worth mentioning. No precomputed and stored table of values of a function could be useful in a subroutine for the computation of this function. The tabular values are not sufficient, and to compute the function's value for any nontabular value of its argument an interpolation is necessary. But an interpolation is a much more time- and space-consuming operation than a direct generation with the aid of an approximate expression. Nevertheless, for some functions, for instance for the logarithm, a very short table of values, in combination with an approximate expression applied in a very small reduced range, can be used to attain a prescribed accuracy with the aid of a relatively small number M of operations.

In this chapter we describe first the subroutines, binary and decimal, of pure type in which no tables of the function's values are used. The subroutines of mixed type will be mentioned only incidentally.

This discussion shows that the construction of an economical and short subroutine is not an easy task. It necessitates a long and thorough study of the conflicting effects of many factors and their delicate balancing. The generation of elementary functions in digital computers is a new and rapidly progressing art. No claims to final results can be made and the subroutines presented in this chapter are not to be considered as the most economical or the best ones. It is hoped that in applying the methods discussed and described here the reader will be able to improve them.

This is why in discussing the subroutines of elementary functions our attention will be directed more towards methods and ideas than to the practical results and the numerical computation of the constants to store. In our opinion, principles are more important, and when methods and general formulas of approximation are well understood the practical construction of a subroutine is a relatively simple and easy task.

PART A

MATHEMATICAL TOOLS

The subroutines of elementary functions reflect in their structure the special properties of each particular function. They are dissimilar and this necessitates a separate study for each one of the elementary functions. Nevertheless, the fundamental mathematical tools used in building the approximations are the same for all functions. Therefore it seems worthwhile to describe first the various mathematical expressions used as approximations before applying them in Part B to the individual elementary functions.

1. Maclaurin's Power Series

Polynomial approximations were the first to be used and, at the beginning of the use of electronic computers, almost all subroutines were based on the partial sums $s_m(x)$ of Maclaurin's power series

$$f(x) = \sum_{n=0}^{\infty} f^{(n)}(0) \cdot x^n/n! \qquad (1)$$

the order m of which was chosen sufficiently large to insure the required accuracy.

To take an example, let us consider $\sin x$ in the interval $(0, \pi/4)$. The absolute error made in approximating it by the partial sum $s_m(x)$ of its expansion (1) is less in absolute value than the first neglected term* and

* In what follows, when an exact estimate of the upper bound of the error cannot be found, then an approximate estimate of the upper bound is considered. After the approximation to the function has been found, the upper bound of the error must then be checked.

therefore, for $|x| \leq \pi/4$, is less than $(\pi/4)^{2m+1}/(2m+1)!$. To insure that there are Dg correct digits after the dot, the order m of $s_m(x)$ must justify the inequality

$$(2m + 1) \log (4/\pi) + \log (2m + 1)! > Dg + 0.30103$$

and this inequality for $Dg = 10$ gives $m > 5$.

A subroutine for fixed point computation of $\sin x$ in the range $|x| \leq \pi/4$ based on the approximation

$$\sin x \approx s_5(x) = x - x^3/3! + x^5/5! - x^7/7! + x^9/9! - x^{11}/11! \quad (1a)$$

is a very short and simple one, but it is too slow. To obtain $Dg = 10$, as many as seven multiplications are needed: one for computing $z = x^2$, five for evaluating the polynomial of fifth degree in z, and one more for the multiplication by x. In the range $\pi/4 \leq N \leq \pi/2$, $\sin N$ is computed as $\cos (\pi/2 - N) = \cos t$, where $0 \leq t \leq \pi/4$. Using $s_m(t)$ as an approximation to $\cos t$ with

$$2m \cdot \log (4/\pi) + \log (2m)! > 10.30103$$

gives $m > 6$. We see that to compute

$$s_6(t) = 1 - t^2/2! + t^4/4! - t^6/6! + t^8/8! - t^{10}/10! + t^{12}/12! \quad (1b)$$

seven multiplications are again necessary. Thus, the subroutines for $\sin x$ and $\cos x$, yielding $Dg = 10$ for $|x| \leq \pi/4$, allow the computation of these functions in $0 \leq |x| \leq \pi/2$ in seven multiplications, if Maclaurin's power series is used, the number of precomputed and stored constants being ten for a binary and eleven for a decimal computer.

2. Chebyshev Expansion

Another source of polynomial approximations is the Fourier cosine series. It is used in the form of the Chebyshev expansion of $f(z)$ in the interval $-a \leq z = ax \leq a$

$$f(ax) = \tfrac{1}{2}c_0(a) + \sum_{n=1}^{\infty} c_n(a) \cdot T_n(x) \qquad (-1 \leq x \leq 1) \quad (2)$$

Here $x = \cos \theta$, and $T_n(x) \equiv \cos n\theta$ is a polynomial of nth degree in x orthogonal in $(-1, +1)$, the weight function being $(1 - x^2)^{-1/2}$. The coefficients $c_n(a)$ of expansion (2) are functions of the parameter a:

$$\pi c_n(a) = 2 \cdot \int_{-1}^{1} f(ax) \cdot T_n(x) \cdot (1 - x^2)^{-1/2} \cdot dx$$

The partial sums $\sigma_n(a, x)$ of (2) are much better for our purpose than the partial sums $s_m(x)$ of the power series (1). The reason is very simple. Since the coefficients of (1) depend on the behavior of the expanded function in the *immediate* neighborhood of the origin $x = 0$ only, the series (1) converges rapidly and represents the function well for small values of the argument x only. The truncation error $f(x) - s_m(x)$ increases very rapidly when $|x|$ increases. But the coefficients $c_n(a)$ of the Fourier series (2) depend on all the values $f(z)$ takes in the interval $(-a, a)$. In general, they rapidly decrease when n increases, the truncation error $f(ax) - \sigma_n(a, x)$ being fairly well approximated by the first neglected term, i.e., by $c_{n+1}(a) \cdot T_{n+1}(x)$. It is obvious that $|T_{n+1}(x)| \leq 1$ and $T_{n+1}(x) = \cos (n + 1)\theta$ oscillates between $+1$ and -1, when x runs through its range $(-1, 1)$, and vanishes $n + 1$ times in the interior of the interval $(-1, +1)$. Therefore the truncation error, the behavior of which is essentially that of $T_{n+1}(x)$, has $n + 1$ zeros and its extrema are approximately equal in absolute value and alternate in sign.

This behavior of $\sigma_n(a, x)$ is very similar, though not identical, to the behavior of the classical (Chebyshev [1]) best approximating polynomial of degree $\leq n$, $P_n^*(x)$, which realizes the least possible deviation Δ_n from $f(ax)$, this deviation Δ_n being defined by

$$\Delta_n = \max_{-1 \leq x \leq 1} |f(ax) - P_n(x)|$$

The idea of choosing $P_n^*(x)$ to represent $f(ax)$ is natural, but the computation of its coefficients is so difficult and time-consuming that it is worthwhile to inquire whether other approximating polynomials of nth degree more easy to determine can replace $P_n^*(x)$ and yield the same Dg in the same number M of operations. The answer is yes: using $\sigma_n(a, x)$, the same practical accuracy is attained as with $P_n^*(x)$ ([2], p. 9). This is important from the practical point of view since the computation of coefficients $\gamma_m(a)$ in

$$\sigma_n(a, x) = \tfrac{1}{2}C_0(a) + \sum_{m=1}^{n} C_m(a) \cdot T_m(x) = \sum_{m=0}^{n} \gamma_m(a) \cdot x^m$$

is much easier and faster than the evaluation of coefficients of $P_n^*(x)$, the difficulty of computing the latter increasing very rapidly with the number Dg of prescribed correct digits. The expression of $\sigma_n(a, x)$ defines $\gamma_m(a)$ and they are obtained by replacing $T_m(x)$ by their explicit expressions

$$T_m(x) = \tfrac{1}{2} \cdot m \cdot \sum_{s=0}^{2s \leq m} \frac{(-1)^s}{m - s} \binom{m-s}{s} (2x)^{m-2s} \qquad (2^*)$$

To illustrate this point let us consider the best polynomial approximation $P_5^*(x)$ of fifth degree to the function arctan x in the interval $(-1, 1)$:

$$P_5^*(x) = 0.995\,3580x - 0.288\,6902x^3 + 0.079\,3390x^5$$

the maximum of $|\arctan x - P_5^*|$ being equal to 0.000 6086 ([2], p. 18). No polynomial of degree $\leqq 5$ can yield a better approximation, in this sense, to arctan x in $(-1, +1)$. Its coefficients were not easy to compute by the method of successive approximations, and it was possible to do it uniquely because of the small accuracy ($Dg = 2$) of the approximation. For larger values of Dg the amount of computation rapidly becomes prohibitive.

To compare $P_5^*(x)$ with $\sigma_n(1, x)$ we will use the Chebyshev expansion of arctan x, which is a particular case for $\theta = \pi/8$ of the expansion ([3], p. 44, form. II):

$$\arctan (x \tan 2\theta) = 2 \sum_{m=0}^{\infty} \frac{(-1)^m}{2m + 1} \cdot \tan^{2m+1} \theta \cdot T_{2m+1}(x) \qquad (3)$$

In our case $\tan \pi/8 = \sqrt{2} - 1 = \tau$ and the partial sum $\sigma_2(1, x) = \sigma_2(x)$ of the first three terms is

$$\sigma_2(x) = 2 \sum_{0}^{2} (-1)^m \tau^m \cdot T_{2m+1}(x)/(2m + 1)$$

Replacing the polynomials $T_{2m+1}(x)$ by their expressions, we have ([3], p. 51, form. 20)

$$\sigma_2(x) = A_{30} \cdot x - A_{31} \cdot x^3/3 + A_{32} \cdot x^5/5$$

where $A_{30} = 1 - \tau^6$, $A_{31} = (1 + 3\tau^2)(1 - \tau^2)^3$, $A_{32} = (1 - \tau^2)^5$. Thus,

$$\sigma_2(x) = 0.994\,9494x - 0.287\,0605x^3 + 0.078\,0372x^5$$

The derivative $f'(x)$ of the difference $f(x) = P_5^*(x) - \sigma_2(x)$ is positive in (0, 1), so that $|f(x)| \leqq \max |f(x)| = f(1) = 0.000\,0807$. Thus, the upper bound of the truncation error $|\arctan x - \sigma_2(x)|$, which is necessarily larger than 0.000 6086, is equal to 0.000 6086 + 0.000 0807 = 0.000 6893 $\sim$ 0.0007. It is clear that both approximations $P_5^*(x)$ and $\sigma_2(x)$ yield the same number ($Dg = 2$) of correct digits and are therefore equivalent from our point of view.

The approximating polynomial

$$0.995\,354x - 0.288\,679x^3 + 0.079\,331x^5$$

found by Hastings ([4], p. 132, sheet 8) is another example of an almost best polynomial approximation of fifth degree to arctan x in $(-1, +1)$. As $P_5^*(x)$, it was obtained by successive approximations. Its deviation from arctan x does not exceed in $(-1, +1)$ 0.000 6094.

Readers interested in the numerical solutions of the famous Chebyshev problem of best polynomial approximation should consult the fundamental work of Remez [5]. A report by Murnaghan and Wrench [2] gives an excellent presentation of an iterative procedure for the determination of $P_n^*(x)$ by successive approximations to its coefficients.

In our opinion, for the practical purposes considered here, the expansion (2) is sufficient and nothing is gained replacing $\sigma_n(a, x)$ by $P_n^*(x)$.

To show how the use of (2) instead of (1) improves a subroutine, we return to sine and cosine in the range $0 < x \leqq \pi/4$ and discuss the subroutine with $Dg = 10$ (fixed point computation).

The expansions (2) for sine and cosine are (see [6], p. 22, form. (3) and (4)):

$$\sin ax = 2 \sum_{0}^{\infty} (-1)^n J_{2n+1}(a) \cdot T_{2n+1}(x) \qquad (4)$$

$$\cos ax = J_0(a) + 2 \sum_{1}^{\infty} (-1)^n J_{2n}(a) \cdot T_{2n}(x) \qquad (5)$$

where $J_n(t)$ denotes the Bessel function of the first kind:

$$J_n(t) = \sum_{s=0}^{\infty} (-1)^s (t/2)^{n+2s}/s!(s + n)!$$

Now $J_{11}(\pi/4) \approx 0.847 \times 10^{-12}$ and $J_{10}(\pi/4) \approx 0.237 \times 10^{-10}$, which proves that the truncation errors in (4) and (5) are less in absolute value than 5×10^{-11} if their partial sums of first five terms are used as polynomial approximations to $\sin(\pi x/4)$ and $\cos(\pi x/4)$ respectively. Therefore $\sin N$ is approximated by an odd polynomial of ninth degree and $\cos N$ by an even polynomial of eighth degree in the range $0 \leq N \leq \pi/4$ with the first ten digits after the dot correct ($Dg = 10$). The use of (1) necessitated seven multiplications for both functions, but now, using (2), $\cos N$ is computed in five and $\sin N$ in six multiplications. Better subroutines for these functions are described in Part B.

3. Relaxation of Power Series

Instead of using (2), polynomial approximations with the aid of partial sums of (1) can be improved by reducing their degree without diminishing the accuracy of approximation. This relaxation of $s_m(x)$ is based on the approximation of integral powers of x by polynomials of lesser degree and the same parity. The best polynomial approximation to x^{2n} and x^{2n+1} is obtained using the explicit expressions of $T_{2n}(x)$ and $T_{2n+1}(x)$. Thus solving the identity $2^{1-2n} \cdot T_{2n}(x) = x^{2n} - P_{2n-2}(x)$, where

$$P_{2n-2}(x) = (-1)^{n-1} \cdot n \cdot 2^{1-2n} \cdot \sum_{m=0}^{n-1} \frac{(-1)^m}{n+m} \binom{n+m}{n-m} (2x)^{2m}$$

for x^{2n}, we obtain

$$x^{2n} = P_{2n-2}(x) + e_{2n}(x) \tag{6}$$

with $|e_{2n}(x)| = 2^{1-2n} \cdot |T_{2n}(x)| \leqq 2^{1-2n}$ in $(-1, 1)$. Therefore, the error $e_{2n}(x)$ made in replacing x^{2n} by $P_{2n-2}(x)$ is less than 2^{1-2n} in absolute value.

Likewise, the identity $2^{-2n}T_{2n+1}(x) = x^{2n+1} - P_{2n-1}(x)$ yields the approximation

$$x^{2n+1} \approx P_{2n-1}(x) = (-1)^{n-1} \cdot (n + \tfrac{1}{2})2^{-2n} \cdot \sum_{m=0}^{n-1} \frac{(-1)^m}{n+m+1} \binom{n+m+1}{n-m} (2x)^{2m+1} \tag{7}$$

the error being less than 2^{-2n} in absolute value.

These approximations allow us to economize on one or more terms in a polynomial approximation. If the coefficient c_{2n} of the last term $c_{2n}x^{2n}$ in the approximating polynomial of degree $2n$, $Q_{2n}(x)$ is so small that $2^{1-2n}c_{2n}$ can be neglected without affecting the value of Dg, then the relaxation can be applied and the modified approximation $Q_{2n}(x) - c_{2n}x^{2n} + c_{2n} \cdot P_{2n-2}(x) = \bar{Q}_{2n-1}$ is a polynomial of degree $2n - 1$. If the last term is $c_{2n+1} \cdot x^{2n+1}$, then the relaxation is conditioned by the smallness of $2^{-2n} \cdot c_{2n+1}$.

For $\sin x$ and $\cos x$ the relaxation of their approximations (1*a*) and (1*b*) yields polynomials of ninth and tenth degree respectively, thus giving the same result as the use of (2) for $\sin x$, but a lesser improvement for $\cos x$. This example indicates that the relaxation of power series is not equivalent to the use of (2).

Another form of the relaxation consists in the following transformation of $s_n(x) = \sum_{m=0}^{n} a_m x^m$. An integral power of x can be expanded as a Chebyshev series (2):

$$(2x)^{2m} \equiv \binom{2m}{m} + 2\sum_{k=1}^{m} \binom{2m}{m-k} \cdot T_{2k}(x) \tag{8}$$

$$(2x)^{2m+1} \equiv 2\sum_{k=0}^{m} \binom{2m+1}{m-k} \cdot T_{2k+1}(x) \tag{9}$$

Replacing in $s_n(x)$ the powers of x by these expansions, we obtain for it an equivalent form

$$s_n(x) = \sum_{m=0}^{n} a_m x^m = \frac{b_0}{2} + \sum_{m=1}^{n} b_m \cdot T_m(x)$$

the coefficients b_m of which are defined by

$$b_{2k} = 2 \cdot \sum_{i=k}^{2i \leqq n} \binom{2i}{i-k} 2^{-2i} \cdot a_{2i}$$

and

$$b_{2k+1} = \sum_{i=k}^{2i+1 \leqq n} \binom{2i+1}{i+k} 2^{-2i} \cdot a_{2i+1}$$

It happens very often that one or more among the last coefficients b_m, $m = n, n-1, \cdots$, are so small that omitting the corresponding terms does not diminish the accuracy of approximation. In such a case $s_n(x)$ can be relaxed. Suppose, for instance, that b_{n-1} and b_n can be omitted. Then the relaxed form is a polynomial $\bar{s}_{n-2}(x)$ of degree $n - 2$ obtained by replacing in the sum $\sum_{m=1}^{n-2} b_m \cdot T_m(x)$ the polynomials $T_m(x)$, $1 \leq m \leq n - 2$, by their explicit expressions in terms of powers of x.

4. τ-Method

For a function to have polynomial or rational approximations it is not at all necessary that it be represented by a convergent expansion (1) or (2). If a function is the solution of a linear differential equation, the coefficients of which are rational functions of the independent variable, the so-called *τ-method*, due to Lanczos ([7], Ch. VII, § 12), allows the formation of a convergent sequence of polynomial (as well as rational) approximations of increasing degree, the accuracy of which increases with the degree of the approximating polynomial. This method works even when the power series (1) diverges everywhere or does not exist at all, as, for instance, for the function $x \log x$. When the power series (1) converges too slowly it accelerates the convergence. In general, the τ-method is an important source of approximations and in many cases, for instance in the case of Bessel functions, it yields very useful approximations. It is worthwhile to study this method from our point of view of the most economical procedure for the computation of elementary functions. We do it on the example of the function arctan x, but the result and the conclusion are the same for all elementary functions.

By definition of the τ-method, a polynomial approximation $P(x)$ to a given $f(x)$ in the interval (0, 1) is constructed as follows. Let $L(y) = 0$ denote the differential equation satisfied by $y = f(x)$, the coefficients of which are made polynomials in x by multiplying them, if necessary, by their common denominator. This equation $L(y) = 0$ is slightly distorted by adding to the right-hand member a new term $\tau \cdot T^*_{n+1}(x)$, where τ is a small constant and $T^*_{n+1}(x) \equiv T_{n+1}(2x - 1) \equiv T_{2n+2}(\sqrt{x})$ is the Chebyshev polynomial orthogonal in the interval (0, 1). The modified equation

$$L^*(y) \equiv L(y) - \tau \cdot T^*_{n+1}(x) = 0$$

has now a polynomial solution, $P(x)$ and this exact solution of the modified equation is an approximation to $f(x)$ in (0, 1).

For instance, the even function x^{-1} arctan x of x, considered as a function $y = y(z)$ of the variable $z = x^2$, satisfies the equation

$$L(y) = (1 + z)\left(2z\frac{dy}{dz} + y\right) - 1 = 0$$

This equation $L(y) = 0$ does not have polynomial solutions at all. But the approximate equation $L^*(y) = L(y) - \tau \cdot T^*_{n+1}(z) = 0$ does and the polynomial of nth degree in z satisfying $L^*(y) = 0$ and the initial condition $y(0) = 1$ is considered as an approximation to x^{-1} arctan x. The degree of this approximation in x is equal to $2n + 1$.

To compare the τ-method with the method based on the use of (2) we consider the case $n = 2$. Since $T^*_3(z) = T_6(\sqrt{z}) = 32z^3 - 48z^2 + 18z - 1$, the solution of the equation $L(y) = \tau \cdot T^*_3(z)$ is a quadratic polynomial $q_2(z)$. Replacing z by x^2 and multiplying by x, we obtain the approximation

$$\begin{aligned}\arctan x \approx Q_5(x) &= x \cdot q_2(x^2)\\ &= (1470x - 400x^3 + 96x^5)/1485\end{aligned}$$

The maximum of the absolute error of arctan $x - Q_5(x)$ is reached at $x_0 = \frac{1}{2}\sqrt{2 - \sqrt{3}}$ and it is equal to 0.00164 since $Q_5(x_0) = 0.25163$, while arctan $x_0 = 0.25327$.

The deviation 0.00164 is larger than that 0.00069 obtained using (2). Therefore nothing is gained using the τ-method, which seems to be useful mostly in the cases when the power series (1) diverges or does not exist at all.

We will now discuss the method which yields rational approximations.

5. The Padé Table ([8]; [9], Ch. XX)

Some seventy years ago the French mathematician H. Padé formulated a general method with the aid of which any power series (convergent or divergent) can be transformed into a table of rational approximations to the function represented by this power series. The coefficients a_k, b_k of a rational function $R_{mn}(x) = P_m(x)/Q_n(x)$, where $P_m(x) = \sum_0^m a_k x^k$ and $Q_n(x) = 1 + \sum_1^n b_k x^k$, are deduced from those c_k of the power series $\sum_0^\infty c_k x^k$ with the aid of the definition:

$$Q_n(x) \cdot \sum_0^\infty c_k x^k - P_m(x) \equiv x^{n+m+1} \cdot \sum_0^\infty \gamma_k x^k \quad (10)$$

This definition yields $n + m + 1$ linear equations for $n + m + 1$ unknowns a_k, b_k $(b_0 = 1)$.

As we will see below, rational approximations give much better subroutines than polynomials since they almost halve the number of operations. Padé approximations in particular are most useful if $m = n$ (diagonal of the Padé table), or $m = n + 1$.

To take an example, we consider the case $m = n$ in which $b_0 = 1$, so that $a_0 = c_0$ and, for $1 \leqq s \leqq n$,

$$\sum_{m=0}^{n} b_m \cdot c_{n-m+s} = 0; \quad a_s = \sum_{m=0}^{s} b_m \cdot c_{s-m} \quad (11)$$

while

$$\gamma_k = \sum_{m=0}^{n} b_m \cdot c_{2n+k+1-m}$$

The coefficients γ_k in the right-hand member of (10) in general decrease so rapidly that its first term $\gamma_0 x^{n+m+1} = \gamma_0 x^{2n+1}$ divided by $Q_n(x)$ is an excellent approximation to the absolute error made in considering $R_{nn}(x) = P_n(x)/Q_n(x)$ instead of $\sum_0^{\infty} c_k x^k$. Moreover, for $|x| < 1$ the value of the denominator $Q_n(x)$ differs very little from one since in general the b_k rapidly decrease and b_1 is already small enough. Thus, before computing the coefficients a_k, b_k it is advisable to obtain a crude estimate of the absolute error E_n corresponding to the value of n. This E_n is a decreasing function of n and its estimate helps in determining that value of n which will yield the required accuracy. Since $E_n \approx \gamma_0 x^{2n+1}$, where the range $0 \leqq x \leqq x_0$ is known, we have $|E_n| \leqq |\gamma_0| \cdot x_0^{2n+1}$ and it is sufficient to compute γ_0. The system of $n + 1$ equations

$$\sum_{m=0}^{n} b_m \cdot c_{n-m+s} = 0 \qquad (1 \leqq s \leqq n)$$

$$\gamma_0 = \sum_{m=0}^{n} b_m \cdot c_{2n+1-m}$$

gives the expression $\gamma_0 = \Delta_n/\delta_n$, where Δ_n is defined by

$$\Delta_n = \begin{vmatrix} c_1 & c_2 & \cdots c_{n+1} \\ c_2 & c_3 & \cdots c_{n+2} \\ \cdot & & \cdot \\ \cdot & & \cdot \\ \cdot & & \cdot \\ c_{n+1} & c_{n+2} & \cdots c_{2n+1} \end{vmatrix}$$

and δ_n is the principal minor of Δ_n obtained by omitting the last row and column in Δ_n.

Applying this method to cos x, we consider

$$\cos x = \cos(\sqrt{z}) = \sum_{k=0}^{\infty} \frac{(-1)^k}{(2k)!} z^k \quad (12)$$

where $z = x^2$. Let $m = n = 3$. Computing Δ_3, δ_3, γ_0 for $c_k = (-1)^k/(2k)!$, we find that

$$\gamma_0 = \frac{\Delta_3}{\delta_3} = \frac{45{,}469}{59 \times 660 \times 14!} < 1.34 \times 10^{-11}$$

Here $b_k > 0$ for $k = 1, 2, 3$, so that $Q_3(z) > 1$ and $E_3 < 1.34 \times 10^{-11}$, if $z = x^2 \leqq 1$. Therefore, the case $m = n = 3$ should yield for cos x ten correct digits after the dot in the range $0 < x \leqq 1$. Cos $x \geqq \cos 1 = 0.54 \cdots$, if $x \leqq 1$ and these ten digits are all significant digits.

The approximation $R_{33}(z) = P_3(z)/Q_3(z)$ is now formed by computing the coefficients a_k, b_k and is then transformed into a continued fraction

$$\cos x \approx C_0 + \sum_{k=1}^{3} \frac{C_k|}{|z + B_k}$$

$$= C_0 + \sum_{k=1}^{3} \frac{C_k|}{|x^2 + B_k} \quad (13)$$

the coefficients C_k, B_k of which are easily found from the identity

$$\left\{\prod_{k=1}^{3}(z + B_k) + C_2(z + B_3) + C_3(z + B_1)\right\}$$
$$\cdot \{P_3(z) - C_0 \cdot Q_3(z)\}$$
$$\equiv C_1[(z + B_2)(z + B_3) + C_3] \cdot Q_3(z)$$

It seems that for $0 < |x| \leqq 1$ a subroutine based on (13) should yield $Dg \doteq 10$, but an important complication arises which is a common feature of many continued-fraction approximations whatever their origin. If a small number is computed as a difference of two large numbers the accuracy is lost (unless extra digits are carried in the computation). Thus, in (13), where the value of the right-hand member is less than one, $C_0 = a_3/b_3 = -14{,}615/127 \approx -115$ and, using (13) as it is, $Dg = 7$ would be obtained instead of $Dg = 10$. A slight modification of the continued-fraction expansion eliminates the trouble. It consists of the addition of a rapidly computed term ξz, where the parameter ξ is chosen in such a way that $C_0 = C_0(\xi)$ is small [10]. To satisfy the requirement of speed, ξ is adapted to particular features of the computer. Thus, for instance, in the IBM type 709, the duration of multiplication depends on the binary form of the factor ξ. If the number of significant bits is small, the multiplication takes a negligible time. In other computers the multiplication by an integral power of 2 is a shift.

This modification of (13) is studied in Part B, where we consider cos N and sin N in the whole range $(0, \infty)$. A reduction of this range to the small range $(0, \pi/2)$ leads to the argument $\sqrt{z} = \pi f$, in which f is known. To avoid the multiplication by π, $f \leqq \frac{1}{3}$ is taken as the argument of continued-fraction approximation to $\cos \pi f$. Therefore, in $\xi z = \xi\pi^2 \cdot f^2$ it is the factor $\xi\pi^2$ which is adapted to the computer.

Thus, a subroutine for $\cos x$ in $(0, \pi/3)$ based on the modified (13) does yield $Dg = 10$ in four multiplications, the number of stored constants being seven. We stress the point that a rational approximation is useful only in the equivalent form of a finite continued fraction. The $R_{33}(z)$, if computed as an ordinary fraction $P_3(z)/Q_3(z)$, necessitates eight operations instead of four.

Returning to our example of $\cos x$, we observe that in the interval $\pi/3 < x \leqq \pi/2$, $\cos x$ can be computed as $\sin t = \sin(\pi/2 - x)$ with $0 < t \leqq \pi/6$. The accuracy of a rational approximation depends on the range, and in general decreasing the range increases the accuracy. Since the range $(0, \pi/6)$ for the sine is smaller than the range $(0, \pi/3)$ for the cosine, let us try the case $m = n = 2$, applying a Padé approximation to $\sin x$.

Again letting $z = x^2$, we form $R_{22}(z)$ for the function

$$x^{-1} \sin x = \varphi(z) = \sum_{k=0}^{\infty} c_k z^k$$

$$= \sum_{k=0}^{\infty} \frac{(-1)^k}{(2k+1)!} z^k \tag{14}$$

An easy computation gives $\delta_2 = -11/302{,}400$ and $\Delta_2 = 121/(63 \times 5!7!10!)$ so that this time γ_0 is negative and $|\gamma_0| = 11/(136 \times 10!)$. But the absolute error $|E_2| = x \cdot (x^2)^5 |\gamma_0| = x^{11} \cdot |\gamma_0|$ depends on the range and for $x \leqq \pi/6$ it is of the order of $(\pi/6)^{11} \cdot |\gamma_0| < 2 \times 10^{-11}$. Therefore, we conclude that using $m = n = 2$, we obtain $Dg = 10$. It is important to observe that the relative error equal to $x^{10} \cdot |\gamma_0| (\sin x/x)^{-1}$ is less than 4×10^{-11}. Therefore, a subroutine for $\sin x$ in the range $0 < x \leqq \pi/6$ based on $R_{22}(z)$ will have the first ten significant digits correct. Transforming $xR_{22}(z)$ into the equivalent continued fraction, we have the approximation,

$$\sin x \approx x \cdot \left(C_0 + \sum_{k=1}^{2} \frac{C_k|}{|x^2 + D_k} \right) \tag{15}$$

which can be computed in four operations. Here again $C_0 = 23.07/\pi$ and a modification is needed.

Combining (13) and (15), we state that Padé approximations for $\sin x$ and $\cos x$ allow us to compute either function in the range $0 < x \leqq \pi/2$ in *four* multiplications and divisions with the first *ten* correct significant digits. The approximations (13) and (15) seem to be new and in Part B we will discuss their use in the generation of $\sin x$ and $\cos x$ for $0 < x < \infty$.

The Padé method can be applied also to functions represented by a quotient $C(x)/D(x)$ of two convergent power series. Let c_k and d_k denote their known coefficients. The coefficients a_k, b_k ($b_0 = 1$) of the numerator $P_m(x)$ and denominator $Q_n(x)$ of a rational approximation $P_m(x)/Q_n(x)$ to $C(x)/D(x)$ are defined by the identity

$$Q_n(x)C(x) - P_m(x)D(x) = x^{m+n+1} \cdot \sum_{0}^{\infty} \gamma_k x^k$$

and they satisfy the system

$$\sum_{0}^{n} b_k c_{N-k} = \sum_{0}^{m} a_k \cdot d_{N-k} \quad (0 \leqq N \leqq m + n)$$

where c_k and d_k with negative subscripts vanish. The error is of the order $\gamma_0 x^{m+n+1}$ and

$$\gamma_0 = \sum_{k=0}^{n} b_k c_{m+n-k+1} - \sum_{k=0}^{m} a_k d_{m+n-k+1}$$

6. Maehly's Method ([11]; [12])

The continued-fraction approximations obtained by the Padé method are characterized —as are the power series from which they are derived—by excellent accuracy near the origin $x = 0$ since the expression of the error involves the factor x^{m+n+1}. But when x increases, the accuracy rapidly diminishes. In most cases their usefulness is due much more to the smallness of the reduced range in which they are used than to the factor γ_0 in the error $\gamma_0 x^{m+n+1}$. The same holds for the classical infinite continued fractions representing $\arctan x$, $\log x$, etc., deduced from the Gauss continued-fraction expansion of the hypergeometric function which we will use in Part B.

Though the accuracy of polynomial approximations deduced from the Chebyshev expansions also depends on the magnitude of the reduced range, in the interior of the range the truncation error does not increase when x increases but oscillates between extrema of almost equal absolute value. In this case the accuracy is due to the rapidity with which the coefficients of Chebyshev expansion decrease.

It is to be expected that for the continued-fraction approximations deduced from the Chebyshev expansions the behavior of the error will be the same as for those expansions which would allow—in the case of slowly convergent power series representations—a much better subroutine than those based on the use of Padé approximations. And, indeed, these expectations are fulfilled.

We owe to Dr. H. Maehly [11] the fundamental idea with the aid of which any Chebyshev expansion (2) of $f(ax)$ in $-a \leq ax \leq a$ can generate a rational approximation to $f(ax)$ of the following type

$$f(ax) \approx R_{MN}(x; a) = \sum_{0}^{M} a_m \cdot T_m(x) / \sum_{0}^{N} b_m \cdot T_m(x)$$

the error of which is of the order of $A_0 \cdot T_{M+N+1}(x)$. Since the coefficients of expansion (2), $c_n = c_n(a)$, are functions of the parameter a, the coefficients a_m, b_m, as well as the factor A_0 in the expression of the error, also depend on a. The most economical values of M are N and $N + 1$ as in the case of the Padé approximations.

If the coefficients $c_n(a) = c_n$ are known, the $M + N + 1$ unknowns a_m, b_m ($b_0 = 1$) are defined by as many conditions $h_m = 0$, $0 \leq m \leq M + N$, h_m being the coefficient of the expansion (2) of the following function $H(x)$:

$$H(x) = \left[\sum_{0}^{N} b_m T_m(x)\right] \cdot \left[\sum_{0}^{\infty} c_n \cdot T_n(x)\right] - \sum_{0}^{M} a_m T_m(x) = \sum_{0}^{\infty} h_m \cdot T_m(x) \quad (16)$$

To find the expressions of h_m we observe that

$$T_{m+n}(x) + T_{m-n}(x) = 2T_m(x) \cdot T_n(x) \quad (m \gtrless n)$$

Therefore, performing the multiplications in (16), applying this identity, and using the conditions $h_m = 0$ for $0 \leq m \leq M + N$, we obtain the following system of linear equations ($j = 1, 2, \cdots, M + N - 1, M + N$):

$$a_0 = c_0 + \frac{1}{2} \sum_{i=1}^{N} b_i c_i \quad (17)$$

$$a_j = c_j + \frac{c_0}{2} b_j + \frac{1}{2} \sum_{i=1}^{N} b_i (c_{j+i} + c_{|j-i|})$$

where $a_j = 0$ for $j > M$ and $b_j = 0$ for $j > N$. We denote by A_0 the coefficient of the first non-zero term in the expansion of $H(x)$, i.e., h_{M+N+1}, so that

$$h_{M+N+1} = A_0 = c_{M+N+1} + \frac{1}{2} \sum_{i=1}^{N} b_i (c_{M+N+i+1} + c_{M+N-i+1}) \quad (18)$$

The error of the approximation $R_{MN}(x; a)$ is equal to $H(x) / \sum_{m=0}^{N} b_m T_m(x)$ and it is of the order $A_0 \cdot T_{M+N+1}(x)$ because h_m and b_m decrease very rapidly so that $H(x)$ is of the order of its first term, while the denominator is of the order of one ($b_0 = 1$). Therefore, $|A_0|$ can be considered as a crude approximation to the upper bound of the absolute error in the range $a \cdot |x| \leq a$. For a prescribed accuracy the order of magnitude of $|A_0|$ should be studied for various combinations of parameters M, N, and a.

To estimate the value of A_0 for given values of M, N, and a it is not necessary to solve the system (17) and use (18). Consider, for instance, the case $M = N$. Eliminating the unknowns b_m, $1 \leq m \leq N$, from (18) and the last N equations of (17) with $N + 1 \leq j \leq 2N$, we can express A_0 as a ratio D/Δ of two determinants, Δ being the principal minor of D. Their elements, except those in the last column of D, are the sums $c_{N+j-i} + c_{N+j+i}$. Since the sequence c_m is a rapidly decreasing one—otherwise we would not use the Chebyshev expansion (2)—omitting the second term c_{N+j+i}, a good approximation D^*/Δ^* to A_0 is easier to compute than $D/\Delta = A_0$.

When the choice of M, N, and a (reduced range) is fixed, replacing $T_m(x) = T_m(z/a)$ by their explicit expressions, the approximation $R_{MN}(x; a)$ is transformed into a continued fraction of the same argument z as in $f(z)$.

If $f(x)$ is odd and vanishes at $x = 0$, so that $\lambda = \lim_{x \to 0} x^{-1} \cdot f(x)$ is finite and does not vanish, applying Maehly's method to $f(ax)$ we have a choice between two Chebyshev expansions:

$$f(ax) = \sum_{n=0}^{\infty} c_n(a) \cdot T_{2n+1}(x) \qquad (19)$$

and

$$x^{-1} \cdot f(ax) = \frac{c_0^*}{2} + \sum_{0}^{\infty} c^*(a) \cdot T_{2n}(x) \quad (20)$$

where naturally $c_n^* = 2\sum_{j=0}^{\infty} (-1)^j c_{n+j}$ and $2c_n = c_n^* + c_{n+1}^*$. The upper bound of the *relative* error, if (20) is used, is of the order of $|A_0^*|/a\lambda$, while if (19) is used, it is $|A_0| \cdot (2M + 2N + 3)/a\lambda$, since in the neighborhood of the origin we have to use the inequality

$$|T_{2M+2N+3}(x)| \leqq (2M + 2N + 3) \cdot |x| \qquad (|x| \leqq 1)$$

Sometimes the neighborhood of the origin is to be considered as a separate interval in which a rational approximation is replaced by a short polynomial because near to the origin the rational approximation does not give the required accuracy. To illustrate Maehly's method let us consider a subroutine for arctan x in $0 < x < \infty$ which gives the first six correct significant digits ($Dg = 6$) in four operations. The importance of such a special subroutine is clear: in many cases, when three or four correct digits only are asked in the final result, using this shorter and more rapid subroutine instead of one giving eight or ten first correct digits means a big saving of machine time.

Applying Maehly's method to arctan x in $0 < x \leq 1$, we choose to use the expansion (20) with $a = 1$. To find the coefficients $G_m = c_m^*(1)$ in

$$x^{-1} \cdot \arctan x = \tfrac{1}{2}G_0 + \sum_{m=1}^{\infty} G_m \cdot T_{2m}(x)$$

we let $a = \tan 2\theta$ (in our case $\theta = \pi/8$) and observe that in (3)

$$c_n(a) = 2(-1)^n \tan^{2n+1}\theta/(2n + 1)$$

This gives $c_n(1) = 2(-1)^n(\sqrt{2} - 1)^{2n+1}/(2n + 1)$, so that $\tfrac{1}{2}G_0 = \log_e \tan(\pi/4 + \theta) = \log_e(\sqrt{2} + 1)$, the other G_m's being computed by recurrence: $G_{n+1} = 2c_n(1) - G_n$. The table of first nine coefficients G_m is as follows:

$$\begin{aligned}
\tfrac{1}{2}G_0 &= 0.881\ 373\ 587\ 019\ 543\\
G_1 &= -0.105\ 892\ 924\ 546\ 706\\
G_2 &= 0.011\ 135\ 842\ 059\ 403\\
G_3 &= -0.001\ 381\ 195\ 003\ 598\\
G_4 &= 0.000\ 185\ 742\ 973\ 276\\
G_5 &= -0.000\ 026\ 215\ 196\ 110\\
G_6 &= 0.000\ 003\ 821\ 036\ 590\\
G_7 &= -0.000\ 000\ 569\ 918\ 604\\
G_8 &= 0.000\ 000\ 086\ 488\ 727
\end{aligned}$$

Solving the system [using $M = N = 2$ and (17)]

$$\begin{aligned}
(G_2 + G_4)b_1 + (G_1 + G_5)b_2 &= -2G_3\\
-(G_3 + G_5)b_1 - (G_2 + G_6)b_2 &= 2G_4
\end{aligned}$$

we have

$$\begin{aligned}
b_1 &= 0.373\ 605\ 4075\ 3\cdots\\
b_2 &= 0.013\ 854\ 1097\ 2\cdots
\end{aligned}$$

The formulas for a_0, a_1, a_2

$$\begin{aligned}
2a_0 &= 2G_0 + b_1G_1 + b_2G_2\\
2a_1 &= 2G_1 + b_1(2G_0 + G_2) + b_2(G_1 + G_3)\\
2a_2 &= 2G_2 + b_1(G_1 + G_3) + b_2(2G_0 + G_4)
\end{aligned}$$

give

$$\begin{aligned}
a_0 &= 0.861\ 669\ 641\ 0\\
a_1 &= 0.224\ 730\ 125\ 2\\
a_2 &= 0.003\ 308\ 679\ 5
\end{aligned}$$

and therefore

$$\arctan x \approx x[a_0 + a_1T_2(x) + a_2T_2(x)] \cdot [1 + b_1T_2(x) + b_2T_4(x)]^{-1}$$

Replacing T_2 and T_4 by $2x^2 - 1$ and $8x^4 - 8x^2 + 1$, we obtain the rational approximation

$$\arctan x = x[\alpha_0 + \alpha_1x^2 + \alpha_2x^4] \times [\beta_0 + \beta_1x^2 + \beta_2x^4]^{-1} \qquad (21)$$

where

$$\begin{aligned}
\alpha_0 &= a_0 - a_1 + a_2 &&= 0.640\ 248\ 1953\\
\alpha_1 &= 2a_1 - 8a_2 &&= 0.422\ 990\ 8144\\
\alpha_2 &= 8a_2 &&= 0.026\ 469\ 4361\\
\beta_0 &= 1 - b_1 + b_2 &&= 0.640\ 248\ 7022\\
\beta_1 &= 2b_1 - 8b_2 &&= 0.636\ 377\ 9373\\
\beta_2 &= 8b_2 &&= 0.110\ 832\ 8778
\end{aligned}$$

Replacing x by $1/N$, we have also

$$\arctan(1/N) = [\alpha_2 + \alpha_1N^2 + \alpha_0N^4] \times [\beta_2 + \beta_1N^2 + \beta_0N^4]^{-1}/N \qquad (22)$$

To study the error we compute

$$\begin{aligned}
A_0^* &= G_5 + \tfrac{1}{2}(G_4 + G_6)b_1 + \tfrac{1}{2}(G_3 + G_7)b_2\\
&= 3 \times 10^{-7}
\end{aligned}$$

Here $b_1 = 0.373 \cdots$ is not small enough to neglect the denominator. So for small x we divide A_0^* by $\beta_0 = 1 - b_1 + b_2$ and obtain 4.5×10^{-7} as the order of magnitude of the upper bound of relative error (here $a = \lambda = 1$, so $A_0^*/a\lambda = A_0^*$). To check the true value of this upper bound we form $\alpha_0/\beta_0 = 0.999\ 9992$ which is an approximation to $\lambda = 1$ and thus find 8×10^{-7} as the extreme value of the relative error. This result shows that in the neighborhood of the origin our approximation (21) will yield for some values of x not six but five correct digits. To improve the subroutine based on (21), obtaining six correct digits for all values of $x \leq 1$, it is necessary to isolate the interval (0; tan 7°30′) and apply (21) in the interval (tan 7°30′; 1) only. In the interval (1; ∞) arctan x is computed as $\pi/2$ − arctan (x^{-1}), so that the approximation (22) yields six correct digits. Since in the neighborhood of the origin for $t = x^{-1}$, arctan x exceeds one, the relative error as expected is bounded by 4.5×10^{-7} and (22) yields always $Dg = 6$.

In the first interval (0; tan 7°30′) the expansion (3) with $2\theta = 7°30'$

$$\arctan (x \cdot \tan 7°30') = 2 \sum_{0}^{\infty} (-1)^n \tan^{2n+1} 3°45' \cdot T_{2n+1}(x)/(2n+1)$$

yields six correct digits if we approximate arctan t for $t = x \cdot \tan 7°30'$ by the sum of first three terms. We have indeed $\tan 3°45' \approx 0.06554$, so that the upper bound of the relative error is $2 \cdot (0.06554)^7 \cdot 24/\pi < 10^{-7}$. This polynomial approximation is, in $0 < N < \tan 7°30'$:

$$P_5(N) = 2 \sum_{\infty}^{2} (-1)^n \tan^{2n+1} 3°45' \cdot T_{2n+1}(x)/(2n+1) = N \cdot [C_0 - N^2(C_1 - C_2N^2)] \quad (23)$$

with

$$C_0 = 0.999\ 999\ 9207 = 1 - 793 \times 10^{-10}$$
$$C_1 = 0.333\ 296\ 6338 = 1/3 - 703.3662 \times 10^{-6}$$
$$C_2 = 0.195\ 740\ 8066 = 1/5 - 4.259\ 1934 \times 10^{-3}$$

In computing the polynomial $P_5(N)$, the variable x was replaced by $N \cdot \text{cotan } 7°30'$. Finally, (21) and (22) are transformed into continued fractions and used in the form:

$$\arctan N \approx N \cdot \left\{ A_0^* + \frac{A_1^*|}{|N^2 + B_1^*} - \frac{A_2^*|}{|N^2 + B_2^*} \right\} \quad (\tan 7°30' \leq N \leq 1) \quad (24)$$

$$\arctan N \approx \pi/2 - \left\{ A_0 - \frac{A_1|}{|N^2 + B_1} - \frac{A_2|}{|N^2 + B_2} \right\} \cdot N^{-1} \quad (1 < N < \infty) \quad (25)$$

The eleven constants in (24) and (25) to be stored are $\pi/2$ and

$A_0^* = 0.238\ 822\ 9612;$	$A_0 = 0.999\ 999\ 2083$
$A_1^* = 2.445\ 205\ 396;$	$A_1 = 0.333\ 287\ 0775$
$B_1^* = 3.943\ 529\ 798;$	$B_1 = 0.598\ 599\ 8078$
$A_2^* = 1.314\ 747\ 223;$	$A_2 = 0.063\ 550\ 0089$
$B_2^* = 1.798\ 249\ 626;$	$B_2 = 0.395\ 354\ 4718$

To locate N one more constant, namely $c = \tan 7°30' = 0.131\ 652\ 497$, is needed.

The flow chart of the subroutine is as follows:

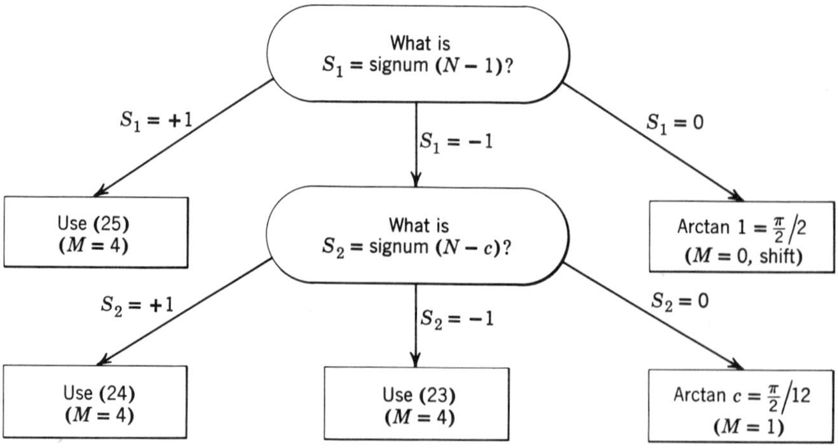

In all there are fifteen constants to store.

This example shows clearly the importance of Maehly's method in the case of a slowly convergent power series. The sum of the first three terms $x - x^3/3 + x^5/5$ of the power series for arctan x considered in the interval $0 < x < 1$ is a very poor approximation to arctan x with a maximum relative error (at $x = 1$) equal to 2/15. The sum of the same number of terms of the Chebyshev expansion is better: it yields two to three correct digits and for it the maximum truncation error is 0.0007. These two polynomial approximations necessitate four multiplications and in the same number of operations the continued fraction deduced from Maehly's rational approximation yields six correct digits instead of two or three.

PART B

In discussing the subroutines for individual elementary functions one by one we can describe in detail—because of limitations of space—only those which seem to be the most economical. The functions will be considered in the order: e^N together with sinhyp N and coshyp N; sin N and cos N together with e^{iN}; tan N and cotan N; log N; arctan N; arcsin N; and $\sqrt{N}$. For all of them, except arcsin N, the infinite range $(-\infty, +\infty)$ of the argument is considered.

Although Bessel functions could be considered as the least elementary among the elementary functions, they are not included: the question of the most economical subroutines for them is still in a state of transition and would necessitate too much space.

e^N in $(-\infty, \infty)$ [13]

To reduce the infinite range of N, the binary representation N_2 of N is multiplied by $\log_2 e$ and the fractional part of the product is denoted by F. Thus, $0 < F < 1$ in $N_2 \cdot \log_2 e = N^* + F$ and $e^N = 2^{N^*} \cdot e^{F \cdot \log 2}$, where 2^{N^*} is accounted for by a shift, if a binary machine is used. If the computer is a decimal one, $e^N = 10^{N^*} \cdot e^{F \cdot \log 10}$ with $0 < F < 1$, where $N_{10} \cdot \log_{10} e = N^* + F$. In the following the symbol G denotes $\log_2 e$ for a binary and $\log_{10} e$ for a decimal machine.

Since $e^{-N} \cdot d(e^N) = dN$, an *absolute* error in the given value N generates an equal *relative* error in the computed value of e^N. In a binary computer the conversion of the decimal form N_{10} of N into its binary form N_2 introduces a truncation error $DN_2 = DF \neq 0$. In a single precision fixed point computation DF can reach $2^{-\beta} \cdot N$ and, in a double precision, $2^{-2\beta} \cdot N$, if the computer is a β-bit machine. Suppose $\beta = 35$ and N large: $10^q > N > 10^{q-1}$, $q > 1$. The absolute error $D(F \log 2)$ can reach $3 \times 10^{q-11}$, so that there will be $Dg = 10 - q$ correct significant digits only in the final value of e^N though the mathematical method used in the subroutine may insure $Dg = 10$.

In a decimal machine N_{10} is given and considered as the exact value of the exponent, but there is another important cause of possible loss of accuracy, also unrelated to the method of generating e^N used in the construction of a subroutine and which affects both binary and decimal machines. It is the error in the value of $F \log 2$ (or $F \log 10$) caused by the multiplication of N by G. Suppose that a binary 35-bit machine is used. The truncation error in the stored value of $G = \log_2 e$ is of the order of 2^{-35} and, if $10^q > N > 10^{q-1}$, then the error in $F \cdot \log 2$ can reach $2^{-35} 10^q \log 2 \approx 7 \times 10^{q-10}$. Thus, instead of the theoretical number Dg related to the method used, $9 - q$ first correct significant digits will be obtained in the computed approximation to e^N.

To avoid the loss of accuracy which may be caused by these two sources of error it is advisable to store the double precision representations of N_2 (or N_{10}) and G and use, if N is large, double precision multiplication in computing F. Thereafter, single precision will be used. This can help only for single precision subroutines.

The partial sums of the expansion (2)

$$e^{ax} = I_0(a) + 2 . \sum_{n=1}^{\infty} I_n(a) \cdot T_n(x) \qquad (26)$$

yield the best practical polynomial approximations $Q_{n-1}(a, x)$, for which the upper bound of absolute error E_n is:

$$|E_n(a, x)| \leq 2I_n(a) \cdot [1 - \tfrac{1}{2}a/n + 1]^{-1}$$

The diagonal terms of the Padé table are the most economical rational approximations $P_m(z)/P_m(-z)$ for e^z. The polynomial $P_m(z)$ is:

$$(2m)!P_m(z) = m! \sum_{j=0}^{m} (2m - j)! z^j / [j!(m - j)!]$$

The *relative* error $R_m(z)$ is bounded by

$$|R_m(z)| < (2m+1)^{-1} \cdot [m!/(2m)!]^2 \cdot |z|^{2m+1} \cdot \exp\left[-\tfrac{1}{2}z + \frac{z^2}{8(2m+3)}\right]$$

We omit the proofs (see [13]).

These rational approximations are not new. They are a very particular case of Darboux's two-point expansion (1876) ([14]). Rediscovered in 1949 in America [15] they were also formed again in 1955 by Lanczos ([7], Chapter VI, §19, pp. 419–427). Since

$$e^x = [1 + \text{tanhyp}\,(x/2)][1 - \text{tanhyp}\,(x/2)]^{-1}$$

the same sequence of approximations $P_m(x)/P_m(-x)$ to e^x is obtained for $t = x/2$ from the sequence of convergents $A_n(t)/B_n(t)$ of the well-known Lambert's continued fraction

$$\text{tanhyp}\ t = \frac{t|}{|1} + \sum_{m=1}^{\infty} \frac{t^2|}{|2m+1}$$

The rational function $[B_m(\frac{1}{2}x) + A_m(\frac{1}{2}x)]/[B_m(\frac{1}{2}x) - A_m(\frac{1}{2}x)]$ is indeed equal to $P_m(x)/P_m(-x)$.

Let us consider first a binary subroutine. The range (0; log 2) of $F \cdot \log 2$ can be reduced to as small a range $(-2^{-k} \cdot \log 2;\ 2^{-k} \cdot \log 2)$ as needed, where k is *any* integer. Choosing a fixed k, we subdivide the interval (0; 1) into 2^k subintervals $[2^{-k}j;\ 2^{-k}(j+1)]$, $0 \leqq j \leqq 2^k - 1$. Beginning with $f_1 = F - \frac{1}{2}$, k numbers f_1 are computed successively in k additions by the rule

$$f_{i+1} = f_i - \sigma_i/2^{i+1} \qquad (1 \leqq i \leqq k-1)$$

where $\sigma_i = \text{signum}\,(f_i) = \pm 1$ is the sign of f_i. Denoting the last f_k simply by f, we have (with $\sigma_0 = 1$)

$$f = f_k = F - \sum_{i=0}^{k-1} \sigma_i/2^{i+1} = F - a(k)$$

The exceptional $k - 1$ cases when some f_i, $i < k$, vanish should be included in the subroutine. Such cases correspond to stored values of $e^{F \cdot \log 2}$. In general, there are 2^{k-1} possible values of the sum $a(k) = \sum_{0}^{k-1} \sigma_i/2^{i+1}$. The 2^{k-1} constants $2^{a(k)}$ are to be stored also since

$$e^N = 2^{N^*} \cdot e^{F \cdot \log 2} = 2^{N^*} \cdot 2^{a(k)} \cdot e^{f \cdot \log 2} = 2^{N^*} \cdot 2^{a(k)} \cdot e^z$$

Thus, the computation of e^N is reduced to that of e^z, where the range of $z = f \cdot \ln 2$ is precisely $|z| \leqq 2^{-k} \cdot \log 2$.

The upper bound $\bar{R}(m, k)$ of the error in the approximation $P_m(z)/P_m(-z)$ to e^z

$$\bar{R}(m, k) = (m!)^2(2^{-k} \cdot \log 2)^{2m+1}/(2m)!(2m+1)!$$

is a decreasing function of m and k. Since the number of operations M is equal to $m + 1$, we consider only four cases $2 \leqq m \leqq 5$. On the other hand, the logical part of the subroutine and the number PC of stored constants increase very rapidly with k, so that the same values $2 \leqq k \leqq 5$ for k as for m are of interest to us.

Omitting the details (see [13]), we give in Table 1 the values of M, PC, and Dg for various combinations (k, m). We would advise using the five combinations shown in Table 2.

Table I

		$k = 2$		$k = 3$		$k = 4$		$k = 5$	
m	M	Dg	PC	Dg	PC	Dg	PC	Dg	PC
2	3	6	7	7	11	9	19	10	35
3	4	10	8	12	12	14	20	16	36
4	5	14	9	16	13	19	21	22	37
5	6	18	8	21	10	24	14	28	22

Table 2

	m	k	Dg
Floating point	2	2	6
	2	4	9
	5	2	18
Fixed point	3	2	10
	5	3	21

Naturally, the final form of the rational approximation $\Pi_m = 2^{a(k)} \cdot P_m(z)/P_m(-z)$ is a continued fraction in terms of $f = f_k$. Thus,

$$\Pi_2 = C + a_2 \cdot [f - c_2 + b_2/f]^{-1}$$

$$\Pi_3 = -C + a_3 \cdot [b_3 - f - c_3(f + d_3/f)^{-1}]^{-1} \tag{27}$$

$$\Pi_4 = C + a_4 \cdot [b_4/f - c_4 + d_4 f + h_4(f + b_4/f)^{-1}]^{-1}$$

$$\Pi_5 = 2C\{\tfrac{1}{2} + f \cdot [b_5 - f - c_5(f^2 + d_5 - h_5 \cdot [f^2 + r_5]^{-1})^{-1}]^{-1}\}$$

The constants are ($G = \log_2 e$ or $G = \log_{10} e$): $C = 2^{a(k)}$; $a_2 = 12GC$; $b_2 = 12G^2$; $c_2 = 6G$; $a_3 = 2a_2$; $b_3 = 12G$; $c_3 = 50G^2$; $d_3 = 10G^2$; $a_4 = 42GC$; $b_4 = 42G^2$; $c_4 = 21G$; $d_4 = 1.05$; $h_4 = 1029G^2/10$; $b_5 = 30G$; $c_5 = 9240G^3$; $d_5 = 4116G^2/11$; $h_5 = 244{,}944G^4/121$; and $r_5 = 504G^2/11$.

For a decimal machine, using the same procedure, we have

$$e^N = 10^{N*} \cdot 10^{a(k)} \cdot e^{f \cdot \log 10}$$

so that $z = f \cdot \log 10$, which entails somewhat greater M and PC for the same Dg. We would advise the combinations given in Table 3.

Table 3

		$k = 3$		$k = 4$		$k = 5$	
m	M	Dg	PC	Dg	PC	Dg	PC
2	3	—		—		8	35
3	4	8	12	10	20	—	
5	6	—		18	14	22	22
6	7	19	12	—		—	

Polynomial approximations yield less economical subroutines and they should be used only in the case when the computer is slow in division. The study of their upper bound $B(m, k)$ of relative error $e^{-z} \cdot E_m(z)$, $z = ax$, where $a = 2^{-k} \cdot \log 10$, gave [13] the values of the number Dg of first correct significant digits shown in Table 4. Therefore, we would advise the combinations:

$(k = 3, n = 6)$ for $Dg = 6$

$(k = 4, n = 6)$ for $Dg = 8$

$(k = 4, n = 7)$ for $Dg = 10$

$(k = 4, n = 10)$ for $Dg = 17$

$(k = 5, n = 10)$ for $Dg = 20$

Table 4

VALUES OF Dg (POLYNOMIAL APPROXIMATIONS)

k \ n	5	6	7	8	9	10
2	3	4	6	7	9	10
3	5	6*	8	10	12	14
4	7	8*	10*	13	15	17*
5	8	10	13	15	17	20*

The number of operations is equal to n: $n - 1$ multiplications to compute $Q_{n-1}(z)$ and one more to compute $f_k = f$.

The coefficients c_j of $Q_{n-1}(z) = Q_{n-1}(ax)$ depend on k because $a = 2^{-k} \cdot \log 10$. They are obtained by replacing $T_j(x)$ in

$$e^z \approx Q_{n-1}(ax) = I_0(a) + 2 \sum_{j=1}^{n-1} I_j \cdot (a) \cdot T_j(x)$$
$$= \sum_0^{n-1} c_j z^j$$

by their explicit expressions (2*). They are of the form $c_j = [1 - b_{nj}(a)]/j!$, where the $b_{nj}(a)$ are small, but these small corrections of the coefficients $1/j!$ of the power series expansion for e^z increase the accuracy 2^{n-1} times: the relative error of the partial sums $s_{n-1}(z)$ of power series for e^z is 2^{n-1} times greater than the relative error of the approximation $Q_{n-1}(z)$, the number of operations M as well as the number of stored constants being the same.

The polynomial $Q_{n-1}(z)$ can be relaxed and we illustrate this fact in the example in which $k = 5$, $n = 6$, so that $a = 2^{-5} \cdot \log 10$ and $Dg = 10$. In this case

$$e^z \approx Q_5(ax) = \sum_{j=0}^{5} c_j \cdot (2x)^j$$

with $c_0 = I_0 - 2I_2 + 2I_4$; $c_1 = I_1 - 3I_3 + 5I_5$; $c_2 = I_2 - 4I_4$; $c_3 = I_3 - 5I_5$; $c_4 = I_4$; and $c_5 = I_5$, the argument of Bessel functions I_n being equal to $2^{-5} \cdot \log 10$. The last term, $c_5 \cdot (2x)^5$, can be relaxed as follows. The modified Chebyshev polynomial

$$T_5^*(x) = 512x^5 - 1280x^4 + 1120x^3$$
$$400x^2 + 50x - 1$$

is related to the interval (0, 1) and $|T_5^*(x)| \leqq 1$, if $0 \leqq x \leqq 1$.

Changing the sign of its argument, we have another modified Chebyshev polynomial related to the interval $(-1, 0)$

$$T_5^{**}(x) = -(512x^5 + 1280x^4 + 1120x^3 + 400x^2 + 50x + 1)$$

which justifies the inequality $|T_5^{**}(x)| \leqq 1$, if $-1 \leqq x \leqq 0$. Solving for x^5, we have in $(-1, +1)$

$$x^5 = (5x^4/2 + 25x^2/32 + 1/512)\sigma(x) - 35x^3/16 - 25x/256 + \epsilon(x)$$

where $\sigma(x) = \text{signum}\,(x)$ and $|\epsilon(x)| < 2^{-9}$ for $-1 \leqq x \leqq 1$.

Now since $a = 2^{-5} \cdot \log 10$, we find that

$2^5 \cdot I_5(a) \leqq 10^{-7.79125}$ and therefore $2^5 I_5 \cdot |\epsilon(x)| \leqq 10^{-10.5}$ which proves that we can omit the term $2^5 I_5 \cdot \epsilon(x)$ without affecting the value of $Dg = 10$. Thus, replacing x^5 by the approximating polynomial of the fourth degree and collecting the like terms, we obtain, instead of $Q_5(ax)$,

$$e^z \approx \bar{Q}_4(x) = \sum_{j=0}^{4} c_j^* \cdot (2x)^j$$

The relaxed coefficients c_j^* are defined by: $c_0^* = c_0 \pm I_5/16$; $c_1^* = c_1 - 25I_5/16$; $c_2^* = c_2 \pm 25I_5/4$; $c_3^* = c_3 - 35I_5/4$; $c_4^* = c_4 \pm 5I_5$, where in c_0^*, c_2^*, c_4^* the double sign $\pm$ means the sign of x: plus, if $1 > x > 0$, and minus, if $-1 < x < 0$. The relaxation of the term in x^4 is impossible, the resulting error being of the order of 10^{-6}.

Thus, e^z can be computed with the aid of *two* approximating polynomials of the fourth degree in *five* multiplications:

$$e^z \approx d_0 + f \cdot \{d_1 + f \cdot d_2 + f \cdot (d_3 + d_4 \cdot f)\}$$

where $d_j = 2^{(k+1)j} \cdot c_j^* = 64^j \cdot c_j^*$. Adding to the eight coefficients d_j, the $2^{k-1} = 16$ constants $2^{a(k)}$, and $\log_2 e$, we obtain in all $PC = 26$ constants to store.

To conclude we mention a rather curious adaptation of an old method, invented in 1624 by Briggs for compiling his famous table of logarithms, for the generation of exponentials in an electronic computer. It is flexible and allows us to obtain any required accuracy. To fix our ideas, let us consider only the case where the first ten correct significant digits are required ($Dg = 10$) and the computer is a binary machine. To compute by Briggs' method the factor 2^F in $e^N = 2^{N^*} \cdot 2^F$, $0 < F < 1$, with the first ten correct digits we need seventeen stored constants $c_k = \log_2 (1 + 2^{-k})$, $1 \leqq k \leqq 17$. Let $f_0 = F$ and define recursively the numbers f_i, a_i for $1 \leqq i \leqq 17$ by the rule:

$$f_{i+1} = f_i \text{ and } a_{i+1} = 0, \text{ if } f_i < c_{i+1}$$

but

$$f_{i+1} = f_i - c_{i+1} \text{ and } a_{i+1} = 1, \text{ if } f_i > c_{i+1}$$

In the exceptional cases, when for some $k < 16$, $f_k = c_{k+1}$, we have

$$F = \sum_{i=1}^{k} a_i c_i + c_{k+1}$$

so that the value of 2^F is obtained by at most k shifts and additions:

$$2^F = \left[\prod_{i=1}^{k} (1 + 2^{-i})^{a_i}\right](1 + 2^{-(k+1)})$$

A multiplication by $1 + 2^{-i}$ in fact is performed as follows: $A(1 + 2^{-i}) = A + 2^{-i}A$. But, in general, shifts and additions are not sufficient and, no f_i vanishing, we arrive at

$$F = \sum_{i=1}^{17} a_i c_i + f^*$$

where $0 < f^* = f_{17} < c_{17} = 1.1 \times 10^{-5}$. Therefore,

$$2^F = 2^{f^*} \cdot \prod_{i=1}^{17} (1 + 2^{-i})$$

and we have to compute 2^{f^*}, f^* being less than 1.1×10^{-5}. Now $\frac{1}{2}(f^* \cdot \log 2)^2 < 3 \times 10^{-11}$ and the sum of the first two terms in

$$2^{f^*} = e^{f^* \cdot \log 2} = 1 + \sum_{1}^{\infty} (f^* \cdot \log 2)^n/n!$$

i.e., $1 + f^* \cdot \log 2$, is a good approximation to 2^{f^*}. In all two multiplications only are used: one to form F, another to compute the product $f^* \cdot \log 2$. The number of additions varies between 18 and 35, depending on the value of N. The number of constants is $PC = 19$: $\log_2 e$, $\log 2$, and seventeen c_i's.

If the first six or eight correct significant digits are required, then $f^* = f_{12}$ and f_{14} respectively. The Briggs method was also adapted to a decimal machine (see [13]).

Maehly's method has not been studied yet for e^z. It would seem that the rapidity with which the power series for e^z converges precludes the success of Maehly's method because it favors Padé convergents, but the question remains open.

The most economical method for generating $\text{sinhyp}\, x = \frac{1}{2}(e^x - e^{-x})$ and $\text{coshyp}\, x = \frac{1}{2}(e^x + e^{-x})$ is without any doubt to compute e^x and then e^{-x}, as its reciprocal value, in one division. The reduction to small range—necessary with any method by which sinhyp x and coshyp x are to be computed—is an obstacle in the direct generation of these functions for all values of their argument in $(-\infty, +\infty)$. Tanhyp $x = 1 - 2 \cdot (1 + e^{2x})^{-1}$ also is generated by computing e^{2x}.

Sin *N*, Cos *N*, e^{iN} in (0, ∞)

One multiplication is sufficient to reduce the infinite range (0, ∞) to a small range. In counting the number M of operations (multiplications and divisions) involved in various subroutines, we have in mind always the whole range (0, ∞) and include in M the multiplication used in reducing the range.

Usually, the same reduced range $(0, \pi/4)$ is used for both sin N and cos N, so that with the aid of $\sin(\pi/2 - x) = \cos x$ and $\cos(\pi/2 - x) = \sin x$ the approximations cover the interval $(0, \pi/2)$. Much better subroutines are obtained if the reduced range for $\sin x$ is smaller than that for $\cos x$, since the generation of $\sin x$ in the same range as $\cos x$ necessitates one operation more. Therefore, we will form the approximation to $\sin x$ in $(0, \pi/6)$, while the corresponding approximation to $\cos x$ will be applied in $(0, \pi/3)$. Together they cover the range $(0, \pi/2)$.

The partial sums of power series in $(0, \pi/4)$ were used for the first subroutines and they were very slow: to obtain $Dg = 8$ seven multiplications were necessary for either function, and if $Dg = 10$ was required eight multiplications were needed. Relaxing their last term, one multiplication was eliminated: $M = 6$ for $Dg = 8$ and $M = 7$ for $Dg = 10$. It was possible to eliminate one more multiplication for $\cos x$, using the partial sum of the expansion (5), but no improvement for $\sin x$ was obtained using the similar expansion (4). As we saw in Part A, the generation of $\sin x$ and $\cos x$ in $(0, \pi/4)$ by this method with $Dg = 10$ necessitates respectively six and five operations which correspond to $M = 7$ for $\sin x$ and $M = 6$ for $\cos x$. But both approximations are necessary to generate either function in $(0, \pi/2)$, so that $M = 7$ remains as for power series with relaxation.

We now consider the same range $(0, \pi/6)$ for both functions and apply Maehly's method, discussing the case of $M = 5$ operations in the whole range (0, ∞), so that $\sin x$ in $(0, \pi/6)$ should need $M = 4$ operations. $\cos x$ in $(0, \pi/6)$ will be obtained in three operations and in the range $(\pi/6, \pi/3)$ it will be computed as $2\cos^2(x/2) - 1$, so that $\cos x$ will need in all four operations in $(0, \pi/3)$. $\sin x$ in $(\pi/6, \pi/2)$ and $\cos x$ in $(\pi/3, \pi/2)$ are computed with the aid of $t = \pi/2 - x$.

To deduce the approximation for $x^{-1}\sin x$ we need the coefficient c_n in

$$\sin\frac{\pi x}{6} = x \cdot \left[\frac{c_0}{2} + \sum_{n=1}^{\infty} c_n \cdot T_{2n}(x)\right]$$

We know the coefficients $\gamma_n = 2(-1)^n \cdot J_{2n+1}(\pi/6)$, of the Chebyshev expansion (4) for $\sin(\pi x/6)$ so that $c_n + c_{n+1} = 2\gamma_n$ and

$$c_n = 2(-1)^n \sum_{m=n}^{\infty} (-1)^m \gamma_m$$

$$= 4(-1)^n \cdot \sum_{m=n}^{\infty} J_{2m+1}\left(\frac{\pi}{6}\right)$$

Forming the function $H(x)$ and applying the conditions $h_m = 0$, $0 \leqq m \leqq 3$, we have the approximation

$$\sin z = \sin\frac{\pi x}{6} \approx \frac{x[A + BT_2(x)]}{1 + \alpha T_2(x) + \beta T_4(x)} \tag{28}$$

where α and β are defined by the system

$$(c_1 + c_3)\alpha + (c_0 + c_4)\beta + 2c_2 = 0$$

$$(c_2 + c_4)\alpha + (c_1 + c_5)\beta + 2c_3 = 0$$

and

$$A = \tfrac{1}{2}(c_0 + \alpha c_1 + \beta c_2)$$

$$B = c_1 + \alpha(c_0 + c_2)/2 + \beta(c_1 + c_3)/2$$

The *relative* error of the approximation (28) in the interval $(0, \pi/6)$ has $(6/\pi)|A_0| \approx 2|A_0|$ as its upper bound, where

$$2A_0 = 2c_4 + \alpha(c_3 + c_5) + \beta(c_2 + c_6)$$

Computing $2|A_0|$ as the ratio of two determinants (see Part A), 1.6×10^{-9} is obtained and this upper bound of the relative error justifies $Dg = 8$: the rational approximation (28) yields eight correct *significant* digits for $\sin z$ in $0 \leqq z \leqq \pi/6$. An equivalent form of (28) is

$$\sin\frac{\pi x}{6} \approx p_1 x \cdot [x^2 + q_1 + p_2(x^2 + q_2)^{-1}]^{-1}$$

To find p_1, q_1, p_2, q_2 we write (28) as $x(a_0 + a_1x^2)/(b_0 + b_1x^2 + b_2x^4)$, where $a_0 = A - B$; $a_1 = 2B$; $b_0 = 1 - \alpha + \beta$; $b_1 = 2\alpha - 8\beta$; and $b_2 = 8\beta$. This gives $p_1 = a_1/b_2$; $q_1 = (p_1b_1 - a_0)/a_1$; $p_2 = (p_1b_0 - a_0q_1)/a_1$; and $q_2 = a_0/a_1$.

Since the reduction to the range $(0, \pi/6)$ leads to the computation of $\sin \pi f$, where the number $f < \frac{1}{6}$ is known, the final form of (28) is

$$\sin\frac{\pi x}{6} = \sin \pi f = 216 p_1 \cdot f \cdot \left[f^2 + \frac{q_1}{36} + \frac{p_2}{1296}\cdot\left(f^2+\frac{q_2}{36}\right)^{-1}\right]^{-1} \qquad (29)$$

Applying the same procedure to $\cos(\pi x/6)$, we obtain the approximation

$$\cos\frac{\pi x}{6} = \cos \pi f = C_0 \cdot [f^2 + C_1 + C_2(f^2 + C_3)^{-1}]^{-1} \qquad (30)$$

but this time the absolute error is of the order of 2×10^{-8}, so that only $Dg = 7$ correct digits are obtained.

We see therefore that for $\sin x$ and $\cos x$ in $(0, \infty)$ only an accuracy of $Dg = 7$ is obtained in $M = 5$ operations with the aid of Maehly's method. Using the modification described in Part A, it is possible to improve the approximation to $\cos x$ and obtain $Dg = 8$.

It remains to describe the beginning of the flow chart for the subroutine based on (29) and (30). Denoting the integral part of the product $N \cdot \pi^{-1}$ by n, so that in $N\pi^{-1} = n + f_0$ we have $0 < f_0 < 1$, we define $f_1 = \frac{1}{2} - \sigma_1 \cdot (\frac{1}{2} - f_0)$, where $\sigma_1 = \text{signum}\ (\frac{1}{2} - f_0)$. If $f_0 = \frac{1}{2} = f_1$ ($\sigma_1 = 0$), $\sin N = (-1)^n$ and $\cos N = 0$, since $\sin N = (-1)^n \sin \pi f_1$, $\cos N = (-1)^n \cdot \sigma_1 \cdot \cos \pi f_1$. Now, if $0 < f_1 < \frac{1}{2}$, we define $\sigma_2 = \text{signum}\ (\frac{1}{3} - f_1)$. If $\sigma_2 = -1$, then $f_2 = \frac{1}{2} - f_1 < \frac{1}{6}$ and $\sin \pi f_1 = \cos \pi f_2$, $\cos \pi f_1 = \sin \pi f_2$ are computed with the aid of (29) and (30). If $\sigma_2 = 0$, i.e., $f_1 = \frac{1}{3}$, the stored values $\sin \pi/3 = \sqrt{3}/2$ and $\cos \pi/3 = \frac{1}{2}$ are used. For $\sigma_2 = +1$, $f_1 < \frac{1}{3}$ and $\cos \pi f_1$ is computed as $2\cos^2(\pi f_1/2) - 1$ since $f_1/2 < \frac{1}{6}$. To generate $\sin \pi f_1$ in the case $\sigma_2 = +1$ we let $\sigma_3 = \text{signum}\ (\frac{1}{6} - f_1)$ and compute $\sin \pi f_1$ by (29), if $\sigma_3 = +1$. If $\sigma_3 = 0$, i.e., $f_1 = \frac{1}{6}$, then $\sin \pi f_1 = \frac{1}{2}$. Finally, if $\sigma_3 = -1$ we define $f_2 = \frac{1}{2}(\frac{1}{2} - f_1) < \frac{1}{6}$ since $\frac{1}{3} > f_1 > \frac{1}{6}$ and compute $\sin \pi f_1$ as $2\cos^2 \pi f_2 - 1$. This part of the flow chart looks as follows:

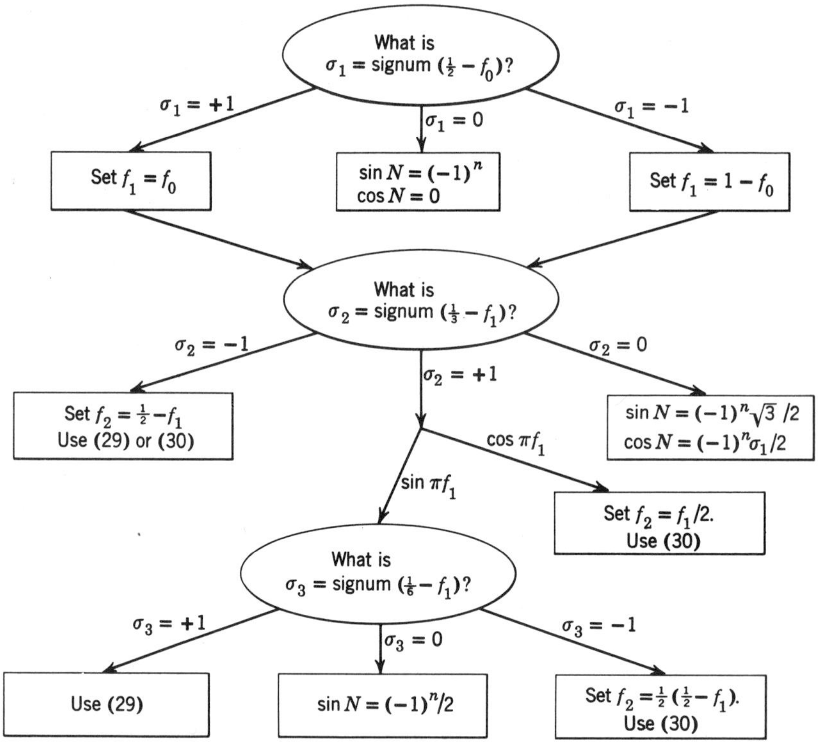

In Part A we saw that the Padé approximations (13) and (15) to $\sin N$, $\cos N$ yield $Dg = 10$ in $M = 5$ operations, $0 < N < \infty$, if modified to eliminate the differences of large numbers. In our opinion the use of the Padé table for $\sin x$, $\cos x$ is the easiest way to obtain the most economical subroutines for these functions for whatever accuracy is required.

We have now to show, for the example of (13)

and (15), how to modify the Padé approximations. Letting $x = \sqrt{z}$ we approximate $\cos \sqrt{z}$ by $(1 + a_1 z + a_2 z^2 + a_3 z^3 - \xi b_3 z^4)(1 + b_1 z + b_2 z^2 + b_3 z^3)^{-1}$ as before, considering ξ as a parameter. This yields $\lambda b_3 = 6!127$, $\lambda b_2 = 8!(297 - 42\xi)$, $\lambda b_1 = 10!(229 - 76\xi)$ with $\lambda = 12!(59 - 34\xi)$. The first nonvanishing coefficient A_0, which determines the accuracy, and the first constant C_0 in (13) become functions of ξ. Since now $C_0 = (a_3 + \xi \cdot b_2)/b_3$, we have

$$127 C_0 = -2{,}352\xi^2 + 33{,}264\xi - 14{,}615$$

while

$$660 \times 14! A_0 = (45469 + 9336\xi)/(59 - 34\xi)$$

We can fix the value of ξ, thus prescribing C_0. But it is important not to destroy the accuracy by keeping $(\pi/3)^{14} \cdot |A_0|$ less than 5×10^{-11}.

The constant C_0 varies with extreme rapidity: if $\xi = \xi_0 = 4.5/\pi^2$

$$\Delta C_0 \approx 245 \Delta \xi \qquad (\xi = 4.5/\pi^2)$$

For $\xi = \xi_0$, $C_0(\xi_0) = 0.493048 \cdots$. The value 4.5 of $\pi^2 \xi$ is good for all binary computers since $4.5 = (100.1)_2$.

It remains to check the corresponding value of A_0. We find that $A_0(\xi_0) \approx 2 \times 10^{-11}$, so that the absolute error $(\pi/3)^{14} \cdot A_0(\xi_0) \approx 10^{-11}$.

Now the approximation (13) for $\cos \pi f$ takes the final form [10]:

$$\cos(x) = \cos \pi f = C_0 + \sum_{k=1}^{3} \frac{C_k|}{|f^2 + B_k} - \eta f^2 \qquad (13^*)$$

where $\eta = 4.5 = (100.1)_2$, the values of C_k and B_k depending on η.

We do not describe the corresponding modification of (15) since it is exactly the same as the procedure just explained. To terminate our example, we give the details of the reduction of range in the case of (13) and (15). It begins as before by $N\pi^{-1} = n + f_0$, $\sigma_1 = $ signum $(\tfrac{1}{2} - f_0)$ and $f_1 = \tfrac{1}{2} - (\tfrac{1}{2} - f_0) \cdot \sigma_1$.

To compute $\sin \pi f_1$, let $\sigma_2 = $ signum $(\tfrac{1}{3} - 2 f_1)$. If $\sigma_2 = 0$, then $\sin \pi f_1 = \tfrac{1}{2}$. For $\sigma_2 = +1$, $f = f_1 < \tfrac{1}{6}$ and $\sin \pi f$ is computed, using the modified (15). But for $\sigma_2 = -1$, $f = \tfrac{1}{2} - f_1 < \tfrac{1}{3}$ and $\sin \pi f_1 = \cos \pi f$ is computed, using (13*).

To compute $\cos \pi f_1$, let $\sigma_2 = $ signum $(\tfrac{1}{3} - f_1)$. If $\sigma_2 = 0$, then $\cos \pi f_1 = \tfrac{1}{2}$. For $\sigma_2 = +1$, $\cos \pi f_1$ is computed directly, using (13*), but for $\sigma_2 = -1$, $\cos \pi f_1 = \sin \pi f$, where $f = \tfrac{1}{2} - f_1 < \tfrac{1}{6}$ and $\sin \pi f$ is computed using the modified (15), i.e.,

$$\sin \pi f = f \cdot \left(C_0 - \eta^* \cdot f^2 + \sum_{k=1}^{2} \frac{C_k|}{|f^2 + D_k} \right) \qquad (15^*)$$

Thus, the described part of flow chart looks as follows:

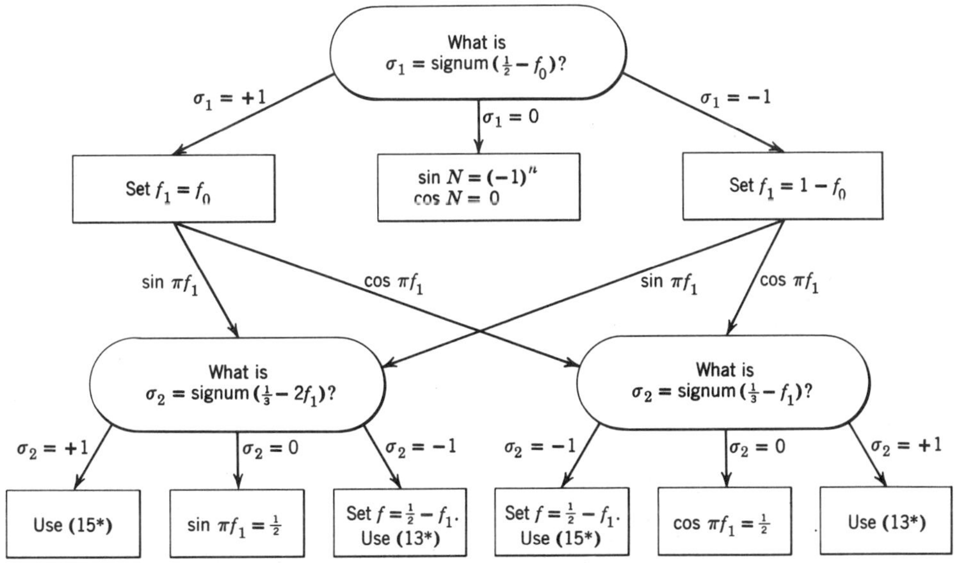

The new method just explained on the example of $Dg = 10$ can be applied to construct subroutines of any required accuracy, the number of operations M increasing with Dg. When both functions $\cos N$ and $\sin N$ for the same N are to be computed, as in the case when e^{iN} is to be generated, f and f^2 are the same in (13*) and (15*), so that $M = 8$ operations suffices to obtain both functions cos N and sin N with $Dg = 10$.

Tan N, Cotan N

The successive convergents $K_n(x)$ of the continued fraction

$$\frac{x}{2}\cdot \text{cotan}\frac{x}{2} = 1 + \sum_{n=1}^{\infty} \frac{-(x/2)^2|}{|2n+1} \quad (31)$$

yield a sequence of approximations to tan $(x/2)$ of increasing accuracy. To illustrate this method, we discuss the approximations $K_2(x)$ and $K_3(x)$

$$K_2(x) = x(60 - x^2)(10 - x^2)^{-1}/12$$
$$(32)$$
$$K_3(x) = 20x(42 - x^2)(1680 - 180x^2 + x^4)^{-1}$$

To find the correct expression of the errors we now deduce $K_n(x)$ from the classical Darboux two-point expansion ([14]):

$$[f(z) - f(a)]\cdot p^{(n)}(0) = \sum_{m=1}^{n} (-1)^m \cdot [p^{(n-m)}(0) \times f^{(m)}(a) - p^{(n-m)}(1) f^{(m)}(z)]\cdot (z-a)^m + h_n(z)$$

with

$$h_n(z) = (-1)^n (z-a)^{n+1}\cdot \int_0^1 p(t) \cdot f^{(n+1)}[a + t\cdot(z-a)]\cdot dt$$

which for $p(t) = t^n\cdot(t-1)^n$, $a = 0$, and $f(z) = e^{iz}$ yields the expansion

$$e^{iz} = 1 + \sum_{m=1}^{n} c_{nm}\cdot[1 - (-1)^m\cdot e^{iz}] \cdot (iz)^m + h_n(z)$$

where

$$h_n(z) = (n!)^2 \cdot i \cdot z^{2n+1} e^{i\theta_n z}/(2n)!(2n+1)! \quad (0 < \theta_n < 1)$$

and

$$c_{nm} = (2n-m)!n!/(2n)!(n-m)!m!$$

Separating the imaginary part, we have

$$P_n(z)\cdot \sin z = Q_n(z)(1 + \cos z) + R_n(z) \quad (33)$$

with

$$P_n(z) = 1 + \sum_{m=1}^{2m\leq n} c_{n,2m}\cdot(-1)^m z^{2m}$$

$$Q_n(z) = \frac{z}{2} + \sum_{m=1}^{2m+1\leq n} c_{n,2m+1}\cdot(-1)^m z^{2m+1}$$

$$R_n(z) = (n!)^2\cdot z^{2n+1}\cdot \cos(\theta_n z)/(2n)!(2n+1)!$$

Dividing (33) by $P_n(z)(1 + \cos z)$, we obtain the $(n-1)$st convergent $K_{n-1}(z)$ of (31) and the error $r_{n-1}(z)$

$$\tan(z/2) = K_{n-1}(z) + r_{n-1}(z)$$

where

$$K_{n-1}(z) = Q_n(z)/P_n(z)$$

and

$$r_{n-1}(z) = R_n(z)/(1 + \cos z)P_n(z)$$

The upper bound B_{n-1} of the relative error E_{n-1} in $0 < |z| \leqq z_0$ is therefore equal to:

$$|E_{n-1}| = |r_{n-1}\cdot \text{cotan}\, z/2| \leqq |z_0|^{2n}\cdot|z_0/\sin z_0| \cdot (n!)^2/P_n(z_0)(2n)!(2n+1)! = B_{n-1}(z_0)$$

The reason why we used Darboux's expansion is clear: now we can estimate the accuracy of the approximation $K_{n-1}(z)$, which depends on the reduced range $0 < |z| \leqq z_0$ and on the value of n.

The cases $n = 3$ and $n = 4$ give (32) and the corresponding upper bounds B_2, B_3 of the relative error in the interval $(-\pi/8, \pi/8)$ are:

$$B_2(\pi/8) = 3.75 \times 10^{-8}$$

and

$$B_3(\pi/8) = 2.4 \times 10^{-11}$$

Therefore, $Dg = 7$ and $Dg = 10$ are obtained when $K_2(x)$ and $K_3(x)$ are used to compute tan $(x/2)$ in the interval $(0 < |x| \leqq \pi/8)$.

Reducing the infinite range $0 < N < \infty$ to $(0, \pi/2)$, we have $N\cdot\pi^{-1} = n + f_1, 0 < f_1 < 1$. Let $\sigma = \text{signum}\,(\frac{1}{2} - f_1)$ and define $f_2 = |\frac{1}{2} - f_1|$; then $\tan N = \tan \pi f_1 = \sigma\cdot\text{cotan}\,\pi f_2$, $0 < f_2 < \frac{1}{2}$. Now, if the binary form of f_2 is $(0.0\alpha_1\alpha_2\beta_1\beta_2\cdots)_2$, we have $(0.0\alpha_1\alpha_2)_2 = i/8$, $0 \leqq i \leqq 3$. Let $f_3 = f_2 - i/8$. If $f = f_3 - \frac{1}{16}$, we have $|f| < 2^{-4}$ and tan πf can be computed directly as $K_2(2\pi f)$ or $K_3(2\pi f)$. We need eight stored constants: $a_i = \tan[(2i+1)\pi/16]$, $b_i = 1 + a_i^2$ for $i = 0, 1, 2, 3$. Then, $f_2 = (2i+1)/16 + f$ and

$$\tan N = \sigma\,\text{cotan}\,\pi f_2 = \sigma\left\{-a_i + \frac{b_i}{a_i + \tan \pi f}\right\}$$

while

$$\text{cotan}\, N = \sigma \tan \pi f_2 = \sigma\left\{-a_{3-i} + \frac{b_{3-i}}{a_{3-i} - \tan \pi f}\right\}$$

Now, for $f < \frac{1}{16}$,

$$\tan \pi f \approx K_2(2\pi f) = f\cdot\left[\frac{\pi}{6} + \frac{25/12\pi}{2.5/\pi^2 - f^2}\right] \quad (34)$$

can be computed in three operations: one multiplication by f, another to square f, and a division.

Therefore, $\tan N$ or $\operatorname{cotan} N$ with $Dg = 7$ can be obtained in $M = 5$ operations. But suppose we use the approximation

$$\tan \pi f \approx K_3(2\pi f) = \tfrac{1}{2} f/D^*(f) \qquad (35)$$

with

$$D^*(f) = 1.725\pi^{-1} - \pi f^2/20 - 12.8625\pi^{-3}(10.5\pi^{-2} - f^2)^{-1}$$

Then $Dg = 10$, $D^*(f)$ can be computed in four operations, and $M = 6$ is necessary to compute $\tan N$ or $\operatorname{cotan} N$ in $0 < N < \infty$.

If we want a subroutine with an accuracy characterized by $Dg = 8$ but more rapid than $K_3(x)$, we have to shorten the reduced range. In the range $0 < |f| < \frac{1}{24}$ the approximation $K_2(2\pi f)$ has the required accuracy: $B_2(\pi/12) = 3.1 \times 10^{-9}$.

Let us denote by I_k the five intervals of length $\pi/12$: $I_k = [(2k-1)(\pi/24);\ (2k+1)(\pi/24)]$ for $k = 1, 2, 3, 4, 5$, while $I_0 = [0; \pi/24]$ and $I_6 = [11\pi/24; \pi/2]$ are half as long. If $\theta = k\pi/12 + \pi f$ belongs to I_k, $0 \leq k \leq 6$, then $-\pi/24 < \pi f < \pi/24$ and $|f| < \frac{1}{24}$. Using the same addition formula for tangent, we have for $1 \leq k \leq 5$

$$\tan \theta = -a_k + \frac{b_k}{a_k \dot{-} \tan \pi f} \qquad (1 \leq k \leq 5)$$

where $a_k = \operatorname{cotan}(k\pi/12)$ and $b_k = a_k^2 + 1$. In I_0, $\tan \theta = \tan \pi f$ and in I_6

$$\tan \theta = \operatorname{cotan} \pi f \approx [K_2(2\pi f)]^{-1} = \left(\frac{6}{\pi} - \frac{75\pi^{-3}}{15\pi^{-2} - f^2}\right) \Big/ f$$

where $\pi f = \pi/2 - \theta$.

Increasing the number of subintervals necessitates more precomputed and stored constants. Thus in our last example we have twenty constants: π^{-1}, $(2k-1)\pi/24$ for $1 \leq k \leq 6$, $a_1 = 2 + \sqrt{3}$, $b_1 = 4(2 + \sqrt{3})$, $a_2 = \sqrt{3}$, $a_4 = \sqrt{3}/3$, $b_4 = 4/3$, $a_5 = 2 - \sqrt{3}$, $b_5 = 4(2 - \sqrt{3})$, $6/\pi$, $75\pi^{-3}$, $15\pi^{-2}$, $\pi/6$, $25/12\pi$, and $5/2\pi^2$.

The approximations to $\tan \pi f$ just studied can be also used in generating both $\cos x$ and $\sin x$. Suppose that their argument was already reduced to the range $0 < x \leq \pi/2$ and therefore they are to be computed in $(0, \pi/4)$. Consider first $0 < x \leq \pi/8 = 2\pi f$, so that $f < \frac{1}{16}$ and use for $\tan \pi f = t$ the approximation $K_2(2\pi f) = f/D(f)$ with

$$D = D(f) = 6/\pi - 75\pi^{-2}(15\pi^{-2} - f^2)^{-1}$$

equivalent to (34) and which insures $Dg = 7$. Since $\sin 2\pi f = 2t(1 + t^2)^{-1}$ and $\cos 2\pi f = (1 - t^2)(1 + t^2)^{-1}$, we have the rules

$$\sin x = 2f \cdot (D + f^2/D)^{-1}$$

and

$$\cos x = (D - f^2/D) \cdot (D + f^2/D)^{-1}$$

Naturally, t is not computed; only f and $D(f)$ in three operations. Thus, in $0 < x \leq \pi/8$ we can compute both functions in $M = 6$ operations.

But if $\pi/8 < x \leq \pi/4$ one more operation is needed. Letting $x = \pi/4 - 2\pi f$, we have $f \leq \frac{1}{16}$ so that $t = \tan \pi f = f/D$ as before. Substituting f/D instead of t in

$$\sin x = \frac{1}{\sqrt{2}}(1 - t^2 - 2t)(1 + t^2)^{-1}$$

$$\cos x = \frac{1}{\sqrt{2}}(1 - t^2 + 2t)(1 + t^2)^{-1}$$

we finally have the rules:

$$\sin x = \frac{\sqrt{2}}{2} - \sqrt{2}(f + f^2/D)/(D + f^2/D)$$

$$\cos x = \frac{\sqrt{2}}{2} + \sqrt{2}(f - f^2/D)/(D + f^2/D)$$

Having computed f, f^2, and D in three operations, we need three more for $\sin x$ and one for $\cos x$ which gives $M = 7$.

Using $K_3(2\pi f)$ and computing D^* instead of D we can construct a subroutine which gives as before both $\cos x$ and $\sin x$ in $M = 8$ operations with $Dg = 10$.

The question of the best subroutines for generating the trigonometric functions is not yet solved. As one of interesting possibilities, we call attention to the series

$$x \cdot \operatorname{cotan} x = 1 - 2 \cdot \sum_{n=1}^{\infty} c_n x^{2n} \qquad (|x| \leq \pi) \qquad (36)$$

with $c_n = \pi^{-2n} \cdot \zeta(2n)$. Padé's and Maehly's methods could be applied to this series and the resulting rational approximations compared to those deduced from the continued fraction

(31). The series (36) can be transformed into a series for sine and cosine:

$$\frac{\sin(\pi z/4)}{\pi z/4} = \left[1 - 2\sum_{1}^{\infty} \zeta(2n)\left(\frac{z}{8}\right)^{2n}\right] \times \left[1 + 2\sum_{1}^{\infty}(2n-1)\zeta(2n)\left(\frac{z}{8}\right)^{2n}\right]^{-1}$$

$$\cos(\pi z/4) = 1 - \frac{\pi^2 z^2}{8}\cdot\left[1 + 2\sum_{1}^{\infty}(2n-1)\cdot\zeta(2n)\left(\frac{z}{8}\right)^{2n}\right]^{-1}$$

Rational approximations to sine and cosine in the interval $(0, \pi/4)$ can be deduced from these expressions by both Padé's and Maehly's methods.

Log *N* in $(0, \infty)$ and Argtanhyp in (0, 1)

Our tools are the Gauss continued fraction for the hypergeometric function $F(\alpha, 1; \gamma; x)$ which for $\alpha = \frac{1}{2}$, $\gamma = \frac{3}{2}$ yields the expansion of $\log[(1 + x)/(1 - x)]$:

$$\log\frac{1+x}{1-x} = \frac{2x|}{|1} + \sum_{m=1}^{\infty}\frac{-m^2x^2|}{|4m^2-1} \tag{37}$$

and the Chebyshev expansion

$$\log f^* = \log\frac{1 + 2ax + a^2}{1 - 2ax + a^2} = 4\cdot\sum_{n=0}^{\infty}\frac{a^{2n+1}}{2n+1}\cdot T_{2n+1}(x) \tag{38}$$

where $|x| \leqq 1$ and the choice of the constant a depends on the range assigned to the argument f^* of $\log f^*$, i.e., on the required accuracy.

Omitting all the proofs and details, we give here the final results only (see our IBM Report #2, Feb. 1958).

To obtain $\log N$ with $Dg = 6$ we use the second convergent $K_2(x) = 6x(3 - x^2)^{-1}$ in the reduced range $2^{-1/8} < f^* < 2^{1/8}$, where $N = 2^{\nu}\cdot f$, $1 < f < 2$, and $f^* = f/r$ with $r = 2^{1/2 + \sigma_1/4 + \sigma_2/8}$, $\sigma_1 = \text{signum}(f - 2^{1/2})$, $\sigma_2 = \text{signum}(f - 2^{1/2+\sigma_1/4})$. It is sufficient to consider $1 < f^* < 2^{1/8}$, since $\log A^{-1} = -\log A$. Now the relative error of the approximation

$$\log f^* \approx K_2[(f-r)/(f+r)] = 3(f^2 - r^2) \times (f^2 + 4rf + r^2)^{-1} = R_2(f)$$

is 3.2×10^{-7}, if $1 < f^* \leqq 2^{1/8}$, and therefore $\log N \approx (\nu + \frac{1}{2} + \sigma_1/4 + \sigma_2/8)\cdot\log 2 + R_2(f)$ yields $Dg = 6$.

$$R_2(f) = a_0 + \sum_{m=1}^{2}\frac{-a_m|}{|b_m + f}$$

and $f = 2^{-\nu}\cdot N$ is defined by a shift (binary computer) so that $Dg = 6$ is obtained in $M = 3$ operations only. The four sets of constants a_m, b_m $(m = 1, 2)$ are $(a_0 = 3)$:

	$\sigma_1 = \sigma_2 = -1$	$\sigma_1 = -1,\ \sigma_2 = +1$	$\sigma_1 = +1,\ \sigma_2 = -1$	$\sigma_1 = \sigma_2 = +1$
$a_1 =$	13.086 092 784	15.562 074 660	18.506 529 900	22.008 097 032
$b_1 =$	3.816 777 062	4.538 938 442	5.397 737 887	6.419 028 301
$a_2 =$	0.891 905 336	1.261 344 622	1.783 810 672	2.522 689 245
$b_2 =$	0.545 253 866	0.648 419 777	0.771 105 412	0.917 004 043

They are defined by $a_1 = 12r$, $b_1 = 7r/2$, $a_2 = 3r^2/4$, and $b_2 = r/2$, and r has one of four values $2^{(2i+1)/8}$; $i = 0, 1, 2, 3$. Three constants $2^{1/4}$, $2^{1/2}$, $2^{3/4}$ are needed to locate f, so that in all $PC = 21$, $M = 3$, and $Dg = 6$.

Applying Maehly's method in the same range, we have in (38) $a = (2^{1/16} - 1)(2^{1/16} + 1)^{-1}$. To simplify the computation of coefficients we choose for this parameter a slightly larger value: 0.02166. Defining the approximation by

$$a_0\cdot T_1(x)\cdot[1 + b_1T_2(x)]^{-1} = Q_2(f) = p_0 + \sum_{m=1}^{3}\frac{-p_m|}{|f + q_m}$$

where $2ax = (1 + a^2)(f - r)(f + r)^{-1}$, we find that $p_0 = 2.954\ 462\ 815$, while

	$\sigma_1 = \sigma_2 = -1$	$\sigma_1 = +1,\ \sigma_2 = -1$	$\sigma_1 = -1,\ \sigma_2 = +1$	$\sigma_1 = \sigma_2 = +1$
$p_1 =$	12.893 50609	3.819 079 867	0.892 184 1794	0.544 998 1062
$q_1 =$	15.333 04918	4.541 676 950	1.261 738 966	0.648 115 6256
$p_2 =$	18.234 17117	5.400 994 540	1.784 368 359	0.770 743 7129
$q_2 =$	21.684 20609	6.422 901 135	2.523 477 933	0.916 573 9072

The upper bound of relative error of the approximation $Q_2(f)$ is equal to 10^{-7}, so that Maehly's $Q_2(f)$, which is characterized by the same values $M = 3$, $Dg = 6$, and $PC = 21$, seems to be slightly more accurate than $R_2(f)$ deduced from (37).

Applying Maehly's method in the range $2^{-1/2} < f^* < 2^{1/2}$, we take $a = 0.0866$ and in this case

$$Q_3(f) = [a_0T_1(x) + a_1T_3(x)]\cdot[1 + b_1T_2(x)]^{-1}$$

gives again $Dg = 6$ since the relative error is less than 7×10^{-8}. The number of operations is $M = 4$, but the subroutine based on $Q_3(x)$ is more economical since $r = \sqrt{2}$ has only one value and the logical part of the subroutine is much shorter. Here

$$Q_3(f) = p_0 + \sum_{m=1}^{3} \frac{-p_m|}{|f + q_m}$$

with a unique set of seven coefficients: $p_0 = 3.681\ 656\ 603$; $p_1 = 34.410\ 69291$; $q_1 = 10.379\ 67214$; $p_2 = 8.126\ 503\ 834$; $q_2 = 2.051\ 212\ 813$; $p_3 = 0.266\ 519\ 5666$ and $q_3 = 0.424\ 995\ 2497$. Here $\log N \approx (v + \frac{1}{2}) \log 2 + Q_3(f)$ yields $Dg = 6$ in $M = 4$ and $PC = 8$. It is interesting to observe that, using $K_3(x) = R_3(f)$ in the same range so that $M = 4$ and $PC = 8$, we cannot insure $Dg = 6$ but only $Dg = 5$. This shows clearly that Maehly's method is more economical for $\log N$ than the convergents of (37). With increasing Dg its superiority becomes more pronounced.

Table 5. VALUES OF Dg FOR $T = T_k$ AND m

T \ m	3	4	5	6	7	8	9	10
T_1	5	8	10	12	14	16	18	20
T_2	7	10	13	15	18	21	24	26
T_3	9	12	16	19	22	26	29	32
M	4	5	6	7	8	9	10	11

A detailed study of both methods in their application to the generation of $\log N$ gave the following results. Denoting the order of the convergent of (37) by m and the three ranges $(1, 2^{1/2})$, $(1, 2^{1/4})$, $(1, 2^{1/8})$ for f^* by T_1, T_2, T_3 respectively, the accuracy of a combination (m, T_k) is as shown in Table 5.

Approximating $\log f^*$ by the rational function

$$\sum_{0}^{M} a_m T_{2m+1}(x) \Big/ \sum_{0}^{N} b_m T_{2m}(x) \qquad \begin{pmatrix} N = M \text{ or} \\ N = M + 1 \end{pmatrix}$$

we consider six cases in Table 6, where, for $i = 1, 2, 3, 4, 5, 6$, $M = [2i + 1 - (-1)^i]/4$ and $N = [2i + 3 + (-1)^i]/4$, and we denote them by I–VI respectively.

Table 6. VALUES OF Dg AND THE NUMBER OF OPERATIONS M

T \ Case	I	II	III	IV	V	VI
T_1	7	10	12	15	18	20
T_2	9	12	15	18	22	—
T_3	11	14	18	22	26	—
M	4	5	6	7	8	9

Comparing Table 6 with Table 5, it can be seen that with increasing accuracy (increasing Dg) the superiority of Maehly's method is more pronounced. Thus, if $Dg = 20$ first significant correct digits is required, they can be obtained with the reduction to the range $2^{-1/8} < f^* < 2^{1/8}$, in eight operations, when K_7 is used, and in seven when (29) is used with $M = 2$, $N = 3$ (case IV). Reducing to the range $2^{-1/4} < f^* < 2^{1/4}$, $Dg = 20$ for K_8 and for (29) with $M = N = 3$ (case V) so that again one division is eliminated using Maehly's method. Finally, $Dg = 20$ for the range $2^{-1/2} < f^* < 2^{1/2}$, if (29) is used with $M = 3$, $N = 4$ (case VI) so that $\log N$ is computed by Maehly's method in nine operations, while if K_{10} is used the number of operations is equal to eleven.

In floating point, double precision computation ($Dg = 18$), the number M of operations is:

Range of f^*	$(2^{-1/2}, 2^{1/2})$	$(2^{-1/4}, 2^{1/4})$	$(2^{-1/8}, 2^{1/8})$
Maehly's method	8	7	6
$K_m(t)$	10	8	7

Concluding the comparative analysis of these two methods, we can state that Maehly's method should be recommended as a basis for subroutines dealing with $\log N$.

Briggs computed his famous table of logarithms approximating $\log (1 + x)$ by $x \cdot (1 - x/2)$ in a sufficiently small interval to insure the required accuracy. The reduction of range was achieved by multiplications, factors of the type 1.1, 1.11, etc., being chosen so that in fact additions were used.

R. W. Bemer [16] gave to Briggs' method the following form. Let $N = 2^n \cdot f$ with $\frac{1}{2} \leq f < 1$ and locate f in one of eight equal subintervals $I_k[\frac{1}{2} + (k - 1)/16;\ \frac{1}{2} + k/16]$, $1 \leq k \leq 8$. To compute $\log_2 N$ (which at the end is converted into $\log N$ or $\log_{10} N$ in one multiplication) eight multipliers A_k are defined such that multiplication of f by not more than three of the A_k results in a product $(\prod_k A_k)f$ which has the property $0 < t = (\prod_k A_k)f - 1 < 2^{-7}/5$. The first multiplier is chosen from Table 7 according to the subinterval I_k in which f is. If the resulting product

Table 7

k	I_k	A_k (binary)	$\text{Log}_2\, A_k$ (octal)
1	1/2– 9/16	1.110 11	.703 723 450 561
2	9/16–10/16	1.101 01	.564 543 765 447
3	10/16–11/16	1.100 00	.453 400 321 640
4	11/16–12/16	1.011 00	.353 165 174 015
5	12/16–13/16	1.010 00	.244 647 411 363
6	13/16–14/16	1.001 10	.176 740 553 440
7	14/16–15/16	1.000 11	.102 142 610 633
8	15/16–1	1.000 01	.026 565 575 654

is less than one, the next A_k is chosen according to which subinterval the product is in, etc. It is guaranteed that after not more than three multiplications t will be in the range indicated above. Moreover, the A_k are so chosen that multiplication by A_k is very rapid. This is easy to achieve in a computer with variable execution time multiplication such as, for instance, the IBM type 709, for which Bemer's subroutine was constructed. When this is done, $\log_2 (1 + t)$ is computed by

$$\log_2 (1 + t) \approx P_3(t) = t \cdot [c_0 - t(c_1 - c_2 t)]$$

in three multiplications. Bemer's $P_3(t)$ is the relaxed partial sum $t - t^2/2 + t^3/3 - t^4/4$ of the power series, multiplied by $\log_2 e$, and the octal coefficients are

$$c_0 = (1.342\ 520\ 342)_8$$
$$c_1 = (0.560\ 7625)_8$$
$$c_2 = (0.35)_8$$

Here $c_2 = (0.011101)_2$ is again such that the multiplication of t by it is very rapid. The absolute error for a natural logarithm is less than $2^{-27} \approx 7 \times 10^{-9}$. The multipliers and their logarithms (for a binary computer) are stored constants. For a decimal computer the range (0.1, 1) is subdivided in ten intervals and the factors A_k, $1 \leq k \leq 10$, are two digit numbers. The final formula is

$$\log_2 N \approx n - \sum_k \log_2 A_k + P_3(t)$$

where the sum includes one, two, or three terms. This shows that the natural logarithm $\log N$ is computed in five to seven multiplications of which two to four are very rapid. For computers with variable time multiplication this subroutine is a very economical one. Replacing $P_3(t)$ by Maehly's approximation to $\log (1 + t)$, one more long multiplication can be eliminated and in this case Bemer's form of Briggs' method can yield any required accuracy.

The function $\text{arctanhyp}\, x = \frac{1}{2} \log [(1 + x)(1 - x)^{-1}]$ can be computed most economically as a natural logarithm.

Arctan N in $(0, \infty)$

The Chebyshev expansion (3) and the continued fraction

$$\arctan t = \frac{t|}{|1} + \sum_{m=1}^{\infty} \frac{m^2 \cdot t^2|}{|2m + 1} \tag{37*}$$

formed in 1812 by Gauss [17] give the most economical subroutines for generating arctan N in $0 < N < \infty$. The methods are the same as for other functions and to avoid repetition we will list the results of a detailed study [3], omitting all the proofs and discussions.

Subdividing the infinite range of $N = \tan \theta$ into $(q/2) + 1$ intervals I_k by points $\theta_0 = 0$, $\theta_k = (k - \frac{1}{2})\pi/q$, $0 < \theta_k \leqslant \pi/2$, and using the addition theorem, we have—if $N \subset I_k$—

$$\arctan N = k\pi/q + \arctan \left(\alpha_k - \frac{\beta_k}{N + \alpha_k}\right)$$

where $\alpha_k = \text{cotan}\,(k\pi/q)$ and $\beta_k = 1 + \alpha_k^2$ are precomputed and stored constants.

The accuracy of an approximation depends on two parameters q and m, where m denotes the degree of a polynomial deduced from (3), while for a rational approximation based on (37*) m is the order of the convergent $K_m(t)$.

Maehly's method when applied to (3) gives for arctangent less economical subroutines than the subroutines based on the convergents $K_m(t)$.

Polynomial approximations with $Dg = 6, 8, 10, 18$, and 20 are obtained, using the combinations (m, q) given in Table 8 (see [3], Table 4, p. 51). They are only useful for computers with a slow division.

The corresponding results for rational approximations based on (37*) are given in Table 9 (see [3], Table 3, p. 49).

It is possible to obtain $Dg = 6, 8$, and 10 in $M = 3$, using $K_3^*(t)$ for $q = 10, 20$, and 45 respectively. Likewise, $Dg = 18$ and 20 is obtainable, using $K_{12}^*(t)$ for $q = 27$ and 45. Naturally, by increasing q it becomes necessary to store more and more constants.

Table 8

q	m	$M = m + 1$	PC	Dg
5	4	5	7	6
9	3	4	11	
12	3	4	9	
6	5	6	8	8
9	4	5	12	
12	4	5	10	
6	6	7	9	10
9	5	6	13	
12	5	6	11	
12	8	9	14	18
12	9	10	15	20

Table 9

q	m	M	PC	Dg
10	3	3	19	6
6	4	4	11	
4	5	5	7	
9	4	4	17	8
6	5	5	9	
9	5	5	14	10
4	8	6	11	
12	8	6	18	18
12	8	7	15	
15	8	6	30	20
12	9	7	16	

We find it necessary to point out that all the values of Dg in these tables mean the number of first *significant* correct digits in the decimal notation of the result. Very often the number Dg cited in the literature on subroutines means the number of first correct digits only and in such a case, if there is a string of zeros at the beginning, the number of significant correct digits may be very small: two, three, one, or even zero. This point is an important one since when an integrand, for instance, is a ratio of two small numbers the computation of a definite integral loses its accuracy completely, unless it is the *relative* error which is accounted for in the expression of Dg. Naturally, two subroutines with the same Dg should be compared from the point of view of economy of machine time only when they belong to the same class. For a function vanishing at the origin such as $\sin x$, $\log(1 + x)$, $\arctan x$, or $\tan x$ a subroutine yielding Dg first *significant* digits is a must and the subroutines which do not guarantee an upper bound of the *relative* error should never be used. Unfortunately, most of subroutines currently used for such functions ignore this important point.

Arcsin *N* in (0, 1)

All known subroutines for arcsine are based on the relation $\arcsin N = \arctan f(N)$, where $f(N) = N \cdot (1 - N^2)^{-1/2}$. Therefore, arcsine is not computed as such, but as an arctangent. The generation of $f(N)$, N being given, causes a big loss of machine time: at least five multiplications and divisions are necessary to compute $f(N)$. It seems that a direct generation of arcsine is worthwhile studying.

A new rational approximation $R(x)$ to arcsin N is obtained by applying Maehly's method to the expansion

$$\arcsin N = \arcsin(x \sin 2\theta) = \sum_{n=0}^{\infty} c_n(\theta) \cdot T_{2n+1}(x) \qquad (0 < \theta \leq \pi/4) \quad (39)$$

$N = 1$ is a singular point of our function. Therefore, $R(x)$ yields the required accuracy only for $N \leq N_0 = \sin 2\theta_0$, where $\theta_0 < \pi/4$. We take $\theta_0 = \pi/8$, but this choice is not unique and we cannot say that it is the best. In the range $0 < N \leq N_0 = 2^{-1/2}$, arcsin N is approximated by $R(x)$, but if $2^{-1/2} < N \leq \sin 3\pi/8$ then $2N^2 - 1 \leq 2^{-1/2}$, so that

$$\arcsin N = \pi/4 + \tfrac{1}{2} \arcsin(2N^2 - 1) \quad (40)$$

Finally, in $\sin 3\pi/8 \leq N \leq 1$ we use the relation

$$\arcsin N = \pi/2 - 2 \arcsin \sqrt{(1 - N)/2} \quad (41)$$

This new approach to the generation of arcsine is fully discussed in [18] and therefore it suffices here to describe the results. The coefficients $c_n(\theta)$ in (39) satisfy, for a given value of θ, the recurrence relation

$$m(2m + 3)^2 \cdot c_{m+1} = (2m + 1) \cdot [2m(m + 1)(\tan^2 \theta + \cot^2 \theta) + 1] \cdot c_m - (m + 1)(2m - 1)^2 c_{m-1}$$

and the first two are expressed in terms of

complete elliptic integrals $E = E(k)$ and $K = K(k)$, $k = \sin 2\theta$:

$$c_0 = 4(E - k'^2 \cdot K)/\pi k$$

$$c_1 = 4[(1 + 7k'^2) \cdot E - k'^2(5 + 3k'^2) \cdot K]/9\pi k^3$$

In particular, for $\theta = \pi/8$ when $2E - K = \pi/2K$ and $K = \Gamma^2(\frac{1}{4})/4\Gamma(\frac{1}{2})$, we have $c_0 = \sqrt{2}/K$ and $c_1 = c_0 - 8K\sqrt{2}/9\pi$. Computing c_m for $0 \leqq m \leqq 7$ and applying Maehly's method, we have the approximation

$$\arcsin N \approx N \cdot \left(A_0 + \frac{A_1|}{|B_1 - N^2} + \frac{A_2|}{|B_2 - N^2}\right) \quad (42)$$

where $A_0 = 0.52499\,78317$, $A_1 = 1.57834\,29040$, $B_1 = 3.55743\,40883$, $A_2 = 0.33215\,85891$, and $B_2 = 1.41569\,02913$. The maximum of the relative error is at $N = 0$ and is equal to 6.4×10^{-7}, if $0 \leqq N \leqq 2^{-1/2}$, so that $Dg = 6$ except in the neighborhood of the origin. The relative error decreases when N increases and for $0.1 \leqq N \leqq 2^{-1/2}$ it is less than 3.8×10^{-7}. For some exceptional values of N in (0; 0.1) the sixth significant digit may exceed by one the corresponding digit in the exact value of arcsin N. To avoid this loss of accuracy and to guarantee $Dg = 6$ for all values of N in $(0, 2^{-1/2})$ the partial sum of first three terms of the power series, namely $N \cdot [1 + N^2 \cdot (\frac{1}{6} + 3N^2/40)]$, could be used in the range (0; 0.1) instead of (42). It yields the first seven significant digits correct.

If this polynomial approximation is used in a binary, variable time multiplication computer the coefficients $\frac{1}{6}$ and $\frac{3}{40}$ can be replaced by $(0.00101\,01010\,10101\,0101)_2$ and $(0.00010\,01100\,11)_2$ without affecting the accuracy of approximation.

In the range $\sin 3\pi/8 < N \leqq 1$, arcsin N exceeds $3\pi/8 = 1.178 \cdots$ so that six significant digits are guaranteed, if the *absolute* error is less than 5×10^{-6}. The first two terms of the expansion (39) with $\theta \geqq \pi/32$ and $\tan \theta \geqq 0.0985$ do yield a polynomial approximation with an absolute error less than 2×10^{-6}. To simplify the numerical computations we take $\tan \theta = 0.1$ so that $c_0 = 0.199\,004\,9628$, $c_1 = 0.000\,330\,8457$, and $c_2 = 0.000\,001\,4875 < 1.5 \times 10^{-6}$. Therefore, we can use the approximation

$$\arcsin (2x/10.1) \approx c_0T_1(x) + c_1T_2(x)$$

which, for $x = 5.05 \cdot [(1 - N)/2]^{1/2}$, becomes

$$\arcsin [(1 - N)/2]^{1/2} \approx \varphi(N) = (A - B \cdot N) \cdot [(1 - N)/2]^{1/2}$$

with $A = 1.08518\,04210$ and $B = 0.08521\,76716$. We have now to compute the square root. It can be done in two divisions and one multiplication as follows. Let $B^2(1 - N)/2 = 2^{-2h} \cdot f$, $0.25 < f < 1$, so that $2^{-2h} \cdot f \leqq 0.000275$ gives $h \geqq 6$. Take as a first approximation to $\sqrt{f}$

$$R_1 = \alpha\left(\beta + f - \frac{\gamma}{f + \delta}\right)$$

with $\alpha = 0.3343\,1261$, $\beta = 2.7691\,3454$, $\gamma = 1.1903\,1245$, and $\delta = 0.5316\,4106$. The relative error $|1 - R_1/\sqrt{f}|$ is less than 6×10^{-4}, if $0.25 \leqq f \leqq 1$. The relative error of the second approximation $R_2 = \frac{1}{2}(R_1 + f/R_1)$ is therefore less than 1.8×10^{-7}, so that in

$$B \cdot [(1 - N)/2]^{1/2} \approx 2^{-h} \cdot R_2$$

the first six significant digits are correct. Thus, arcsin N in the range $\sin 3\pi/8 < N \leqq 1$ is computed by (41), where

$$\arcsin [(1 - N)/2]^{1/2} \approx (12.734\;21816 - N) \cdot 2^{-h}R_2 \quad (43)$$

in three multiplications and two divisions. Four operations suffice in the first range $0 < N \leqq 2^{-1/2}$, but if $2^{-1/2} \leqq N \leqq \sin 3\pi/8$ one more operation is necessary to compute $2N^2 - 1$.

The choice of range (0.25; 1) for f presupposes that a binary machine is used. For a decimal machine which cannot perform the extraction of square roots directly, as, for instance, the IBM type 610 can, we let $B^2(1 - N)/2 = 10^{-2h} \cdot f$ with $0.01 \leqq f \leqq 1$, $h \geqq 2$. Two sets of constants are needed here to insure $Dg = 6$: one for the range $0.01 \leqq f \leqq 0.1$, another for $0.1 \leqq f \leqq 1$ (see [18], p. 222).

Thus, with $M = 5$ the first six correct significant digits can be generated for arcsin N in $0 < N < 1$.

Arctan x can be generated for $Dg = 6$ in three operations, so that, when arcsin N is computed as arctan $[N(1 - N^2)^{-1/2}]$, in all $M = 8$ operations are needed. Thus, for $Dg = 6$ the direct computation of arcsin N as such economizes almost 40% of machine time, reducing $M = 8$ to $M = 5$. The difference of two operations between the direct subroutine for arcsine and the subroutine for arctangent we found for $Dg = 6$ does not depend on the accuracy. Therefore, the gain of three operations between the new and the old methods of generating arcsine exists for any Dg.

$\sqrt{N}$ in $(0, \infty)$

This is one of the most important elementary functions. It is generated usually with the aid of Heron's method of successive approximations. In this almost two-millenium-old method the $(n+1)$st approximation x_{n+1} to $x = \sqrt{N}$ is related to the nth, x_n, by

$$x_{n+1} = \tfrac{1}{2}(x_n + N/x_n) \qquad (44)$$

which is a particular case of the Newton-Raphson method to find the roots of $f(x) = 0$ by

$$x_{n+1} = x_n - f(x_n)/f'(x_n)$$

The relative error $e_n = |1 - x_n/\sqrt{N}|$ decreases very rapidly since $e_{n+1} \approx e_n^2/2$, but if the first guess x_1 is not sufficiently near to x as many as five iterations may be necessary to obtain $Dg = 10$: letting, in a binary computer, $N = 2^{2m} \cdot f$, $\sqrt{f}$ is to be computed, where $0.25 < f < 1$. Taking as a first guess f_1 the mid-point of the range, $f_1 = 0.625$, we may have $e_5 \approx 1 \times 10^{-10}$ if the value of f is near to 1.0. If $f_1 = f$, then $e_5 \approx 5 \times 10^{-10}$ if f is near to 0.25 and again four iterations are not enough for $Dg = 10$. Things are much worse in a decimal computer where the range of f is $0.01 \leq f \leq 1$.

We will now construct a Padé approximation to $\sqrt{f}$ in a given interval (b, c), $b > 0$

$$\sqrt{f} \approx d_0 + \frac{d_1|}{|f + d_2} + \frac{d_3|}{|f + d_4} \qquad (b \leq f \leq c) \quad (45)$$

which is computed in two divisions. It can be used also as a first approximation to $\sqrt{f}$ and then Heron's method applied to it. The five coefficients d_k are functions of a, where a belongs to the interval (b, c), $b < a < c$, and its choice influences the accuracy of (45). This accuracy depends on the ratio c/b of the range and increases when this ratio diminishes.

Letting $b = a(1 - r_1)$, $c = a(1 + r_2)$, where $0 < r_1 < 1$ and $0 < r_2 < 1$, we use a variable t related to f by $f = a(1 + t)$. The range of t, $-r_1 \leq t \leq t_2$, is interior to the interval $(-1, +1)$. Expanding $\sqrt{f} = \sqrt{a}(1 + t)^{1/2}$ into a power series $\sum_{n=0}^{\infty} c_n \cdot t^n$ with $c_n = \binom{\frac{1}{2}}{n} \cdot \sqrt{a}$, we approximate it by $(A + Bt + Ct^2)/(1 + \alpha t + \beta t^2)$:

$$(1 + \alpha t + \beta t^2) \cdot \sum_0^{\infty} c_n t^n - (A + Bt + Ct^2)$$
$$= \sum_{m=0}^{\infty} H_m \cdot t^{m+5} = t^5 \cdot H(t)$$

Here $A = c_0$, $B = c_1 + c_0\alpha$, $C = c_2 + c_1\alpha + c_0\beta$ as well as

$$\beta c_1 + \alpha c_2 + c_3 = 0$$
$$\beta c_2 + \alpha c_3 + c_4 = 0$$
$$\beta \cdot c_{m+3} + \alpha c_{m+4} + c_{m+5} = H_m$$

Thus, $\alpha = \frac{3}{4}$, $\beta = \frac{1}{16}$, $A = \sqrt{a}$, $B = (\frac{5}{4}) \cdot \sqrt{a}$, $C = 5 \cdot \sqrt{a}/16$, and $H_0 = 2^{-9} \cdot \sqrt{a}$. The ratio H_{m+1}/H_m rapidly approaches -1, when m increases, so that the function $H_0(1 + t)^{-1}$ is a good approximation to $H(t)$.

Therefore, the relative error is closely approximated by $H_0 \cdot t^5 \cdot (1 + \alpha t + \beta t^2)^{-1} \cdot (1 + t)^{-3/2} \cdot /\sqrt{a} = 2^{-9} \cdot \varphi(t)$. To study its behavior in $-1 \leq t \leq 1$, we observe that $2t \cdot (16 + 12t + t^2)\varphi'(t)/\varphi(t) = 3t^2 + 66t^2 + 208t + 160 \geq 15$ for $t \geq -1$. Thus, $\varphi'(t) > 0$ and the extrema of the relative error are at $t = -r_1$ and $t = r_2$.

Equating their absolute values, we have the condition

$$K(r_1) = r_1^5(1 - \alpha r_1 + \beta r_1^2)^{-1}(1 - r_1)^{-3/2}$$
$$= r_2^5(1 + \alpha r_2 + \beta r_2^2)^{-1}(1 + r_2)^{-3/2}$$
$$= G(r_2) \qquad (46)$$

Since the range of f is prescribed, the ratio $(1 + r_2)/(1 - r_1) = c/b$ is known, so that r_1 and r_2 can be found. To take an example, consider the whole range (0.25; 1). Then, $(1 + r_2)/(1 - r_1) = 4$ and combining it with (46) we find $r_1 = 0.511$, $r_2 = 0.956$, so that $1.956a = 1$. Taking $a = 1/1.96$, so that $r_2 = 0.96$ and $r_1 = 0.51$, we find $H_0 = 2^{-9}\sqrt{a} = 2^{-9}/1.4$, while the left- and right-hand members of (46) are equal to 0.155 and 0.166 respectively. Thus, the upper bound of the relative error in the interval (0.25; 1) is $2^{-9}/8.4 \approx 2.325 \times 10^{-4}$. Computing the constants $A = (1.4)^{-1}$, $B = 1.25/1.4$, and $C = 5/22.4$, transforming the rational function into a continued fraction, and replacing t by $t = (f - a)/a = 1.96f - 1$, we finally obtain the approximation (45) where $d_0 = C/\beta = 5\sqrt{a}$, $d_1 = -40a\sqrt{a}$, $d_2 = 9.4a$, $d_3 = -16a^2/25$, and $d_4 = 3a/5$.

Thus, using $a = (1.96)^{-1}$, we have

$$\sqrt{f} \approx 25/7 - \frac{5000 \times 7^{-3}|}{|f + 235 \times 7^{-2}} - \frac{400 \times 7^{-4}|}{|f + 15 \times 7^{-2}} \tag{47}$$

For $f = 1$ our approximation has a relative error equal to $2/7777 < 3 \times 10^{-4}$. If (47) is considered as a first approximation x_1 to $\sqrt{f}$ and x_2 is computed by (44), then the relative error of x_2 is less than 4.5×10^{-8}, so that $Dg = 7$ in the whole range (0.25; 1).

To improve the accuracy we now consider (45) in the interval (0.25; 0.50) for which $r_2 + 2r_1 = 1$. Combining this relation with (46), where now $(1 + r_2)/(1 - r_1) = 2$, we find $r_1 = 0.3$, $r_2 = 0.4$, and $a = 5/14$:

$$\sqrt{f} \approx 5\sqrt{70}/14 - \frac{50\sqrt{70}/49|}{|f + 47/14} - \frac{4/49|}{|f + 3/14} \qquad (\tfrac{1}{4} \leq f \leq \tfrac{1}{2}) \tag{48}$$

The upper bound of the relative error, $2^{-9} \cdot \varphi(-r_1) = 2^{-9} \cdot \varphi(r_2)$, is equal this time to 10^{-5}, so that, applying to the rational approximation (48) Heron's method (44) only once, we obtain $\sqrt{f}$, and therefore $\sqrt{N}$ for any N in $(0, \infty)$, in three operations with $Dg = 10$. The approximation (48) holds in the interval (0.25; 0.5) only. In (0.5; 1) it takes the following form:

$$\sqrt{f} \approx 5\sqrt{35}/7 - \frac{200\sqrt{35}/49|}{|f + 47/7} - \frac{16/49|}{|f + 3/7} \qquad (\tfrac{1}{2} \leq f \leq 1) \tag{48*}$$

The relative error remains the same and $Dg = 10$ with the aid of (44). Applying (48) and (48*), two sets of five constants are to be stored.

It is natural to try to apply the same method to the interval (0.01; 1) if the computer is a decimal machine. But the method does not work for too large values of the ratio $\rho = c/b$, which in this case is equal to 100. Since the expression for the relative error is based on the convergence of the series $\sum_0^\infty (-t)^n$, the necessary condition $r_2 < 1$ appears as a limitation of the ratio ρ. From $1 - r_1 = (1 + r_2)/\rho < 2/\rho$, we deduce that $r_1 > 1 - 2/\rho$. To satisfy condition (46), we must have the inequality $K(1 - 2/\rho) < G(1)$, i.e.,

$$F(\rho) = \rho^{3/2} \cdot [5\rho^2 + 20\rho + 4] - 29 \cdot (\rho - 2)^5 > 0$$

We have $F(4) = 394$, but $F(5) = -4486.78$.

Therefore we have to subdivide the range (0.01; 1) into at least four intervals since $F(\sqrt[3]{100}) = -684.5$, while $F(\sqrt[4]{100}) > 0$. Letting $k^2 = 10$, we consider the intervals (0.01; 0.01k), (0.01k; 0.1), (0.1; 0.1k), (0.1k; 1). Combining $(1 - r_1)\sqrt{10} = 1 + r_2$ with (46), we have $a^{-1} = 55.02$, $r_1 = 0.44976$, and $r_2 = 0.74$. This gives 1.3×10^{-4} as an upper bound of the relative error, so that applying to the rational approximation

$$\sqrt{f} \approx 0.674055 - \frac{0.098\,002|}{|f + 0.170836} - \frac{2.11388 \times 10^{-4}|}{|f + 0.0109044} \qquad (1 < 100f < k) \tag{49}$$

Heron's method, we can obtain $Dg = 7$, the relative error of the last approximation being less than 10^{-8}. The coefficients in the other three intervals are obtained by observing that if the range of f is $\lambda^2 \cdot b \leq f \leq \lambda^2 \cdot c$, then $f \cdot \lambda^{-2}$ belongs to the range (b, c). Applying this to (45), we obtain in the new range the approximation:

$$\sqrt{f} \approx \lambda d_0 - \frac{\lambda^3 d_1|}{|f + \lambda^2 d_2} - \frac{\lambda^4 d_3|}{|f + \lambda^2 d_4}$$

It is clear that our method can be used to generate roots of any integral order, or even any power of x. Thus, approximating $f^{1/m}$, where m is any positive integer, by $[A(m) + B(m) \cdot t + C(m)t^2] \cdot [1 + \alpha(m) \cdot t + \beta(m) \cdot t^2]^{-1}$, we obtain for $f = a(1 + t)$, $2^{-m} < a < 1$: $A_m = A(m) = a^{1/m}$, $B_m = (2m + 1)A_m/2m$, $C_m = (m + 1)B_m/6m$, $\alpha_m = (2m - 1)/2m$, $\beta_m = (m - 1)(2m - 1)/12m^2$, and $H_0(m) = (m^2 - 1)(4m^2 - 1)a^{1/m}/6!m^5$.

On the other hand, to improve the accuracy of our approximations it suffices to increase the degree of numerator and denominator.

Each unit added to this degree increases the number of operations M by one. If—for greater accuracy—a loss of digits occurs because of too large a constant term d_0, the same modification we used for (13) and (15) helps to eliminate this cause of loss of accuracy. Thus, approximating $\sqrt{f}$ by the ratio of two cubic polynomials $(A + Bt + Ct^2 + Dt^3)/(1 + \alpha t + \beta t^2 + \gamma t^3)$, we find $\alpha = \frac{5}{4}$, $\beta = \frac{3}{8}$, $\gamma = \frac{1}{64}$, $A = \sqrt{a}$, $B = 7A/4$, $C = 7A/8$, and $D = 7A/64$, so that $d_0 = D/\gamma = 7\sqrt{a}$. Computing H_0, we have

$$H_0 = -21 \times 2^{-15}\sqrt{a}$$

instead of $2^{-9}\sqrt{a}$. Choosing $r_1 = 0.3$ and $r_2 = 0.4$ we obtain, for $b \leqq f \leqq 2b$, $Dg = 6$ since the upper bound of relative error is now 4×10^{-7}, instead of 2.3×10^{-4}, when quadratic polynomials were used. The final form of the approximation is now

$$\sqrt{f} \approx 7\sqrt{a} - \frac{112a\sqrt{a}|}{|19a + f} - \frac{24a^2/7|}{|f + 5a/3} - \frac{8a^2/63|}{|f + a/3}$$

where $a = b \cdot (1 - r_1)^{-1} = 2b \cdot (1 + r_2)^{-1}$ Thus, in $M = 3$ operations $\sqrt{N}$ can be computed with the first six correct significant digits.

Concluding the discussion of methods studied in this chapter, we would like to stress that these methods can be applied to any function possessing a power series, convergent or divergent, or represented by a Chebyshev expansion.

REFERENCES

1. P. L. Chebyshev, *Oeuvres*, Vol. I, St. Petersburg, 1899, pp. 271–378.
2. F. D. Murnaghan and J. W. Wrench, Report No. 1175, David Taylor Model Basin, Apr. 1958.
3. E. G. Kogbetliantz, Computation of Arctan N for $-\infty < N < +\infty$ Using an Electronic Computer, *IBM J. Research and Devel.*, vol. 2, no. 1, Jan. 1958, pp. 43–53.
4. C. Hastings, *Approximations for Digital Computers*, Princeton University Press, 1955.
5. E. Remez, *General Computation Methods for Chebyshev Approximation. Problems with Real Parameters Entering Linearly*, Izdat. Akad. Nauk. Ukrainsk. SSR, Kiev, 1957.
6. G. N. Watson, *Theory of Bessel Functions*, The Macmillan Co., New York, 1944.
7. Cornelius Lanczos, *Applied Analysis*, Prentice-Hall, New York, 1956.
8. H. Padé, Sur la répresentation approchée d'une fonction par des fractions rationnelles, *Ann. sci. Éc. norm. sup., Paris*, vol. 9, 1892, pp. 1–93; vol. 16, 1899, pp. 395–426.
9. H. S. Wall, *Continued Fractions*, D. Van Nostrand Co., New York, 1948.
10. E. G. Kogbetliantz, Computation of Sin N, Cos N, and $\sqrt[m]{N}$ Using an Electronic Computer, *IBM J. Research and Devel.*, vol. 3, no. 2, Apr. 1959.
11. H. Maehly, Monthly Progress Report, Institute for Advanced Study, Princeton, Oct. 1956.
12. H. Maehly, First Interim Progress Report on Rational Approximations, Project NR 044–196, Princeton University, June 23, 1958.
13. E. G. Kogbetliantz, Computation of e^N for $-\infty < N < +\infty$ Using an Electronic Digital Computer, *IBM J. Research and Devel.*, vol. 1, no. 2, Apr. 1957, pp. 110–115.
14. G. Darboux, *Jour. de Math.* (3), vol. II, 1876, p. 271.
15. P. M. Hummel and C. L. Seebeck, A Generalization of Taylor's Expansion, *Amer. Math. Monthly*, vol. 56, Apr. 1949, pp. 243–247.
16. R. W. Bemer, A Subroutine Method for Calculating Logarithms, *Communs. Assoc. Comp. Mach.*, vol. 1, no. 5, May 1958, pp. 5–7.
17. C. F. Gauss, *Werke*, Vol. 3, 1876.
18. E. G. Kogbetliantz, Computation of Arcsin N for $0 < N < 1$ Using an Electronic Computer, *IBM J. Research and Devel.*, vol. 2, no. 3, July 1958, pp. 218–222.

PART II | MATRICES AND LINEAR EQUATIONS

Matrix inversion and related topics by direct methods

2

Alex Orden
University of Chicago

I. FUNCTION

In preparing a matrix inversion program for an electronic digital computer on the basis of a direct (i.e., noniterative) method, it is desirable to have the same program also serve for the solution of linear algebraic equations. The function of such a program would be:

$$\text{Solve } AX = B \tag{1}$$

where A is an $m \times m$ real matrix, B is a given $m \times r$ matrix which implies that r sets of "right-hand sides" are given, and X is an $m \times r$ matrix of solutions—each column being the solution which goes with the corresponding column of B. By setting $B = I$, the $m \times m$ identity matrix, a digital computer program for (1) provides $X = A^{-1}$, i.e., a matrix inversion program is available.

The programming of matrix inversion and simultaneous equations implied by (1) is a commonly used level of generality in the application of digital computers to linear algebraic systems. The approach which will be presented here proceeds to further generality. As an intrinsic and proper aspect of digital computer programming for matrix inversion by a direct method, it is recommended that the program have the generality to solve any of the following closely related problems (or combinations of these problems):

$$\left.\begin{array}{lll} \text{Compute } A^{-1} & & \\ \text{Solve} & AX & = B \text{ for } X \\ \text{Find} & \delta(A) & = \text{determinant of } A \\ \text{Find} & \rho(A) & = \text{rank of } A\text{, where } A \\ & & \quad \text{is allowed to be a} \\ & & \quad \text{rectangular matrix as} \\ & & \quad \text{well as a square} \\ & & \quad \text{matrix} \end{array}\right\} \tag{2}$$

In fact, the plan of attack which will be presented here appears to yield almost any of the results which a direct method can provide in matrix problems which arise in common applications. For example, if A is singular, and if the problem $Ax = b$ (a single right-hand side vector) has solutions, a general solution can be obtained in the form $x = x_0 + \Sigma \alpha_j y_j$, where x_0 is a particular solution, the y_j are linearly independent vectors, and the α_j are arbitrary parameters. In the case of a rectangular matrix, the program can find the rank, designate the rows and columns which provide a nonsingular square submatrix of that rank, and give the value of the determinant of that submatrix.

The computation method to be described is basically the classical "Gauss-Jordan" approach, which lends itself very well to the general concept of the function, given by (2), for a digital computer matrix inversion program.

2. MATHEMATICAL DISCUSSION

a. Choice of Method

Direct methods of matrix inversion and solution of linear equations have been derived in a considerable number of ways, involving the handling of matrices, equations, and determinants, and with the aid of various geometrical images. To some extent the availability of a variety of mathematical representations has facilitated the derivation and expression of alternative computational techniques. With the development of the digital computer field, however, it has been not so much the mathematical forms of the algorithms but rather a number of considerations in the storage and transfers of sequences of coefficients that have led to formulation of the computation process in a considerable variety of ways. Beyond that the digital computer field has also brought forth the need for error analysis in computations on large matrices, and the choice of computational method or development of new methods which minimize the errors that are intrinsic to the computations.

Before the advent of digital computers the methods usually presented were standard Gaussian elimination, use of determinants with Cramer's rule, and—relatively recently—several techniques to facilitate hand computation, which were developed by Doolittle, Choleski, Crout, and others. Looking over the latter techniques nowadays, one might say that they are directed toward optimizing the data-handling aspects of one particular "mechanization" of the solution of the problem, namely the situation in which the facilities available are a man with a pencil, paper, and a desk calculator. The digital computer field has introduced, not just one, but rather a variety of forms of mechanization. The proper choice of computational technique and some associated aspects of mathematical description have come to depend on such matters as the size and speed of the main memory of a digital computer and the relative speed of auxiliary memory devices. Two illustrations follow:

1. In machines with relatively small central memory, but with relatively high-speed auxiliary memory (magnetic drums or magnetic tapes), large matrices can be handled with very good efficiency by partitioning them into small square submatrices.
2. In machines with relatively slow auxiliary memory, in particular, where only the basic input-output is available as auxiliary memory, it is desirable to handle problems as large as possible without intermediate storage on auxiliary devices. A technique is available for solving linear equations in which the number of data cells needed in memory is approximately one-fourth of the number of terms in the original equations. The coefficients are entered a row at a time. When the kth row is entered, its first k terms are reduced to zero by use of multiples of the preceding rows. The kth row is then divided through by the kth term on the diagonal, and the kth row used to reduce the terms in column k in preceding rows to zero. The next row is then brought in and the process repeated. The configuration of matrix terms after, for example, the third stage is:

$$\begin{matrix} 1 & 0 & 0 & x & x & \\ 0 & 1 & 0 & x & x & x \\ 0 & 0 & 1 & x & x & x \end{matrix}$$

where the x's are matrix terms which have been altered by preceding calculations. There is a 3×3 unit matrix on the left and, in general, a $k \times k$ unit matrix after the kth row has been entered and processed. The successive unit matrices need be present by implication only, i.e., they need no memory space, and it can be seen that, as successive rows are entered, the memory space required for matrix coefficients builds up to about one-fourth the number of terms in the problem to be solved, and then decreases.

Considerations such as those in the above two illustrations have become more a matter of computer programming than of mathematical representation. One deals primarily with the order in which matrix rows or columns, terms, or submatrices are to be processed, and any one of several mathematical representations can provide a suitable basis for the work. This is somewhat in contrast to earlier work where the mathematical representation, such as by determinants or decomposition into upper and lower triangular matrices, was used to indicate a hand-in-hand connection between a mathematical solution and a computation algorithm.

We will not specialize, in the program to be described here, to the extent indicated by the two particular cases just described. However, choice of method should always involve consideration of the characteristics of the particular computing machine which is to be programmed. A fairly general situation to consider is that of programming for a machine with a fairly sizable main memory without recourse to auxiliary memory during the course of the computations. The fact that the memory size is fixed (never the case for the man with pencil and paper) becomes a factor in laying out the computational scheme. In particular, with some care in programming, the entire memory, except for the program, can be made available for a matrix which is to be inverted, rather than just half of that memory.

The classic Gauss-Jordan approach and other well-established techniques are satisfactory for handling fairly large matrices on digital computers, provided the matrices are well behaved. However, with very large matrices and with "ill-conditioned" matrices, the cumulative effect of arithmetic round-off becomes a very serious problem. This problem became apparent as soon as the development of digital computers made the handling of large matrices technically feasible, and appeared at first to offer what might be a limiting factor on matrix inversion and related problems. As a first step von Neumann and Goldstine [1] carried through an intricate analysis of the problem, attempting both to establish a method which would minimize the round-off error problem and to develop mathematical aspects of this type of error analysis.

In practice the error build-up problem has generally not been quite as serious as at first seemed a possibility, and approaches are available for extending the range of solvability. In the first place, one can take the brute force approach of double precision storage and arithmetic. This necessitates the use of subroutines for double precision arithmetic, which increases the computation time in the order of five to ten times. Secondly, direct methods can be followed by one or two iterative refinement stages.

The outlook with regard to enlarging the range of solvability is still developing, with double precision direct methods, combination of direct methods with iterative refinements, and successive approximation methods under study. In the material presented here, the outlook is that of providing a digital computer approach, considering the factors which are characteristic of the computing machines, but leaving the question of extending the range to very large matrices or ill-conditioned matrices out of direct consideration. Of course, the approach which will be described can be set up for double precision arithmetic, if desired, and this is at present the most widely used approach for extending accuracy and the size of matrix that can be handled.

b. Consideration of Computing Machine Characteristics

The reasons for presenting a matrix inversion scheme which includes the various subsidiary capabilities lie in the versatility of stored program digital computers with regard to variations in programs, and in the fact that direct method programs for the various problems in (2) would be almost identical.

A few branch points are required in a computer program to permit the choice of function indicated by (2). The amount of additional program to permit these choices is quite minor. In the class of problems at hand—direct methods of solution of matrix problems—the point of view here is that it is as desirable to make use of digital computer capabilities to provide generality of function as it is to provide the usual generality with regard to size of matrix.

In addition to defining the problem, as in (2), in a very general way with regard to function, which takes advantage of the inherent capabilities of stored program digital computers, the approach to be presented involves several aspects of digital computer characteristics from the point of view of memory capacity and computation time. The aspects of memory usage and computation time which are taken into account are rather broadly applicable, as compared with some other considerations which may arise in programming for specific machines.

The aspect of memory usage which is to be treated involves making the distinction in (2) between obtaining A^{-1} and solving $AX = B$. If, by setting $B = I$, we consider solving the problem $AX = B$ to obtain $X = A^{-1}$, the usual implication in computer programming is that memory cells in the computer are to be allocated for two square matrices, A and I, and at the end of the computation A^{-1} will be available in the cells originally used for I. This means that in matrix inversion only half of the memory available for matrix coefficients can be used to store A. A more careful examination of the computations shows that the full section of memory which is available for matrix coefficients can be used to store A, and thus larger matrices can be inverted without resorting to transfers to auxiliary memory devices during the computations. Questions of programming strategy rather than of mathematical method are involved, but these questions are considered here to be important to the digital computer point of view.

Two aspects of computation time will be taken into account. They are side considerations, which do not affect the basic structure of the approach, but they are of sufficient generality to be shown as simple options which may well be desirable. They are:

1. "Sparse matrices," i.e., matrices with a considerable number of zeros, offer opportunities to reduce the number of arithmetic steps.
2. Direct methods permit the reduction of a matrix A to an array which can be saved in place of A^{-1} for use at any time for solving sets of equations $Ax = b$, with different right-hand sides. Obtaining the reduced array requires only about half as much computation as obtaining A^{-1}. If the terms of A^{-1} do not in themselves have significance in the work at hand, i.e., if the purpose of obtaining A^{-1} is only to solve a number of cases of $Ax = b$ for different b's, then it is really rather pointless to spend the computation time to obtain A^{-1}, when the special array will do just as well. Generation of the array involves no additional aspect of programming: it emerges from the computation by calling for $AX = B$, but with no B matrix present. The array which is obtained has been termed the "product form of the inverse" of A in the literature of linear programming (see Chapter 25), and that term will be used here.

Scaling, word length, and form of mechanization of the arithmetic operations are interrelated primary factors in considering the effects of round-off error. These factors cannot be covered here, except to note some scaling considerations.

There has been a general tendency in scientific and engineering computations to use "floating point numbers," i.e., representation of each number by an exponent (power of 10 or power of 2 depending on whether the machine is decimal or binary), and a fractional part, which is less than one, and whose leftmost digit is not zero. The exponent and fractional part are stored in the same word. Some of the newer machines provide direct mechanization of arithmetic operations on numbers in floating point form. In machines without this capability, floating point arithmetic can be provided by means of subroutines.

Despite the appearance of optimum scaling and consequent optimum preservation of significant figures which use of floating point numbers appears to offer, the matrix inversion problem can be handled as well, *or perhaps better*, in fixed point arithmetic. In matrix operations such as the method which will be described here, the approach which can be used is to determine, and keep in memory, a scale factor for each column of the matrix such that the largest term in the column is "normalized," i.e., has a nonzero leftmost digit. This requires some coding of the arithmetic steps to permit holding the terms of a column temporarily in double length form, determination of whether a change in scale factor is required, and shifting of each term to correspond to the new scale factor before dropping the "lower half" of the term. This can be done at relatively little cost in computation time, and has the advantage over floating point arithmetic that more significant figures are carried along, since part of each word is not used by an exponent.

The computation plan which will be described below can be considered as basically applicable to either floating point or fixed point number representation. Certain questions of finding the largest terms in matrix columns and associated permutations of the matrix during the computations are desirable in either case, and are fully taken into account below. Beyond that, the question can be considered as one of detailed coding for particular machines. Coding for fixed point operation would require bringing in scale control over each column. Similarly, the computational plan can be considered as applicable with other coding aspects such as storage of double length numbers throughout the computations and corresponding use of double length arithmetic subroutines.

c. Mathematical Description

In most of the mathematical description which follows it is assumed that the successive "pivot terms" in successive matrix transformations lie on the main diagonal of a square matrix, as this simplifies the presentation. A detailed description is then given of the permutations of rows and columns which are associated with off-diagonal pivots. Although the latter is to some extent a matter of detail, it is included because it has a general character in the sense that it should be included in computer programs regardless of whether the number representation is fixed point or floating point, and because careful choice of a permutation algorithm is a factor in optimizing storage space and computation time.

The memory of the computing machine is assumed to have an area for matrix storage which holds a matrix A, to which may be adjoined a matrix B, both sequenced by columns. The memory space available for the matrix or matrices is fixed; it may be filled by a square or rectangular matrix A, as well as by two matrices A and B.

(1) General Description

The method which, as noted earlier, is basically the Gauss-Jordan reduction will be expressed here in the form of matrix products, an approach which yields various desired aspects of the computation quite conveniently.

Consider the problem $AX = B$, where A is an $m \times m$ square matrix, and X and B are $m \times r$ matrices, i.e., the solution of linear equations for a matrix A and r different right-hand sides which are the columns of B.

The computational process will be considered to represent the generation of a succession of matrix products:

$$
\begin{aligned}
[A, B] = I[A, B] &= V^{(0)}W^{(0)} \\
&= V^{(1)}W^{(1)} \\
&= V^{(2)}W^{(2)} \\
&\quad \cdot \\
&\quad \cdot \\
&\quad \cdot \\
&= V^{(m)}W^{(m)} = A[I, X]
\end{aligned} \tag{3}
$$

where $[A, B]$ means that A and B adjoined are to be treated as a single matrix.

Each matrix product in (3) is of the form $D = V^{(k)}W^{(k)}$. D is a matrix of original data, in this case $[A, B]$, $V^{(k)}$ is a sequence of square matrices, initially the identity matrix I, and at the final stage $V^{(m)} = A$. The $W^{(k)}$ are rectangular matrices which satisfy the matrix product expression. Each stage of calculation involves replacing a unit vector in $V^{(k)}$ by a vector in A so that at the final stage the V matrices have changed from $V^{(0)} = I$ to $V^{(m)} = A$. Computationally it is only the matrix W that is operated on, since the left side of (3) is always the same, and the successive matrices $V^{(k)}$ merely require a bookkeeping process to keep track of which columns of A are in $V^{(k)}$. The computational algorithm on W will be set up in the next section.

In terms of real vector spaces the matrices $V^{(k)}$ form a succession of bases for an m-dimensional space, and the succession of products in (3) are the representations of $[A, B]$ in terms of successive bases $V^{(k)}$ until A becomes the basis, which yields the solution to $AX = B$. The process is such that the matrices $V^{(k)}$ are always nonsingular, i.e., always form a basis for an m-dimensional vector space.

To invert a matrix A in the notation of (3), with $B = I$, we write

$$
\begin{aligned}
[A, I] &= I[A, I] \\
&\quad \cdot \\
&\quad \cdot \\
&\quad \cdot \\
&= A[I, A^{-1}]
\end{aligned} \tag{4}
$$

If A is singular, the process stops at a stage where there is a nonsingular matrix $V^{(k)}$ and the corresponding matrix product, as in (3), but no further columns from A can be introduced into $V^{(k)}$ to form a nonsingular $V^{(k+1)}$.

(2) Basic Computational Algorithm

The computations to transform $W^{(k-1)}$ to $W^{(k)}$ are the familiar pivotal transformations. This section will set up the algorithm in terms of the algebraic notations that we are using and will serve as a basis for facets of the problem which appear in later sections.

At stage $k - 1$ we have:

$$D = [A, B] = V^{(k-1)} W^{(k-1)} \tag{5}$$

with

$$V^{(k-1)} = [A_1, A_2, \cdots, A_{k-1}, I_k, I_{k-1}, \cdots, I_m]$$

where I_k is the kth column of the identity matrix. We wish to introduce $V^{(k)}$, which differs from $V^{(k-1)}$ in having A_k in place of I_k. By (5) we have for all columns in D:

$$D_j = w_{1j}A_1 + w_{2j}A_2 + \cdots + w_{k-1,\,j}A_{k-1} + w_{kj}I_k + \cdots + w_{mj}I_m \tag{6}$$

where w_{ij} for $i = 1 \cdots m$ are the terms in column j of $W^{(k-1)}$. In particular, $D_k = A_k$, so that

$$A_k = w_{1k}A_1 + \cdots + w_{k-1,\,k}A_{k-1} + w_{kk}I_k + \cdots + w_{mk}I_m \tag{7}$$

We assume in this section that $w_{kk} \neq 0$; then solving (7) for I_k and substituting for I_k in (6), we get:

$$D_j = \left(w_{1j} - w_{kj}\frac{w_{1k}}{w_{kk}}\right)A_1 + \left(w_{2j} - w_{kj}\frac{w_{2k}}{w_{kk}}\right)A_{2j} + \cdots + \frac{w_{kj}}{w_{kk}}A_k + \left(w_{k+1,j} - w_{kj}\frac{w_{k+1,k}}{w_{kk}}\right)I_{k+1} + \cdots + \left(w_{mj} - w_{kj}\frac{w_{mk}}{w_{kk}}\right)I_m \tag{8}$$

The basis vectors in (8) differ from those in (6) only in the replacement of I_k by A_k, i.e., the new basis is the desired matrix $V^{(k)}$. Equation (8) thus gives the desired computation algorithm:

$$\begin{aligned} w'_{ij} &= \left(w_{ij} - w_{kj}\frac{w_{ik}}{w_{kk}}\right) \quad \text{for all } i \text{ except } i = k \text{ and for all } j \\ w'_{kj} &= \frac{w_{kj}}{w_{kk}} \quad \text{for all } j \end{aligned} \tag{9}$$

where w'_{ij} are the terms of $W^{(k)}$.

The quotients w_{ik}/w_{kk} in (9) are used in the transformation of every column. Therefore the computation scheme is written:

$$\begin{aligned} c_i &= \frac{w_{ik}}{w_{kk}} \quad \text{for all } i \text{ except } i = k \\ w'_{ij} &= w_{ij} - w_{kj}c_i \quad \text{for all } i \text{ except } i = k \text{ and for all } j \\ w'_{kj} &= \frac{w_{kj}}{w_{kk}} \quad \text{for all } j \end{aligned} \tag{10}$$

In (10), if $w_{kj} = 0$, then $w'_{ij} = w_{ij}$ for the entire jth column. Since large matrices which occur in practice are fairly often "sparse," i.e., have many zeros, it is worthwhile in a digital computer to test whether $w_{kj} = 0$ in each column, and if so to bypass computation on that column.

Let $C^{(k)}$ be a square matrix, which differs from a unit matrix only in the kth column, and is of the form:

$$C^{(k)} = \begin{bmatrix} 1 & & -c_1 & & \\ & 1 & -c_2 & & \\ & & 1 & 0 & \\ & & \frac{1}{w_{kk}} & & \\ 0 & & \cdot & 1 & \\ & & \cdot & & \\ & & \cdot & & 1 \\ & & -c_m & & \end{bmatrix} \tag{11}$$

Then (10) is equivalent to the matrix product

$$W^{(k)} = C^{(k)} W^{(k-1)} \tag{12}$$

(3) Solution of Linear Equations

As noted earlier, it is only the successive matrices $W^{(k)}$ which need be in the computer memory. At the kth stage we have:

$$[A, B] = V^{(k)} W^{(k)}$$

where the first k columns of the basis $V^{(k)}$ are $A_1 \cdots A_k$; consequently the first k columns of $W^{(k)}$ are unit vectors. At each stage another unit vector enters W and those already present remain unchanged during all succeeding transformations. Consequently, at each stage a computer program should operate only on the columns from k on to $(n + r)$, rather than on all columns of W.

Rather than generate and store a unit vector in the kth column the program should at each stage store the column $C_k^{(k)}$ which is given in (11).

This has functions, which will appear below, in connection with compact memory arrangement for matrix inversion, and in connection with the reduced array which can serve in place of the inverse. This is, however, a matter of efficiency of computer storage. For mathematical purposes the expression $[A, B] = VW$ should be thought of as having the appropriate unit vectors in W, with the columns $C_k^{(k)}$ held separately.

(4) THE INVERSE

For matrix inversion the product representation is:

$$[A, I] = I[A, I] = V^{(0)}[A, I]$$
$$\vdots$$
$$= V^{(k)}W^{(k)}$$
$$\vdots$$
$$= V^{(m)}[I, A^{-1}]$$

where at each stage

$$V^{(k)} = [A_1, A_2, \cdots, A_k, I_{k+1}, \cdots, I_m]$$

Since the basis, $V^{(k)}$, always consists of m columns of the left-hand side $[A, I]$, $W^{(k)}$ always has m unit vectors in it. It can be written:

$$W^{(k)} = [I_1, I_2, \cdots, I_k, W_{k+1}^{(k)}, \cdots, W_{k+m}^{(k)}, I_{k+1}, \cdots, I_m] \quad (13)$$

The proper mechanization of this is to store only the columns $W_{k+1}^{(k)}, \cdots, W_{k+m}^{(k)}$, which at the start means storing A instead of $[A, I]$. Thus a fixed memory size can either handle a larger matrix or, alternatively, half of the available memory can be used to save A, while the other half is used for computing A^{-1}.

Equations (12) and (13) show that $W_{k+m}^{(k)}$ in (13) is just column $C_k^{(k)}$ given in (11) [since column $(k + m)$ in $W^{(k-1)}$ is I_k]. Thus the storing of $C_k^{(k)}$ at each stage, as called for in the preceding section, is intrinsic to the matrix inversion case. For simultaneous equations the $C_k^{(k)}$ are computational entities for the kth stage which need no further processing. For matrix inversion they become part of the matrix W, and must be transformed in all succeeding stages. The difference between handling simultaneous equations and matrix inversion from the digital computer programming point of view is the very minor branching aspect that for the former the number of columns to be transformed decreases by one at each stage, while in the latter all columns are transformed at each stage.

(5) PRODUCT FORM OF THE INVERSE

Consider the left-hand side of (3) to be a nonsingular matrix A. Then using the notation of (12) the process may be expressed in the form:

$$W^{(0)} = A$$
$$W^{(1)} = C^{(1)}W^{(0)} = C^{(1)}A$$
$$W^{(k)} = C^{(k)}W^{(k-1)}$$

The final result is:

$$I = W^{(m)} = C^{(m)}C^{(m-1)} \cdots C^{(1)}A$$

Thus,

$$C^{(m)}C^{(m-1)} \cdots C^{(1)} = A^{-1} \quad (14)$$

Since each $C^{(k)}$ differs from a unit matrix in only one column, it requires just m storage cells, and the product form of the inverse requires m^2 cells, just as the true inverse does. Proceeding at each stage as described in Section (3), i.e., storing column $C^{(k)}$ in column k and transforming only the succeeding columns, the product form of the inverse requires $(m^3 + m^2)/2$ multiplications or divisions, as compared with m^3, to compute A^{-1}. Thus the amount of computation to obtain the product form of the inverse is about one-half of that for obtaining the true inverse.

The product form of the inverse is as efficient for solving linear equations as is the true inverse. In both cases the solution of $Ax = b$ requires m^2 multiplications. The solution using the product form is:

$$x = A^{-1}b = C^{(m)}C^{(m-1)} \cdots C^{(1)}b$$

Set $x^{(0)} = b$. Then m successive matrix multiplications

$$x^{(k)} = C^{(k)}x^{(k-1)} \qquad k = 1 \cdots m \quad (15)$$

give the final solution $x^{(m)}$, where each of the matrix by vector multiplications, (15), requires just m multiplications because of the form of the matrices $C^{(k)}$, namely at each stage:

$$x_i^{(k)} = x_i^{(k-1)} + c_i^{(k)}x_k^{(k-1)} \qquad i = 1 \cdots m \text{ except } i = k \quad (16)$$

$$x_k^{(k)} = c_k^{(k)}x_k^{(k-1)}$$

A minor change permits simpler expression of (16). Replace each of the diagonal terms, $c_k^{(k)}$, in the product form of the inverse by $(c_k^{(k)} - 1)$. Then (16) becomes simply

$$x_i^{(k)} = x_i^{(k-1)} + c_i^{(k)}x_k^{(k-1)} \qquad i = 1 \cdots m \quad (17)$$

(6) THE DETERMINANT

Our process will always deal with determinants of the matrices $V^{(k)}$ in (3). The successive $V^{(k)}$ always consist of linearly independent columns from the original matrix A, supplemented by unit vectors to make up an $m \times m$ nonsingular matrix. It is straightforward to confirm that at each transformation

$$V^{(k)}C^{(k)} = V^{(k-1)}$$

and that

$$\delta(C^{(k)}) = \frac{1}{w_{kk}}$$

Consequently

$$\delta(V^{(k)}) = w_{kk}\delta(V^{(k-1)}) \qquad (18)$$

Thus starting with $V^{(0)} = I$ and $\delta(V^{(0)}) = 1$, the determinant of $V^{(k)}$ is obtained from the determinant of $V^{(k-1)}$ by multiplying by the "pivot term" w_{kk}. Assuming that A is square and nonsingular, the final stage gives $\delta(V^{(m)}) = \delta(A)$.

Since merely one multiplication per transformation stage is involved, it is desirable to carry the determinant evaluation along in all of the problems to which the over-all process can be applied. In the particular case where only the determinant of a square matrix A is wanted, the process is exactly the same as for solution of linear equations, but with the left-hand side of (3) simply the matrix A, instead of $[A, B]$.

$\delta(V^{(k)})$ is at each stage the determinant of a submatrix of A. This will be noted further in connection with determination of the rank of A.

(7) THE RANK

Assume that the process has proceeded as described in Section (2) to the kth stage, specifically that up to that stage the pivot terms have not been zero. Since the columns of B in (3) can be any m-dimensional vectors, $V^{(k-1)}$ is at that point an m-dimensional basis. The first $(k - 1)$ columns of $V^{(k-1)}$ are at that stage $A_1 \cdots A_{k-1}$; therefore these columns of A are linearly independent and the rank of A is at least $(k - 1)$.

At that stage if the terms $w_{kk} \cdots w_{mk}$ in (7) are all zero, A_k is a linear combination of $A_1 \cdots A_{k-1}$, and A_k cannot be introduced into the basis $V^{(k)}$. One option at that point is for the computation process to stop with an indication that A is singular.

Alternatively, the process can be allowed to continue by considering A_{k+1} in place of A_k. If any term $w_{k,\,k+1} \cdots w_{m,\,k+1}$ differs from zero, the vector A_{k+1} permits the formation of a new basis $V^{(k)}$ in place of $V^{(k-1)}$; otherwise A_{k+1} is also a linear combination of $A_1 \cdots A_{k-1}$. Similarly, one proceeds to subsequent columns if necessary. (Assume here, in order to avoid questions of row or column permutation, that the next pivot term is in row k of W regardless of whether it is in column W_k, W_{k+1}, or a subsequent column.)

Proceeding as above, each transformation stage replaces a basis $V^{(k-1)}$ by a basis $V^{(k)}$ which contains one more linearly independent column from A. Each column of A gets only one chance to enter the basis, at which time either a nonzero pivot term in the corresponding column of W permits the column of A to enter the basis or else that column has been shown to be a linear combination of columns of A which are already in the basis.

Thus by going through essentially the identical process needed for simultaneous equations or matrix inversion, the process can tally up the number of columns of A which enter the basis, and give the rank of A. A can be a rectangular matrix, as well as a square one.

As in the case of obtaining the determinant, if the only result wanted is the rank of A the process should be considered to be the same as that for linear equations, except that the left side of (3), and consequently $W^{(0)}$, is A instead of $[A, B]$. A decreasing number of columns are transformed at each stage.

At each stage the basis V consists of columns of A and unit vectors, and the determinant $\delta(V)$ is available. It is clear that $\delta(V^{(k)})$ is the determinant of a submatrix of A whose columns are those which have entered the basis $V^{(k)}$, and whose rows are those in which the pivot terms have been found. A digital computer can of course develop and store a record of the rows and columns of A which are involved.

(8) UNDERDETERMINED SYSTEMS

By including the steps for determination of rank, the process described in Sections (2) and (3) for linear equations covers the general solution of underdetermined sets of linear equations.

The process starts with $[A, B] = W^{(0)}$. Consider B to be a single column. After all columns of A have had a chance to enter the basis, as described in Section (7), the results are available. Just as for a vector of A as described in the section on rank, the positions of zero and nonzero terms in the last column of the final matrix W show whether the rank of $[A, B]$ is greater than the rank of A. If so, the system of equations $AX = B$ is inconsistent. If the rank of $[A, B]$ is seen to be the same as the rank of A, the equations are consistent. A particular solution and general solution are then available from the final matrix W as follows:

Let A be $m \times n$ and of rank ρ. The process stops after ρ transformations with a basis $V^{(\rho)}$ which consists of ρ columns of A and $(m - \rho)$ unit vectors. We denote the final situation by $[A, B] = V^{(\rho)}W^{(\rho)}$ and assume for simplicity that the first ρ columns, $A_1 \cdots A_\rho$, of A are linearly independent and have entered the basis, and that no row permutations have been involved. Then (dropping the superscript ρ from the w_{ij}):

$$B = \sum_{i=1}^{\rho} w_{i,n+1}A_i$$

i.e., the $(n + 1)$st column of W gives a particular solution:

$$x_i = w_{i,n+1} \qquad i = 1 \cdots \rho$$
$$x_i = 0 \qquad i = (\rho + 1) \cdots n$$

For a general solution we have:

$$A_j = \sum_{i=1}^{\rho} w_{ij}A_i \qquad j = (\rho + 1) \cdots n$$

Let λ_j for $j = (\rho + 1) \cdots n$ be arbitrary parameters. Then

$$B = \sum_{i=1}^{\rho} w_{i,n+1}A_i + \sum_{j=\rho+1}^{n} \lambda_j(A_j - \sum_{i=1}^{\rho} w_{ij}A_i)$$
$$= \sum_{i=1}^{\rho} (w_{i,n+1} - \sum_{j=\rho+1}^{n} \lambda_j w_{ij})A_i + \sum_{j=\rho+1}^{n} \lambda_j A_j$$

i.e., the general solution is obtained essentially from the last column of W and arbitrary multiples of the columns of W which correspond with columns of A which are not in the final basis. Explicitly, the general solution is:

$$x_i = w_{i,n+1} - \sum_{j=\rho+1}^{n} \lambda_j w_{ij} \qquad i = 1 \cdots \rho \qquad (19)$$
$$x_i = \lambda_i \text{ (arbitrary)} \qquad i = (\rho + 1) \cdots n$$

(9) PERMUTATION PROCEDURE

In order to guard against $w_{kk} = 0$ [Eqs. (7) and (8)], the process has to be modified, leading to permutation of rows and columns.

Let A be a nonsingular matrix. Then at the kth transformation stage A_k can always be introduced into the basis. Consider the terms w_{ik}, which are the coefficients of unit vectors on the right side of (7). At least one of these must be nonzero, since otherwise the matrix would be singular. Let the largest, in absolute value of the coefficients of unit vectors, be $w_{\phi k}$. Substitute A_k into the basis in place of I_ϕ; the transformation $W^{(k-1)}$ to $W^{(k)}$ then takes the form [in place of (10)]:

$$c_i = \frac{w_{ik}}{w_{\phi k}} \quad \text{for all } i \text{ except } i = \phi$$
$$w'_{ij} = w_{ij} - w_{\phi j}c_i \quad \text{for all } i \text{ except } i = \phi \text{ and for all } j \qquad (20)$$
$$w'_{\phi j} = \frac{w_{\phi j}}{w_{\phi k}} \quad \text{for all } j$$

If A is singular or rectangular, the column which enters the basis at transformation k need not be A_k, since at that stage or preceding stages linear dependence may have been uncovered, so that at stage k it may be a vector A_{q_k}, with $q_k > k$ which is to be substituted into the basis. Let q_k be the index of the vector which enters the basis at the kth transformation stage, where $q_k \geq k$. In the computer program (as detailed in the flow chart) the index q_k replaces the index k in (20). In this section we will, however, consider A as nonsingular, and consequently denote the pivotal column as A_k.

Selection of the largest w_{ik}, of those in column k which are coefficients of unit vectors, is a convenient approach in computer programming to the selection of a nonzero pivot term. It has the further virtue that the use of the largest eligible term as the pivot term is favorable with regard to minimizing error propagation, since, in general, small terms are likely to contain larger percentage errors, and the use of small terms as pivot terms would have a tendency to propagate their error levels into all terms of the matrix.

In the earlier discussion, which did not call for selecting the largest eligible terms in the pivotal column, (8) indicates that after a transformation, $W^{(k-1)}$ to $W^{(k)}$, the coefficients of A_k appear in the kth row of $W^{(k)}$. Use of (20) implies that the coefficients of A_k in the relations equivalent to (8) appear in $W^{(k)}$ in row ϕ instead of row k. Rows ϕ and k can be interchanged after each stage, or the successive ϕ values can be recorded and the required over-all permutation applied to the rows of the final matrix W. Assume the latter procedure, and designate the sequence of ϕ values which arise in the transformations from $W^{(0)}$ to $W^{(m)}$ by f_k, $k = 1 \cdots m$. Then the permutation to be carried out on the final matrix W is that row f_k is to be moved to row k for $k = 1 \cdots m$.

For the solution of simultaneous equations the row permutation puts the final results into proper order. For matrix inversion, in addition to the row permutation, the inverse permutation must be applied to the columns, i.e., column k is to be moved to column f_k for $k = 1 \cdots m$.

The need for column permutation (as well as rows) in the case of matrix inversion is associated with the memory reduction scheme which was described in Section (4). Consider again that A_k replaces I_ϕ in the basis at stage k. If W were stored as $2m$ columns [see (13)], then at that point the basic computational algorithm would replace a unit vector in column $(m + \phi)$ of $W^{(k-1)}$ by new terms in the same column of $W^{(k)}$, and after further transformation stages, the ϕth column of A^{-1} would appear, in proper order, in column $(m + \phi)$. However, the memory compacting scheme for matrix inversion calls for inserting the terms which replace I_ϕ at stage k in column k. Denoting the sequence of ϕ values for all the stages by f_k, the effect is that at the end of the process the permutation of columns k to f_k, i.e., the inverse of the row permutation, puts the columns into the proper order for A^{-1}.

If we were dealing with hand computation, no more would have to be said about the permutation aspects of the process. On a digital computing machine, however, it becomes fairly important to consider how to handle the permutations from the point of view of programming efficiency and economy of memory usage. A good digital computer approach will therefore be described here. The approach permits permuting the rows alone, or, for the matrix inversion case, carrying out the column permutation concurrently.

The method calls for a series of $(m - 1)$ pairwise exchanges of rows (and columns, if necessary). Consider the rows as being identified by their sequential positions at the time the transformation calculations have been completed, but the permutations not yet carried out. The sequence f_k, $k = 1 \cdots m$, is available and the permutation required is that each row vector f_k is to be transferred to row k.

Because digital computers operate efficiently by means of table look-up (as compared with scanning a sequence to find a specific value), the permutation by means of pairwise exchanges will be set up on the basis of a table which permits each step to be based on table look-ups. Each exchange of rows will place a row vector f_k in its proper final position, but in doing so will move the row replaced to a position which may, in general, be neither its original nor final position. For each row exchange we will want to know from where to obtain a row vector f_k, since it may have been moved by previous exchanges. A record of the positions of the f_k will be kept and denoted by $l(f_k)$. We also have to know which of the row vectors is currently in row k, since it may have been moved into row k by a previous exchange, and will be moved again by the current exchange. The contents of row k in terms of original positions will be denoted by g_k.

A position table which starts out as shown in Table 1 will be used to keep track of the positions of all rows.

In a computer, in order to economize in storage, Table 1 can be stored in m words, one word for each row of the table, and the quantities extracted as needed. k would not be stored as it would simply correspond to the position of each word in the sequence of m words. Values of the f_k, l_k, and g_k represent row numbers in the matrix, and since three

Table I

k	f_k	l_k	g_k
1	f_1	1	1
2	f_2	2	2
3	.	.	.
.	.	.	.
.	.	.	.
.	.	.	.
m	f_m	m	m

decimal digits are enough to designate row numbers, practically all digital computers could readily hold all three quantities in a single word.

Use of the table will involve compound look-up. Subscript k notation below will indicate simple look-up, e.g., f_k involves finding the f value in the kth row of the table. Functional notation will imply compound look-up, namely, $l(f_k)$ means, after finding f_k in row k of the table, look in row f_k to find the l value. Similarly, $g[l(f_k)]$ will mean the value of g in row $l(f_k)$ of the table.

With the aid of Table 1 the permutation process takes the following form:

1. Starting with $k = 1$ and proceeding sequentially to $k = m$, put row vector f_k into row k by exchanging rows $l(f_k)$ and row k.

2. Along with each of the pairwise exchanges of rows make the following changes in the table, with primed values meaning new values in the table:

(a) Changes in column l of the table:

$l'(f_k) = k$—the new location of vector f_k is row k, i.e., f_k has been moved to its correct final position.

$l'(g_k) = l(f_k)$—the vector in row k before the exchange is g_k. As a result of the exchange it is moved to a new location $l'(g_k)$, which is the position $l(f_k)$ from which f_k has just been removed.

(b) Changes in column g of the table:

$g_k' = f_k$—the kth row of the matrix contains its proper final row vector f_k after the kth exchange.

$g'[l(f_k)] = g_k$—row $l(f_k)$ in the matrix, which contained vector f_k before the exchange, contains vector g_k after the exchange.

For column permutation, which is required in the case of matrix inversion, the inverse permutation, k to f_k, is obtained with the same table simply by redesignating the quantities involved as follows. k in the table should be thought of as giving the positions before permutation and providing name tags for the column vectors; the f_k are the desired final positions, the g_k are the current positions of vector k, and the $l(f_k)$ give the vectors currently in positions f_k. Applying these designations to the table developed for row permutation, the column permutation is accomplished by pairwise exchanges of columns g_k and f_k for $k = 1 \cdots m$. The table changes required during the permutation process are taken care of by the relations given above for the row permutation process. Thus, by making a column exchange along with each row exchange the column permutation process is taken care of with little extra effort since the processing of the auxiliary table is shared by the two permutations.

3. FLOW CHART

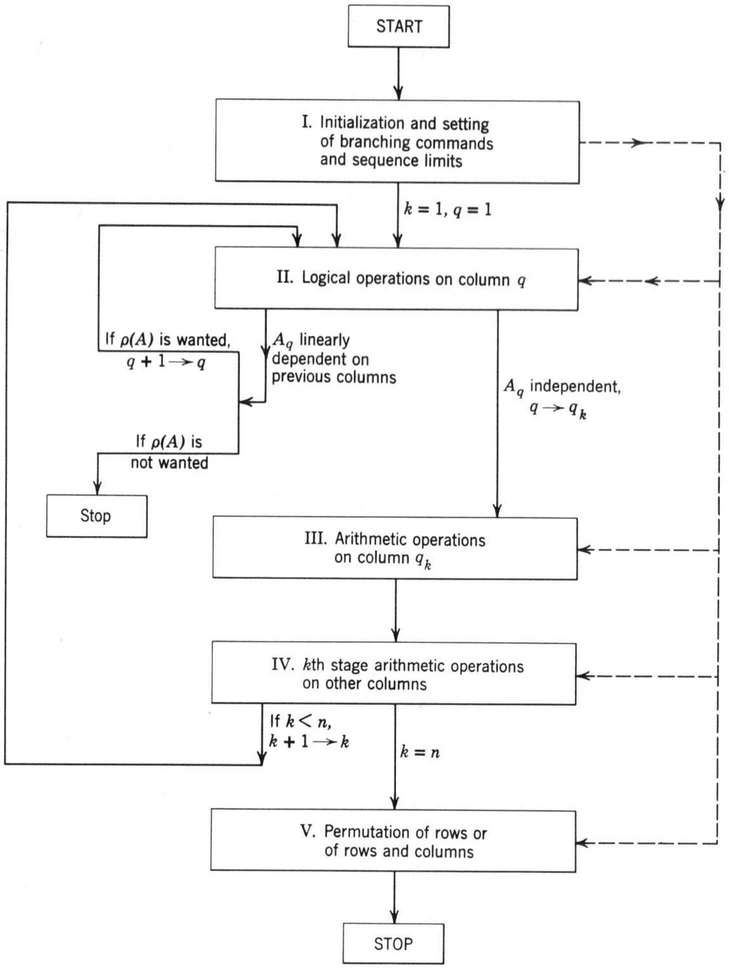
START
I. Initialization and setting of branching commands and sequence limits
k = 1, q = 1
II. Logical operations on column q
If ρ(A) is wanted, q + 1 → q
A_q linearly dependent on previous columns
A_q independent, q → q_k
If ρ(A) is not wanted
Stop
III. Arithmetic operations on column q_k
IV. kth stage arithmetic operations on other columns
If k < n, k + 1 → k
k = n
V. Permutation of rows or of rows and columns
STOP

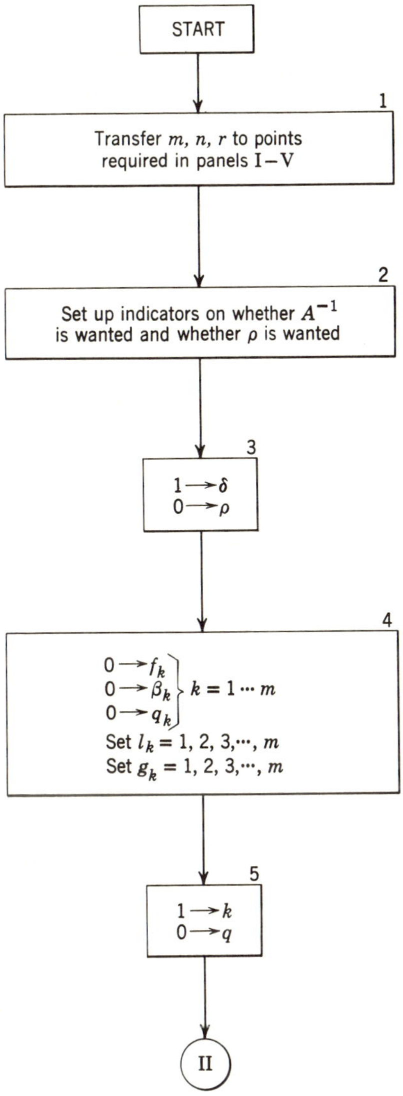

Panel I. Initialization.

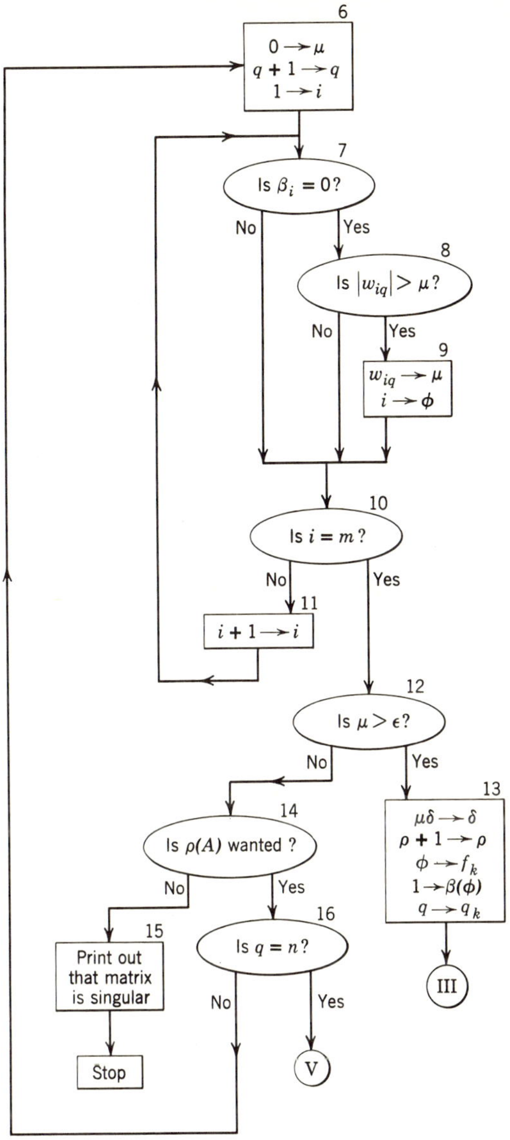

Panel II. Logical operations on column k.

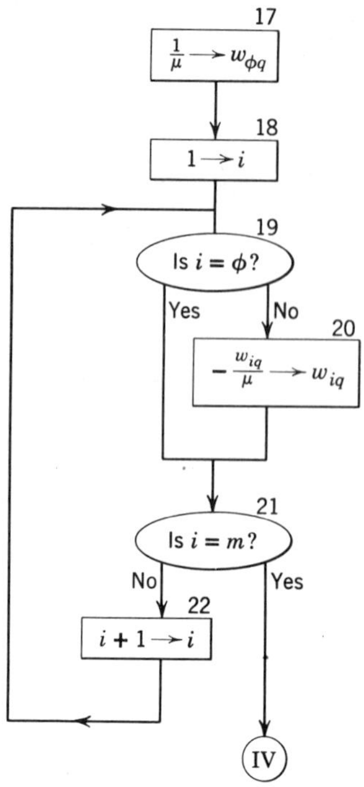

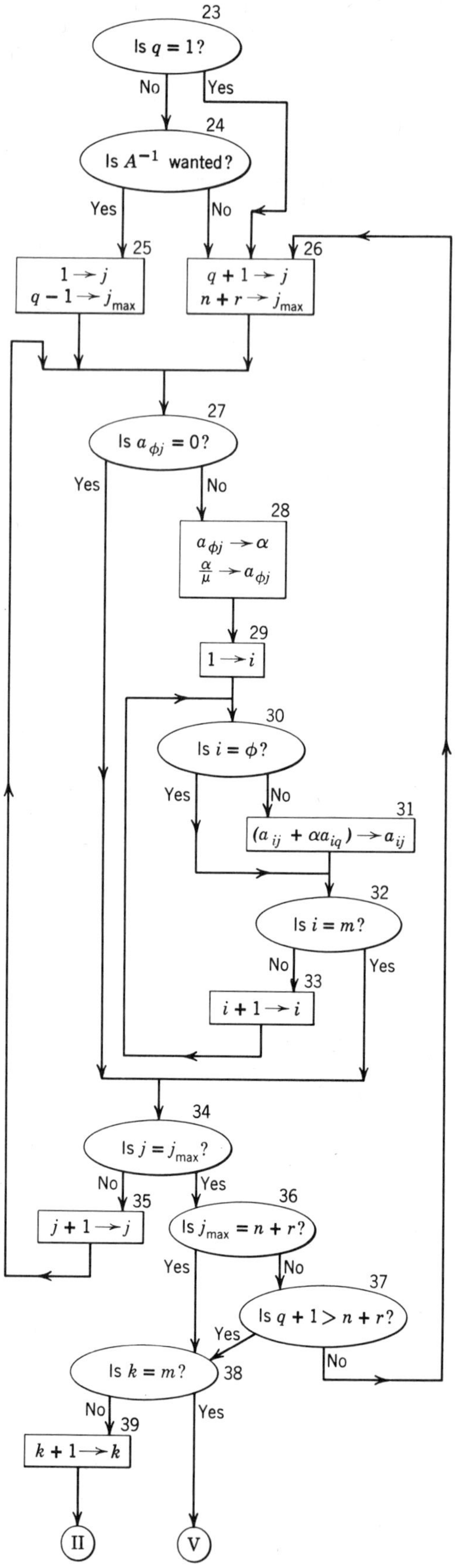

Panel III (above). Arithmetic operation on column q_k.

Panel IV (right). kth stage arithmetic operation on other columns.

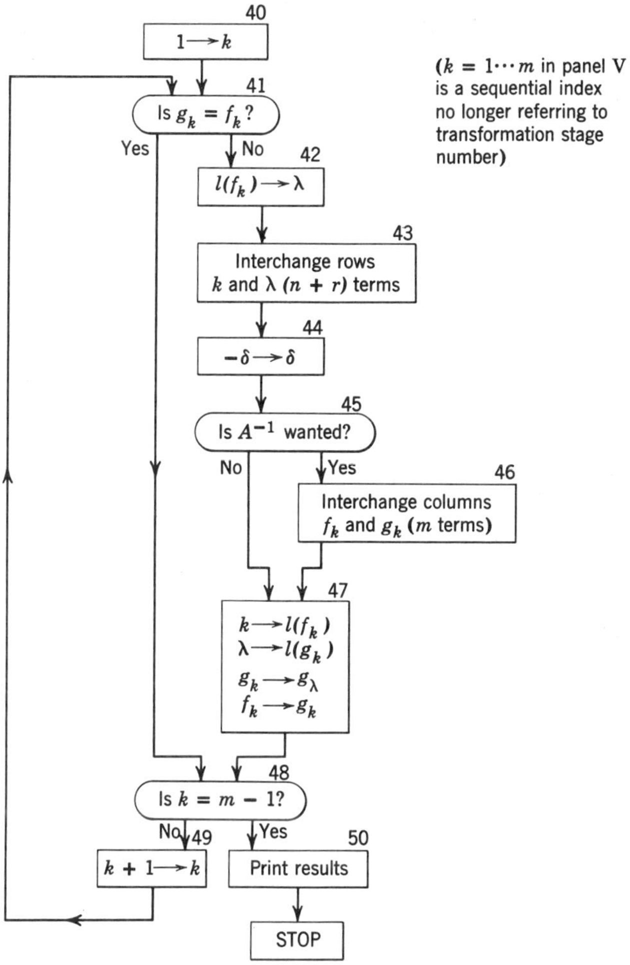

Panel V. Permutation.

4. DESCRIPTION OF THE FLOW CHART

As is generally the case in digital computer programming, the flow chart quickly shows that the logical aspects of organizing the process as a computer program are considerably dominant over the arithmetic steps. This is in fact the underlying reason why computer programming is much more time-consuming than is expected by the uninitiated.

Since the flow chart turns out to be of such length, one might suspect that the variety of capabilities of the program have considerably increased its complexity. Actually, of some 50 boxes in panels I–V, only a few would be eliminated if the only capability of the program were to invert a square matrix.

The flow chart structure panel gives the function and interconnection of each of five flow chart sections, panels I–V. The boxes in panels I–V are identified by a single sequence of box numbers. Since the flow chart, in conjunction with the text of Section 2c should be generally self-explanatory, we merely comment here on some of the flow chart steps rather than give a full step-by-step annotation.

Box 4: For the definition of β see the comments for box 7.

Box 6: μ is to be the value of the pivot term for the kth transformation. Set $\mu = 0$, and, in the course of panel II on the basis of box 8, replace it by successively larger eligible terms, thus finding the largest eligible term in the column.

q is the index of the column under consideration as the pivotal column for the kth transformation. If column A_q is found to be linearly dependent on previous columns of A, column $q + 1$ is tried. The first linearly independent column thus found is denoted by q_k, and A_{q_k} goes into the basis.

Box 7: β_i for $i = 1 \cdots m$ is an indicator as to whether the ith unit vector is still in the basis. $\beta_i = 0$ indicates that the ith unit vector is in the basis (row i is eligible for providing the pivot term); $\beta_i = 1$ indicates that the ith unit vector was replaced by a vector from A in a previous stage (row i is not eligible).

Box 9: ϕ serves for successive interim recording of the row in which the largest eligible w_{iq} is located.

Box 12: ϵ is an empirical tolerance parameter. If the largest term found in the column for use as a pivot term is $<\epsilon$, column A_k is considered to be linearly dependent on earlier columns of A to within this tolerance level.

Box 16: A is considered to be an $m \times n$ matrix and all columns up to the nth are to be taken in sequence as vectors A_q which may enter the basis. If m columns from A have entered the basis before all n columns of A have been tested, all the β_i will be 1, and the program will circulate in panel II until $q = n$.

Box 17: Compute and store the transformed term in the pivot position.

Box 20: Compute and store the transformed terms in column q in all the positions other than the pivot position.

Boxes 24, 25, 26: If A^{-1} is not wanted the program proceeds to 26 rather than 25, i.e., columns 1 to $(q - 1)$ are not transformed. Box 25 sets up the range of columns 1 to $(q - 1)$, and box 26 the range from $q + 1$ to the end of the matrix.

Box 27: If the term $a_{\phi j}$, which is in the same row as the pivot term, is zero, the jth column remains unaltered, so the program bypasses the calculations on that column.

Box 28: Compute and store the transformed term in the pivotal row in column j.

Box 31: Compute and store the transformed value of the general term a_{ij}.

Boxes 36, 37: 36 determines whether it is the first group of columns, set up by 25, or the second group, set up by 26, which has been completed. If the former, ordinarily the second group has to be handled, so the program goes from 36 back to 26, via 37. 37 takes care of an exception that if there is no B matrix involved $(r = 0)$, and if the last column of A is the pivotal column $(k = n)$, then there is no second group of columns to be transformed.

Box 41: If $g_k = f_k$, row vector f_k is already in the correct position, either because it was properly located before the permutations started, or as a by-product of preceding row exchanges.

Box 44: Each exchange of rows implies a change of sign of the determinant.

Box 47: Changes in the permutation table per Section 2c(9).

Box 48: Limit is $k = m - 1$ since the requirement, vector f_k to row k for $k = m$, is taken care of by a pairwise exchange prior to $k = m$.

5. SAMPLE PROBLEM

Invert A and solve $Ax = b$ for $[A,b]$ given by $W^{(0)}$ in Table 2.

Table 3 shows the rearrangement of the permuted inverse back to natural ordering.

Final columns 1, 2, 3 are A^{-1} and column 4 is the solution of $Ax = b$ properly ordered.

6. MEMORY REQUIREMENTS

The program, including parameters, constants, and temporary storage cells, should take no more than about 400 words. To this should be added the memory requirements for arithmetic subroutines if they are involved, e.g., double precision arithmetic, or programmed floating point arithmetic on a fixed point machine.

The program should be written in such a way that the remaining memory can be used to store rectangular matrices, of any proportions, as a sequence of columns. The matrix domain stores the original matrix A or $[A, B]$, and the successive transformations $W^{(1)}$ to $W^{(m)}$. If the program is in double precision, of course, two words are required per term of the matrix.

The only other memory requirements are for the permutation table and for records of two other sets of quantities, q_k, $k = 1 \cdots m$ (pivotal columns), and β_i, $i = 1 \cdots m$, which are indicators, 0 or 1, as to whether row i is eligible to be the pivotal row (see flow chart). The word length of the computer may be great enough to hold a set of permutation table quantities, as well as a q_k, and a β_i value in a single word (in a binary machine each β_i

Table 2. TRANSFORMATION CALCULATIONS

	Matrix Terms				Permutation Table k	f	l	g	β	δ	Remarks
$W^{(0)}$	2	−1	1	3	1	0	1	1	0	1	Row 3 of column 1 contains the largest term of $W^{(0)}$, showing that f_1 will be changed to 3 and β_3 changed to 1.
	1	0	3	10	2	0	2	2	0		
	5	2	−1	6	3	0	3	3	0		
$W^{(1)}$	$\frac{-2}{5}$	$\frac{-9}{5}$	$\frac{7}{5}$	$\frac{3}{5}$	1	3	1	1	0	5	$\delta_1 = 5 \times 1$
	$\frac{-1}{5}$	$\frac{-2}{5}$	$\frac{16}{5}$	$\frac{44}{5}$	2	0	2	2	0		
	$\frac{1}{5}$	$\frac{2}{5}$	$\frac{-1}{5}$	$\frac{6}{5}$•	3	0	3	3	1		
$W^{(2)}$	$\frac{2}{9}$	$\frac{-5}{9}$	$\frac{-7}{9}$	$\frac{-3}{9}$	1	3	1	1	1	−9	$\delta_2 = \frac{-9}{5} \times 5$
	$\frac{-1}{9}$	$\frac{-2}{9}$	$\frac{26}{9}$	$\frac{78}{9}$	2	1	2	2	0		
	$\frac{1}{9}$	$\frac{2}{9}$	$\frac{1}{9}$	$\frac{12}{9}$	3	0	3	3	1		
$W^{(3)}$	$\frac{5}{26}$	$\frac{-16}{26}$	$\frac{7}{26}$	2	1	3	1	1	1	−26	
	$\frac{-1}{26}$	$\frac{-2}{26}$	$\frac{9}{26}$	3	2	1	2	2	1		
	$\frac{3}{26}$	$\frac{6}{26}$	$\frac{-1}{26}$	1	3	2	3	3	1		

Table 3. PERMUTATION OF ROWS AND COLUMNS

	Matrix Terms				Permutation Table k	f	l	g	δ	Remarks
First Permutation	$\frac{-1}{26}$	$\frac{6}{26}$	$\frac{3}{26}$	1	1	3	3	3	26	Using Table 2 with $k = 1$, we have: $f_1 = 3$
	$\frac{9}{26}$	$\frac{-2}{26}$	$\frac{-1}{26}$	3	2	1	2	2		$l(f_1) = 3$, $k = 1$ } rows exchanged
	$\frac{7}{26}$	$\frac{-16}{26}$	$\frac{2}{26}$	2	3	2	1	1		$g_1 = 1$, $f_1 = 3$ } columns exchanged
										Changes in table (unprimed values are from Table 2): $g'_k = f_k$: $g'_1 = 3$; $g'[l(f_k)] = g_k$: $g'_3 = 1$; $l'(f_k) = k$: $l'_3 = 1$; $l'(g_k) = l(f_k)$: $l'_1 = 3$. Sign of δ is changed since rows are exchanged
Second permutation	$\frac{6}{26}$	$\frac{-1}{26}$	$\frac{3}{26}$	1	1	3	2	3	−26	Using Table 2 with $k = 2$: $f_2 = 1$
	$\frac{-16}{26}$	$\frac{7}{26}$	$\frac{5}{26}$	2	2	1	3	1		$l(f_2) = 3$, $k = 2$ } rows exchanged
	$\frac{-2}{26}$	$\frac{9}{26}$	$\frac{-1}{26}$	3	3	2	1	2		$g_2 = 2$, $f_2 = 1$ } columns exchanged

requires just one bit). Otherwise two words would be used per set of these quantities. Allocation of storage by the program should be thought of as though one or two columns were being added to the matrix, thus still allowing the storage domain, outside the program itself, to be completely flexible.

7. ESTIMATION OF THE RUNNING TIME

The time for matrix inversion is approximated by $T \simeq \alpha m^3 \mu$, where μ is the multiplication time of the computer, and α is a factor, usually between 2 and 3, which takes care of addition and bookkeeping (for the most part one addition is associated with each multiplication). μ is meant to represent the time for whatever form of multiplication is used, whether the built-in multiplication of the machine, or a subroutine such as for double precision. When arithmetic subroutines are used, the relative time for bookkeeping operations decreases, but the time for programmed addition becomes about the same as for programmed multiplication so that α remains in the range of about 2 to 3.

If the matrix is sparse (mostly zeros), the time will be decreased, but is not likely to fall below about half the time for a "full" matrix of the same size, even if the original matrix looks very sparse.

For simultaneous equations, $AX = B$, with A $m \times m$, and B $m \times r$, the time estimate formula becomes $T \simeq \alpha[(m^3) + m^2 r]\mu$ (ignoring terms lower than third degree in m and r). Details can be worked out if they are desired for other cases discussed in Section 2c on the basis of the number of columns which have to be transformed.

8. REFERENCE

1. J. von Neumann and H. Goldstine, The Numerical Inverting of Matrices of High Order, *Bull. Amer. Math. Soc.*, *vol.* 53, 1947, pp. 1021–1099.

3 The solution of linear equations by the Gauss-Seidel method

R. Van Norton
Institute of Mathematical Sciences
New York University

1. FUNCTION

The function of this program is to solve N linear algebraic equations in N unknowns by the method of Gauss-Seidel.

2. MATHEMATICAL DISCUSSION

A well-known linear, iterative process for approximating the solution of a set of N simultaneous linear algebraic equations in N unknowns is the method of Gauss-Seidel. Crudely speaking, the process is useful for equations which can be so arranged that the diagonal elements of the coefficient matrix dominate the other elements. Some conditions for the convergence of the process will be stated explicitly in the next section. (For example, if the equations can be so arranged that the coefficient matrix is symmetric and positive definite, then we will see that the process is convergent.)

Many generalizations and specializations of the method have been investigated. These investigations have resulted in the availability of more swiftly convergent methods for some special cases. Some of these researches have been directed to the important application of solving boundary value problems for second-order elliptic partial difference equations.

References [1] and [2] are interesting and important surveys of this field. The reader's attention is urgently directed to these papers as well as to Chapter 13 of this book.

a. Symbols Used

Matrices are denoted by capital English letters, vectors by lower case English letters. The matrix of coefficients is called A, the solution vector x, the inhomogeneous vector b. Scalars are denoted by Greek letters except for various indices. Superscripts on vectors or their components are reserved for the count of iterations. The superscript counts individual component corrections. Some ambiguities may exist, but these will either be easily interpreted or explained in the text.

b. Derivation of the Method

Briefly, one wishes to solve

$$Ax = b \tag{1}$$

by starting with a trial vector x^0 and successively improving this guess to vectors $x^1, x^2, \cdots$ which converge to the solution vector x if one exists. The improvement is made by cycling through the equations, replacing only the ith ($i = 1, 2, \cdots, N$) component of the trial vector by the value necessary to satisfy the ith equation. In terms of the matrix A, let A be decomposed by writing

$$A = L + D + R \tag{2}$$

where L is lower triangular, D is diagonal, and R is upper triangular. Then, in the Gauss-Seidel method, we have after $k + 1$ passes through all the equations:

$$x^{N(k+1)} = D^{-1}(b - Lx^{N(k+1)} - Rx^{Nk})$$

or

$$x^{N(k+1)} = (I + D^{-1}L)^{-1}D^{-1}(b - Rx^{Nk})$$

or, finally,

$$x^{N(k+1)} = Mb + Hx^{Nk}$$

where $M = (D + L)^{-1}$

$$H = -(D + L)^{-1}R$$

I is the identity matrix of order N

One condition necessary for success of the process is now apparent. The equations must be ordered so that D^{-1} exists, i.e., $a_{ii} \neq 0$ for any i. This will be assumed in what follows. We will also assume that A is nonsingular.

THEOREM: For arbitrary x^0 the process yields

$$\lim_{k\to\infty} x^{Nk} = A^{-1}b$$

if and only if all the eigenvalues of H are less than one in absolute value.

Proof: If x^0 is the initial trial vector, then

$$x^{Nk} = H^k x^0 + (I + H + \cdots + H^{k-1})Mb$$

If all the eigenvalues of H are less than one in absolute value, then

$$\lim_{k\to\infty} H^k x^0 = 0$$

and

$$\lim_{k\to\infty} \sum_{i=0}^{k-1} H^i = (I - H)^{-1}$$

The sum is convergent. Therefore,

$$\begin{aligned}\lim_{k\to\infty} x^{Nk} &= (I - H)^{-1}Mb \\ &= [I + (D + L)^{-1}R]^{-1}(D + L)^{-1}b \\ &= (D + L + R)^{-1}b \\ &= A^{-1}b\end{aligned}$$

On the other hand, these limits exist only if all the eigenvalues of H are less than one in absolute value.

The next two theorems provide some theoretical, but none too practical, information about the application of the Gauss-Seidel method.

THEOREM: If A is a nonsingular, symmetric matrix with positive diagonal elements and if the iteration scheme is convergent for every initial vector, then A is positive definite.

Proof: Let

$$e^{Nk} = x - x^{Nk} \tag{3}$$

be the error vector after the kth sweep of the equations. We define the associated quadratic form

$$Q(e^{Nk}) = (e^{Nk}, Ae^{Nk}) \tag{4}$$

In order to simplify the analysis, let the superscript t be increased by one for the correction of each equation.

Define the residual vector at stage t as

$$r^t = Ae^t = b - Ax^t \tag{5}$$

The correction of the ith component during the kth sweep is given by

$$\xi_i^{Nk+i} = \xi_i^{Nk+i-1} + \alpha_{ii}^{-1}\rho_i^{Nb+i-1}$$

where the ξ_i^{Nk+i} are the components of x^{Nk+i}, ρ_i^{Nk+i-1} is the ith component of r^{Nk+i-1}, and α_{ii} is the ith diagonal element of A.

At the same step one has

$$\epsilon_i^{Nk+i} = \epsilon_i^{Nk+i-1} - \alpha_{ii}^{-1}\rho_i^{Nk+i-1} \tag{6}$$

while the other components of e remain unchanged.

Assume, without loss of generality, that the initial error vector yields

$$Q(e^0) < 0$$

Then

$$Q(e^1) - Q(e^0) = Q(e^0 - \alpha_{11}^{-1}\rho_1^0) - Q(e^0)$$

When a component appears as the argument of a function of vectors, it denotes a vector with only the indicated component nonzero. So, using (5),

$$Q(e^1) - Q(e^0) = -\alpha_{11}^{-1}(\rho_1^0)^2$$

and after a sweep

$$Q(e^{N(k+1)}) - Q(e^{Nk}) = -\sum_{i=1}^{N} \alpha_{ii}^{-1}(\rho_i^{Nk+i})^2 \tag{7}$$

Therefore, the $Q(e^{Nk})$ form a decreasing sequence. By the hypothesis of convergence for all initial vectors, the limit of this sequence must be zero. This is contrary to the assumption that for some error vector the quadratic form could be negative. Thus, A is positive definite.

The next theorem is in the nature of a converse.

THEOREM: If the matrix A is symmetric and positive definite, then the iteration scheme converges for every initial vector.

Proof: Since A is positive definite, it has positive diagonal elements. In the preceding theorem it was shown that the $Q(e^{Nk})$ form a decreasing sequence. Since A is positive definite, this sequence is bounded below by zero, and hence convergent. Thus, for arbitrary positive δ, there exists a K so large that

$$Q(e^{Nk}) - Q(e^{N(k+1)}) < \delta$$

for all $k > K$.

So from (7)

$$\sum_{i=1}^{N} \alpha_{ii}^{-1}(\rho_i^{Nk+i})^2 < \delta \qquad (8)$$

and

$$(\rho_i^{Nk+i})^2 < \delta[\max_{j=i,\cdots,N} \alpha_{jj}] \qquad (i = 1, \cdots, N)$$

So the ρ_i^{Nk+i} converge to zero. But $[\rho_i^{Nk+i}]$ is only a subsequence of $[\rho_i^t]$ $(t = 0, 1, 2, \cdots)$. However, the error vector satisfies

$$e^{N(k+1)} - e^{Nk} = \sum_{i=1}^{N} (e^{Nk+i} - e^{Nk+i-1})$$

and from (6) and (8)

$$|e^{N(k+1)} - e^{Nk}|^2 = \sum_{i=1}^{N} (\alpha_{ii})^{-2}(\rho_i^{Nk+i})^2$$

$$< \left[\max_{i=1,\cdots,N} \left(\frac{1}{\alpha_{ii}}\right)\right] \delta$$

for $k > K$.

That is to say, the error vectors, computed at the end of each sweep, form a Cauchy sequence. Since only the pertinent component of the error vector is disturbed by the correction of its equation, the error vector sequence is convergent. Since A is nonsingular and

$$r^t = A^{-1}e^t$$

the residual vectors, r^t, are convergent. But each component of r^t has a subsequence convergent to zero and one concludes that

$$\lim_{k\to\infty} r^{Nk} = 0$$

and

$$\lim_{k\to\infty} e^{Nk} = 0$$

A practical, sufficient criterion for convergence of the method is given in [3]. A slightly weaker version of this theorem is given below.

Let S be the set of integers $1, 2, \cdots, N$. The matrix A is said to be irreducible if, for every disjoint and exhaustive partition of S into two sets T_1 and T_2, there exists at least one nonzero element α_{ij} of A such that i is an element of T_1 and j is an element of T_2. That is to say, the solution of the equations cannot be reduced to the solution of two or more sets of lower order.

THEOREM: If the matrix A is irreducible and nonsingular, if the diagonal elements of A are all unity, and if

$$\sum_{\substack{j=1\\ j\neq i}}^{N} |\alpha_{ij}| \leq 1 \qquad (i = 1, 2, \cdots, N)$$

and for at least one value of i

$$\sum_{\substack{j=1\\ j\neq i}}^{N} |\alpha_{ij}| \leq \mu < 1$$

then the iterates converge from any starting vector to the solution.

Proof: The error in the ith component at the kth cycle is

$$\epsilon_i^{N(k+1)} = \xi_i - \xi_i^{N(k+1)} = \sum_{j=1}^{i-1} \alpha_{ij}(\xi_j^{N(k+1)} - \xi_j)$$

$$+ \sum_{j=i+1}^{N} \alpha_{ij}(\xi_j^{Nk} - \xi_j)$$

Let γ^{Nk} be the maximum absolute error component at the kth cycle. Assume, without loss of generality, that the first row sum is less than one. Then

$$|\epsilon_1^1| \leq \gamma^0(|\alpha_{12}| + |\alpha_{13}| + \cdots + |\alpha_{1,n}|) \leq \mu\gamma^0$$

Perhaps the first variable does not appear in the second equation; then

$$|\epsilon_2^2| \leq \gamma^0(0 + |\alpha_{23}| + |\alpha_{24}| + \cdots + |\alpha_{2,n}|) \leq \gamma^0$$

Because A is irreducible, there is at least one other equation in which ξ_1 appears, say the jth equation. Then

$$|\epsilon_j^1| \le \gamma^0(|\alpha_{j1}|\mu + |\alpha_{j2}| + \cdots + |\alpha_{j,N}|) = \mu^1\gamma^0 < \gamma^0$$

One continues in this manner. Thus, after $N - 1$ sweeps through the equations one concludes that

$$\gamma^{(N-1)k} \le \mu''\gamma^0$$

where

$$\mu \le \mu'' < 1$$

The irreducibility of the matrix has assured that the decrease in $|\epsilon_1|$ can eventually be spread through all the error components. The estimate

$$\gamma^{kN(N-1)} \le (\mu'')^N\gamma^0$$

implies convergence of the iterates to the solution as $k \to \infty$.

c. Error Analysis

Only the case of a real, symmetric, positive definite coefficient matrix will be considered under this heading. It has already been shown in this case (7) that a certain measure of the error is monotone decreasing. There is the possibility that rounding could disturb this situation.

Reference [4] analyzes this possibility for a special set of circumstances. Assume that a fixed point computer uses numbers less than one in absolute value. Define δ such that the largest number the computer uses is 1–2δ. If the symbol p^* denotes the number which is the computer's closest approximation to p, then, for $p < 1 - \delta$,

$$|p^* - p| < \delta$$

Assume further that the diagonal elements of the matrix are all unity. (In this case these elements would not be stored, so the assumption about the size of the computer's numbers is not violated.)

The vector x which satisfies (1) is also the unique vector which minimizes the quadratic form

$$(x, Ax - 2b) = -(x, b + r)$$

Therefore, one may consider x^{t+1} a better approximation to the solution than x^t if

$$(x^{t+1}, b + r^{t+1}) > (x^t, b + r^t)$$

The correction of the ith equation replaces ξ_i^t by $\xi_i^t + \rho_i^{t*}$. This yields

$$(x^{t+1}, b + r^{t+1}) = (x^t + \rho_i^{t*}, 2b - Ax^t - A\rho_i^{t*})$$

where, as before, the use of a component of a vector as an argument of a function of vectors denotes a vector with zero for all other components. Consider

$$(x^{t+1}, b + r^{t+1}) - (x^t, b + r^t) = (x^t + \rho_i^{t*}, 2b - Ax^t - A\rho_i^{t*}) - (x^t, b + r^t) = 2(r^t, \rho_i^{t*}) - (\rho_i^{t*}, A\rho_i^{t*}) = 2\lambda\lambda^* - (\lambda^*)^2$$

where

$$\lambda = \rho_i^t \quad \text{and} \quad \lambda^* = \rho_i^{t*}$$

For this expression to be greater than zero we must have

$$|\lambda^*| > 2|\lambda^* - \lambda|$$

But

$$|\lambda^* - \lambda| < \delta$$

If $|\lambda^*| > 2\delta$, the result follows. However, every nonzero computer number is $\ge 2\delta$ in absolute value. If $|\lambda^*| = 2\delta$, then the correction does not make the vector any worse, and if $\lambda^* = 0$ the new vector is unchanged from the old. Therefore, in the sense of this error measure, the error is nonincreasing.

3. SUMMARY OF THE CALCULATION PROCEDURE

(a) Choose an arbitrary vector x^0. In many cases $x^0 = 0$ or $x^0 = D^{-1}b$ are reasonable choices.

(b) Successively alter the ith component of the trial vector so that the ith equation is satisfied. ($i = 1, 2, \cdots, N$.)

(ç) At the same time accumulate an error estimate; e.g., $\sum_{i=1}^{N} |\xi_i^{N(k-1)} - \xi_i^{Nk}| = E^{Nk}$, an error estimate at the kth iteration. Here ξ_i^{Nk} is the ith component of x^{Nk}.

(d) Test if E^{Nk} is less than some error criterion. This might be an absolute criterion; e.g., $E^{Nk} < 0.001N$.

(e) If the error is not sufficiently small, then return to step (b). Otherwise, accept x^{Nk} as an estimate of the solution and proceed to print or make other use of it.

4. FLOW CHART

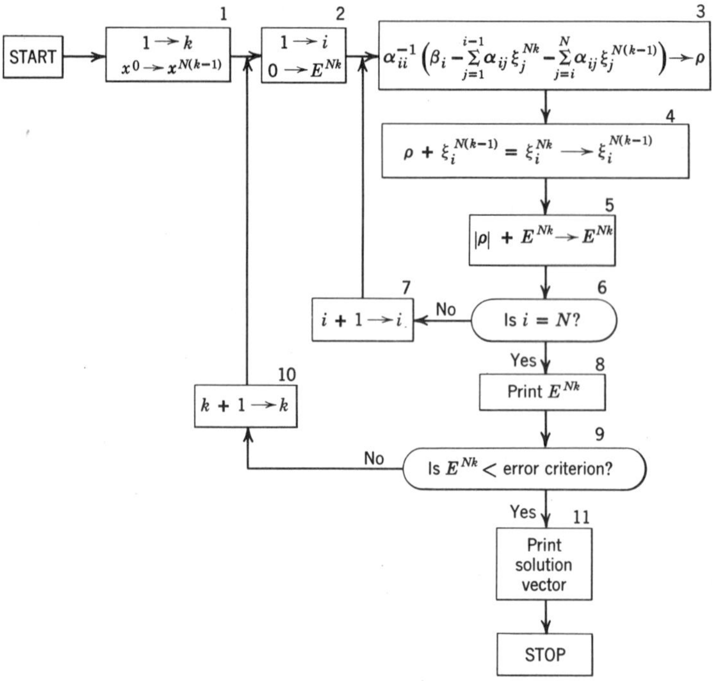

5. DESCRIPTION OF THE FLOW CHART

Box 1: The counter k, which counts cycles through all N equations, is set to one. The initial guess vector, x^0, is put into the $x^{N(k-1)}$ locations.

Box 2: The counter i, which counts component corrections, is set to one. The error estimater, E^{Nk}, is initialized to zero.

Box 3: The ith residual component is computed.

Box 4: The next approximation to the ith component of the solution is computed.

Box 5: The error estimate is progressively accumulated.

Box 6: If i has reached N, all N components have been altered, and the error estimate should be checked. In this case proceed to box 8, otherwise to box 7.

Box 7: The component counter is advanced, and control is returned to box 3.

Box 8: The error estimate is printed.

Box 9: If the error estimate is less than the predetermined error criterion, the problem is finished. Otherwise proceed to box 10.

Box 10: The cycle counter is advanced, and control is returned to box 2.

Box 11: The results are printed.

6. SUBROUTINES

Although no subroutines are explicitly used in the flow chart, it is often the case that subroutines may be useful for computing the elements of the coefficient matrix for systems of large order. For example, in the solution of elliptic boundary value problems the foregoing flow chart is too general. Only a small fraction of the coefficients will be nonzero and frequently these do not vary much.

7. SAMPLE PROBLEM

$$A = \begin{pmatrix} 3 & 1 \\ 1 & 2 \end{pmatrix} \qquad b = \begin{pmatrix} 5 \\ 5 \end{pmatrix}$$

$$x^0 = (0, 0)$$
$$\rho_1^1 = 5/3$$
$$\xi_1^1 = 5/3$$
$$\rho_2^1 = 5/3$$
$$\xi_2^1 = 5/3$$
$$E^1 = 10/3$$

Some further iterations are compressed in Table 1. In the last column of the table are the sums of the absolute values of the error vector's components. The reader is warned that the example is favorable to the method.

Table I

Cycle	ξ_1	ξ_2	E	
2	10/9	35/18	5/9	1/6
3	55/54	215/108	5/36	1/36
.				
.				
.				
∞	1	2	0	0

8. MEMORY REQUIREMENTS

A storage table might include:

x^t	N words
A	N^2 words in general, $[N(N+1)]/2$ if A is symmetric
b	N words
Program	About 50 words

9. ESTIMATION OF THE RUNNING TIME

Excluding bookkeeping operations, there will be at most about $N^2/2$ multiplications and additions per sweep of the equations. So the time for an iteration is about

$$N^2(\mu + \nu)/2$$

where μ and ν are the multiply and add times of the computer.

The number of iterations will depend on the matrix A and the accuracy desired. There is too great a latitude here to permit a reasonable estimate for this number.

10. REFERENCES

1. George E. Forsythe, Solving Linear Equations Can Be Interesting, *Bull. Amer. Math. Soc.*, vol. 59, 1953, pp. 299–329.
2. David Young, On the Solution of Linear Systems by Iteration, *Proceedings of Symposia in Applied Mathematics*, Vol. VI, McGraw-Hill Book Co., New York, 1956.
3. H. Geiringer, On the Solution of Systems of Linear Equations by Certain Iteration Methods, *Reissner Anniversary Volume*, J. W. Edwards, Ann Arbor, Mich., 1949.
4. A. S. Householder, *Principles of Numerical Analysis*, McGraw-Hill Book Co., New York, 1953.
5. S. Schechter, Relaxation Methods for Linear Equations, Report NYO-8673, AEC Computing and Applied Mathematics Center, Institute of Mathematical Sciences, New York University, Sept. 1, 1958.

The solution of linear equations by the conjugate gradient method

4

F. S. Beckman
International Business Machines Corporation

I. FUNCTION

The method of conjugate gradients may be used for solving a system of simultaneous linear algebraic equations,

$$Ax = k \quad (1)$$

where A is an $N \times N$ matrix of coefficients, x is an $N \times 1$ vector of unknowns, and k is an $N \times 1$ vector of constants. The method will provide a solution where it exists.* In the derivation of the basic algorithm of the method, the matrix A is assumed to be positive definite and symmetric, but the method can easily be extended to handle the case in which A is an arbitrary matrix and this variation of the algorithm also appears below.

The method is an N-step iterative one; i.e., an algorithm is applied to give successive approximations to the solution of the given linear system and, if computations are done with complete accuracy, a solution is obtained (where it exists) after M iterations where $M \leq N$ (the order of the system). The basic algorithm applied is an elegant one and has the appeal of simplicity. We shall later review some of the advantages of the procedure that are claimed in [1]. These are principally:

(a) The simplicity of the computational procedures.

(b) The preservation of the original matrix of coefficients during the computation and the consequent ability to exploit the advantage of having many zeros in this matrix.

(c) The need for a small amount of storage space to hold intermediate results.

(d) The ability to start anew at any point in the computation.

(e) The superiority of each approximation to the preceding ones in the sense of being closer to the true solution.

These appear to be cogent arguments for the superiority of the method to that of elimination in many cases. However, we offer some arguments below for not arriving at this conclusion in general.

* If no solution exists, it is of theoretical interest to note that the method can be used to provide a least squares solution, i.e., one which minimizes the sum of the squares of the residuals, or the square of the length of the vector, $k - Ax$. However, this is of limited practical value because (cf. below) round-off errors make the recognition of a singular system very difficult.

2. MATHEMATICAL DISCUSSION

a. Symbols Used

A represents a symmetric and positive definite matrix, x a vector of unknowns to be determined, and h the solution vector in the system $Ax = k$. Vectors are indicated by other lower case English letters and scalars by Greek letters. (p, q) represents the inner product of the vectors p, q, or the value of p^Tq. Note that, if A is symmetric, $(p, Aq) = (Ap, q) = (q, Ap) = (Aq, p)$. $|r|$ stands for the length of the vector r or the value of $(r, r)^{1/2}$.

b. Derivation of the Method

The method of conjugate gradients is a special case of the method of conjugate directions which is described below. We shall assume in what follows that a solution vector h of the system $Ax = k$ exists.

METHOD OF CONJUGATE DIRECTIONS

Let us suppose that a set of N "A-conjugate," or "A-orthogonal," vectors $\{p_i\}$, $i = 0, \cdots, N-1$, is available to us. This means that the inner product $(Ap_i, p_j) = 0$ where $i \neq j$. If A is positive definite, then $(Ap_i, p_i) > 0$. In this case, since the $\{p_i\}$ are necessarily linearly independent and span the N-dimensional space, the solution vector h can be written as

$$h = c_0p_0 + c_1p_1 + \cdots + c_{N-1}p_{N-1}$$

If we can determine the $\{c_i\}$ the solution can be quickly calculated. This is easy, however, for from the above it follows that

$$(Ah, p_i) = (k, p_i) = c_i(Ap_i, p_i)$$

and consequently

$$c_i = \frac{(k, p_i)}{(Ap_i, p_i)}$$

and

$$h = \frac{(k, p_0)}{(Ap_0, p_0)}p_0 + \frac{(k, p_1)}{(Ap_1, p_1)}p_1 + \cdots + \frac{(k, p_{N-1})}{(Ap_{N-1}, p_{N-1})}p_{N-1} \qquad (2)$$

In the method of conjugate gradients, a particular set of A-conjugate vectors, $\{p_i\}$, is developed and a solution formed in terms of these. In general, the "Gram-Schmidt" orthogonalization procedure can be used to produce such a set starting with any set of N linearly independent vectors. This procedure is not, per se, a part of the conjugate gradient method. We describe it here briefly because of its importance in allied methods and because it will be used below to derive the basic algorithm of the conjugate gradient method.

GRAM-SCHMIDT ORTHOGONALIZATION PROCEDURE

Let $\{v_i\}$ be a set of N linearly independent vectors. It is desired to produce a set of vectors $\{t_i\}$ which is A-conjugate.

Let $t_1 = v_1$. We can choose t_2 as a linear combination of v_2 and t_1, $t_2 = v_2 + \alpha_{21}t_1$, which will be A-conjugate to t_1 if $(At_2, t_1) = A(v_2, t_1) + \alpha_{21}(At_1, t_1) = 0$ or if $\alpha_{21} = -[(Av_2, t_1)/(At_1, t_1)]$. In general, having selected $t_1, t_2, \cdots, t_k$ we can choose t_{k+1} of the form

$$t_{k+1} = v_{k+1} + \alpha_{k+1,1}t_1 + \alpha_{k+1,2}t_2 + \cdots + \alpha_{k+1,k}t_k$$

It will be A-conjugate to $t_r\,(r \le k)$ if $(At_{k+1}, t_r) = (Av_{k+1}, t_r) + \alpha_{k+1,r}(At_r, t_r) = 0$ or if $\alpha_{k+1,r} = -[(Av_{k+1}, t_r)/(At_r, t_r)]$. Thus, the set $\{t_i\}$ can be developed from the set $\{v_i\}$ by application of the recursion formula

$$t_{k+1} = v_{k+1} - \left[\frac{(Av_{k+1}, t_1)}{(At_1, t_1)}t_1 + \cdots + \frac{(Av_{k+1}, t_k)}{(At_k, t_k)}t_k\right] \qquad (3)$$

A set of orthogonal vectors will be produced if, in the above, A is set equal to the identity matrix. None of the divisors will vanish if A is positive definite.

Note that since, from (3), v_k is a linear combination of $t_1, \cdots, t_k$, it follows that

$$(v_k, At_i) = 0 \quad \text{for } i > k \qquad (4)$$

It will be convenient in the discussion which follows to note at this point that the algorithm defined by (3) can be modified slightly if we choose t^*_{k+1} to be of the form:

$$t^*_{k+1} = t^*_k + \beta_{k+1,1}t^*_1 + \beta_{k+1,2}t^*_2 + \cdots + \beta_{k+1,k-1}t^*_{k-1} + \beta_{k+1,k}v_{k+1}$$

We shall do this for the case in which A is the identity matrix and, therefore, a set of orthogonal vectors will be produced. By paralleling what is done above, the β's are easily found and a different set $\{t_1^*\}$ can be developed by use of the formula

$$t_{k+1}^* = t_k^* + \frac{(t_k^*, t_k^*)}{(v_{k+1}, t_k^*)}\left[\frac{(v_{k+1}, t_{k-1}^*)}{(t_{k-1}^*, t_{k-1}^*)} t_{k-1}^* + \cdots + \frac{(v_{k+1}, t_1^*)}{(t_1^*, t_1^*)} t_1^* - v_{k+1}\right] \quad (5)$$

and $t_1^* = v_1$, if we assume $(v_{k+1}, t_k^*) \neq 0$. This condition will be satisfied in the application of (5) which is discussed below. As before, (4) holds (with A equal to the identity matrix).

We shall show that the method of conjugate directions can also be considered a variational method (where the iterative procedure is designed to minimize at each step, subject to certain restraints, a certain nonnegative quadratic function of the unknowns that vanishes at the solution). The computation defined by (2) can be described as the following iterative scheme where $p_0, p_1, \cdots, p_{N-1}$ is the given set of N A-conjugate vectors

$$x_0 = \beta_0 p_0$$

$$x_{i+1} = x_i + \beta_i p_i$$

where

$$\beta_i = \frac{(k, p_i)}{(Ap_i, p_i)} \quad (6)$$

$x_M = h$ where $M \leq N$ (h can be a linear combination of the first M p's with $M < N$). The approximation x_{i+1} is seen to be obtained by traveling a certain distance along the vector p_i from the preceding approximation (point) x_i. We show that the distance traveled is such as to minimize a function $H(x)$, whose minimum occurs at $x = h$, over all points on the line of action of the vector p_i passing through x_i. In this sense, then, we are choosing the "best" possible point on this line as the next approximation.

Consider the function

$$H(x) = (A(h - x), h - x) \quad (7)$$

Since A is positive definite, $H(x) \geq 0$. Given the vectors x_i, p_i, let us minimize $H(x)$ over all x of the form $x = x_i + \lambda_i p_i$. We have by expansion,

$$H(x_i + \lambda_i p_i) = H(x_i) + \lambda_i^2(Ap_i, p_i) - 2\lambda_i(r_i, p_i) \quad (8)$$

where r_i represents the residual vector $k - Ax_i$ (note that $k = Ah$). By setting $dH/d\lambda_i = 0$ we find that the minimum value of $H(x_i + \lambda_i p_i)$, with respect to λ_i, occurs when

$$\lambda_i = \frac{(r_i, p_i)}{(Ap_i, p_i)} \quad (9)$$

However, $(r_i, p_i) = (k - Ax_i, p_i) = (k, p_i) - (x_i, Ap_i)$. From (6) it follows that $x_i = \beta_{i-1}p_{i-1} + \beta_{i-2}p_{i-2} + \cdots + \beta_0 p_0$ and, since the p's are A-conjugate, that $(x_i, Ap_i) = 0$. Consequently (9) can be written as

$$\lambda_i = \frac{(k, p_i)}{(Ap_i, p_i)} \quad (10)$$

which proves the assertion that (6) defines a variational method based on the minimization of $H(x_i + \lambda_i p_i)$. The function $H(x)$ can be interpreted as the square of the "A-metric length" of the "error vector" $h - x$.

Basic Algorithm of the Conjugate Gradient Method

Let x_0 be an arbitrary starting approximation to the solution vector h. Then, the following formulas define the fundamental conjugate gradient iterative procedure leading to a solution of (1):

$$p_0 = r_0 = k - Ax_0 \quad (11)$$

$$\alpha_i = \frac{|r_i|^2}{(p_i, Ap_i)} \quad (12)$$

$$x_{i+1} = x_i + \alpha_i p_i \quad (13)$$

$$r_{i+1} = r_i - \alpha_i Ap_i \quad (14)$$

$$\beta_i = \frac{|r_{i+1}|^2}{|r_i|^2} \quad (15)$$

$$p_{i+1} = r_{i+1} + \beta_i p_i \quad (16)$$

After M iterations, with $M \leq N$, x_M will be equal to the solution h if all computations are done with no loss of accuracy. Many relations hold among the quantities appearing in (11)–(16) and these are derived in [1]. In place of (12) and (15) one may use:

$$\alpha_i = \frac{(p_i, r_i)}{(p_i, Ap_i)} \quad (17)$$

$$\beta_i = -\frac{(r_{i+1}, Ap_i)}{(p_i, Ap_i)} \quad (18)$$

which the authors of [1] report have led to greater accuracy in the limited numerical tests of the method conducted by them and their colleagues. Although the use of (12) and (15) involves somewhat fewer arithmetic operations than (17) and (18), the saving is not sufficiently great to offset this apparent advantage of greater accuracy.

Some other salient relations that hold among the quantities appearing, in these equations are:

$$(r_i, r_j) = 0 \qquad (i \neq j) \tag{19}$$

$$(p_i, Ap_j) = 0 \qquad (i \neq j) \tag{20}$$

$$\begin{aligned}(p_i, r_j) &= 0 \qquad (i < j),\\ (p_i, r_j) &= |r_i|^2 \qquad (i \geq j)\end{aligned} \tag{21}$$

$$(r_i, Ap_i) = (p_i, Ap_i) \tag{22}$$

$$(r_i, Ap_j) = 0 \qquad (i \neq j, j + 1) \tag{23}$$

These relations will follow quickly from what is proved below.

We shall derive the equations (11)–(16) which define the method by showing that they basically describe the formation of two sequences of vectors by the Gram-Schmidt process. The residual vectors r_i and direction vectors p_i are formed in the order: $r_0, p_0, r_1, p_1, r_2, \cdots, r_{N-1}, p_{N-1}$. If we form the $\{r_i\}$ by applying an orthogonalization procedure to the vectors $r_0, Ap_0, Ap_1, \cdots, Ap_{N-2}$ and form the $\{p_i\}$ by producing a set of A-conjugate vectors from $r_0, r_1, \cdots, r_{N-1}$, we shall show that the formulas describing these Gram-Schmidt processes reduce to (11)–(18). That this will lead to a solution $h = x_M$ $(M \leq N)$ follows from the fact that r_N, if calculated using Ap_{N-1}, would necessarily be zero since it would be orthogonal to all the N orthogonal vectors $\{r_i\}$ $(i \leq N - 1)$ (r_N would be expressible as a linear combination of these N linearly independent vectors and, since it is orthogonal to all of them, the coefficients in this linear expression must all vanish).

Before writing in detail the formulas describing these two orthogonalization procedures going on in parallel, let us observe some of the relations among the $\{r_i\}$ and $\{p_i\}$ which result from their means of formation. At the time that r_{i+1} is formed from $r_0, Ap_0, Ap_1, \cdots, Ap_i$, the set $p_0, p_1, \cdots, p_i$ has already been formed from the A-orthogonalization of $r_0, r_1, \cdots, r_i$. If this is done by application of (3), then (4) implies that

$$(r_k, Ap_i) = 0 \quad \text{for } i > k \tag{24}$$

On the other hand, at the time that p_{i+1} is formed from the A-orthogonalization of $r_0, r_1, \cdots, r_{i+1}$, this latter set has been formed from the orthogonalization of $r_0, Ap_0, Ap_1, \cdots, Ap_i$. If this is done by using (5) then, as noted earlier, (4) is true, with A replaced by the identity matrix, and this implies (Ap_{k+1} corresponds to v_k and r_i to t_i) that

$$(Ap_k, r_i) = (Ar_i, p_k) = 0 \quad \text{for } i > k + 1 \tag{25}$$

Now, let us apply (5) to develop the $\{r_i\}$ from $r_0, Ap_0, Ap_1, \cdots, Ap_{N-2}$. This leads to

$$r_{i+1} = r_i - \frac{|r_i|^2}{(Ap_i, r_i)} Ap_i \tag{26}$$

since all the terms on the right-hand side of (5) vanish because of (24) except for the first and last.

When (3) is used to develop the $\{p_i\}$ from the A-orthogonalization of the $\{r_i\}$ this becomes

$$p_{i+1} = r_{i+1} - \frac{(Ar_{i+1}, p_i)}{(Ap_i, p_i)} p_i \tag{27}$$

since, because of (25), all terms vanish on the right-hand side of (3) except for the first and last. We see that (27) is the same as (16) with the value of β_i given in (18). Also, since

$$p_i = r_i + \beta_{i-1} p_{i-1}$$

it follows that $(p_i, Ap_i) = (r_i, Ap_i)$, because p_{i-1} is A-conjugate to p_i, and consequently (26) is the same as (14) with the value of α_i given by (12). The equivalence of (12) and (17) and of (15) and (18) are easily shown by making use of the foregoing relations.

That r_{i+1}, defined by (14), represents the residual vector immediately follows by induction since

$$\begin{aligned} r_{i+1} = r_i - \alpha_i Ap_i &= k - (Ax_i + \alpha_i Ap_i)\\ &= k - Ax_{i+1}\end{aligned}$$

The Gradient

The locus of the equation

$$H(x) = \text{constant}$$

is an ellipsoid in N-dimensional space. If x is a point on the surface of this ellipsoid it is known from the calculus that the directional derivative of H at x, or the rate of change of H when moving in a particular direction from x, assumes the greatest absolute value when moving from

x in the direction of the gradient [the vector whose components are $\partial H/\partial x^{(i)}$ $(i = 1, 2, \cdots, N)$ where $x^{(i)}$ $(i = 1, 2, \cdots, N)$ represent the N independent variables; the symbol ∇H will be used to represent the gradient vector of H]. This is the direction of the normal to the surface of the ellipsoid at x. If $\{\alpha_{ij}\}$ represents the elements of the matrix A and if $x^{(i)}$ represents the ith component of the vector x, then $H(x) = (A(h - x), h - x)$ is the sum of terms of the form $\alpha_{ij}(h^{(i)} - x^{(i)})(h^{(j)} - x^{(j)})$ over all possible values of i, j. Since $\alpha_{ij} = \alpha_{ji}$ (A is symmetric), it follows by differentiation that

$$\begin{aligned}\frac{\partial H}{\partial x^{(i)}} &= -2[\alpha_{i1}(h^{(1)} - x^{(1)}) + \alpha_{i2}(h^{(2)} - x^{(2)}) \\ &\qquad + \cdots + \alpha_{iN}(h^{(N)} - x^{(N)})] \\ &= -2(\alpha_{i1}h^{(1)} + \alpha_{i2}h^{(2)} + \cdots + \alpha_{iN}h^{(N)}) \\ &\qquad + 2(\alpha_{i1}x^{(1)} + \alpha_{i2}x^{(2)} \\ &\qquad + \cdots + \alpha_{iN}x^{(N)}) \\ &= -2(k^{(i)} - \alpha_{i1}x^{(1)} - \alpha_{i2}x^{(2)} \\ &\qquad - \cdots - \alpha_{iN}x^{(N)}) \\ &= -2r^{(i)}\end{aligned}$$

Therefore, we see that the negative residual vector has the same direction as the gradient vector ∇H. This means that, starting at some approximation x_0 to the solution h, we can achieve the greatest "instantaneous" reduction in $H(x_0)$ by traveling along a line whose direction is that of the residual vector r_0. The point $x_0 + \lambda r_0$ on this line where the minimum value of H occurs can be found, as proven above, by choosing λ in accordance with (9). This procedure is an example of the "gradient methods" which have widespread applications in numerical analysis [2].

However, although a choice of ∇H_i [i.e., $\nabla H(x_i)$] as the direction vector p_i in an iterative procedure such as (6) would have the "advantage" mentioned above, there is no assurance that the procedure would yield the solution in N, or less, steps. It is clear that choosing the direction which maximizes the instantaneous rate of change of H does not necessarily lead to a "better" approximation than by using the direction p_i assumed in the conjugate gradient method. The direction vector p_i is not ∇H_i (except for $i = 0$). It is, however, a scalar multiple of the projection of ∇H_i in the linear space spanned by $p_i, p_{i+1}, \cdots, p_{N-1}$ [i.e., the space consisting of all points (vectors) of the form $\alpha_i p_i + \alpha_{i+1}p_{i+1} + \cdots + \alpha_{N-1}p_{N-1}$]. This assertion, which will be proved below, together with the fact that the $\{p_i\}$, $i = 0, 1, \cdots, N - 1$, are A-conjugate, accounts for the name given to the conjugate gradient method.

Geometric Interpretation and Restatement of the Conjugate Gradient Method

We shall first prove the following theorem which was mentioned above.

THEOREM: The direction vector p_i, along which we move from x_i in order to arrive at x_{i+1}, is a scalar multiple of the projection of the gradient vector ∇H_i, in the linear space spanned by $p_i, p_{i+1}, \cdots, p_{N-1}$.

Proof: We shall show that p_i can be expressed as $\gamma_i r_i + \delta_i q_i$, where q_i is a vector that is orthogonal to the hyperplane, or linear space, spanned by $p_i, p_{i+1}, \cdots, p_{N-1}$. This will imply the conclusion that p_i is a scalar multiple of the projection of $-r_i$ or of ∇H in this hyperplane.

We shall prove this result by induction. From (11), $p_0 = r_0$, and so the theorem is true for $i = 0$. Since the N independent vectors $\{p_i\}$, $i = 0, 1, \cdots, N - 1$, are A-conjugate, it follows that the space orthogonal to the hyperplane spanned by $p_i, p_{i+1}, \cdots, p_{N-1}$ is spanned by the vectors $Ap_0, Ap_1, \cdots, Ap_{i-1}$. If we assume the theorem to be true for $i = k$, this implies that p_k can be written in the form

$$p_k = \gamma_k r_k + \delta_k q_k = \gamma_k r_k + (\delta_{k,0}Ap_0 + \delta_{k,1}Ap_1 + \cdots + \delta_{k,k-1}Ap_{k-1}) \quad (28)$$

From (16) p_{k+1} is a linear combination of r_{k+1} and p_k. Also, (14) implies that r_k is a linear combination of r_{k+1} and Ap_k. These two statements, together with (28), imply that p_{k+1} is a linear combination of $r_{k+1}, Ap_0, Ap_1, \cdots, Ap_k$, i.e.,

$$p_{k+1} = \gamma_{k+1}r_{k+1} + (\delta_{k+1,0}Ap_0 + \delta_{k+1,1}Ap_1 + \cdots + \delta_{k+1,k}Ap_k) \quad (29)$$

This completes the proof by induction since the expression in parentheses in (29) represents a vector orthogonal to the hyperplane spanned by $p_{k+1}, p_{k+2}, \cdots, p_{N-1}$.

Now, we can give the following geometric interpretation of the conjugate gradient method. We start at an arbitrary point x_0, and consider the hyperellipsoid $H(x) = H(x_0)$ passing through this point. The center of this hyperellipsoid is the desired solution h. We travel along the chord of this hyperellipsoid which is normal to the surface at x_0 (i.e., in the direction of p_0, which has the same direction as the gradient for $i = 0$) in order to arrive at x_1. It is easily shown that $H(x_0) = H(x_0 + 2\alpha_0 p_0)$, where α_0 is obtained from (12), and consequently, that $x_1 = x_0 + \alpha_0 p_0$ is the mid-point of this normal chord. Having found x_1 we then consider the projection of the hyperellipsoid $H(x) = H(x_1)$ in the $(N - 1)$-dimensional space that is A-conjugate to p_0 and we repeat the above procedure with this hyperellipsoid projection of one lower dimension. The iteration continues so that at each step the dimension of the space in which we are seeking the solution is decreased by one. After not more than N steps the solution is found.

An important property of the conjugate gradient method is given by the following theorem.

THEOREM: The approximation x_j is closer to the solution h than x_i, $i < j$, i.e., $|h - x_j| < |h - x_i|$.

This result indicates that if we stop the iterative process at any step, the last obtained approximation to the solution is the best, in the sense of being the closest to the true solution. This does not imply, and it is not necessarily true, that the residual vector corresponding to this last approximation is less in length than the residual vectors corresponding to the preceding approximations. We shall need the following lemma to prove the theorem.

LEMMA: $(p_i, p_j) > 0$.

There is no loss of generality in assuming $j \geq i$. The result, of course, is immediate if $j = i$.

From (16), we have

$$p_i = r_i + \beta_{i-1} p_{i-1} = r_i + \beta_{i-1} r_{i-1} + \beta_{i-1}\beta_{i-2} p_{i-2} = \cdots = r_i + \beta_{i-1} r_{i-1} + \beta_{i-1}\beta_{i-2} r_{i-2} + \cdots + \beta_{i-1}\beta_{i-2}\cdots\beta_0 r_0$$

But, because of (15), this can be written as

$$p_i = |r_i|^2\left(\frac{r_i}{|r_i|^2} + \frac{r_{i-1}}{|r_{i-1}|^2} + \frac{r_{i-2}}{|r_{i-2}|^2} + \cdots + \frac{r_0}{|r_0|^2}\right) \tag{30}$$

If, now, we form (p_i, p_j), $j \geq i$, by using the values of p_i, p_j given by (30) we get [because from (19) $(r_i, r_j) = 0$ for $i \neq j$]:

$$(p_i, p_j) = |r_i|^2\,|r_j|^2\left(\frac{1}{|r_0|^2} + \frac{1}{|r_1|^2} + \cdots + \frac{1}{|r_i|^2}\right) \tag{31}$$

Since all the terms on the right of (31) are positive, the lemma is seen to be true.

Before proceeding with the proof of the theorem we observe that (13) implies that

$$x_i = x_0 + \alpha_0 p_0 + \alpha_1 p_1 + \cdots + \alpha_{i-1} p_{i-1} \tag{32}$$

If $x_N = h$, we have

$$h = x_0 + \alpha_0 p_0 + \alpha_1 p_1 + \cdots + \alpha_{N-1} p_{N-1}$$

Let us denote the vector $h - x_i$ by d_i. We shall now show that $(d_{i+1}, d_{i+1}) < (d_i, d_i)$ which implies the theorem to be proved (because of the definition of $|h - x_i|$). From (13), we have $d_i = d_{i+1} + \alpha_i p_i$ and therefore

$$(d_i, d_i) = (d_{i+1}, d_{i+1}) + 2\alpha_i (p_i, d_{i+1}) + \alpha_i^2 (p_i, p_i) \tag{33}$$

Since the third term on the right of (33) is positive and since α_i, by (12), is positive, our result will follow if $(p_i, d_{i+1}) > 0$. To show this, we have, using (32) and assuming that $x_N = h$:

$$\begin{aligned}(p_i, d_{i+1}) &= (p_i, h - x_{i+1}) = (p_i, \alpha_{i+1} p_{i+1} + \alpha_{i+2} p_{i+2} + \cdots + \alpha_{N-1} p_{N-1}) \\ &= \alpha_{i+1}(p_i, p_{i+1}) + \alpha_{i+2}(p_i, p_{i+2}) + \cdots + \alpha_{N-1}(p_i, p_{N-1})\end{aligned} \tag{34}$$

Because of our lemma and because $\alpha_i > 0$, it follows from (34) that $(p_i, d_{i+1}) > 0$ and, thus, the theorem is proved.

GENERALIZATION OF THE METHOD TO HANDLE AN ARBITRARY MATRIX OF COEFFICIENTS, A

In what has preceded, the assumption that A was positive definite and symmetric was vital to the discussion. However, the method can be easily extended to handle the case in which A is arbitrary. For, in this event, the system (1) is equivalent, by premultiplying both sides by A^T, to the system

$$(A^T A)x = A^T k \tag{35}$$

in which the matrix of coefficients $A^T A$

satisfies the necessary requirements of symmetry and positive definiteness if A is nonsingular. If A is singular, A^TA is nonnegative definite $[(Ap, p) \geq 0$ for all $p]$ and we shall show that in this event a least squares solution can be found.

If we apply (11)–(16) to the system (35) the following iterative scheme can be obtained:

$$r_0 = k - Ax_0, \qquad p_0 = A^T r_0 \tag{36}$$

$$\alpha_i = \frac{|A^T r_i|^2}{|Ap_i|^2} \tag{37}$$

$$x_{i+1} = x_i + \alpha_i p_i \tag{38}$$

$$r_{i+1} = r_i - \alpha_i A p_i \tag{39}$$

$$\beta_i = \frac{|A^T r_{i+1}|^2}{|A^T r_i|^2} \tag{40}$$

$$p_{i+1} = A^T r_{i+1} + \beta_i p_i \tag{41}$$

In the above, r_i denotes the residual vector corresponding to the system (1) while $A^T r_i$ is the residual vector corresponding to (35). Note that in using (36)–(41) it is not necessary to form A^TA in solving (35).

Termination of the Iteration Procedure

Let us assume that in (1) A is nonnegative definite, rather than positive definite. Then it is clear that the iteration defined by (11)–(16) can only terminate at $i = M$ if either

$$r_M = 0 \tag{42}$$

or

$$(p_M, Ap_M) = 0 \tag{43}$$

If (42) holds, then x_M is a solution of (1). If (43) is true, we shall show that a least squares solution to (1) (i.e., one which minimizes the length of the residual vector $k - Ax$) can be written by forming a certain linear combination of the approximations $x_0, x_1, \cdots, x_M$.

We recall that a least squares solution to (1) is obtained by solving the system of normal equations given by (35) ([3], p. 72; [4], p. 452). If A is symmetric, this becomes

$$A^2 x = Ak \tag{44}$$

The vector $\bar{x}_M$ will be a solution of (44) and therefore a least squares solution of (1) if

$$A(k - A\bar{x}_M) = A\bar{r}_M = 0 \tag{45}$$

We shall now show that $\bar{x}_M$, satisfying the following equation, is a linear combination of $x_0, x_1, \cdots, x_M$ which satisfies (45) if (43) is true.

$$\bar{x}_M\left(\frac{1}{|r_0|^2} + \frac{1}{|r_1|^2} + \cdots + \frac{1}{|r_M|^2}\right) = \frac{x_0}{|r_0|^2} + \frac{x_1}{|r_1|^2} + \cdots + \frac{x_M}{|r_M|^2} \tag{46}$$

If we compute $\bar{r}_M = k - A\bar{x}_M$ using (46) we obtain

$$\bar{r}_M\left(\frac{1}{|r_0|^2} + \frac{1}{|r_1|^2} + \cdots + \frac{1}{|r_M|^2}\right) = \frac{r_0}{|r_0|^2} + \frac{r_1}{|r_1|^2} + \cdots + \frac{r_M}{|r_M|^2} \tag{47}$$

However, because of (30) this can be written as

$$\bar{r}_M\left(\frac{1}{|r_0|^2} + \frac{1}{|r_1|^2} + \cdots + \frac{1}{|r_M|^2}\right) = \frac{p_M}{|r_M|^2} \tag{48}$$

Now, if (43) holds, then $Ap_M = 0$; for, if we assume A to be of the form B^TB, then $(p_M, Ap_M) = (p_M, B^TBp_M) = (Bp_M, Bp_M) = 0$ which implies that $Bp_M = 0$ and

$$Ap_M = B^TBp_M = 0 \tag{49}$$

Therefore, $A\bar{r}_M = 0$ because of (48) and (49); and $\bar{x}_M$ is a least squares solution of (1).

c. Error Analysis and Salient Properties of the Computational Procedure

First, we observe that it is difficult to set up, beforehand, optimal bounds on the intermediate computed quantities that appear in (11)–(16) and, therefore, it is desirable to use floating point arithmetic during the course of the calculation. Because of the round-off errors that occur during an application of the computational procedure, much of the preceding discussion is of theoretical interest only. It is very unlikely, for example, that the iterative procedure will end after M $(M < N)$ steps because of condition (42) or (43) being satisfied. The probability of the computed values of either r_M or (p_M, Ap_M) being zero is very small. It is also unlikely that the conditions of orthogonality given in (19) and (20) will be satisfied exactly. This not only decreases the effectiveness of these relationships as checks that can be used in an application of the procedure on a fallible computer, but it also, of course, affects the accuracy of the result. It is true that the method permits one to restart the iteration with x_N as the initial approximation if this last approximation is not sufficiently accurate. This can be done with the assurance that the error vector $h - x_i$ will decrease in length. However, one cannot exploit the full advantages of an N-step method without executing all N steps.

As pointed out earlier, although each approximation x_i is closer to the true solution h_1 than the preceding one, there is no assurance that the residual vector will be smaller. In "solving" a linear system, the objective may be not to arrive at an approximation which is very "close" to the true solution, but to satisfy the system well—or to make the residual vector small. These two standards for the accuracy of an approximate solution are quite different. There is a variation of the conjugate gradient method described in [1], in which both the error vector and residual vector diminish at each step. It involves a negligible increase in computational labor and may be preferable in some applications.

The residual vector calculated from (14) will, in general, differ from its defined value $k - Ax_i$. If the procedure is restarted at any point using the last obtained approximation as the starting one, it is better, because of this, to recalculate the residual vector from its definition rather than to assume the last computed value obtained from (14).

When the matrix A is positive definite, symmetric, and has no zero elements, about (neglecting terms of lower order) $N^3 + 5N^2$ multiplications or divisions and $N^3 + 6N^2$ additions or subtractions are needed to arrive at x_N. If the matrix is arbitrary, about twice this number of arithmetic operations is needed. Thus, the method involves about *six* times the number of arithmetic operations involved in some of the efficient elimination schemes. The method shows to greater advantage when the matrix of coefficients has many zero elements. Of course, however, there are variations of the elimination scheme which can also exploit this situation in many instances—as in many finite difference linear systems. It is true that the computer program for a general elimination scheme would be more complex than that for the conjugate gradient method, but this is a specious disadvantage of the elimination scheme. It is worthwhile to put a substantial effort into a library program of general applicability if substantial use of it is anticipated.

Since so many of the linear problems that arise in practice stem from finite difference approximations to differential equations, many of the problems with which we are concerned involve matrices with a large number of zero coefficients. To apply the elimination scheme effectively to these problems will often mean changing it to take advantage of a particular pattern of nonzero coefficients. The conjugate gradient method can be used with great economy in some of these situations, and the savings resulting from the zero coefficients will not be dependent upon their particular locations within the coefficient matrix. In practice it has been noticed that, frequently, the $(N + 1)$st approximation obtained by the method is significantly better than the Nth approximation and this may compensate, to some extent, for the growth of round-off errors in arriving at the Nth approximation. As mentioned earlier, the comparatively small storage requirements and the simplicity of the computational procedure in applying the method are quite attractive. The ideas involved in the derivation of the conjugate gradient method are capable of more general application [2], [5].

No complete error analysis for the procedure is known to us at this time. In [1] there is some discussion of the growth of round-off errors, but this is limited to an indication, which appears below, of some of the factors that influence this growth. No bound on the error of the final approximation is given.

In view of the large number of multiplications that are required to arrive at the solution, it is not surprising that experience in applying the method to some large ($N > 50$) linear systems indicates that the method compares unfavorably with elimination (accompanied by positioning for size) in the accuracy achieved (see Chapter 2). Similar experiences have been reported by Olle Karlquist of the Swedish Board for Computing Machinery in applications on the BESK computer. The elimination method can be applied *using double precision arithmetic* to a linear system with few zero coefficients without involving more elementary arithmetic (single precision) operations than the conjugate gradient method.

A FACTOR INFLUENCING THE GROWTH OF ROUND-OFF ERRORS

The following relations can be proved using (11)–(16) (cf. [1]):

$$\frac{(r_i, r_{i+1})}{|r_i|^2} = \frac{\alpha_i}{\alpha_{i-1}} \frac{(r_{i-1}, r_i)}{|r_{i-1}|^2} \tag{50}$$

$$\frac{(Ap_i, p_{i+1})}{(Ap_i, p_i)} = \frac{\alpha_i}{\alpha_{i-1}} \frac{(Ap_{i-1}, p_i)}{(Ap_{i-1}, p_{i-1})} \tag{51}$$

These show that if, at the beginning of any iteration, $(r_{i-1}, r_i) \neq 0$ and $(Ap_{i-1}, p_i) \neq 0$, then the computed values of (r_i, r_{i+1}) and (Ap_i, p_{i+1}) will deviate from zero also, and by an amount which depends upon the size of α_i/α_{i-1}. The bigger this ratio is the greater will be the disturbance of the orthogonality relations. In [1] it is shown that

$$\frac{1}{\lambda_{max}} < \alpha_i < \frac{1}{\lambda_{min}} \tag{52}$$

where λ_{max}, λ_{min} denote, respectively, the greatest and least characteristic roots of the matrix A. It follows from (52) that $\lambda_{max}/\lambda_{min}$ is an upper bound of the ratio α_i/α_{i-1} which affects the sensitivity of the process to round-off errors. It is interesting to note that this same ratio, $\lambda_{max}/\lambda_{min}$, is the critical "condition number" of a matrix which plays such an important role in the von Neumann-Goldstine analysis of the growth of round-off errors during the elimination process (see [6]). The authors in [1] show that it is possible to construct examples of the conjugate gradient method in which the ratios α_i/α_{i-1}, $i = 1, 2, \cdots, N - 1$, take on any preassigned set of values and, consequently, that the stability with respect to the growth of round-off errors can be quite low.

3. SUMMARY OF THE CALCULATION PROCEDURE

We restate the iterative procedure using (17) and (18) in the place of (12) and (15) because of the more favorable results obtained with their use in the numerical tests reported in [1].

(a) Choose an arbitrary vector x_0 (the initial approximation to the solution).

(b) Set $p_0 = r_0 = k - Ax_0$.

(c) Determine successively

$$\alpha_i = \frac{(p_i, r_i)}{(p_i, Ap_i)}$$

$$x_{i+1} = x_i + \alpha_i p_i$$

$$r_{i+1} = r_i - \alpha_i Ap_i$$

$$\beta_i = -\frac{(r_{i+1}, Ap_i)}{(p_i, Ap_i)}$$

$$p_{i+1} = r_{i+1} + \beta_i p_i$$

(d) Repeat step (c) with $i + 1$ replacing i and continue until $i = N - 1$ or until the residual vector becomes sufficiently small, whichever condition may be satisfied first.

4. FLOW CHART

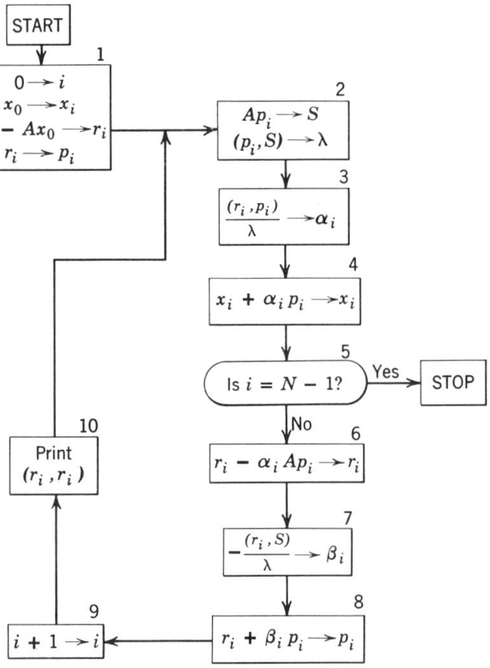

5. DESCRIPTION OF THE FLOW CHART

Box 1: The counter i is set to zero. The initial guess vector is put into the x_i locations. The initial residual vector is computed and stored in the r_i and p_i locations.

Box 2: The vector Ap_i, which occurs twice, is computed and set aside in temporary storage. The scalar $\lambda_i = (p_i, Ap_i)$ is computed.

Box 3: α_i is computed.

Box 4: The next approximation to the solution vector x_{i+1} is calculated.

Box 5: If i has reached $N - 1$, the problem is over and x_i holds the solution vector. Otherwise proceed to box 6. (The termination of the procedure can also be based on the adequacy of the last obtained approximation as indicated by the smallness of the residual vector r_i. In this case, the test would occur after box 10.)

Box 6: The next residual vector is computed.

Box 7: The scalar β_i is calculated.

Box 8: The next member of the conjugate set p_i is calculated.

Box 9: The index i is increased to start the next iterative cycle.

Box 10: The magnitude of the residual vector is computed and printed out to provide some measure of how well the last obtained approximation satisfies the solution. Control is then changed to box 2 for the next cycle.

6. SUBROUTINES

(a) SR1—Dot product of two vectors. Input is first location of each vector involved and location of output scalar. Output is the scalar product of the given vectors. Used in boxes 2, 3, 7, 10. In this computation, it is desirable to take advantage of the facility, available on many computers, of obtaining the full double precision product of single precision operands. These double precision products can be accumulated, with varying difficulty and increase in time, depending upon the command structure of the computer, to give a more accurate value of the sum or scalar product.

(b) SR2—Matrix by vector multiplication. Input is first location of input vector and first location of output vector. Output is product of matrix A by vector. Used in boxes 1 and 2. This routine can, of course, use SR1 above as a subroutine, since each element in the output vector is the scalar product of a row of the matrix and the input vector.

(c) SR3—Sum and difference of two vectors. Input is first location of vectors involved and first location of output vector. Also a signal word to indicate whether addition or subtraction is being called for. Output is the sum or difference of the given vectors. Used in boxes 1, 4, 6, 8.

7. SAMPLE PROBLEM

$$A = \begin{pmatrix} 3 & 1 \\ 1 & 2 \end{pmatrix} \qquad k = \begin{pmatrix} 5 \\ 5 \end{pmatrix}$$

$$\begin{aligned}
x_0 &= (0, 0) \\
p_0 &= r_0 = (5, 5) \\
Ap_0 &= (20, 15) \\
(p_0, Ap_0) &= 175 \\
\alpha_0 &= 2/7 \\
x_1 &= (2/7)(5, 5) \\
r_1 &= (5/7)(-1, 1) \\
\beta_0 &= 1/49 \\
p_1 &= (1/49)(-30, 40) \\
Ap_1 &= (1/49)(-50, 50) \\
(p_1, Ap_1) &= 3500/2401 \\
\alpha_1 &= 7/10 \\
x_2 &= (1, 2)
\end{aligned}$$

(For more compact format, all the vectors involved in the computation have been listed as row vectors. They should, of course, be interpreted as column vectors.)

Note that even in as simple a problem as this one the numbers can grow quite large. This illustrates the fact that round-off error can become a serious problem when N is large. It also indicates the desirability of using floating arithmetic.

8. MEMORY REQUIREMENTS

A table of the quantities which must be stored and the memory space required for each is as follows:

x_i	N words
p_i	N words
α_i	1 word
β_i	1 word
r_i	N words
A	$N(N + 1)/2$ words for a symmetric matrix
k	N words

The program requires about 300 words (depending upon the availability of built-in floating point operations.)

After the calculations of box 1, the N words of k can be used for, say, the residual vector. The total memory requirement is thus about $N(N + 7)/2$ words.

9. ESTIMATION OF THE RUNNING TIME

Excluding the initial guesses, there will be about $N(N + 6)$ additions and $N(N + 5)$

multiplications per iteration. The total calculation time required for N steps is therefore given by

$$T \sim 2(N^3 + 5N^2)\mu + 2(N^3 + 6N^2)\nu$$

where μ and ν are the multiply and add times on the computer in question, and we have doubled the arithmetic time in the over-all estimate to account for "bookkeeping" operations.

10. REFERENCES

1. M. Hestenes and E. Stiefel, Method of Conjugate Gradients for Solving Linear Systems, Report 1659, *Nat. Bur. Standards*, 1952.
2. J. W. Fischbach, Some Applications of Gradient Methods, *Proceedings of Symposia in Applied Mathematics*, Vol. VI, McGraw-Hill Book Co., New York, 1956, pp. 59–72.
3. A. S. Householder, *Principles of Numerical Analysis*, McGraw-Hill Book Co., New York, 1953.
4. F. B. Hildebrand, *Methods of Applied Mathematics*, Prentice-Hall, New York, 1952.
5. M. R. Hestenes, The Conjugate-Gradient Method for Solving Linear Systems, *Proceedings of Symposia in Applied Mathematics*, Vol. VI, McGraw-Hill Book Co., New York, 1956, pp. 83–102.
6. J. von Neumann and H. H. Goldstine, Numerical Inverting of Matrices of High Order, *Bull. Amer. Math. Soc.*, vol. 53, no. 11, Nov. 1947, pp. 1021–1099.

5 Matrix inversion by the method of rank annihilation*

Herbert S. Wilf
The University of Illinois

1. FUNCTION

To invert a nonsingular square matrix.

2. MATHEMATICAL DISCUSSION

Let A be a square matrix of known inverse, and let u and v be column vectors. The inversion formula

$$(A + uv^T)^{-1} = A^{-1} - \frac{(A^{-1}u)(v^T A^{-1})}{1 + v^T A^{-1}u} \tag{1}$$

can be immediately verified, and provides a method for finding the inverse of a matrix which differs from a matrix of known inverse by a matrix of rank unity. Equation (1) was originally derived by Sherman and Morrison [1], used by Bartlett [2], and generalized by Woodbury [3], as reported by Householder [4] to a wider class of modifications of a matrix of known inverse.

Now, since an arbitrary matrix can obviously be written as a finite sum of matrices of rank one, it is perfectly clear that one can, by repeated application of (1), invert an arbitrary nonsingular matrix.

* Portions of this chapter are reprinted here with the kind permission of the *Journal of the Society for Industrial and Applied Mathematics* (see [5]).

To be specific, suppose that the given matrix B has been written in the form

$$B = D + \sum_{i=1}^{I} u_i v_i^T \tag{2}$$

where D is a nonsingular diagonal matrix. Defining the partial sum

$$C^{(k)} = \left\{ \sum_{i=1}^{k} u_i v_i^T + D \right\}^{-1} \tag{3}$$

we find easily the algorithm

$$C^{(1)} = D^{-1} \tag{4}$$

$$\begin{aligned} C^{(k+1)} &= \left\{ \sum_{i=1}^{k+1} u_i v_i^T + D \right\}^{-1} \\ &= \left\{ D + \sum_{i=1}^{k} u_i v_i^T + u_{k+1} v_{k+1}^T \right\}^{-1} \\ &\qquad (k = 1, \cdots, I) \quad (5) \\ &= \{(C^{(k)})^{-1} + u_{k+1} v_{k+1}^T\}^{-1} \\ &= C^{(k)} - \frac{(C^{(k)}u_{k+1})(v_{k+1}^T C^{(k)})}{1 + v_{k+1}^T C^{(k)} u_{k+1}} \end{aligned}$$

which terminates with

$$C^{(I+1)} = B^{-1} \tag{6}$$

The question which arises, then, is how to

choose, among all the possible decompositions (2) of the given matrix B, the one which minimizes the total labor required to find B^{-1}. Without claiming to have solved this optimization problem, we will give below a particular decomposition of B which is interesting and quite efficient.

We now suppose that B is an $N \times N$ matrix, and that $B_{11} - 1 \neq 0$. Writing

$$A = B - I \tag{7}$$

consider the sequence of matrices whose elements are determined by

$$A_{ij}^{(1)} = A_{ij} \tag{8}$$

$$A_{ij}^{(n+1)} = A_{ij}^{(n)} - \frac{A_{in}^{(n)}A_{nj}^{(n)}}{A_{nn}^{(n)}} \qquad (n = 1, 2, \cdots, N) \tag{9}$$

We have

THEOREM 1: $A_{ij}^{(n)}$ vanishes for $i < n$ or $j < n$, for each $n = 1, \cdots, N$.

Proof: The result is clearly true for $n = 1$. If true for any n, then (9) shows the desired result immediately for $i < n$ or $j < n$, and it suffices to consider $i = n$ or $j = n$. Substitution of these values of i or j into (9) again proves the theorem for $n + 1$ and hence for all n.

COROLLARY:

$$A^{(N+1)} = 0 \tag{10}$$

THEOREM 2: The given matrix B can be written in the form (2) by

$$(B)_{ij} = \delta_{ij} + \sum_{n=1}^{N} \frac{A_{in}^{(n)}A_{nj}^{(n)}}{A_{nn}^{(n)}} \tag{11}$$

Proof: Summing both sides of (9) from $n = 1$ to N,

$$\sum_{n=1}^{N} \frac{A_{in}^{(n)}A_{nj}^{(n)}}{A_{nn}^{(n)}} = \sum_{n=1}^{N} A_{ij}^{(n)} - \sum_{n=1}^{N} A_{ij}^{(n+1)}$$
$$= A_{ij}^{(1)} - A_{ij}^{(N+1)}$$
$$= A_{ij} = B_{ij} - \delta_{ij}$$

which was to be shown.

We have found, therefore, an expansion of the type (2), in which

$$D_{ij} = \delta_{ij} \tag{12}$$
$$(u_i)_j = A_{ji}^{(i)}/A_{ii}^{(i)} \tag{13}$$
$$(v_i)_j = A_{ij}^{(i)} \tag{14}$$

and the number of terms, I, in (2) is equal to the order of B.

3. SUMMARY OF THE CALCULATION PROCEDURE

Given a nonsingular matrix B, take

$$A_{ij}^{(1)} = B_{ij} - \delta_{ij} \qquad C_{ij}^{(1)} = \delta_{ij} \tag{15}$$

Then for each $n = 1, 2, \cdots, N$ calculate

$$A_{ij}^{(n+1)} = A_{ij}^{(n)} - \frac{A_{in}^{(n)}A_{ij}^{(n)}}{A_{nn}^{(n)}} \qquad (i, j > n) \tag{16}$$

$$u_i = \sum_{k=n}^{N} C_{ik}^{(n)}A_{kj}^{(n)} \qquad (i = 1, 2, \cdots, N) \tag{17}$$

$$v_j = \sum_{k=n}^{N} A_{nk}^{(n)}C_{kj}^{(n)} \qquad (j = 1, 2, \cdots, N) \tag{18}$$

$$\lambda = \sum_{j=n}^{N} v_j A_{jn}^{(n)} + A_{nn}^{(n)} \tag{19}$$

$$C_{ij}^{(n+1)} = C_{ij}^{(n)} - \frac{1}{\lambda} u_i v_j \qquad (i, j = 1, 2, \cdots, N) \tag{20}$$

It follows that

$$B^{-1} = C^{(N+1)} \tag{21}$$

4. FLOW CHART

The flow chart appears on page 75.

5. DESCRIPTION OF THE FLOW CHART

Box 1: The input matrix B has been put into the locations A_{ij}. The identity matrix will now be subtracted from B, and entered in C_{ij}. The counter i is set to one.

Box 2: The ith diagonal element of B is reduced by one and the result put into A. A one is put into the ith diagonal element of C, whose other elements are assumed to be zero.

Box 3: The index i is tested to see if all diagonal elements have been done.

Box 4: If not, i is increased by one, and control is returned to box 2 to do the next element.

Box 5: The iterative cycle is started with n set to one.

Box 6: The elements of the matrix of rank one on the right-hand side of (16) will now be stored for future use. The index i is set to $n + 1$.

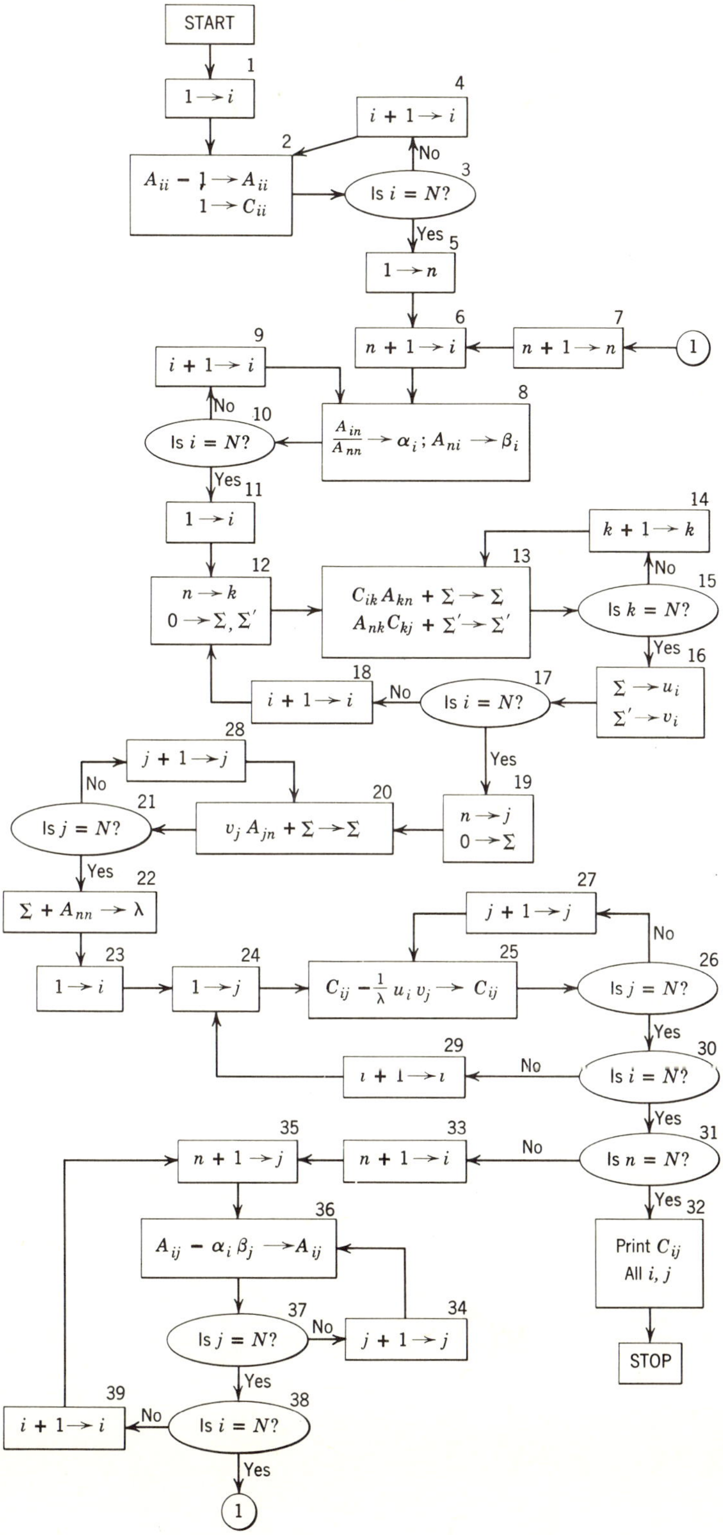
START
1 → i
A_ii − 1 → A_ii
1 → C_ii
Is i = N?
i + 1 → i
No
Yes
1 → n
n + 1 → i
n + 1 → n
1
A_in / A_nn → α_i ; A_ni → β_i
Is i = N?
i + 1 → i
1 → i
n → k
0 → Σ, Σ′
C_ik A_kn + Σ → Σ
A_nk C_kj + Σ′ → Σ′
Is k = N?
k + 1 → k
Σ → u_i
Σ′ → v_i
Is i = N?
i + 1 → i
n → j
0 → Σ
v_j A_jn + Σ → Σ
Is j = N?
j + 1 → j
Σ + A_nn → λ
1 → i
1 → j
C_ij − (1/λ) u_i v_j → C_ij
Is j = N?
j + 1 → j
Is i = N?
i + 1 → i
Is n = N?
Print C_ij
All i, j
STOP
n + 1 → i
n + 1 → j
A_ij − α_i β_j → A_ij
Is j = N?
j + 1 → j
Is i = N?
i + 1 → i
1

Box 7: Returning from the end of a cycle, the iteration counter n is increased by one.

Box 8: The ith elements of the vectors α_i and β_i are stored. Their products will later be subtracted from A as in (16).

Box 9: The index i is increased by one and control is returned to box 8 to do the next elements.

Box 10: The index i is tested to see if all elements have been done.

Box 11: We now calculate the vectors u and v given by (17) and (18). The index i is set to one.

Box 12: To start the sums indicated in (17) and (18), two locations denoted by Σ and Σ' are cleared, and the summation index k is set to n.

Box 13: The kth terms of the sums defining u and v are calculated and accumulated.

Box 14: The index k is increased by one to accumulate the next term of the sum.

Box 15: The summation index k is tested to see if all terms have been done.

Box 16: Having done the entire sum, the location Σ is stored as u_i and the location Σ' as v_i.

Box 17: The index i is tested to see if all components of u and v have been calculated.

Box 18: If not, the i index is increased by one and control is returned to box 12 to calculate the next components.

Box 19: We now prepare the calculation of λ given by (19) by clearing a sum location and setting the summation index j to n.

Box 20: The jth term of the sum defining λ is accumulated.

Box 21: The summation index j is tested against N to see if all terms have been done.

Box 22: The sum is added to $A_{nn}^{(n)}$ as in (19) to give λ.

Box 23: Preparatory to calculating the new C matrix from (20), the row counter i is set to one.

Box 24: Similarly, the column counter is set to one.

Box 25: The i, j element of the new C matrix is calculated as in (20) and stored.

Box 26: The column index j is tested to see if the complete row has been done.

Box 27: If not, the column counter is increased by one and control is returned to box 25 to do the next element in the same row.

Box 28: The index j is increased by one to calculate the next term in the sum.

Box 29: The row index i is increased by one and control is returned to box 24 to do the next row.

Box 30: The row index is tested against N to see if all rows have been done.

Box 31: The grand iteration counter n is tested against N to see if the entire inverse has been calculated.

Box 32: If so, the entire C matrix, which is the required inverse of B, is typed out, and the program stops.

Box 33: Otherwise, the new A matrix must be calculated, and we set the row counter i to $n + 1$.

Box 34: The column counter is increased by one to do the next column.

Box 35: The column counter for the new A matrix is set to $n + 1$.

Box 36: The i, j element of the new A matrix is calculated from (16).

Box 37: The column counter j is tested to see if all columns of A have been calculated.

Box 38: The row counter i is tested to see if all rows of A have been calculated.

Box 39: The row counter is increased by one and control is returned to box 35 to do the next row.

6. SUBROUTINES

A subroutine which can be given the location of any two indices and the first location of a matrix in storage, and which will then pick out the element in question, will be found very useful.

7. SAMPLE PROBLEM

$$B = \begin{pmatrix} 5 & 2 \\ 3 & 1 \end{pmatrix}$$

$$A^{(1)} = \begin{pmatrix} 4 & 2 \\ 3 & 0 \end{pmatrix} \qquad C^{(1)} = \begin{pmatrix} 1 & 0 \\ 0 & 1 \end{pmatrix}$$

$$u = (4, 3) \qquad v = (4, 2)$$

$$\lambda = 26$$

$$A^{(2)} = \begin{pmatrix} 0 & 0 \\ 0 & -\frac{3}{2} \end{pmatrix} \qquad C^{(2)} = \tfrac{1}{13}\begin{pmatrix} 5 & -4 \\ -6 & 10 \end{pmatrix}$$

$$u = \tfrac{1}{13}(6, -15) \qquad v = \tfrac{1}{13}(9, -15)$$

$$\lambda = \tfrac{3}{13}$$

$$C^{(3)} = B^{-1} = \begin{pmatrix} -1 & 2 \\ 3 & -5 \end{pmatrix}$$

8. MEMORY REQUIREMENTS

Memory requirements are for two matrices, four vectors, a few scalars and the program, for a total of $2N^2 + 4N + 300$ words, approximately.

9. ESTIMATION OF THE RUNNING TIME

If B is not symmetric, the number of multiplications and divisions required for the calculation of B^{-1} is $\frac{7}{3}N^3$, for large N. If B is symmetric, the program can be modified slightly to recognize the fact that in this case $u = v$, and the matrices $A^{(n)}$ and $C^{(n)}$ are symmetric at every stage of the cycle. In this way the amount of work can be cut about in half. The running time for the nonsymmetric case is about

$$T = 5N^3(\mu + \nu)$$

for large N, where μ and ν are the multiply and add times, respectively, of the computer.

10. REFERENCES

1. J. Sherman and W. J. Morrison, Adjustment of an Inverse Matrix Corresponding to Changes in a Given Column or a Given Row of the Original Matrix, *Ann. Math. Stat.*, vol. 21, p. 124, 1949.
2. M. S. Bartlett, An Inverse Matrix Adjustment Arising in Discriminant Analysis, *Ann. Math. Stat.*, vol. 22, pp. 107–111, 1951.
3. Max Woodbury, Inverting Modified Matrices, Memo. Report 42, Statistical Research Group, Princeton, 1950.
4. A. S. Householder, *Principles of Numerical Analysis*. McGraw-Hill Book Co., New York, 1953, pp. 79, 83.
5. H. S. Wilf, Matrix Inversion by the Annihilation of Rank, *J. Soc. Indust. Appl. Math.*, vol. 7 (2), pp.149–151, 1959.

6 Matrix inversion by Monte Carlo methods

Florence Jeanne Oswald
Nuclear Development Corporation of America

1. FUNCTION

The Monte Carlo method provides a simple computational approach to the statistical estimation of the elements of the inverse of a given matrix. The technique to be described will give any one element, a single row, or all of the elements of a matrix when certain conditions are met.

Assume that the inverse of a matrix A of order N is desired and let $D = I - A$, where I is the unit matrix. Choose quantities p_{ij} and v_{ij} so that the elements of D are

$$d_{ij} = p_{ij}v_{ij} \tag{1}$$

with the restrictions that all $p_{ij} \geq 0$, and $\sum_{j=1}^{N} p_{ij} < 1$. Define the matrix B with elements

$$b_{ij} = p_{ij}v_{ij}^2 \tag{2}$$

For this method to be applicable it is necessary that

$$\max_i |\lambda_i(D^*)| < 1 \tag{3}$$

where D^* is the matrix with elements $|d_{ij}|$ and $\lambda_i(D^*)$ is the ith eigenvalue of the matrix D^*.

Furthermore, to insure that the estimate of the elements of A^{-1} will have a finite variance one must have

$$\max_i |\lambda_i(B)| < 1 \tag{4}$$

2. MATHEMATICAL DISCUSSION

a. Description of the Method

The method consists of taking a random walk in N points. Let $1, 2, 3, \cdots, N$ be the domain of the walker. The walk begins at any specified point i and moves from point to point with the probabilities p_{ij} defined by (1). The walk terminates after h steps at some point k (possibly immediately) with the stop probability $p_k = 1 - \sum_{j=1}^{N} p_{kj}$. When the walk stops, a score is tallied for every element in the row i of the inverse matrix. The score G_{ij} is zero if $j \neq k$. For $j = k$ the score is the product of the weights or transition values v_{ij} associated with each step in the walk and defined by (1), divided by the probability of stopping at point k, so that

$$G_{ij} = \begin{cases} 0 & \text{if } j \neq k \\ v_{ii_1}v_{i_1i_2}\cdots v_{i_{h-1}j}p_j^{-1} & \text{if } j = k \end{cases}$$

where $i \rightarrow i_1 \rightarrow \cdots \rightarrow i_{h-1} \rightarrow j$ is the route ρ from i to j. If condition (3) is satisfied, then

$$A^{-1} = (I - D)^{-1} = I + D + D^2 + \cdots + D^h + \cdots = \sum_{h=0}^{\infty} D^h$$

Therefore, $(A^{-1})_{ij} = \sum_{k=0}^{\infty} (D_k)_{ij}$. Now we can prove the

THEOREM [1]: The expected value of G_{ij} is $(A^{-1})_{ij}$.

Proof: The probability of following a route ρ and then stopping at j is

$$P_\rho p_j = p_{ii_1} p_{i_1 i_2} \cdots p_{i_{h-1} j} p_j$$

The expected score after a walk is

$$E(G_{ij}) = \sum_\rho (P_\rho p_j)(V_\rho p_j^{-1}) = \sum_\rho P_\rho V_\rho$$

where $V = v_{ii_1} v_{i_1 i_2} \cdots v_{i_{h-1} j}$ and where the sum is over all routes from i to j. Since $p_{ij} v_{ij} = d_{ij}$,

$$E(G_{ij}) = \delta_{ij} + \sum_{h=1}^{\infty} \sum_{i_1=1}^{N} \cdots \sum_{i_{h-1}=1}^{N} d_{ii_1} d_{i_1 i_2} \cdots d_{i_{h-1} j}$$
$$= I_{ij} + \sum_{h=1}^{\infty} (D^h)_{ij} = (A^{-1})_{ij}$$

which proves the theorem. The δ_{ij} term arises from those walks which terminate immediately.

A fairly simple sufficient condition [2] exists for testing whether condition (3) is satisfied. Let

$$S_j = \sum_{i=1}^{\infty} d_{ij}$$

If $S_j < 1$ for all j then $\max_i |\lambda_i(D^*)| < 1$. If the matrix A is such that this test fails, it is possible in some cases to divide A by a constant factor f so that $A = fA'$ and using $I - A' = D'$, $\max_i |\lambda_i(D'^*)| < 1$. Then $(A')^{-1}$ is obtained by Monte Carlo and $(A^{-1})_{ij} = 1/f\,(A')_{ij}^{-1}$.

b. Error Analysis

In a Monte Carlo calculation the problem of round-off and truncation error has little effect on the accuracy. The statistical variation of the result is a more important factor. Accordingly, a measure of the precision of the answer is the mean square deviation, the variance. It should be noted that the first few walks will tend to improve the results markedly while many additional random walks may be necessary to refine them. Therefore, this technique may be particularly useful in obtaining a rough estimate quickly which can be improved by some other technique, such as the Hotelling-Bodewig iteration [3].

The variance of the score G_{ij} about its expected value [1] is given by $\sigma_{ij}^2 = H_{ij} p_j^{-1} - (A^{-1})_{ij}^2$, where $H = (I - B)^{-1}$. The variance depends on the choice of the p_{ij} and v_{ij}. In some cases one of two simple procedures can be used to choose these quantities.

The first is to let all the $v_{ij} = 1$. The variance will then correspond to the variance of the binomial distribution [1]. This choice is possible only when all the elements $d_{ij} \geq 0$. The other choice is to let all the p_{ij} equal a constant $p < 1/N$. Consider $H = (I - B)^{-1} = [I - (1/p)W]^{-1}$, where W is the matrix whose elements are d_{ij}^2 since $b_{ij} = p_{ij} v_{ij}^2 = d_{ij} v_{ij} = d_{ij}^2 / p_{ij}$. The condition $\max_i |\lambda_i(B)| < 1$ implies

$$\max_i \left| \lambda_i \left(\frac{1}{p} W \right) \right| = \frac{1}{p} \max_i |\lambda_i(W)| < 1$$

or $\max_i |\lambda_i(W)| < p$. Therefore, to insure a finite variance the constant value of p must lie in the range

$$\max_i |\lambda_i(W)| < p < \frac{1}{N}$$

if possible.

3. SUMMARY OF THE CALCULATION PROCEDURE

(a) Choose the quantities p_{ij} and v_{ij}, C' the number of walks executed before calculating the variance, i the row of elements to be computed, U the tolerance on the variance.

(b) Compute p_i^{-1}.

(c) Input to the code: steps (a) and (b) and N, the dimension of the matrix.

(d) Choose a random number ξ between 0 and 1.

(e) If j is the point to which the walk has proceeded, find the smallest k for which the cumulative probability $\left(\sum_{r=1}^{k} p_{jr}\right)$ of going from j to k is greater than ξ. Calculate the contribution to the score of the step from j to k. If

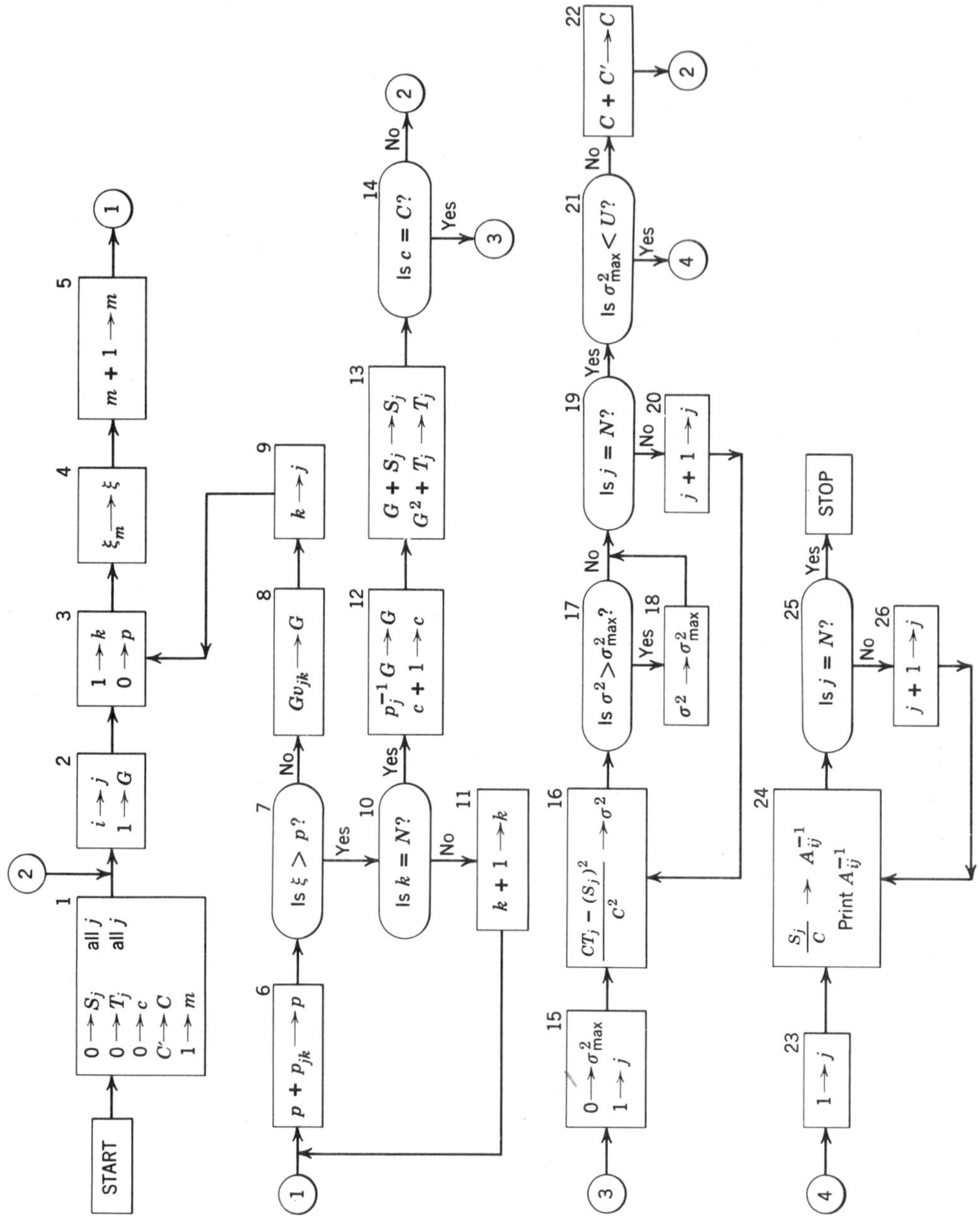
START
$0 \to S_j$ all j
$0 \to T_j$ all j
$0 \to c$
$C' \to C$
$1 \to m$
$i \to j$
$1 \to G$
$1 \to k$
$0 \to p$
$\xi_m \to \xi$
$m + 1 \to m$
$p + p_{jk} \to p$
Is $\xi > p$?
Yes
No
$Gv_{jk} \to G$
$k \to j$
Is $k = N$?
$k + 1 \to k$
$p_j^{-1} G \to G$
$c + 1 \to c$
$G + S_j \to S_j$
$G^2 + T_j \to T_j$
Is $c = C$?
$0 \to \sigma^2_{max}$
$1 \to j$
$\frac{CT_j - (S_j)^2}{C^2} \to \sigma^2$
Is $\sigma^2 > \sigma^2_{max}$?
$\sigma^2 \to \sigma^2_{max}$
Is $j = N$?
$j + 1 \to j$
Is $\sigma^2_{max} < U$?
$C + C' \to C$
$1 \to j$
$\frac{S_j}{C} \to A_{ij}^{-1}$
Print A_{ij}^{-1}
Is $j = N$?
$j + 1 \to j$
STOP

$\sum_{r=1}^{N} p_{jr} \leq \xi$ the walk stops. It is easy to see that in this way the conditions of the walk defined in Section 2a are satisfied.

(f) Repeat steps (d) and (e) until a stop occurs.

(g) When a stop occurs, compute the score for the walk.

(h) Repeat steps (d) through (g) for C' walks.

(i) Calculate the variance of the score for each element in the row.

(j) If the variance is sufficiently small, print the results; otherwise repeat steps (d) through (i).

4. FLOW CHART

The flow chart appears on page 80.

5. DESCRIPTION OF THE FLOW CHART

Box 1: The locations S_j that will contain the running score and the locations T_j that will contain the sum of the squares of the score for all walks ending at j are set equal to zero for all j. The walk counter c is set equal to zero. The test quantity C is set equal to the increment C'. The random number counter m is set equal to one.

Box 2: The counter j is set equal to i, the point at which the random walk is to start. The score for the walk, G, is set equal to the starting value one.

Box 3: The counter k is set equal to one. The cumulative probability of taking a step from point j is initially set to zero.

Box 4: The mth random number ξ_m is calculated and set aside in temporary storage.

Box 5: The random number counter is increased by one.

Box 6: The cumulative probability of going from point j to point k is calculated.

Box 7: If the random number is greater than the cumulative probability of going to point k, then the step does not end in k; proceed to box 10. Otherwise the step ends at k; proceed to box 8.

Box 8: The score for the walk thus far is multiplied by the weight associated with the step j to k.

Box 9: The origin of the next step j is set equal to the terminus of the preceding step k and the walk continues by proceeding to box 3 to calculate the next step.

Box 10: If the counter k is not equal to the dimension N, proceed to box 11. Otherwise, the walk ends at the point j; proceed to box 12.

Box 11: The counter k is increased by one.

Box 12: The score for the walk thus far is multiplied by the inverse of the probability of stopping at point j. The walk counter is increased by one.

Box 13: The score for the walk is added to the sum of the scores for walks ending at j. The score for the walk is squared and added to the sum of the squares of the scores for walks ending at j.

Box 14: If the walk counter is not equal to a multiple of a predetermined number of walks, the random walk process is continued by returning to box 2. Otherwise proceed to box 15 to calculate the variance of the results so far.

Box 15: Set the location which will contain the largest variance to zero. Set the counter j equal to one.

Box 16: Calculate the variance of $(A^{-1})_{ij}$.

Box 17: If the variance of box 16 is greater than $\sigma^2_{\max}$, proceed to box 18. Otherwise, procced to box 19.

Box 18: Replace $\sigma^2_{\max}$ with σ^2.

Box 19: If j is not equal to the dimension N, go to box 20. Otherwise, proceed to box 21.

Box 20: Increase j by one and return to box 16 to calculate the variance of the next element.

Box 21: If the largest variance is greater than some preassigned tolerance U, proceed to box 22. Otherwise, the calculation has achieved the desired accuracy; proceed to box 23.

Box 22: Increment the test quantity C and go to box 2 to continue the random walk process.

Box 23: Set the counter j equal to one.

Box 24: Calculate and print the inverse matrix element $(A^{-1})_{ij}$.

Box 25: If $j = N$, the elements of one entire row of the inverse matrix have been computed. Otherwise, proceed to box 26.

Box 26: Increase j by one. Proceed to box 24.

6. SUBROUTINES

SR1—Random number generator. No input required. Output is one member of a "random sequence." Used in box 4.

To calculate a pseudo random number ξ_m, let

$$\xi_m = \xi_{m-1} R_0 (\text{mod } q^a)$$

$$\xi_0 = 1$$

where q equals the number base of the computer, a depends on the machine word size, and R_0 depends on both. Specifically, for a ten-decimal digit computer $q \neq 10$, $a = 10$, $R_0 = 8212890627$ is a good choice. For a binary computer $q = 2$, a equals the maximum number of bits in a machine word, and R_0 is equal to the largest odd power of a number relatively prime to the machine base which fits into a machine word. A frequent choice is an odd power of the number 3 or 5; see [4].

7. SAMPLE PROBLEM

$$A = \begin{pmatrix} +.4 & -.2 & -.1 \\ -.2 & +.5 & -.1 \\ -.1 & -.2 & +.6 \end{pmatrix}$$

$$I - A = \begin{pmatrix} +.6 & +.2 & +.1 \\ +.2 & +.5 & +.1 \\ +.1 & +.2 & +.4 \end{pmatrix} = D = P$$

$$V = \begin{pmatrix} 1 & 1 & 1 \\ 1 & 1 & 1 \\ 1 & 1 & 1 \end{pmatrix}$$

$$p = \begin{pmatrix} .1 \\ .2 \\ .3 \end{pmatrix} \qquad \begin{aligned} p_1^{-1} &= 10 \\ p_2^{-1} &= 5 \\ p_3^{-1} &= 3.333333 \end{aligned}$$

$$N = 3, \qquad U = 26$$
$$C' = 2 = C$$
$$i = j = 2$$
$$G = 1$$
$$\xi_1 = \xi = .82$$
$$p = p_{21} = .2$$
$$p = p_{21} + p_{22} = .7$$
$$p = p_{21} + p_{22} + p_{23} = .8$$
$$Gp_2^{-1} = 5 = G$$
$$c = 1$$
$$S_2 = 5$$
$$T_2 = 25$$
$$i = j = 2$$
$$G = 1$$
$$\xi_2 = \xi = .15$$
$$p = p_{21} = .2$$
$$Gv_{21} = 1 = G$$
$$j = 1$$
$$k = 1$$
$$\xi_3 = \xi = .95$$
$$p = p_{11} = .6$$
$$p = p_{11} + p_{12} = .8$$
$$p = p_{11} + p_{12} + p_{13} = .9$$
$$Gp_1^{-1} = 10 = G$$
$$c = 2$$
$$S_1 = 10$$
$$T_1 = 100$$
$$\sigma^2 = \frac{2(100) - (10)^2}{4} \doteq 25$$
$$\sigma^2_{\max} = 25$$
$$\sigma^2 = \frac{2(25) - (5)^2}{4} = 6.25$$
$$\sigma^2_{\max} = 25$$
$$\sigma^2 = \frac{2(0) - (0)^2}{4} = 0$$
$$\sigma^2_{\max} = 25$$
$$U > \sigma^2_{\max}$$
$$(A^{-1})_{21} = \tfrac{10}{2} = 5$$
$$(A^{-1})_{22} = 2.5$$
$$(A^{-1})_{23} = 0$$

This example illustrates the fact that many walks are often required to obtain tolerable accuracy. Obviously setting the quantity U equal to 26 is highly unrealistic. For the above matrix in a test case, the element $(A^{-1})_{22}$ was calculated as 2.994 after 2376 walks. This result was in error by 7 in the last place and the variance was 3.4×10^{-3}. This is a low probability result.

8. MEMORY REQUIREMENTS

The following is a table of quantities that must be stored and the number of words of memory required for each:

S_j	N words
T_j	N words
c	1 word
C'	1 word
C	1 word
i	1 word
j	1 word

G	1 word
k	1 word
p	1 word
ξ_m	1 word
ξ	1 word
p_{jk}	N^2 words
N	1 word
v_{jk}	N^2 words
p_j^{-1}	N words
$\sigma^2_{\max}$	1 word
σ^2	1 word
U	1 word

The program requires approximately 400 words. The total memory requirement is about $N(2N + 3) + 414$ words.

9. ESTIMATION OF THE RUNNING TIME

$$T \sim 4(3N + 2C)\mu + [(W + 2)C + 3N]2\nu$$

where W is the average number of steps per random walk and depends on the magnitude of the stop probabilities, and μ and ν are the multiplication and addition times, respectively, of the computer.

10. REFERENCES

1. G. E. Forsythe and R. A. Leibler, Matrix Inversion by a Monte Carlo Method, *MTAC*, vol. 4, 1950, pp. 127–129.
2. Y. K. Wong, Some Properties of the Proper Values of a Matrix, *Proc. Amer. Math. Soc.*, vol. 6, 1955, pp. 891–899.
3. A. S. Householder, *Principles of Numerical Analysis*, McGraw-Hill Book Co., New York, 1953.
4. H. A. Meyer (editor), *Symposium on Monte Carlo Methods*, John Wiley & Sons, New York, 1956.
5. W. R. Wasow, A Note on the Inversion of Matrices by Random Walks, *MTAC*, vol. 6, 1952, pp. 78–81.
6. W. Feller, *An Introduction to Probability Theory and Its Applications*, Vol. 1, 2nd ed., John Wiley & Sons, New York, 1957.
7. J. H. Curtiss, Monte Carlo Methods for the Iteration of Linear Operators, *Jour. of Math. and Phys.*, vol. 32, 1953, pp. 209–232.
8. A. D. Booth, *Numerical Methods*, Academic Press, New York, 1955, pp. 107–109.
9. R. Courant and D. Hilbert, *Methods of Mathematical Physics*, Vol. 1, New York, 1953.
10. H. P. Edmundson, Monte Carlo Matrix Inversion and Recurrent Events, *MTAC*, vol. 7, 1953, pp. 18–21.
11. G. E. Forsythe and R. A. Leibler, Correction to the Article, "Matrix Inversion by a Monte Carlo Process," *MTAC*, vol. 5, 1951, p. 55.

The determination of the characteristic roots of a matrix by the Jacobi method

7

John Greenstadt
International Business Machines Corporation

I. FUNCTION

The aim of this diagonalization method—originally invented by C. G. J. Jacobi [1] and recently revived by J. von Neumann for modern large computers—is to find the eigenvalues and eigenvectors of a real symmetric matrix.

In order to demonstrate the pertinence of the method, we shall first write down some familiar definitions and relations. Let us denote the matrix in question by A (and its elements by A_{ik}), and assume the symmetry property:

$$A^T = A \tag{1}$$

where A^T denotes the transpose of A. Let the order of A be N. With A we associate a set of scalar eigenvalues $\alpha^{(i)}$ $(i = 1, \cdots, N)$ and eigenvectors $a^{(i)}$ $(i = 1, \cdots, N)$, which satisfy the following relations:

$$Aa^{(i)} = \alpha^{(i)}a^{(i)} \qquad i = 1, \cdots, N \tag{2}$$

The $a^{(i)}$ are column vectors.

Even when not all the eigenvalues are distinct, it is still possible to find N distinct eigenvectors for an Nth-order real symmetric matrix.

2. MATHEMATICAL DISCUSSION

Let us assume that it is feasible to find an Nth-order orthogonal matrix S with elements S_{ik}, with the property that the transform of A by S is a diagonal matrix D. In symbols:

$$S^T AS = D \tag{3}$$

The elements D_{ik} of D have the form:

$$D_{ik} = \epsilon_i \delta_{ik} \tag{4}$$

where δ_{ik} is the well-known Kronecker delta.

We shall now multiply (3) on the left by S, and making use of its orthogonality, obtain:

$$AS = SD \tag{5}$$

Equating the elements of the matrices in (5), we obtain:

$$\sum_{k=1}^{N} A_{mk}S_{ki} = \sum_{k=1}^{N} S_{mk}D_{ki} = \sum_{k=1}^{N} S_{mk}\epsilon_k\delta_{ki} = S_{mi}\epsilon_i \tag{6}$$

Let us now fix our attention on a single value of i and consider the meaning of (6). We can interpret the array of elements S_{ki} for fixed i as a column vector (actually, the ith column of S). This vector, which we may denote by $S^{(i)}$, is premultiplied by the matrix A. The result is, of course, the scalar ϵ_i multiplied by the same vector $S^{(i)}$. Thus:

$$AS^{(i)} = \epsilon_i S^{(i)} \tag{7}$$

Referring back now to (2), we may identify ϵ_i with $\alpha^{(i)}$ and $S^{(i)}$ with $a^{(i)}$. Hence, we may state:

If an orthogonal matrix S has been found which transforms a real symmetric matrix A into a diagonal matrix D, then the i*th diagonal element of D may be adopted as the* i*th eigenvalue of A, and the* i*th column of S is then the* i*th eigenvector of A.*

It is clear that the central difficulty of this approach to the eigenvalue problem is that of finding the orthogonal matrix S. The method we shall consider is based on that due to Jacobi [1], but has been modified in the last ten years.

The fundamental approach of Jacobi is to annihilate, in turn, selected off-diagonal elements of A by "elementary" orthogonal transformations. To illustrate: Let us single out an off-diagonal element of A, namely, $A_{pq}\,(p < q)$ and, by an elementary transformation, annihilate it. The transformation in question (denoted by R) has the following form:

$$\begin{aligned} R_{pp} &= \cos\theta, \qquad & R_{pq} &= \sin\theta \\ R_{qp} &= -\sin\theta, \qquad & R_{qq} &= \cos\theta \end{aligned} \tag{8}$$

$$R_{ii} = 1, \qquad R_{pk} = R_{iq} = R_{ik} = 0 \qquad \begin{matrix} i \neq p, q; \\ k \neq p, q \end{matrix}$$

θ is an angle of rotation to be determined.

Let us denote by B the result of the transformation R^TAR. Then the elements of B are as follows:

$$\left.\begin{aligned} B_{pk} &= A_{pk}\cos\theta - A_{qk}\sin\theta \\ B_{qk} &= A_{pk}\sin\theta + A_{qk}\cos\theta \end{aligned}\right\} \tag{9a}$$

$$\left.\begin{aligned} B_{ip} &= A_{ip}\cos\theta - A_{iq}\sin\theta \\ B_{iq} &= A_{ip}\sin\theta + A_{iq}\cos\theta \end{aligned}\right\} \; i, k \neq p, q \tag{9b}$$

$$B_{ik} = A_{ik} \tag{9c}$$

$$\begin{aligned} B_{pp} &= A_{pp}\cos^2\theta + A_{qq}\sin^2\theta - 2A_{pq}\sin\theta\cos\theta \\ B_{qq} &= A_{pp}\sin^2\theta + A_{qq}\cos^2\theta + 2A_{pq}\sin\theta\cos\theta \end{aligned} \tag{10}$$

$$\begin{aligned} B_{pq} &= (A_{pp} - A_{qq})\sin\theta\cos\theta \\ &\qquad + A_{pq}(\cos^2\theta - \sin^2\theta) \\ &= \tfrac{1}{2}(A_{pp} - A_{qq})\sin 2\theta + A_{pq}\cos 2\theta \end{aligned}$$

Hence, in order to make B_{pq} vanish (annihilation), we must have:

$$\tfrac{1}{2}(A_{pp} - A_{qq})\sin 2\theta + A_{pq}\cos 2\theta = 0 \tag{11}$$

or

$$\tan 2\theta = \frac{-A_{pq}}{\tfrac{1}{2}(A_{pp} - A_{qq})} \tag{12}$$

Actually, it is not necessary in practice to evaluate trigonometric functions, since there are purely algebraic relationships between $\tan 2\theta$, $\sin\theta$, and $\cos\theta$. If we now denote $-A_{pq}$ by λ and $\tfrac{1}{2}(A_{pp} - A_{qq})$ by μ, then the formulas for $\sin\theta$ and $\cos\theta$ may be computed as algebraic functions of λ and μ.

First define (all square roots are positive):

$$\omega \equiv \operatorname{sgn}(\mu)\frac{\lambda}{\sqrt{\lambda^2 + \mu^2}} \tag{13}$$

Then:

$$\sin\theta = \frac{\omega}{\sqrt{2(1 + \sqrt{1 - \omega^2})}} \tag{14}$$

$$\cos\theta = \sqrt{1 - \sin^2\theta} \tag{15}$$

A computational advantage of these formulas is that a vanishing μ will cause no difficulty. If λ vanishes, no rotation should be performed at all.

Let us now suppose that we shall select each off-diagonal element in turn according to some order, and perform on the symmetric matrix a single orthogonal transformation of the type described above appropriate to annihilate the off-diagonal (or *pivotal*) element. The elements A_{pp}, A_{qq}, A_{pq}, and A_{qp} we shall call the *pivotal set* associated with A_{pq}. Now, since each of the elementary orthogonal transformations affects many more elements than just those in the pivotal set, we cannot in general expect that a single pass through all pivotal elements (with the appropriate elementary transformation for each one) will result in a diagonal matrix, but that, in fact, some one of the elementary transformations designed to annihilate a later pivot will undo the annihilation of a previously treated pivot which is in the same row or the same column. A glance at (9) will show that even if, for example, A_{pk} were zero (for some k), the corresponding B_{pk} need not be zero.

For this reason, Jacobi's method is not a finite process, but is an iterative method which is carried on indefinitely until the required accuracy is obtained. Jacobi's original mode of selecting pivots was to choose for the next one that of largest magnitude. This was then annihilated by a transformation. He himself proved that this procedure will converge.

Various modifications have been made to this iterative aspect of Jacobi's technique, the fastest of which, on large computers, seems to be one due to Pope and Tompkins [2]. This consists in examining each pivotal "candidate" in a regular sequence, and performing the elementary transformation only if the magnitude of the pivotal element exceeds a certain threshold value. This value is lowered whenever there are no remaining off-diagonal elements with magnitudes larger than the threshold. The proof of the convergence of this technique follows along the lines of Jacobi's proof and is outlined below. It has also been proved [3] that even if one performs all rotations, regardless of the size of the pivotal element, the process will still converge (subject to rather weak restrictions). Further, results due to Goldstine, Murray, and von Neumann [4] show that the Jacobi method is completely stable against rounding error. However, it is quite important that the square root in (15) be taken as accurately as possible, since the (eigenvalue-preserving) orthogonality of S depends completely on the exactness of the relation:

$$\sin^2\theta + \cos^2\theta = 1 \tag{16}$$

We shall now, in the course of proving the convergence of the Pope-Tompkins scheme, see how the exactness of (16) plays a role.

First, let us evaluate the following combinations from (9*a*) and (9*b*), where $k \neq p, q$:

$$B_{pk}^2 + B_{qk}^2 = A_{pk}^2(\cos^2\theta + \sin^2\theta) + A_{qk}^2(\sin^2\theta + \cos^2\theta) + A_{pk}A_{qk}(-\cos\theta\sin\theta + \sin\theta\cos\theta) \tag{17}$$

$$B_{kp}^2 + B_{kq}^2 = A_{kp}^2(\cos^2\theta + \sin^2\theta) + A_{kq}^2(\sin^2\theta + \cos^2\theta) + A_{kp}A_{kq}(-\cos\theta\sin\theta + \sin\theta\cos\theta)$$

Hence, to the extent to which (16) is accurate, we have:

$$\begin{aligned} B_{pk}^2 + B_{qk}^2 &= A_{pk}^2 + A_{qk}^2 \\ B_{kp}^2 + B_{kq}^2 &= A_{kp}^2 + A_{kq}^2 \end{aligned} \tag{18}$$

Therefore, since all other elements of A of the form A_{ik} $(i, k \neq p, q)$ are unaffected by the transformation, we see that *with the exception of* A_{pq} $(= A_{qp})$ the sum of squares of all off-diagonal elements is invariant.

A parallel evaluation of the sum of squares of elements with subscripts p and q alone, yields the following result:

$$B_{pp}^2 + B_{pq}^2 + B_{qp}^2 + B_{qq}^2 = A_{pp}^2 + A_{pq}^2 + A_{qp}^2 + A_{qq}^2 \tag{19}$$

Now, since we have removed the (p, q) element, we set $B_{pq} = 0$ and obtain:

$$B_{pp}^2 + B_{qq}^2 = A_{pp}^2 + A_{qq}^2 + 2A_{pq}^2 \tag{20}$$

Hence, *we have lost* $2A_{pq}^2$ *from the sum of squares of off-diagonal terms*, and this has been absorbed into the sum of squares of diagonal terms.

Let us now define the initial (off-diagonal) norm ν_I as follows:

$$\nu_I \equiv \left\{ \sum_{\substack{i,k=1 \\ i\neq k}}^{N} A_{ik}^2 \right\}^{1/2} \tag{21}$$

We are here summing $N(N-1)$ off-diagonal terms.

We shall establish a threshold ν_1 by dividing ν_I by a fixed constant σ. In the first stage of the process we will annihilate all the off-diagonal elements for which

$$|A_{ik}| \geq \nu_1 = \nu_I/\sigma \qquad i \neq k \tag{22}$$

If we choose σ = at least N, we see that there would be at least one off-diagonal element with magnitude greater than ν_1, since, if all were equal or smaller, we would have:

$$\sum_{i\neq k} A_{ik}^2 \leq \sum_{i\neq k} \nu_1^2 = N(N-1)\nu_1^2 < N^2\nu_1^2 \leq \sigma^2\nu_1^2 = \nu_I^2 \tag{23}$$

which contradicts (21). For any element whose magnitude is not smaller than ν_1, we perform the appropriate rotation to remove it. Hence, we have decreased the off-diagonal squared norm by at least $2\nu_1^2$ [(20)]. It is conceivable that we might reduce the off-diagonal norm to zero without adjusting ν_1, but this is unlikely. If, after performing all the necessary rotations, we find no more "large" off-diagonal elements, the off-diagonal norm is bounded as follows:

$$\nu_{od}^2 = \nu_I^2 - \sum_{|A_{ik}|\geq \nu_1} 2|A_{ik}|^2 < \nu_I^2 - 2\nu_1^2$$

$$= \left(1 - \frac{2}{\sigma^2}\right)\nu_I^2 \tag{24}$$

We now lower the threshold again:

$$\nu_2 = \nu_1/\sigma \tag{25}$$

and proceed as before. If no off-diagonal elements are found whose magnitude is less than ν_2, we lower the threshold again ($\nu_3 = \nu_2/\sigma$) and continue. At some point we must have $\nu_r \leq \nu_{od}/\sigma$. By the previous argument, with ν_{od} playing the role of ν_I, we shall find at least one off-diagonal element with magnitude greater than ν_{od}/σ and, therefore, greater than ν_r. Then, performing the appropriate rotations on all elements whose magnitude is not smaller than ν_r, we can bound the new off-diagonal norm, $\nu_{od}^{(2)}$, as follows:

$$(\nu_{od}^{(2)})^2 = \nu_{od}^2 - \sum_{|B_{ik}|\geq \nu_r} 2|B_{ik}|^2 < \nu_{od}^2 - 2\nu_{od}^2/\sigma^2$$

$$= (1 - 2/\sigma^2)\nu_{od}^2 < (1 - 2/\sigma^2)^2\nu_I^2 \tag{26}$$

By induction, we can see that if $\nu_{od}^{(m)}$ is the off-diagonal norm after m stages in which at least one rotation has been performed, then, at worst, we shall have:

$$(\nu_{od}^{(m)})^2 \leq \left(1 - \frac{2}{\sigma^2}\right)^m \nu_I^2 \tag{27}$$

For convergence, we want to establish a final threshold ν_F such that:

$$\nu_{od}^2 \leq \rho^2\nu_I^2 \tag{28}$$

where ρ is an accuracy requirement. We have, as before:

$$\nu_{od}^2 = \sum_{i\neq k} A_{ik}^2 \leq N(N-1)\nu_F^2 < N^2\nu_F^2 \tag{29}$$

Hence, if we select

$$\nu_F = (\rho/N)\nu_I \tag{30}$$

we shall have the required accuracy.

To estimate the number of threshold lowerings in the worst case (i.e., when we reduce the off-diagonal norm by the minimum amount with each lowering, through being able to find only one "large" element), we must have, for some m:

$$(\nu_{od}^{(m)})^2 < \left(1 - \frac{2}{\sigma^2}\right)^m \nu_I^2 \leq \rho^2\nu_I^2 \tag{31}$$

Hence, we require:

$$\left(1 - \frac{2}{\sigma^2}\right)^m \leq \rho^2 \tag{32}$$

or

$$m \geq 2 \ln \rho / \ln \left(1 - \frac{2}{\sigma^2}\right) \tag{33}$$

Clearly, m is finite, although it has been estimated with extreme pessimism.

Granted then that the Jacobi iterative process converges, let us see what quantities must be computed and preserved to yield the required final results. In what follows, let the mth elementary orthogonal transformation matrix in the iterative sequence be denoted by $R_{(m)}$ and let the original matrix A be taken as $A_{(1)}$. Then we define:

$$A_{(m+1)} = R_{(m)}^T A_{(m)} R_{(m)} \tag{34}$$

Further, let the result of the sequential compounding of products of m R's be denoted by $S^{(m)}$. We have:

$$S^{(1)} = R_{(1)} \tag{35a}$$

$$S^{(m+1)} = S^{(m)} R_{(m+1)} \tag{35b}$$

The elements of $S^{(m+1)}$ are given in terms of those of $S^{(m)}$ by:

$$S_{ik}^{(m+1)} = S_{ik}^{(m)} \qquad k \neq p, q \tag{36a}$$

$$S_{ip}^{(m+1)} = S_{ip}^{(m)} \cos\theta_{(m+1)} - S_{iq}^{(m)} \sin\theta_{(m+1)} \tag{36b}$$

$$S_{iq}^{(m+1)} = S_{ip}^{(m)} \sin\theta_{(m+1)} + S_{iq}^{(m)} \cos\theta_{(m+1)} \tag{36c}$$

where $\theta_{(m+1)}$ is the rotation angle for the $(m+1)$st transformation.

If these transformations are carried out as indicated, then, since the process converges:

$$\lim_{m\to\infty} A_{(m)} = D \tag{37}$$

$$\lim_{m\to\infty} S^{(m)} \doteq S \tag{38}$$

In a numerical computation, of course, one does not carry this process on indefinitely, but only until all of the off-diagonal elements have become sufficiently small.

There are various good mathematical checks for the Jacobi process, all depending on the orthogonality of the $R_{(m)}$, and finally, of S. We may make use of relations (18), (20), and the following, which may easily be verified:

$$B_{pp} + B_{qq} = A_{pp} + A_{qq} \tag{39}$$

These may be checked at each step, if desired.

A good final check is obtained by premultiplying (3) by S, and postmultiplying by S^T. Then, in virtue of the orthogonality of S ($S^T S = I$), we obtain:

$$A^* = SDS^T \tag{40}$$

and A^* may be compared with A for accuracy.

3. SUMMARY OF THE CALCULATION PROCEDURE

a. Input Quantities

(1) N = order of the matrix.

(2) A_{ik} $(i, k = 1, \cdots, N)$.

b. Order of Calculation

(1) Locate pivotal element.

(2) Compute $\sin\theta$ and $\cos\theta$ by (13), (14), and (15). Note again that it is of extreme importance to verify that $\sin^2\theta + \cos^2\theta = 1$ with maximum accuracy.

(3) Transform A by (9*b*). In practice, it is not necessary to make exceptions of the pivotal set, since these will be computed separately. Also, the row transformations need not be done [(9*a*)], since step (5) accomplishes the same thing.

(4) Compute the new pivotal set B_{pp}, B_{qq}, and B_{qp} (B_{pq}) and place these in the matrix.

(5) Transpose the transformed elements to maintain the symmetry of A.

(6) Transform $S^{(m)}$ according to (36).

(7) Check to see if all of the pivotal elements are less than a preset threshold. If so, stop computation. If not, repeat steps (1)–(6).

c. Output Quantities

(1) The eigenvalues $\alpha^{(k)} (= D_{kk}, k = 1, \cdots, N)$.

(2) The corresponding eigenvectors $a^{(k)}$ $(a_i^{(k)} = S_{ik};\ i, k = 1, \cdots, N)$.

4. FLOW CHART

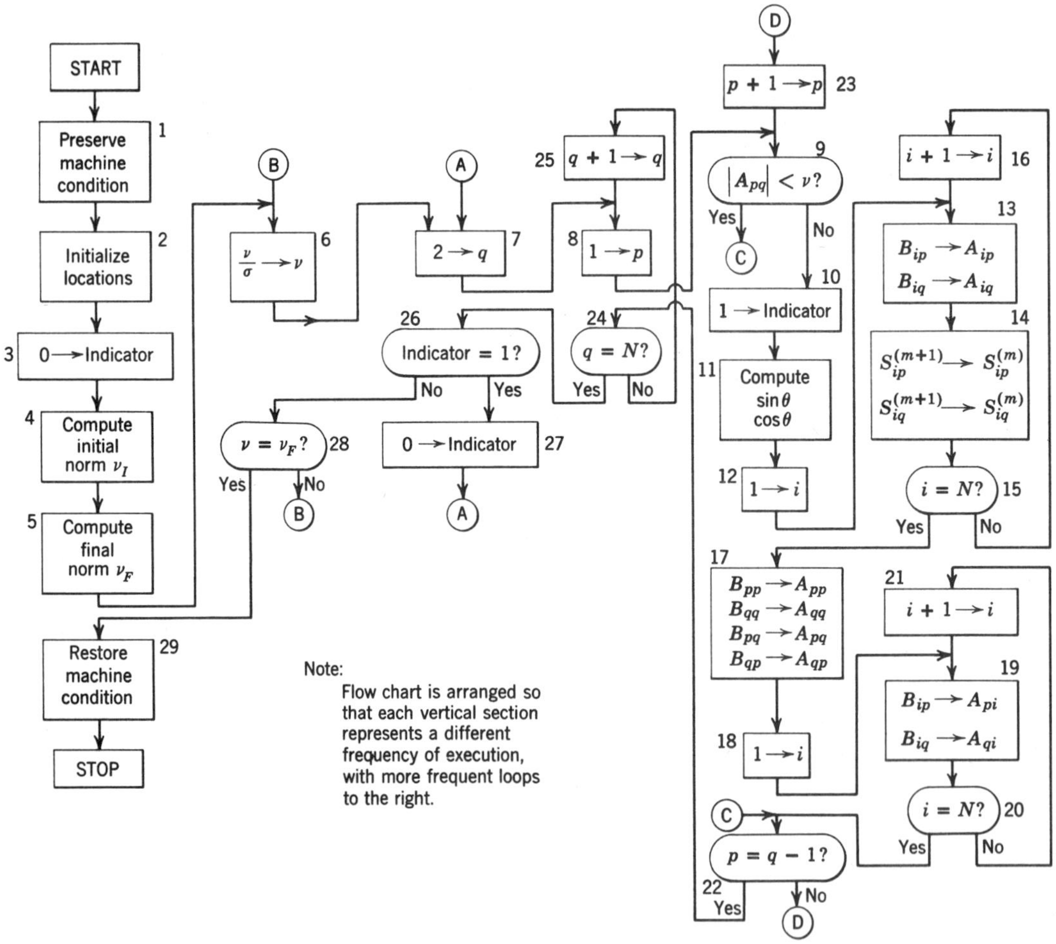

5. DESCRIPTION OF THE FLOW CHART

Box 1: If this program is considered as a subroutine, it is then necessary to preserve the status of various organs in the machine (e.g., index registers, indicator lights, etc.) which are to be used by the subroutine. These organs should then be restored to their conditions before the subroutine was called.

Box 2: The location of the original matrix A and the location of the transformation matrix S (which is to be generated) are presumed to be specified in the sequence of instructions which calls for the Jacobi diagonalization subroutine. Hence, there are many references to elements of these matrices which have to be adjusted properly for each new case. The specified order of the matrix comes into play here. Further, the identity matrix must be generated for the first approximation to S.

Box 3: An indicator (which may be a word in memory or a trigger of some sort) is initialized. This indicator will later be used to tell whether any off-diagonal elements have been found which are greater than the present threshold.

Box 4: The initial norm ν_I of the matrix is computed as follows: $\nu_I = \left\{ \sum_{i \neq k} A_{ik}^2 \right\}^{1/2}$. This norm is repeatedly divided subsequently to produce the threshold.

Box 5: The final norm ν_F is computed. This is set sufficiently small so that the requirement that any element A_{pq} shall be smaller than ν_F in absolute magnitude defines the convergence of the process.

Box 6: The present threshold ν_m at each stage is redivided by the factor σ (e.g., N) to lower the threshold.

Box 7: The initialization of a systematic sweep through off-diagonal elements begins here. The subscript q in the matrix element A_{pq} is set to two.

Box 8: The sweep is through columns of the off-diagonal elements A_{pq} $(p < q)$. For initializing each column, p is set to one.

Box 9: Here the comparison is made between $|A_{pq}|$ and the threshold norm ν_m. If $|A_{pq}| \geq \nu_m$, the $R_{(m)}$ transformation is made.

Box 10: The indicator is set to one to show that at least one element was $\geq \nu_m$, and that a repeat sweep is necessary.

Box 11: Sin θ and cos θ are computed according to (13), (14), and (15).

Box 12: The row index i is initialized for the postmultiplication $AR_{(m)}$.

Box 13: The elements of A in the pth and qth columns are transformed by (9*b*). The new elements replace the old.

Box 14: The elements of $S^{(m)}$ in the pth and qth columns are transformed by (36). The new elements replace the old, thus forming $S^{(m+1)}$.

Box 15: The test for the end of the column.

Box 16: If the end of the column has not been reached, the row index i is stepped up.

Box 17: If the end of the column has been reached, the new pivotal element and its associated diagonal elements are computed by (10) for $k = p$ and q. The new elements replace the old.

Box 18: The row (column) index is initialized for the reflection into the rows, of the new column elements.

Box 19: Since the matrix A was symmetric, and since an orthogonal transformation preserves symmetry, it is only necessary at this point to replace the old row elements by the new column elements. Since $B_{pq} = B_{qp}$ have already been computed, this interchange will not affect them, nor will it affect the diagonal elements B_{pp} and B_{qq}. A more elaborate manner of storing the matrix A together with a more complex address computation in box 13 would have eliminated this step.

Box 20: Test for end of row replacement.

Box 21: If row replacement is not complete, step up index and repeat.

Box 22: If row replacement is complete, test for end of column of pivotal elements. This test is also made when the decision in box 9 to make the transformation is negative.

Box 23: If the column of pivots is not completely examined, step up row index p and repeat box 9 test.

Box 24: If column of pivots has been covered, test whether it was the last column of pivots in the superdiagonal elements.

Box 25: If there are more columns of pivot candidates, step up to the next column and initialize to its first element (box 8).

Box 26: Test indicator to discern whether the entire set of superdiagonal elements was tested without the necessity of transforming the matrix so as to annihilate a pivot.

Box 27: If at least one transformation was performed, reset indicator and begin entire search again (box 7).

Box 28: If no transformation was performed, test whether the threshold was set to its final value. If it was not, divide and begin new search (box 6).

Box 29: If the threshold was set to its final value and all pivot candidates were smaller, then the diagonalization is complete and the process is terminated. At this point, preparation is made to leave the subroutine by restoring the states of all organs of the machine which have been preserved in box 1.

6. SUBROUTINES

The only standard subroutine required for this computation is a square root program. This routine should be capable of extracting a square root of maximum accuracy, so that $\cos\theta$ may be computed from $\sin\theta$ [(15)] as precisely as possible. The square root subroutine would be used in boxes 4 and 11 only.

7. SAMPLE PROBLEM

Let A be the 3×3 symmetric matrix:

$$A = \begin{bmatrix} 1 & 1 & \frac{1}{2} \\ 1 & 1 & \frac{1}{4} \\ \frac{1}{2} & \frac{1}{4} & 2 \end{bmatrix} \tag{41}$$

We compute the initial norm ν_I as indicated in box 4.

$$\nu_I = \{1^2 + 1^2 + (\tfrac{1}{2})^2 + (\tfrac{1}{2})^2 + (\tfrac{1}{4})^2 + (\tfrac{1}{4})^2\}^{1/2} = 1.62019 \tag{42}$$

We may now set the final threshold equal to (for example) $10^{-9}\nu_I$. Thus:

$$\nu_F = 1.62019 \times 10^{-9} \tag{43}$$

as indicated in box 5. Now let us set $\sigma = 3$, for example, and compute the threshold. We have, as in box 6:

$$\nu_1 = 1.62019/3 = 0.40505 \tag{44}$$

After setting $p = 1, q = 2$, we arrive at box 9, where we are to compare A_{12} against ν. The result is, of course, that $|A_{12}| > \nu_1$, so we prepare to annihilate A_{12}.

In box 11, we compute $\sin\theta$ and $\cos\theta$. The calculation is as follows:

$$\begin{aligned} \tfrac{1}{2}(A_{11} - A_{22}) \equiv \mu = 0 \\ -A_{12} \equiv \lambda = -1 \end{aligned} \tag{45}$$

$$\omega = -1/\sqrt{1^2 + 0^2} = -1$$

$$\sin\theta = -1/\sqrt{2(1 + \sqrt{1 - 1^2})} = -1/\sqrt{2} \tag{46}$$

$$\cos\theta = \sqrt{1 - \tfrac{1}{2}} = 1/\sqrt{2}$$

We now prepare to transform A as in Box 13. The results are:

$$AR_{(1)} = \begin{bmatrix} \sqrt{2} & 0 & \frac{1}{2} \\ \sqrt{2} & 0 & \frac{1}{4} \\ \frac{3}{4\sqrt{2}} & \frac{-1}{4\sqrt{2}} & 2 \end{bmatrix} \tag{47}$$

The evaluation of the pivotal elements (box 17) yields:

$$[AR_{(1)}]_{\text{modified}} = \begin{bmatrix} 0 & 0 & \frac{1}{2} \\ 0 & 2 & \frac{1}{4} \\ \frac{3}{4\sqrt{2}} & \frac{-1}{4\sqrt{2}} & 2 \end{bmatrix} \tag{48}$$

We next symmetrize A (box 19). The result is:

$$R^T_{(1)}AR_{(1)} \equiv B = \begin{bmatrix} 0 & 0 & \frac{3}{4\sqrt{2}} \\ 0 & 2 & \frac{-1}{4\sqrt{2}} \\ \frac{3}{4\sqrt{2}} & \frac{-1}{4\sqrt{2}} & 2 \end{bmatrix} \tag{49}$$

In box 22, we notice that $p = q - 1$. Hence, we go to box 25, step up q, and reinitialize p as in box 8. We now compare $|A_{13}|$ with ν_1 and find that we must rotate again. This process continues.

We have not examined box 14 in detail since the compounded matrix $S^{(1)}$ is just $R_{(1)}$ itself, namely:

$$S^{(1)} = R_{(1)} = \begin{bmatrix} \frac{1}{\sqrt{2}} & \frac{-1}{\sqrt{2}} & 0 \\ \frac{1}{\sqrt{2}} & \frac{1}{\sqrt{2}} & 0 \\ 0 & 0 & 1 \end{bmatrix} \tag{50}$$

The final results of repeating the process to 8-decimal-place accuracy are as follows:

$$D = \begin{bmatrix} 2.53652582 & 0 & 0 \\ 0 & -.01664729 & 0 \\ 0 & 0 & 1.48012146 \end{bmatrix} \quad (51)$$

$$S = \begin{bmatrix} .53148338 & -.72120712 & -.44428106 \\ .46147338 & .68634928 & -.56210938 \\ .71032933 & .09372796 & .69760117 \end{bmatrix} \quad (52)$$

$$SDS^T = \begin{bmatrix} .99999991 & .99999992 & .49999992 \\ .99999991 & .99999990 & .24999998 \\ .49999991 & .24999998 & 2.00000014 \end{bmatrix} \quad (53)$$

Equation (53) illustrates the final mathematical check on the results, namely, SDS^T should be equal to the original matrix A.

The output quantities are the diagonal elements of D and the matrix S, each of whose columns is an eigenvector.

8. MEMORY REQUIREMENTS

We now relate the program as outlined above to an actual code, written for the IBM 704 by Y. Bard [5] and denoted by NY EVV. This program is not identical in all details to what has been described here, but it is sufficiently similar that memory and time estimates are almost exact.

The memory requirement for NY EVV is:

Instructions	324 words
Constants	13 words
Working storage	16 words
The matrix A	N^2 words
The matrix S	N^2 words

The original matrix A is destroyed during the computation. In its place is generated the diagonal matrix D.

9. ESTIMATION OF THE RUNNING TIME

The time required for diagonalizing a matrix of order N and generating all eigenvectors is approximately as follows:

$$10(2\nu + \mu)N^3$$

where ν is the addition time and μ is the multiplication time of the computer.

10. REFERENCES

1. C. G. J. Jacobi, Über ein leichtes Verfahren, die in der Theorie der Säkularstörungen vorkommenden Gleichungen numerisch aufzulösen, *J. reine angew. Math.*, vol. 30, 1846, pp. 51–95.
2. D. A. Pope and C. Tompkins, Maximizing Functions of Rotations, *J. Assoc. Comp. Mach.*, vol. 4, 1957, pp. 459–466.
3. G. E. Forsythe and P. Henrici, The Cyclic Jacobi Method for Computing the Principal Values of a Complex Matrix, *Tech. Report No. 74*, Appl. Math. and Statist. Lab., Stanford Univ., 1958.
4. H. H. Goldstine, F. J. Murray, and J. von Neumann, The Jacobi Method for Real Symmetric Matrices, *J. Assoc. Comp. Mach.*, vol. 6, 1959, pp. 59–96.
5. Y. Bard, 704 Program Distributions of SHARE Organization, No. 339, 1957.

PART III ORDINARY DIFFERENTIAL EQUATIONS

Numerical integration methods for the solution of ordinary differential equations

8

Anthony Ralston
Bell Telephone Laboratories

1. FUNCTION

In this chapter we will discuss methods for the numerical integration of a system of N ordinary differential equations

$$\frac{dy}{dx} = f(x, y) \tag{1}$$

with the initial condition $y(x_0) = y_0$, where y, y_0, and f are column vectors with N components. Our discussion will be oriented toward those methods which can best be used to solve large systems of equations. These methods enable the computer to calculate y at a sequence of values $x_i > x_0$, $i = 1, 2, \cdots$. By using the term numerical integration we mean to restrict our attention to those methods which require the evaluation of $f(x, y)$ in (1) only at the points x_i.* This is a basic difference between the methods considered in this chapter and the Runge-Kutta methods of the next chapter in which evaluations of $f(x, y)$ are made both at the points x_i and at values of x in between the successive x_i.

Since any differential equation of higher than first order which can be solved for the highest derivative can be reduced to a system of first-order equations (by suitably defining new variables), the discussions of this chapter and the next are also applicable to higher order equations or systems of higher order equations.

* A slight exception to this rule is considered in Section 2d [(12)] in which $f(x, y)$ is evaluated at values of $x = x_i + h/2$ in order to simplify the computational procedure where $h = x_{i+1} - x_i$.

2. MATHEMATICAL DISCUSSION

a. Introduction

In some areas of numerical analysis the first step in the solution of a problem is to find one technique which can be used to effect the solution. In the area of ordinary differential equations the first step is more often to choose that technique among the many available which will serve the purpose best.

In order to choose the best method with which to effect the solution of a system of ordinary differential equations, it is necessary to keep in mind a number of factors:

1. The accuracy required. The error in the result depends both on the error incurred at each step of the integration (truncation and round-off error) and on how the error incurred at one step propagates in later steps (stability).

2. The ease with which the estimation of the error at each step may be made. In order to know when to change the interval between steps in the integration (see Section 2d) it is very important to be able to estimate the error at each step.

3. The speed with which the computation will be performed. Since a large system of equations ($N = 10$ or more) may require a large amount of machine time on even the fastest computer, it is clearly of importance to consider the time required by the various methods available.

4. The ease with which a method can be programmed for a computer. This is affected by such matters as the ease with which the computation can be started (Section 2c) and the ease with which the interval between steps can be changed (Section 2d).

In the discussion which follows these factors will be kept in mind in order that the methods presented can be fairly evaluated.

b. Derivation of the Formulas

The following symbols will be used in everything that follows:

$$y_i = y(x_i), \qquad y_i' = \left.\frac{dy}{dx}\right|_{x=x_i} \qquad i = 0, 1, \cdots$$

where x_i is the value of x at the ith step in the integration and $x_{i+1} - x_i = h$. The interval between steps, h, will generally be constant over a number of steps but can be changed if necessary (see Section 2d). We note here that since we are considering a system of N equations many of the quantities which are used in succeeding pages must be interpreted as N-dimensional column vectors.

Formulas for the numerical integration of ordinary differential equations can generally be divided into two categories:

1. Forward integration formulas in which y_{n+1} is expressed as a linear combination of values of y_i and y_i' for $i \leq n$.

2. Iterative formulas in which y_{n+1} is expressed as a linear combination of y_{n+1}' and values of y_i and y_i' for $i \leq n$.* Since y_{n+1}' appears in the equation for the evaluation of y_{n+1}, it is in general necessary to use an iterative technique to solve this equation.

The general form of such a numerical integration formula is then

$$\begin{aligned} y_{n+1} = {} & a_n y_n + a_{n-1} y_{n-1} + \cdots + a_{n-p} y_{n-p} \\ & + h(b_{n+1} y_{n+1}' + b_n y_n' + \cdots \\ & + b_{n-p} y_{n-p}') + E_n \end{aligned} \tag{2}$$

where $b_{n+1} = 0$ for a forward integration formula and where some of the other coefficients may be postulated equal to zero. The error E_n is the error incurred at the $(n + 1)$st step of the integration assuming that all the y_i and y_i' on the right-hand side of (2) are known without error. This error has two causes:

1. The right-hand side of (2) is not an exact representation of y_{n+1}. It would require an infinite sum on the right-hand side of (2) to express y_{n+1} exactly for all n. The right-hand side of (2) may be considered a truncated version of this infinite sum so that with it there is associated an error called *truncation error*.

2. The quantities on the right-hand side of (2) must be rounded off to a suitable number of decimal places in performing the computation. This causes another error called *round-off error*.

In general, the quantities on the right-hand side of (2) are not known without error. We must therefore consider the effect of using inexact quantities on the right-hand side of (2) in computing successive values of y_{n+1}. How these errors affect successive values of y_{n+1} is called the stability of the procedure. We will consider stability in Section 2e.

For any given p we may determine the coefficients in (2) by expanding each y_i and y_i' in a Taylor series in h about x_n and then equating the coefficients of $1, h, h^2, \cdots$ on both sides of (2). In general, the number of powers of h

* The linear combination can, in special cases, include values of y_i' for $i > n + 1$. See, for example, [1].

which can be made to vanish is equal to the number of nonzero coefficients in (2). However, in some important cases that we will consider, one higher power of h than the number of nonzero coefficients can be made to vanish. The error E_n in general is of the form $ch^k y^{(k)}(\xi)$, where c is a constant, k is the lowest power of h that does not vanish in the above derivation, and $y^{(k)}(\xi)$ is the kth derivative of y evaluated at a point ξ in the interval occupied by the abscissas used in (2).

Until recently the commonly used numerical integration techniques all had $a_i = 0$ for all i except $n - p$. Instead of using a Taylor series expansion, formulas of this form can be derived directly from Newton's interpolation formula using backward differences [2] expressed in either of the two equivalent forms

$$\begin{aligned} y'_{n+s} = y'(x_n + sh) &= y'_n + s\,\nabla y'_n \\ &+ \frac{s(s+1)}{2!}\nabla^2 y'_n + \cdots \\ &+ \frac{s(s+1)\cdots(s+k-1)}{k!} \\ &\times \nabla^k y'_n + E_s \end{aligned} \tag{3a}$$

or

$$\begin{aligned} y'_{n+s} = y'(x_n + sh) &= y'_{n+1} + (s-1)\,\nabla y'_{n+1} \\ &+ \frac{s(s-1)}{2!}\nabla^2 y'_{n+1} + \cdots \\ &+ \frac{(s-1)s(s+1)\cdots(s+k-2)}{k!} \\ &\times \nabla^k y'_{n+1} + \bar{E}_s \end{aligned} \tag{3b}$$

where the error terms are

$$E_s = h^{k+1}\frac{s(s+1)\cdots(s+k)}{(k+1)!}y^{k+2}(\eta) \tag{4a}$$

$$\bar{E}_s = h^{k+1}\frac{(s-1)s(s+1)\cdots(s+k-1)}{(k+1)!}y^{k+2}(\bar{\eta}) \tag{4b}$$

and where the backward difference operator $\nabla y_k = y_k - y_{k-1}$; η is in the interval occupied by $x_0, x_1, \cdots, x_n, x_n + sh$; and $\bar{\eta}$ is in the interval occupied by $x_0, x_1, \cdots, x_{n+1}, x_n + sh$.

To derive numerical integration formulas from (3) we use the relation

$$y_{n+1} = y_{n-r} + h\int_{-r}^{1} y'_{n+s}\,ds \tag{5}$$

Substitution of (3*a*) into (5) leads to a class of forward integration formulas. The corresponding error terms can be calculated as indicated in [2]. For example, with $r = 0$ we have

$$y_{n+1} = y_n + h(1 + \tfrac{1}{2}\nabla + \tfrac{5}{12}\nabla^2 + \cdots)y'_n + E_n \tag{6}$$

The forward integration method using (6) is known as Adams' method [3]. If only zero-order differences are retained, (6) becomes*

$$y_{n+1} = y_n + hy'_n + h^2 y''(\xi) \tag{7}$$

where ξ is (x_n, x_{n+1}). The term $h^2y''(\xi)$ is the truncation error of (7). The round-off error is always implicit in the form of the equation itself. Equation (7) is known as Euler's method. It has the charm of simplicity and is self-starting. That is, given the initial condition $y(x_0) = y_0$, (7) can be used to calculate all successive values of y_i. This is not true for most numerical integration formulas. The truncation error in Euler's method is, however, much larger than that in most of the formulas we will derive and for this reason it is seldom used in practice.

The formulas we get with r odd and with r differences retained in (3*a*) are of particular interest since in these cases it can be shown [2] that the coefficient of the rth difference is zero so that using $r - 1$ or r differences affords the same accuracy.† The formulas for $r = 1, 3, 5$ are

$$y_{n+1} = y_{n-1} + 2hy'_n + \frac{h^3}{3}y'''(\xi) \tag{8a}$$

$$\begin{aligned} y_{n+1} &= y_{n-3} + \frac{4}{3}h(2y'_n - y'_{n-1} + 2y'_{n-2}) \\ &+ \frac{14}{45}h^5 y^{\text{v}}(\xi) \end{aligned} \tag{8b}$$

$$\begin{aligned} y_{n+1} &= y_{n-5} + \frac{3}{10}h(11y'_n - 14y'_{n-1} + 26y'_{n-2} \\ &- 14y'_{n-3} + 11y'_{n-4}) + \frac{41}{140}h^7 y^{\text{vii}}(\xi) \end{aligned} \tag{8c}$$

where in each formula ξ is in (x_{n-r}, x_{n+1}). Note that (8*a*) has all the simplicity of (7)

* The numerical integration formulas we present will all be expressed in terms of ordinates (Lagrangian form), not differences, since this is the form most convenient for use on a digital computer.

† This corresponds to using Newton-Cotes integration formulas of the open type (see Chapter 22).

but more accuracy. However, neither (8*a*) nor (8*b*) nor (8*c*) is self-starting since all require more than one past value of y to compute y_{n+1}.

Substitution of (3*b*) into (5) leads to a class of iterative integration formulas. For example, taking $r = 0$ and retaining one difference in (3*b*) gives

$$y_{n+1} = y_n + \frac{h}{2}(y'_{n+1} + y'_n) - \frac{h^3}{12}y'''(\xi) \quad (9)$$

with ξ in (x_n, x_{n+1}). Equation (9) is the familiar trapezoidal rule. As before, a more accurate set of formulas can be found taking r odd. In this case it can be shown [2] that the coefficient of the $(r + 2)$nd difference vanishes so that retaining $r + 1$ differences in (3*b*) affords the same accuracy as $r + 2$ differences.* For $r = 1$ and 3 we get

$$y_{n+1} = y_{n-1} + \frac{h}{3}(y'_{n+1} + 4y'_n + y'_{n-1}) - \frac{h^5}{90}y^{\text{v}}(\xi) \quad (10a)$$

$$y_{n+1} = y_{n-3} + \frac{2h}{45}(7y'_{n+1} + 32y'_n + 12y'_{n-1} + 32y'_{n-2} + 7y'_{n-3}) - \frac{8}{945}h^7y^{\text{vii}}(\xi) \quad (10b)$$

Equations (10*a*) and (10*b*), like their analogs (8*b*) and (8*c*), are not self-starting. We note that the error terms in (10*a*) and (10*b*) are significantly smaller than those of (8*b*) and (8*c*), respectively. It is a general rule that iterative formulas have much smaller error terms than their analogous forward integration formulas. For this reason, when high accuracy is required in a solution of ordinary differential equations an iterative formula is generally used.

In order to use an iterative formula it is necessary to perform the following steps:

1. Guess or predict a value of y at $x = x_{n+1}$, call it $\bar{y}_{n+1}$.
2. Using the differential equation, calculate $\bar{y}'_{n+1} = f(x_{n+1}, \bar{y}_{n+1})$.
3. Using the iterative formula, calculate a new y_{n+1}.
4. Repeat steps 2 and 3 until the difference between successive values of y_{n+1} is within some predetermined limit. In [2] it is shown that if $h < 1/(|b_{n+1}|g)$ then the iterative process converges, where b_{n+1} is the coefficient of y'_{n+1} in the iterative formula and g is such that $[|\partial f(x, y)/\partial y|] < g$ in the neighborhood of (x_{n+1}, y_{n+1}).

Methods using the above procedure are called predictor-corrector techniques since step 1 predicts y_{n+1} and step 3 corrects this prediction. It is clear that the better the original prediction the quicker the iteration will converge. The method most often utilized for predicting is to use the forward integration formula which corresponds to the iterative formula being employed. For example, if we were using (9) for the iteration, we would use (8*a*) for the prediction. The complete set of equations would be*

Predictor: $$\bar{y}_{n+1} = y_{n-1} + 2hy'_n \quad (11a)$$

$$\bar{y}'_{n+1} = f(x_{n+1}, \bar{y}_{n+1}) \quad (11b)$$

Corrector: $$y_{n+1} = y_n + \frac{h}{2}(y'_{n+1} + y'_n) \quad (11c)$$

where the accuracy of both the predictor and corrector is $0(h^3)$ (order of h^3). (The importance of the predictor and corrector having the same order of accuracy will be discussed below.) An alternative to (11) which is self-starting uses (7) in place of (11*a*). But we note that in this case the accuracy of the predictor is $0(h^2)$ while that of the corrector is $0(h^3)$.

Another alternative to (11) is the "mid-point" method discussed by Lotkin [4]. Here the equations are

Predictor: $$y_{n+\frac{1}{2}} = y_n + 1/2(y_n - y_{n-1}) \quad (12a)$$

$$y'_{n+\frac{1}{2}} = f(x_{n+\frac{1}{2}}, y_{n+\frac{1}{2}})$$

where $$x_{n+\frac{1}{2}} = x_n + \frac{h}{2} \quad (12b)$$

Corrector: $$y_{n+1} = y_n + hy'_{n+\frac{1}{2}} \quad (12c)$$

Here the predictor is not an integration formula at all but rather an extrapolation formula. Its accuracy is $0(h^2)$ while that of the corrector is $0(h^3)$. The main advantage of this

* This corresponds to using Newton-Cotes integration formulas of the closed type (see Chapter 22).

* In all that follows we will use the equality sign in equations such as (11*a*) and (11*c*) to indicate that the quantity on the left-hand side of the equation is *calculated* (except for round-off error) by evaluating the quantity on the right-hand side. The truncation error is implicit in all these equations and its implications will be considered in Section 2e.

method is that it requires only one computation of $f(x, y)$ at each step. Its main disadvantage, besides the above mentioned difference in accuracy of predictor and corrector, is that the method is not self-starting. We note that this method is the exception mentioned before to the rule that $f(x, y)$ would be evaluated only at values of x_i.

In integrating a system of ordinary differential equations it is often desirable to have a higher order of accuracy than that provided by (11). The classic predictor-corrector method which provides this greater accuracy is that due to Milne [5]. It uses (8*b*) for prediction and (10*a*) for correction and has an accuracy in both predictor and corrector of $0(h^5)$. The complete set of equations is

$$\text{Predictor:}\quad \bar{y}_{n+1} = y_{n-3} + \frac{4h}{3}(2y'_n - y'_{n-1} + 2y'_{n-2}) \tag{13a}$$

$$\bar{y}'_{n+1} = f(x_{n+1}, \bar{y}_{n+1}) \tag{13b}$$

$$\text{Corrector:}\quad y_{n+1} = y_{n-1} + \frac{h}{3}(y'_{n+1} + 4y'_n + y'_{n-1}) \tag{13c}$$

One very desirable feature of both (11) and (13) is that, because the predictor and corrector have the same order of accuracy, we can at each step estimate the truncation error in a very simple fashion. We illustrate this for the case of (13). From (8*b*) the truncation error of (13*a*) is $T_1 = \frac{14}{45}h^5y^{\text{v}}(\xi_1)$, where ξ_1 is in (x_{n-3}, x_{n+1}), and from (10*a*) the truncation error of (13*b*) is $T_2 = -\frac{1}{90}h^5y^{\text{v}}(\xi_2)$, where ξ_2 is in (x_{n-1}, x_{n+1}). Therefore if $y^{\text{v}}(x)$ is continuous we have

$$y_{n+1} - \bar{y}_{n+1} = +\frac{1}{90}h^5y^{\text{v}}(\xi_2) + \frac{14}{45}h^5y^{\text{v}}(\xi_1) = +\frac{29}{90}y^{\text{v}}(\xi) \tag{14}$$

where ξ is in (x_{n-3}, x_{n+1}). If we assume that $y^{\text{v}}(x)$ does not vary much for $x_{n-3} < x < x_{n-1}$, then we may write

$$y_{n+1} - \bar{y}_{n+1} = +\frac{29}{90}h^5y^{\text{v}}(\xi) \approx \begin{cases} \dfrac{29}{28}T_1 \\ -29T_2 \end{cases} \tag{15}$$

Therefore, by calculating $y_{n+1} - \bar{y}_{n+1}$ at each step we can get a measure of the truncation error. This is an invaluable aid in deciding when to change the value of h (see Section 2d). That is, if $y_{n+1} - \bar{y}_{n+1}$ indicates the calculation is more accurate than we require, we can increase the size of h. If $y_{n+1} - \bar{y}_{n+1}$ grows steadily to too large a value we can shorten the interval, while if it changes suddenly this probably indicates a machine error. The corresponding result for (11) is

$$y_{n+1} - \bar{y}_{n+1} = +\frac{5}{12}h^3y'''(\xi) \approx \begin{cases} \dfrac{5}{9}\bar{T}_1 \\ -5\bar{T}_2 \end{cases} \tag{16}$$

where $x_{n-1} < \xi < x_{n+1}$, $\bar{T}_1$ is the truncation error of (11*a*), and $\bar{T}_2$ is the truncation error of (11*c*).

As we have said, one advantage of iterative integration formulas is the high accuracy they achieve at each step. Another advantage is the means for estimation of the truncation error discussed above. One serious disadvantage, however, is the iteration requirement itself. The iterative procedure requires successive recalculation of $f(x, y)$ at each step until the process converges. Often the evaluation of $f(x, y)$ requires more time than any other part of the calculation. Thus, having to calculate $f(x, y)$ many times at each step is a serious disadvantage, especially when the number of equations is large so that the total machine time is large.

One way of avoiding the iteration is to accept the first value of the corrector as y_{n+1}. This is often done, for example, when using (11) or (12). This is more accurate than using a forward integration formula but of course not as accurate as iterating to convergence.

A variation of (13), which incorporates the noniteration feature but is more sophisticated than merely using the first value of the corrector, is the following set of equations [6]:

$$\text{Predictor:}\quad p_{n+1} = y_{n-3} + \frac{4h}{3}(2y'_n - y'_{n-1} + 2y'_{n-2}) \tag{17a}$$

$$\text{Modifier:}\quad m_{n+1} = p_{n+1} - \frac{28}{29}(p_n - c_n) \tag{17b}$$

$$m'_{n+1} = f(x_{n+1}, m_{n+1}) \tag{17c}$$

Corrector: $$c_{n+1} = y_{n-1} + \frac{h}{3}(m'_{n+1} + 4y'_n + y'_{n-1}) \qquad (17d)$$

Final value: $$y_{n+1} = c_{n+1} + \frac{1}{29}(p_{n+1} - c_{n+1}) \qquad (17e)$$

Equations (17*a*) and (17*d*) are merely (13*a*) and (13*c*) with $\bar{y}_{n+1}$ and y_{n+1} replaced by p_{n+1} and c_{n+1}, respectively. Equation (17*b*) is derived from (15), which can be written (replacing $n + 1$ by n and using p_n and c_n instead of $\bar{y}_n$ and y_n)

$$p_n - c_n \approx -\frac{29}{28} T_1 \qquad (18)$$

where T_1 is the truncation error of the predictor. Thus $-\frac{28}{29}(p_n - c_n)$ is a measure of the truncation error at step n. If we assume that this truncation error remains nearly constant from step n to step $n + 1$, then (17*b*) serves the purpose of modifying the predicted value by the estimated truncation error. Similarly, in (17*e*) we modify the corrected value by the estimated truncation error at step $n + 1$, again using (15).

The advantages of this method are

1. Since there is no iteration we compute only two values of $f(x, y)$ at each step.
2. Modifying the predicted and corrected values by an estimation of the truncation error makes up for much of the error caused by not iterating.
3. The difference $c_{n+1} - p_{n+1}$ plays the role of $y_{n+1} - \bar{y}_{n+1}$, thus enabling us to estimate the error at each step.

An obvious application of this same technique can be applied to (11).

Unfortunately, Milne's method (13) or its modified version (17) still has one major disadvantage and this is that the method is unstable. That is, no matter what the form of $f(x, y)$ any error at step n (due to truncation error or round-off error) tends to propagate with increasing magnitude in succeeding steps (see Section 2e).

A method which retains the advantages of (17) and is stable (under the definition of stability in Section 2e) has been devised by Hamming [6]. In place of (13*c*) this method uses the generalized corrector

$$y_{n+1} = a_n y_n + a_{n-1} y_{n-1} + a_{n-2} y_{n-2} + h(b_{n+1} y'_{n+1} + b_n y'_n + b_{n-1} y'_{n-1}) \qquad (19)$$

analogously to (2). Using a Taylor series expansion we determine the six coefficients so that all powers of h through h^4 vanish in (19). Therefore the accuracy is of the order of h^5 as with (13*c*). Since there are six coefficients we could also make the coefficient of h^5 vanish. If we do not do this, then one of the six coefficients in (19) may be chosen arbitrarily. Hamming chooses $a_{n-1} = 0$, in which case (19) becomes

$$y_{n+1} = \frac{1}{8}[9y_n - y_{n-2} + 3h(y'_{n+1} + 2y'_n - y'_{n-1})] - \frac{1}{40} h^5 y^{\text{v}}(\xi) \qquad (20)$$

where ξ is in (x_{n-2}, x_{n+1}). The truncation error is somewhat greater than that for Milne's method. However, the choice $a_{n-1} = 0$ means that the method is stable for virtually all values of h that would be used in practice (see Section 2e). The truncation error of (20) can be made as small as that for (13*c*) by reducing the value of h by 15%.

Using (20) as the predictor, the equations analogous to (17) are

Predictor: $$p_{n+1} = y_{n-3} + \frac{4h}{3}(2y'_n - y'_{n-1} + 2y'_{n-2}) \qquad (21a)$$

Modifier: $$m_{n+1} = p_{n+1} - \frac{112}{121}(p_n - c_n) \qquad (21b)$$

$$m'_{n+1} = f(x_{n+1}, m_{n+1}) \qquad (21c)$$

Corrector: $$c_{n+1} = \frac{1}{8}[9y_n - y_{n-2} + 3h(m'_{n+1} + 2y'_n - y'_{n-1})] \qquad (21d)$$

Final value: $$y_{n+1} = c_{n+1} + \frac{9}{121}(p_{n+1} - c_{n+1}) \qquad (21e)$$

Equations (21*b*) and (21*e*) are derived in just the same way as (17*b*) and (17*e*) by using the truncation errors of (13*a*) and (20). The equation analogous to (15) is

$$c_{n+1} - p_{n+1} = \frac{121}{360} h^5 y^{\text{v}}(\xi_1) \approx \begin{cases} \dfrac{121}{112} T_1 \\[2ex] \dfrac{121}{9} T_2' \end{cases} \tag{22}$$

where T_1 is the truncation error of (13*c*) and T_2' is the truncation of (20). In Sections 3–7 we will discuss the actual use of Hamming's method on a digital computer.

Another stable method which uses an iterative formula is discussed by Wilf [1]. As opposed to Hamming's method, this method is self-starting but it requires the evaluation of $f(x, y)$ three times at each step to obtain an accuracy of h^4.

c. Starting the Solution

We have pointed out that some of the formulas presented in Section 2b are not self-starting. That is, in these methods each step of the integration requires the knowledge of past values of the solution which are not available at the start of the integration. For example, to compute p_{n+1} using (21*a*) requires a knowledge of y_n, y_{n-1}, y_{n-2}, and y_{n-3} but at the start ($n = 0$) only one value, y_0, is known. In this section we will discuss briefly methods of obtaining the values required to start the solution. The points at which the starting values are computed can be:

1. Consecutive values of $x_i > x_0$. For (21*a*) these would be x_1, x_2, and x_3.
2. Values of x_i both greater or less than x_0 although the values of y at $x_i < x_0$ may have no physical significance. For (21*a*) one such choice would be x_{-1}, x_1, and x_2.

The advantage of the second case is that it is generally somewhat easier to compute accurately values near x_0 (e.g., x_{-1}) than values further away (e.g., x_3). The methods we will discuss can be used for either of the above cases. For the sake of simplicity, however, the discussion will consider explicitly only the first case above.

It is important that the values of y_n we calculate in order to start the solution be calculated with at least as high an accuracy as the accuracy of the numerical integration procedure to be used. Otherwise the large errors in the first few values may degrade the later accuracy or increase the tendency toward instability.

One of the most common ways of starting the solution is to use one of the Runge-Kutta integration formulas which are described in the next chapter. These formulas require only the initial condition to start the solution. They can be readily programmed for a digital computer as described in the next chapter.

Another starting procedure which can be programmed for a computer involves the use of Newton's interpolation formula using forward differences to get the required values y_1, $y_2, \cdots, y_s$ as a linear combination of y_0 and $y_0', y_1', \cdots, y_s'$. For example, with $s = 3$ we have

$$y_1 = y_0 + \frac{h}{24}(9y_0' + 19y_1' - 5y_2' + y_3') \tag{23a}$$

$$y_2 = y_0 + \frac{h}{3}(y_0' + 4y_1' + y_2') \tag{23b}$$

$$y_3 = y_0 + \frac{h}{8}(3y_0' + 9y_1' + 9y_2' + 3y_3') \tag{23c}$$

To use these formulas we estimate y_1, y_2, and y_3, calculate y_1', y_2', and y_3' by using the differential equations, and then calculate y_1, y_2, and y_3 using (23). Then we iterate until convergence. The error in all three equations of (23) is $0(h^5)$. This method is discussed more fully in [2] and [7].

There also exist analytic methods for computing the starting values. These are useful when the starting values are to be computed before the problem is put on the digital computer. The most common of the analytic methods is to expand y in a Taylor series about x_0,

$$y(x_0 + rh) = y_0 + rhy_0' + \frac{r^2h^2}{2} y_0'' + \cdots \tag{24}$$

Using the initial condition $y(x_0) = y_0$ and the differential equation we can compute $y_0', y_0'', \cdots$. Then substituting these into (24) we can compute $y_1, \cdots, y_s$ by letting $r = 1, 2, \cdots, s$, if the Taylor series converges.

Another analytic method which can be used is that of Picard, sometimes called the method of successive substitutions. This method is of theoretical importance in the study of differential equations but is generally too tedious to be used in practice. Discussions of this method can be found in [2] and [7].

Using (21) requires values of y_1, y_2, and y_3 which can be computed as above to start the solution and an estimate of $p_3 - c_3$. To get the latter we can estimate the fifth derivative of y by using (23c) with the addition on the right-hand side of its truncation error term which is $-\frac{3}{80}h^5y^{\mathrm{v}}(\xi), x_0 < \xi < x_3$. If we solve this equation for $y^{\mathrm{v}}(\xi)$, then, estimating $p_3 - c_3$ in terms of $y^{\mathrm{v}}(x)$ from (8b) and (20) and assuming $y^{\mathrm{v}}(x)$ is constant in (x_0, x_3), we get

$$p_3 - c_3 = \frac{242}{27}\left[y_3 - y_0 - \frac{3h}{8}(y_3' + 3y_2' + 3y_1' + y_0')\right] \quad (25)$$

An argument against using (25) to estimate $p_3 - c_3$ is that the size of the multiplicative constant, $\frac{242}{27}$, magnifies the errors in y_3 and y_0. That is, if we take the square root of the sum of the squares of the coefficients in (25) we get (neglecting the terms in h) $\frac{242}{27}\sqrt{2} \approx 12.5$. Therefore, the error in $p_3 - c_3$ will be about 12.5 times that of the starting values of y, on the average. Thus, unless the starting values are computed with greater accuracy than that of the integration process to be used, it is probably just as good to set $p_3 - c_3 = 0$ as to use (25) to compute it. We note that if (23) is used to compute the starting values then (25) gives identically $p_3 - c_3 = 0$.

It must be remembered that programming the computation of the starting values for a digital computer can be a tedious process. This may not be vital if the program is being written as a library routine but if the program is being written to solve a single system of ordinary differential equations it may be well to consider self-starting methods such as those discussed in Section 2b or [1].

d. Choosing and Changing the Interval of Integration

There is no formula that can be given for choosing the value of h with which to start the integration. We must first estimate the per step truncation error as a function of h (by estimating the derivative in the truncation error term). Then we choose a value of h which will give an initial per step error sufficiently small so that, over the number of steps required to complete the integration, the total error will not be larger than that which can be tolerated at the end of the integration *if the per step error does not increase*. Unfortunately, the per step error often will increase, requiring us to decrease the interval of integration (see below). Also unfortunately, the per step errors do not necessarily add to produce the final error. Indeed, it generally cannot be predicted a priori just how the per step errors will interact. The final error may be greater than the sum of the per step errors. Thus it is usually wise to choose the initial value of h somewhat conservatively and to decrease the interval when the estimated per step error grows beyond a conservative bound. However, it is not wise to choose a value of h which gives an estimated per step error much less than that required. The main reason for this is that the smaller the value of h, the larger the number of steps in the integration. Thus any tendency in the method toward instability will be increased (see Section 2e). Also, the time of the computation naturally increases when h is decreased. The value of h should never be chosen so small that the per step truncation error is significantly less than that of the per step round-off error because in that case the round-off error determines the final error. In practice the per step truncation error is generally significantly greater than the per step round-off error so that the latter can be neglected.

Thus it is important to decrease the value of h when the estimated per step truncation error grows too large and to increase h if the estimated per step error is much smaller than is required. If the interval is to be increased in size, it is general practice to double it since this is quite easy to do. For example, if, using (21), we desired to double h and thereby go from y_n to y_{n+2}, (21a) would be

$$p_{n+2} = y_{n-6} + \frac{4(2h)}{3}(2y_n' - y_{n-2}' + 2y_{n-4}') \quad (26)$$

and the other equations would be changed in an analogous fashion. To get (26) we skip every other value of n, thereby obtaining values of y and y' spaced at intervals of $2h$. In order to be able to do this on a digital computer it is necessary only to have enough past values of y stored in order to effect the transition from h to $2h$. In the case of (21), equation (26) indicates that six past values of y must be stored at every step instead of the four required by (21).

When decreasing the size of the interval it is customary to halve the value of h. As opposed to doubling the step size, halving the step size requires no extra storage of past values of y. Two general approaches are possible:

1. Use a self-starting integration formula with a value of $h/2$ to compute enough values of y to continue the solution as before. That is, we consider $y_n = y(x_n)$ as a new initial condition and proceed as described in Section 2c. This once again illustrates the advantage of self-starting integration formulas since in using them all that is required is to change h to $h/2$ and continue as before.

2. Using suitable interpolation formulas, calculate enough interpolated values of y so that the solution can be continued with a step size $h/2$.

For example, in the case of (21) we need to know y at $x_n - \frac{1}{2}h = x_{n-1/2}$ and $x_n - \frac{3}{2}h = x_{n-3/2}$. Then (21$a$) becomes

$$p_{n+1/2} = y_{n-3/2} + \frac{4(h/2)}{3}(2y'_n - y'_{n-1/2} + 2y'_{n-1}) \tag{27}$$

and the other equations would be changed in an analogous fashion. It is important of course that the interpolation formulas be at least as accurate as the integration procedure. Suitable interpolation formulas for use with Milne's method or Hamming's method are [8]:

$$y_{n-1/2} = \frac{1}{256}(80y_n + 135y_{n-1} + 40y_{n-2} + y_{n-3}) + \frac{h}{256}(-15y'_n + 90y'_{n-1} + 15y'_{n-2}) \tag{28a}$$

$$y_{n-3/2} = \frac{1}{256}(12y_n + 135y_{n-1} + 108y_{n-2} + y_{n-3}) + \frac{h}{256}(-3y'_n - 54y'_{n-1} + 27y'_{n-2}) \tag{28b}$$

where the error term in (28a) and (28b) is $O(h^7)$. Note that all the quantities required for (28a) and (28b) will be stored in the computer automatically if (21) is being used. In [8] an extrapolation formula is presented which enables the computer to double the interval. This formula also uses only points that are stored in the computer for use by (21).

e. Stability

Stability is a term which appears often in the literature on the numerical solution of ordinary differential equations but is seldom defined. We define it here as follows:*

A numerical integration procedure is said to be stable if, when $f_y = \partial f(x, y)/\partial y < 0$, the error, measured by the difference between the true solution and the numerical solution, decreases in magnitude on the average with increasing n (i.e., as the integration proceeds step by step).*

By "on the average" we mean that over a long series of steps the error will decrease in magnitude. Variations in f_y and in the per step error (truncation and round-off) may cause the error to increase for a given step. If f_y and the per step error are constant, then stability implies a decrease in the error at each step (for $f_y < 0$). This definition of stability says nothing about the situation when $f_y > 0$, in which case the solution itself and generally also the error are increasing exponentially. In this case it is useful to define the term relative stability as follows:

A numerical integration procedure is said to be relatively stable if the rate of change of the error with respect to n is less than the rate of change of the solution with respect to n.

Relative stability is also a useful concept when the solution is decreasing to zero and we are interested in knowing if the error is approaching zero as rapidly as the solution. Most discussions in the literature which treat stability do

* In this section only we consider the case of a single first-order ordinary differential equation instead of a system. We do this solely to simplify the discussion. We emphasize that this discussion carries over in its entirety to a system of equations if we consider each component of the vector f_y and the error vector separately.

not also treat relative stability but it is important to know when a numerical integration procedure is both stable and relatively stable.

A procedure which is not stable by the above definition will always have an error which tends to increase "on the average." In fact, as the illustrations below will indicate, the error in an unstable process tends to grow exponentially with n.

We have been careful to say that the error in an unstable process "tends to increase" rather than to say that it "increases." This is because in considering the stability of a numerical integration procedure as below, we will make certain assumptions (e.g., that f_y is a negative constant) that lead us either to the result that the method is stable or alternatively that it is not stable because the error would grow exponentially with increasing n under the assumptions we will have made. In particular cases, however, variations in f_y might mean that the error in an unstable process would not increase with n. Therefore, we say "tends to increase." However, there is very strong empirical evidence that the error in an unstable process will behave badly.

To test the stability of any numerical integration scheme as expressed by (2), we consider the difference between y_n and the true solution Y_n. We define

$$\epsilon_n = Y_n - y_n \tag{29}$$

At the $(n + 1)$st step of the calculation we have

$$y_{n+1} = a_n y_n + a_{n-1} y_{n-1} + \cdots + a_{n-p} y_{n-p} + h(b_{n+1} y'_{n+1} + \cdots + b_{n-p} y'_{n-p}) + R_n \tag{30a}$$

and

$$Y_{n+1} = a_n Y_n + a_{n-1} Y_{n-1} + \cdots + a_{n-p} Y_{n-p} + h(b_{n+1} Y'_{n+1} + \cdots + b_{n-p} Y'_{n-p}) + T_n \tag{30b}$$

where R_n is the round-off error introduced at the $(n + 1)$st step and T_n is the truncation error introduced by the numerical integration scheme. Subtracting $(30a)$ from $(30b)$ and using (29), we have

$$\epsilon_{n+1} = \sum_{i=0}^{p} a_{n-i}\epsilon_{n-i} + h \sum_{i=-1}^{p} b_{n-i}\epsilon'_{n-i} + E_n \tag{31}$$

where $E_n = T_n - R_n$. Now using the mean value theorem we may write

$$\begin{aligned} \epsilon'_{n-i} &= Y'_{n-i} - y'_{n-i} = (Y_{n-i} - y_{n-i}) f_y(x_{n-i}, \eta_{n-i}) \\ &= \epsilon_{n-i} f_y(x_{n-i}, \eta_{n-i}) \end{aligned} \tag{32}$$

if f_y is continuous and where η_i is between y_i and Y_i. Substituting (32) into (31) we get

$$\epsilon_{n+1}[1 - hb_{n+1} f_y(x_{n+1}, \eta_{n+1})] = \sum_{i=0}^{p} a_{n-i}\epsilon_{n-i} + h \sum_{i=0}^{p} b_{n-i} f_y(x_{n-i}, \eta_{n-i})\epsilon_{n-i} + E_n \tag{33}$$

In order to study the stability of the integration scheme $(30a)$ we wish to study the growth of ϵ_n with n. One way of doing this [2] is to consider a difference equation related to (33) whose solution dominates (33) (i.e., the solution of the related equation is greater than the absolute value of ϵ_n for all n). By solving the related equation it is possible to bound the error very conservatively using only a bound on the magnitude of f_y. However, according to our definition of stability we are interested in the solution of (33) when f_y is negative. We will solve (33) approximately by assuming f_y and E_n to be a negative constant $-K$ and a constant E, respectively. In practice, these quantities tend to vary slowly from step to step so that the solution of (33) under these assumptions does indeed indicate whether or not the method is stable. With these assumptions, (33) is then a linear difference equation with constant coefficients whose solution can be obtained [9] by letting $\epsilon_n = \rho^n$, substituting this into (33), and finding the roots of the resulting polynomial equation.* We now consider this solution for some of the numerical integration formulas we have considered.

For $(11c)$ the solution of (33) with $f_y = -K$ and $E_n = E$ is [7]

$$\epsilon_n = A\left(\frac{1 - hK/2}{1 + hK/2}\right)^n + \frac{E}{hK} \tag{34}$$

where A is a constant determined by the initial condition. For convergence of the iterations we must have (see p. 98) $hK/2 < 1$. Therefore,

* If all the roots lie within the unit circle, the method is stable. In [10] the location of the roots with respect to the unit circle is related in a very useful fashion to the positive definiteness of a matrix whose elements are functions of the a_i's and b_i's.

if $f_y < 0$ (i.e., $K > 0$), then $(1 - hK/2)/(1 + hK/2) < 1$ and ϵ_n decreases with increasing n which indicates the stability of the numerical integration procedure (11).

Now we consider (13c). In [6] it is shown that

$$\epsilon_n = C_1(\rho_1)^n + C_2(\rho_2)^n + C_3 \tag{35}$$

where C_1, C_2, and C_3 are constants and where either ρ_1 or ρ_2 has an absolute value >1 depending on whether f_y is positive or negative. Thus, even if f_y is negative, ϵ_n grows in magnitude with increasing n and, therefore, Milne's method is unstable. However, in [6] it is also shown that if (21d) is used as the corrector instead of (13c), then for $f_y < 0$ the method is both stable and relatively stable if h is chosen such that

$$h < \frac{.75}{|f_y|} \tag{36}$$

Furthermore, for $f_y > 0$ the method is relatively stable if $h < .4/f_y$. In practice the inequality (36) will virtually always be satisfied in order to keep the truncation error small.* Therefore, for all practical purposes, the predictor-corrector method using (21d) as the corrector is stable. If, instead of using the corrector to iterate to convergence, we use the modified version (21), then in place of (36) we have the inequality

$$h < \frac{.65}{|f_y|} \tag{37}$$

which is also nonrestrictive for all practical purposes.

f. How to Choose a Numerical Integration Method

It is impossible to give a general rule for choosing the best numerical integration procedure to solve a particular system of ordinary differential equations. However, some precepts may be stated.

1. For large, complicated systems where the evaluation of $f(x, y)$ is tedious, a method such as Hamming's is to be preferred since it has a high per step accuracy, only two evaluations of $f(x, y)$ are required at each step, and it is stable. Methods analogous to Hamming's with truncation errors of $0(h^7)$ or higher can be developed but we must be very careful in doing this. This is because higher derivatives tend to behave increasingly badly so that, despite the higher power of h, the error may not be less than that of a fifth-order method. An error of $0(h^5)$ is usually sufficient to achieve the required results.

2. For large systems where the evaluation of $f(x, y)$ is not tedious a method such as that discussed in [1] is to be preferred since it is stable and self-starting.

3. When the integration is going to involve a large number of steps (i.e., n large at the end of the integration) a stable method should be used.

4. If the system involves few equations so that the time of computation is not a factor and if stability is not a problem because of the short number of steps of the integration, then the most convenient method should be used. It is easiest to program a self-starting method such as those discussed here or in the next chapter.

A note is in order here on differential equations of second order or higher. Although, as we said, such equations can generally be reduced to a system of first-order equations, it is sometimes advisable, especially in the case of second-order equations, to use methods specially adapted to such equations. In the case of second-order equations such methods are discussed in [2] and [7].

3. SUMMARY OF THE CALCULATION PROCEDURE

We will consider the calculation procedure using Hamming's method (21). Before going any further, however, we must consider the criteria to be used to double or halve the interval. If we denote the components of the vector $p_n - c_n$ by $p_n^{(i)} - c_n^{(i)}, i = 1, \cdots, N$, then we may compute $\sum_{i=1}^{N} a_i|p_n^{(i)} - c_n^{(i)}|$. The coefficients a_i must be specified by the programmer and depend on the relative weighting he wishes to give to errors in the N dependent

* We must remember that in general f_y will not be negative during the whole computation. Therefore, in order to restrict the growth of the error when $f_y > 0$, it is important to have the per step error small since the growth of error depends on the per step error.

variables. Alternatively, we might compute $\sum_{i=1}^{N} a_i(p_n^{(i)} - c_n^{(i)})^2$ but using the absolute value makes the computation simpler, so that is what we will use here. The determination of whether to double or halve the interval is then made by comparing $\sum_{i=1}^{N} a_i|p_n^{(i)} - c_n^{(i)}|$ with two numbers α_1 and α_2 such that if

1. $\sum_{i=1}^{N} a_i|p_n^{(i)} - c_n^{(i)}| < \alpha_1$ the interval is doubled at the next step.
2. $\sum_{i=1}^{N} a_i|p_n^{(i)} - c_n^{(i)}| > \alpha_2$ the interval is halved at the next step.

For checking purposes it is worthwhile to consider one more number α_3 such that if

3. $\sum_{i=1}^{N} a_i|p_n^{(i)} - c_n^{(i)}| > \alpha_3$ the *n*th step is recomputed under the assumption that a machine error has been made.

The input quantities then are:

(a) The initial condition vector $y(x_0) = y_0$.
(b) The initial value of h.
(c) The numbers a_i, $i = 1, \cdots, N$, and α_j, $j = 1, 2, 3$.

The steps in the calculation are:

(a) Compute the starting values (e.g., y_1, y_2, y_3 and $p_3 - c_3$) using one of the methods discussed in Section 2c.
(b) Using (21*a*)–(21*e*) compute successively p_{n+1}, m_{n+1}, m'_{n+1}, c_{n+1}, and y_{n+1}.
(c) Compute

$$p_{n+1} - c_{n+1} \text{ and } \sum_{i=1}^{N} a_i|p_{n+1} - c_{n+1}|.$$

(d) If necessary halve or double the interval or recompute the $(n + 1)$st step.
(e) Repeat steps (b)–(d) as long as desired.

The output quantities are:

(a) Values of y_n and the corresponding x_n which can be printed as often as desired.
(b) Values of $y'_n, p_n - c_n, \cdots$, if desired.

The details of the calculation will be made clearer by the flow chart in the next section. Of course, unforeseen things can happen during the calculation, such as the appearance of a discontinuity in some function so that no matter what h is chosen $p_n - c_n$ exceeds the allowable value. If the programmer feels it necessary to provide for such contingencies he must program the proper checks. For example, the programmer could provide that if h became less than a certain value the computation would be terminated.

If the initial value of h is chosen so that $x_0 + kh$, where k is an integer, is an exact value of x at which the value of y is desired, then the programmer must be careful that, when the interval is doubled, the resulting value of h is such that $x_0 + kh$ will be one of the points of the integration. The flow chart does not provide for this because in many problems it is not necessary to hit an exact end point with a value of $x_0 + kh$.

4. FLOW CHART

The flow chart appears on page 107.

5. DESCRIPTION OF THE FLOW CHART

Box 1: The starting values needed by (21) are computed using one of the methods of Section 2c.

Box 2: The starting value of $p_n - c_n$ is computed using (25).

Box 3: If $|p_3 - c_3|$ is too large, go to box 4. This box should be omitted if a starting formula is used which makes $|p_3 - c_3|$ identically zero.

Box 4: A new starting value of h is tried.

Box 5: Equation (21) is used to compute the quantities for the $(n + 1)$st step.

Box 6: The estimation of the truncation error at step $n + 1$ and the sum to test for halving or doubling of the interval are computed.

Box 7: If $\sum_{i=1}^{N} a_i|p_{n+1}^{(i)} - c_{n+1}^{(i)}|$ is too small, the value of h is doubled for the next step.

Box 8: If $\sum_{i=1}^{N} a_i|p_{n+1}^{(i)} - c_{n+1}^{(i)}|$ is too great, a test is made to determine whether there was a probable machine error or if the interval should be halved.

Box 9: If $\sum_{i=1}^{N} a_i|p_{n+1}^{(i)} - c_{n+1}^{(i)}| > \alpha_3$, the computation at the $(n + 1)$st step is repeated.

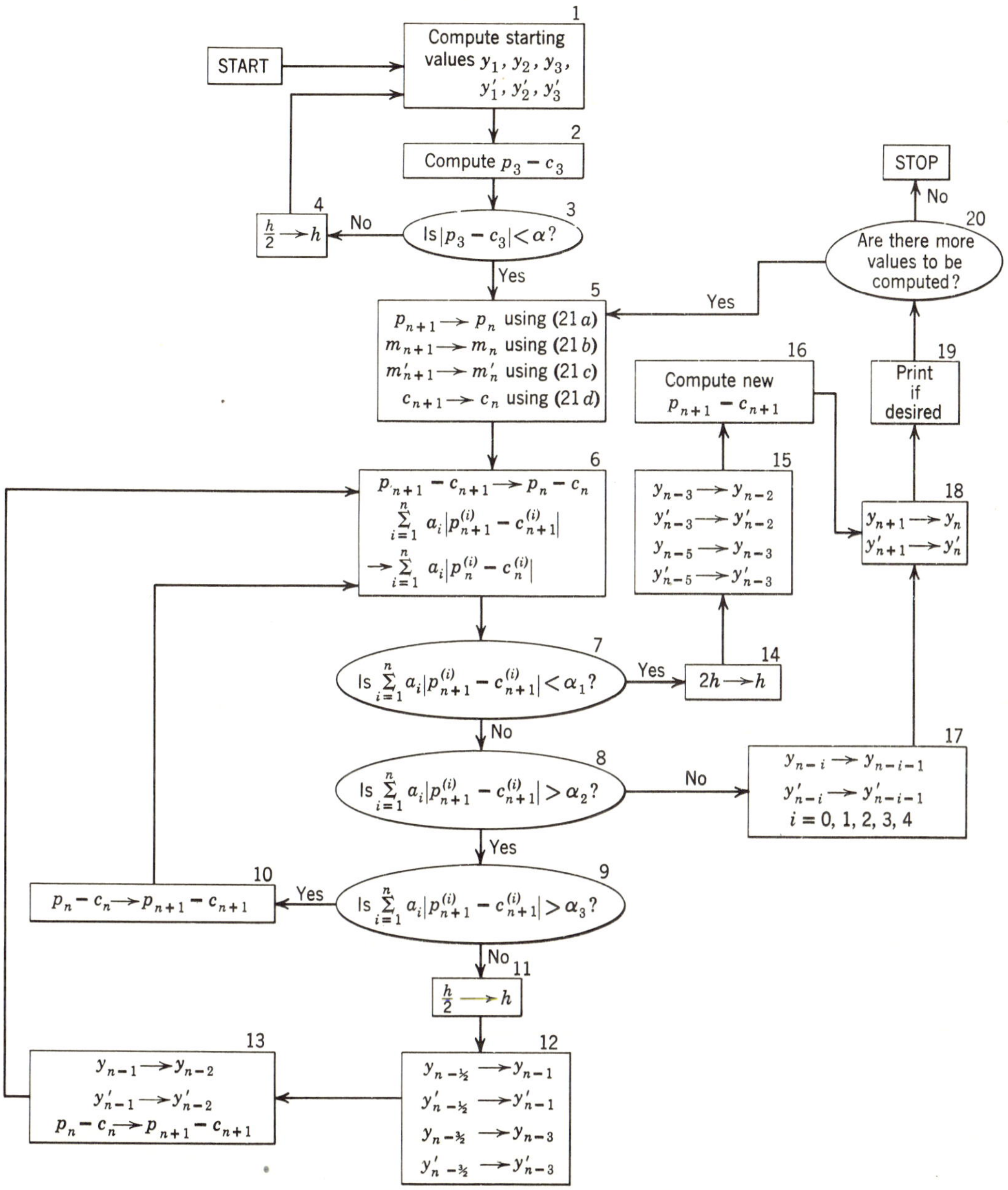

Box 10: In order to repeat the computation certain quantities must be put in their former locations.

Box 11: The interval is halved.

Box 12: The needed interpolations are performed using, for example, (28) and then the derivatives of the interpolated quantities are computed.

Box 13: Data are moved to prepare for the next step in the computation.

Box 14: The interval is doubled. It is also necessary to provide that the interval will not be doubled again until enough past values are available at the new spacing.

Box 15: Data are moved to prepare for the next step in the computation.

Box 16: $p_{n+1} - c_{n+1}$ must be recomputed using (25).

Box 17: Data are moved to prepare for the next step in the computation.

Box 18: y'_{n+1} is calculated as $f(x_{n+1}, y_{n+1})$.

Box 19: The programmer must provide a program here to print the output quantities he wants when he wants them.

Box 20: The programmer must provide a criterion to determine when to stop the computation. This criterion might depend on the value of x or y, etc.

6. SUBROUTINES

(a) The program to compute the starting values in box 1 may be considered a subroutine.

(b) The evaluation of $f(x, y)$ in boxes 5 and 18 may require one or more of the subroutines normally available in libraries.

7. SAMPLE PROBLEM

The example we will use to illustrate Hamming's method is that which appears in Milne [7] to illustrate Milne's method. The problem is that of the single first-order differential equation,

$$y' = x^2 - y^2 \tag{38}$$

with the initial condition $y(-1) = 0$. Using an initial $h = 0.1$ we can compute the starting values from (23). These are

x	y	y'
-1.0	.0	1.0
-0.9	.09004	.80189
-0.8	.16072	.61417
-0.7	.21347	.44443

Since we have used (23) to compute the starting values, we get from (25) $p_3 - c_3 = 0$. Thus we have completed boxes 1 through 4 of the flow chart. Now using (21) we calculate

$$\begin{aligned} p_4 &= .25046 \\ m_4 &= .25046 \\ m_4' &= .29727 \\ c_4 &= .25035 \\ y_4 &= .25036 \\ y_4' &= .29732 \\ p_4 - c_4 &= .00011 \end{aligned}$$

The results for this step and three more steps of the computation are

x	y	y'	$p - c$
-0.6	.25036	.29732	.00011
-0.5	.27376	.17506	.00010
-0.4	.28621	.07808	.00007
-0.3	.29020	.00578	.00004

8. MEMORY REQUIREMENTS

N words are required for each of the following eighteen quantities at each step:

$$p_{n+1},\ p_n - c_n,\ m_{n+1},\ m_{n+1}',\ c_{n+1},\ y_n,\ y_{n-1},\ y_{n-2},\ y_{n-3},\ y_{n-4},\ y_{n-5},\ y_n',\ y_{n-1}',\ y_{n-2}',\ y_{n-3}',\ y_{n-4}',\ y_{n-5}',\ a_i$$

The values of y_{n-4}, y_{n-5}, y_{n-3}', y_{n-4}', y_{n-5}' are required for doubling the interval. If storage is short it might be desirable to recompute y_{n-3}', y_{n-4}', y_{n-5}' when they are needed. Only one set of N registers is needed for m_{n+1}, m_{n+1}', and c_{n+1}. The quantity $p_n - c_n$ cannot be written over during the nth step since it will be needed if the interval has to be halved. Therefore, if y_{n-3}', y_{n-4}', and y_{n-5}' are not stored and one set of N registers is used for m_{n+1}, m_{n+1}', and c_{n+1}, we need $13N$ registers to store the above quantities. Also we require about ten words for the constants used in (21*a*)–(21*e*) h, and the α_j, $j = 1, 2, 3$.

The number of program steps depends so strongly on the form of $f(x, y)$ that it cannot reasonably be estimated. If we call this number P, then the total memory required is

$$P + 13N + 10$$

9. ESTIMATION OF THE RUNNING TIME

If we denote by T the time required to compute $f(x, y)$, then the time required to compute (21*a*)–(21*e*) and y_{n+1}' is about

$$2T + N(16\nu + 5\mu)$$

where ν is the addition time and μ the multiplication time of the computer. Also we need $N(2\nu + \mu)$ to compute $p_{n+1} - c_{n+1}$ and $\sum_{i=1}^{N} a_i(p_{n+1}^{(i)} - c_{n+1}^{(i)})$. Doubling these times to take into account all the bookkeeping operations in the flow chart, we arrive at the estimated time per step of the integration as

$$4T + 2N(18\nu + 6\mu)$$

10. REFERENCES

1. H. S. Wilf, An Open Formula for the Numerical Integration of First Order Differential Equations, *MTAC*, vol. 11, 1957, pp. 201–203.

2. F. B. Hildebrand, *Introduction to Numerical Analysis*, McGraw-Hill Book Co., New York, 1956.
3. F. Bashforth and J. C. Adams, An Attempt to Test the Theories of Capillary Action . . . with an Explanation of the Method of Integration Employed, Cambridge University Press, London, 1883.
4. M. Lotkin, A Note on the Midpoint Method of Integration, *J. Assoc. Comp. Mach.*, vol. 3, 1956, pp. 208–211.
5. W. E. Milne, *Numerical Integration of Ordinary Differential Equations*, *Amer. Math. Monthly*, vol. 38, 1931, pp. 455–460.
6. R. W. Hamming, Stable Predictor Corrector Methods for Ordinary Differential Equations, *J. Assoc. Comp. Mach.*, vol. 6, 1959, pp. 37–47.
7. W. E. Milne, *Numerical Solution of Differential Equations*, John Wiley and Sons, New York, 1953.
8. G. H. Keitel, An Extension of Milne's Three-Point Method, *J. Assoc. Comp. Mach.* vol. 3, 1956, pp. 212–222.
9. F. B. Hildebrand, *Methods of Applied Mathematics*, Prentice-Hall, New York, 1952.
10. H. S. Wilf, A Stability Criterion for Numerical Integration, *J. Assoc. Comp. Mach.*, vol. 6, 1959, pp. 363–365.

The total number of publications on the numerical solution of ordinary differential equations is almost endless. Extensive bibliographies can be found in [7] and in *Numerical Integration of Differential Equations* by A. A. Bennett, W. E. Milne, and H. Bateman, Dover Publications, New York, 1956.

Runge-Kutta methods for the solution of ordinary differential equations

9

Michael J. Romanelli
Ballistic Research Laboratories

1. FUNCTION

The purpose of the Runge-Kutta method is to obtain an approximate numerical solution of an ordinary differential equation. The theory, given here only for a single first-order equation, can readily be extended to a system of such equations. The extension, from one equation to a system, can be achieved by treating y as a vector in the derivation given in Section 2b. The calculation procedure and flow chart are written for a system of first-order equations. Only the fourth-order integration procedure is treated here since it is adequate for most practical applications. Procedures for various orders are given in [1], [2], and [3].

Given the system of first-order ordinary differential equations

$$\frac{dy_i}{dx} = y_i' = f_i(x, y_1(x), y_2(x), \cdots, y_n(x)) \quad (1)$$

with the initial conditions

$$y_i(x_0) = y_{i0} \quad (2)$$

we seek the values

$$y_i(x_0 + h) \quad (3)$$

where h is an increment of the independent variable x, and $i = 1, 2, \cdots, n$. The input consists of:

1. A description of the system to be solved, or, in computer terminology, a sequence of instructions designed to evaluate

$$y_i'(x) = f_i(x, y_0(x), y_1(x), \cdots, y_n(x))$$

2. h, an arbitrary but usually small increment in x.

3. $y_i(x_0)$, the initial values.

The output consists of approximations $\bar{y}_i(x_0 + h)$ to the true values $y_i(x_0 + h)$.

The Runge-Kutta method is an algorithm designed to approximate the Taylor series solutions,

$$y_i(x_0 + h) = y_i(x_0) + hy_i'(x_0) + \frac{h^2}{2!} y_i''(x_0) + \cdots \quad (4)$$

In contrast to the formal Taylor series solution, (4), the Runge-Kutta method does not require explicit definitions of, nor evaluations of, derivatives beyond the first. Approximations are obtained at the expense of several evaluations of the first derivatives. In the classical

Runge-Kutta fourth-order method, four evaluations of the first derivatives are required to obtain agreement with the Taylor series solution through terms of order h^4. Several evaluations of the first derivatives may be time-consuming and hence impractical, particularly if the expressions for the first derivatives are complicated. Another disadvantage of the method is that neither the truncation errors nor estimates of them are obtained in the calculation procedure. On the other hand, the method is stable and "self-starting"; i.e., only the functional values at a single previous point are required to obtain the functional values ahead. Furthermore, it is easy to change the step size h at any step in the calculations. For these reasons the Runge-Kutta method can be used profitably for starting other methods which are not self-starting, even though for speed and efficiency other methods are used for the bulk of the computation.

2. MATHEMATICAL DISCUSSION

a. Symbols Used

To simplify the derivation of the method and truncation error, use is made of certain operators and their algebraic properties. In developing Taylor series expansions for functions of two variables

$$f(x + \alpha_r, y + \beta_r)$$

about the point (x_0, y_0), we use differential operators

$$D_r = \alpha_r \frac{\partial}{dx} + \beta_r \frac{\partial}{\partial y} \qquad (r = 1, 2, 3)$$

where α_r and β_r are scalars.

The expansions are then expressed as

$$f(x_0 + \alpha_r, y_0 + \beta_r) = \left[f(x, y) + D_r f(x, y) + \frac{D_r^2 f(x, y)}{2!} + \frac{D_r^3 f(x, y)}{3!} + \cdots \right]_0$$

where

$$D_r^j f = \left(\alpha_r \frac{\partial}{\partial x} + \beta_r \frac{\partial}{\partial y} \right)^j f$$

and the zero subscript means that the functions are to be evaluated at the point (x_0, y_0). We also use the notation $f(x_0, y_0) = f_0$. The particular operators, D_r, will be defined in the sections where they are required.

b. Derivation of the Method for a Single First-Order Equation*

Given the equation

$$y' = f(x, y)$$

with the initial condition

$$y(x_0) = y_0$$

it is assumed that x_0 is not a singular point, and hence a solution exists and can be expressed in the form of the Taylor series,

$$y(x_0 + h) = y_0 + hy_0' + \frac{h^2}{2!} y_0'' + \frac{h^3}{3!} y_0''' + \cdots \tag{5}$$

where $h = (x - x_0)$ is sufficiently small so that the Taylor series converges. It is further assumed throughout that the higher derivatives and partials exist at points required. Since

$$y' = f(x, y) \equiv f$$

differentiating with respect to x we obtain

$$y'' = f_x + ff_y$$

Similarly,

$$y''' = f_{xx} + 2ff_{xy} + f^2 f_{yy} + f_y(f_x + ff_y)$$

$$\vdots$$

If we let

$$D = \frac{\partial}{\partial x} + f \frac{\partial}{\partial y}$$

we have

$$\begin{aligned} y' &= f(x, y) \equiv f \\ y'' &= f_x + ff_y = Df \\ y''' &= f_{xx} + 2ff_{xy} + f^2 f_{yy} + f_y(f_x + ff_y) \\ &= D^2 f + f_y Df \\ &\vdots \end{aligned}$$

* Portions of the derivation included here are given by numerous authors; see, e.g., [4], [5], or [1], and others listed in the bibliography. The original form is due to Runge [6]; later modifications are due to Kutta [7] and Gill [8].

This substitution in (5) then yields

$$y(x_0 + h) - y(x_0) = k = \Big[hf + \frac{h^2}{2!} Df + \frac{h^3}{3!}(D^2f + f_y Df) + \frac{h^4}{4!}(D^3f + f_y D^2f + f_y^2 Df + 3DfDf_y) + \frac{h^5}{5!}(D^4f + 6DfD^2f_y + 4D^2fDf_y + D^2ff_y^2 + Dff_y^3 + 3(Df)^2f_{yy} + D^3ff_y + 7f_yDfDf_y) + \cdots \Big]_0 \tag{6}$$

where we have retained terms up to order h^5, which are required for the determination of the truncation error in the fourth-order method.

The solution we are seeking, $y(x_0 + h)$, can also be written in the integral form

$$y(x_0 + h) = y(x_0) + \int_{x_0}^{x_0+h} f(x, y(x))\,dx \tag{7}$$

and applying the mean value theorem of integral calculus guarantees the existence of an x such that for $x = x_0 + \theta h$, $0 < \theta < 1$,

$$\int_{x_0}^{x_0+h} f(x, y(x))\,dx = hf(x_0 + \theta h, y(x_0 + \theta h))$$

The substitution of this result in (7) then yields

$$y(x_0 + h) - y(x_0) = hf(x_0 + \theta h, y(x_0 + \theta h)) \tag{8}$$

To avoid evaluations of explicit higher derivatives required in (5), and in the expansion of (8), we define the following:

$$\begin{aligned} k_1 &= hf(x_0, y_0) \\ k_2 &= hf(x_0 + \alpha h, y_0 + \beta k_1) \\ k_3 &= hf(x_0 + \alpha_1 h, y_0 + \beta_1 k_1 + \gamma_1 k_2) \\ k_4 &= hf(x_0 + \alpha_2 h, y_0 + \beta_2 k_1 + \gamma_2 k_2 + \delta_2 k_3) \end{aligned} \tag{9}$$

and then set

$$\bar{y}(x_0 + h) - y_0 = \bar{k} = \mu_1 k_1 + \mu_2 k_2 + \mu_3 k_3 + \mu_4 k_4 \tag{10}$$

The problem now is to determine the α's, β's, γ's, δ_2, and μ's, so that the Taylor expansion of (10) agrees exactly with (6) up to terms of order h^4.

To obtain the Taylor expansion of (10) about the point (x_0, y_0), the operators D_1, D_2, and D_3 are defined as follows:

$$D_1 = \alpha \frac{\partial}{\partial x} + \beta f_0 \frac{\partial}{\partial y}, \text{ for the expansion of } k_2$$

$$D_2 = \alpha_1 \frac{\partial}{\partial x} + (\beta_1 + \gamma_1) f_0 \frac{\partial}{\partial y}, \text{ for the expansion of } k_3 \tag{11}$$

$$D_3 = \alpha_2 \frac{\partial}{\partial x} + (\beta_2 + \gamma_2 + \delta_2) f_0 \frac{\partial}{\partial y}, \text{ for the expansion of } k_4$$

We now proceed to expand, in turn, each of the k's in (10). First we have

$$k_1 = hf_0 \tag{12}$$

Next, to obtain the expansion for k_2, we have

$$hD_1 = \alpha h \frac{\partial}{\partial x} + \beta h f_0 \frac{\partial}{\partial y} = \alpha h \frac{\partial}{\partial x} + \beta k_1 \frac{\partial}{\partial y} \equiv D_{11}$$

hence,

$$\begin{aligned} k_2 &= h\Big[f + D_{11}f + \frac{D_{11}^2 f}{2!} + \frac{D_{11}^3 f}{3!} + \frac{D_{11}^4 f}{4!} + \cdots \Big]_0 \\ &= h\Big[f + hD_1 f + \frac{h^2}{2!} D_1^2 f + \frac{h^3}{3!} D_1^3 f + \frac{h^4}{4!} D_1^4 f + \cdots \Big]_0 \end{aligned} \tag{13}$$

To obtain the expansion of k_3, we have

$$\begin{aligned} \alpha_1 h \frac{\partial}{\partial x} + (\beta_1 k_1 + \gamma_1 k_2) \frac{\partial}{\partial y} &= h\left(\alpha_1 \frac{\partial}{\partial x} + \beta_1 f_0 \frac{\partial}{\partial y}\right) + \gamma_1 k_2 \frac{\partial}{\partial y} \\ &= hD_2 + \gamma_1(k_2 - hf_0) \frac{\partial}{\partial y} \end{aligned}$$

and using (13),

$$= hD_2 + h^2\gamma_1\Big[D_1 f + \frac{h}{2!}D_1^2 f + \frac{h^2}{3!}D_1^3 f + \cdots\Big]_0 \frac{\partial}{\partial y} \equiv D_{21}$$

hence,

$$k_3 = h\Big[f + D_{21}f + \frac{D_{21}^2 f}{2!} + \frac{D_{21}^3 f}{3!} + \frac{D_{21}^4 f}{4!} + \cdots\Big]_0$$

$$= h\Big\{f + hD_2 f + \frac{h^2}{2!}D_2^2 f + \frac{h^3}{3!}D_2^3 f + \frac{h^4}{4!}D_2^4 f + \cdots + h^2\gamma_1\Big[f_y D_1 f + \frac{h}{2!}f_y D_1^2 f + hD_1 f D_2 f_y + \frac{h^2}{3!}f_y D_1^3 f + \frac{h^2}{2}D_1^2 f D_2 f_y + \frac{h^2}{2!}\gamma_1 f_{yy}(D_1 f)^2 + \frac{h^2}{2!}D_1 f D_2^2 f_y + \cdots\Big]\Big\}_0 \tag{14}$$

Finally, to obtain the expansion of k_4, we have

$$\alpha_2 h\frac{\partial}{\partial x} + (\beta_2 k_1 + \gamma_2 k_2 + \delta_2 k_3)\frac{\partial}{\partial y} = hD_3 + \Big[\gamma_2(k_2 - hf_0) + \delta_2(k_3 - hf_0)\Big]_0 \frac{\partial}{\partial y} \equiv D_{31}$$

Using (13) and (14),

$$hD_3 + h^2\Big\{\gamma_2\Big[D_1 f + \frac{h}{2}D_1^2 f + \frac{h^2}{3!}D_1^3 f + \cdots\Big] + \delta_2\Big[D_2 f + \frac{h}{2}(D_2^2 f + 2\gamma_1 f_y D_1 f) + \frac{h^2}{3!}D_2^3 f + \frac{h^2}{2!}\gamma_1 f_y D_1^2 f + h^2\gamma_1 D_1 f D_2 f_y + \cdots\Big]\Big\}_0 \frac{\partial}{\partial y} \equiv D_{31}$$

Hence,

$$k_4 = h\Big[f + D_{31}f + \frac{D_{31}^2 f}{2!} + \frac{D_{31}^3 f}{3!} + \frac{D_{31}^4 f}{4!} + \cdots\Big]_0$$

$$= h\Big\{f + hD_3 f + h^2 f_y\Big[\gamma_2\Big(D_1 f + \frac{h}{2}D_1^2 f + \frac{h^2}{3!}D_1^3 f + \cdots\Big) + \delta_2\Big(D_2 f + h\gamma_1 f_y D_1 f + \frac{h}{2}D_2^2 f + \frac{h^2}{2}\gamma_1 f_y D_1^2 f + h^2\gamma_1 D_1 f D_2 f_y + \frac{h^2}{3!}D_2^3 f + \cdots\Big)\Big] + \frac{1}{2!}\times\Big(h^2 D_3^2 f + 2h^3 D_3 f_y\Big[\gamma_2\Big(D_1 f + \frac{h}{2}D_1^2 f + \cdots\Big) + \delta_2\Big(D_2 f + \frac{h}{2}D_2^2 f + h\gamma_1 f_y D_1 f + \cdots\Big)\Big] + h^4 f_{yy}[\gamma_2^2 D_1^2 f + 2\gamma_2\delta_2 D_1 f D_2 f + \delta_2^2(D_2 f)^2 + \cdots]\Big) + \frac{1}{3!}(h^3 D_3^3 f + 3h^4 D_3^2 f_y[\gamma_2 D_1 f + \delta_2 D_2 f + \cdots]) + \frac{1}{4!}(h^4 D_3^4 f + \cdots) + \cdots\Big\}_0 \tag{15}$$

After the substitution from (12), (13), (14), and (15) into (10), and collection of terms, we obtain the expansion for $\bar{k}$. In this expansion, we equate the coefficients of hf, h^2Df, h^3D^2f, h^3f_yDf, h^4D^3f, $h^4f_yD^2f$, $h^4f_y^2Df$, and h^4DfDf_y to the corresponding coefficients of the Taylor series (6) so as to obtain agreement through terms of order h^4, for all functions f. This leads to eight equations among the thirteen parameters characterizing the algorithm.

$$\begin{aligned}
\mu_1 + \mu_2 + \mu_3 + \mu_4 &= 1\\
\mu_2 D_1 f + \mu_3 D_2 f + \mu_4 D_3 f &= \frac{1}{2!}Df\\
\mu_2 D_1{}^2 f + \mu_3 D_2{}^2 f + \mu_4 D_3{}^2 f &= \frac{2!}{3!}D^2 f\\
\mu_2 D_1{}^3 f + \mu_3 D_2{}^3 f + \mu_4 D_3{}^3 f &= \frac{3!}{4!}D^3 f \qquad (16)\\
\mu_3\gamma_1 D_1 f + \mu_4(\gamma_2 D_1 + \delta_2 D_2 f) &= \frac{1}{3!}Df\\
\mu_3\gamma_1 D_1{}^2 f + \mu_4(\gamma_2 D_1{}^2 f + \delta_2 D_2{}^2 f) &= \frac{2!}{4!}D^2 f\\
\mu_3\gamma_1 D_1 f D_2 f_y + \mu_4(\gamma_2 D_1 f + \delta_2 D_2 f) D_3 f_y &= \frac{3}{4!}Df Df_y\\
\mu_4\gamma_1\delta_2 D_1 &= \frac{1}{4!}Df
\end{aligned}$$

If (16) is to be independent of f, so that the method applies to all functions f, the ratios $D_r^j f / D^j f$, $(r, j = 1, 2, 3)$, must be constant. From the definitions of the operators, the ratios will be constants if

$$\begin{aligned} \alpha &= \beta \\ \alpha_1 &= \beta_1 + \gamma_1 \\ \alpha_2 &= \beta_2 + \gamma_2 + \delta_2 \end{aligned} \tag{17}$$

or

$$D_1 = \alpha D, \qquad D_2 = \alpha_1 D, \qquad D_3 = \alpha_2 D \tag{18}$$

The substitution of (18) into (16), and the division of each equation by its respective homogeneous term in Df, yields the following conditions independent of f:

$$\begin{aligned} \mu_1 + \mu_2 + \mu_3 + \mu_4 &= 1 \\ \mu_2\alpha + \mu_3\alpha_1 + \mu_4\alpha_2 &= \tfrac{1}{2} \\ \mu_2\alpha^2 + \mu_3\alpha_1^2 + \mu_4\alpha_2^2 &= \tfrac{1}{3} \\ \mu_2\alpha^3 + \mu_3\alpha_1^3 + \mu_4\alpha_2^3 &= \tfrac{1}{4} \\ \mu_3\alpha\gamma_1 + \mu_4(\alpha\gamma_2 + \alpha_1\delta_2) &= \tfrac{1}{6} \\ \mu_3\alpha^2\gamma_1 + \mu_4(\alpha^2\gamma_2 + \alpha_1^2\delta_2) &= \tfrac{1}{12} \\ \mu_3\alpha\alpha_1\gamma_1 + \mu_4(\alpha\gamma_2 + \alpha_1\delta_2)\alpha_2 &= \tfrac{1}{8} \\ \mu_4\alpha\gamma_1\delta_2 &= \tfrac{1}{24} \end{aligned} \tag{19}$$

Since (19) is a system of eight equations in ten unknown parameters, μ_1, μ_2, μ_3, μ_4, α, α_1, α_2, γ_1, γ_2, and δ_2, it has 2 degrees of freedom. This circumstance enabled Kutta [7] to impose one arbitrary condition and to express the solution in terms of one of the unknowns. Of his five special solutions, two that are of interest correspond to $\alpha = \frac{1}{2}$ and $\alpha = 1 - \alpha_1$. It is noteworthy that $\alpha_2 = 1$ is a consequence of equations (19), regardless of any arbitrary condition imposed [5].

Solution I ($\alpha = \frac{1}{2}$)	Solution II ($\alpha = 1 - \alpha_1$)
$\alpha = \frac{1}{2}$	$\alpha = \alpha$ (to be specified)
$\alpha_1 = \frac{1}{2}$	$\alpha_1 = 1 - \alpha$
$\alpha_2 = 1$	$\alpha_2 = 1$
$\beta = \frac{1}{2}$	$\beta = \alpha$
$\beta_1 = \frac{1}{2} - (1/2\delta_2)$	$\beta_1 = (1 - \alpha)(2\alpha - 1)/2\alpha$
$\beta_2 = 0$	$\beta_2 = 1 - (3\alpha - 1)/2\alpha[6\alpha(1 - \alpha) - 1]$
$\gamma_1 = 1/2\delta_2$	$\gamma_1 = (1 - \alpha)/2\alpha$
$\gamma_2 = 1 - \delta_2$	$\gamma_2 = (1 - \alpha)(2\alpha - 1)/2\alpha[6\alpha(1 - \alpha) - 1]$
$\delta_2 = \delta_2$ (to be specified)	$\delta_2 = \alpha/[6\alpha(1 - \alpha) - 1]$
$\mu_1 = \frac{1}{6}$	$\mu_1 = [6\alpha(1 - \alpha) - 1)]/12\alpha(1 - \alpha)$
$\mu_2 = (2 - \delta_2)/3$	$\mu_2 = 1/12\alpha(1 - \alpha)$
$\mu_3 = \delta_2/3$	$\mu_3 = 1/12\alpha(1 - \alpha)$
$\mu_4 = \frac{1}{6}$	$\mu_4 = [6\alpha(1 - \alpha) - 1]/12\alpha(1 - \alpha)$

The choice, $\delta_2 = 1$ in Solution I, leads to the classical formulas of Runge [6]:

$$\begin{aligned} k_1 &= hf(x_0, y_0) \\ k_2 &= hf(x_0 + h/2, y_0 + k_1/2) \\ k_3 &= hf(x_0 + h/2, y_0 + k_2/2) \\ k_4 &= hf(x_0 + h, y_0 + h) \\ \bar{k} &= \tfrac{1}{6}(k_1 + 2k_2 + 2k_3 + k_4) \end{aligned} \tag{20}$$

When $y' = f(x)$, Runge's formula reduces to Simpson's rule:

$$\bar{k} = \int_{x_0}^{x_0+h} f(x)\,dx = \frac{h}{6}\left[f(x_0) + 4f\left(x_0 + \frac{h}{2}\right) + f(x_0 + h)\right]$$

The choice, $\alpha = \frac{1}{3}$ in Solution II, leads to the formulas of Kutta [7]:

$$\begin{aligned} k_1 &= hf(x_0, y_0) \\ k_2 &= hf(x_0 + h/3, y_0 + k_1/3) \\ k_3 &= hf(x_0 + \tfrac{2}{3}h, y_0 - k_1/3 + k_2) \\ k_4 &= hf(x_0 + h, y_0 + k_1 - k_2 + k_3) \\ \bar{k} &= \tfrac{1}{8}(k_1 + 3k_2 + 3k_3 + k_4) \end{aligned} \tag{21}$$

To apply the method on high-speed digital computers, Gill [8] developed a calculation procedure which:

1. requires a minimum number of storage registers;
2. gives the highest attainable accuracy (i.e., controls the growth of round-off errors);
3. requires comparatively few instructions.

As given in (9) and (10) the procedure consists of five stages: the first four stages correspond to the calculation of k_1, k_2, k_3, and k_4 respectively; the fifth (final stage) corresponds to the addition of the weighted k's to the initial value y_0. An analysis of the number of storage registers required at each stage shows that four registers are required during the third stage; i.e., at the third stage the following quantities must be available or made available for subsequent stages:

$y_0 + \beta_1 k_1 + \gamma_1 k_2$, an argument of f required for the determination of k_3;

$y_0 + \beta_2 k_1 + \gamma_2 k_2$, an argument of f which must be retained to determine k_4;

$y_0 + \mu_1 k_1 + \mu_2 k_2$, required for the fifth stage;

k_3 is required in both the fourth and fifth stages.

A similar analysis will show that two registers suffice for the first stage and three registers suffice for each of the second, fourth, and fifth stages. Hence, if the number of registers required during the third stage can be reduced to three, then the over-all process will require a maximum of three registers at each stage. Three registers will suffice during the third stage if the quantities to be stored are linearly dependent; i.e., if

$$\begin{vmatrix} 1 & \beta_1 & \gamma_1 \\ 1 & \beta_2 & \gamma_2 \\ 1 & \mu_1 & \mu_2 \end{vmatrix} = 0$$

or equivalently, if

$$\beta_2\mu_2 - \mu_1\gamma_2 - \beta_1\mu_2 + \beta_1\gamma_2 + \mu_1\gamma_1 - \beta_2\gamma_1 = 0 \quad (22)$$

The substitution of Solution I into (22) yields

$$2\delta_2^2 - 4\delta_2 + 1 = 0$$

or

$$\delta_2 = 1 \pm \sqrt{\tfrac{1}{2}}$$

The values of the other parameters corresponding to the choice* $\delta_2 = 1 + \sqrt{\frac{1}{2}}$ are:

$$\begin{aligned} \alpha &= \tfrac{1}{2} & \gamma_1 &= 1 - \sqrt{\tfrac{1}{2}} \\ \alpha_1 &= \tfrac{1}{2} & \gamma_2 &= -\sqrt{\tfrac{1}{2}} \\ \alpha_2 &= 1 & \delta_2 &= 1 + \sqrt{\tfrac{1}{2}} \\ \beta &= \tfrac{1}{2} & \mu_1 &= \mu_4 = \tfrac{1}{6} \\ \beta_1 &= -\tfrac{1}{2} + \sqrt{\tfrac{1}{2}} & \mu_2 &= \tfrac{1}{3}(1 - \sqrt{\tfrac{1}{2}}) \\ \beta_2 &= 0 & \mu_3 &= \tfrac{1}{3}(1 + \sqrt{\tfrac{1}{2}}) \end{aligned} \quad (23)$$

The substitution of (23) into (9) and (10) yields

$$\begin{aligned} k_1 &= hf(x_0, y_0) \\ k_2 &= hf(x_0 + h/2, y_0 + \tfrac{1}{2}k_1) \\ k_3 &= hf(x_0 + h/2, y_0 + (-\tfrac{1}{2} + \sqrt{\tfrac{1}{2}})k_1 + (1 - \sqrt{\tfrac{1}{2}})k_2) \\ k_4 &= hf(x_0 + h, y_0 + (-\sqrt{\tfrac{1}{2}})k_2 + (1 + \sqrt{\tfrac{1}{2}})k_3) \end{aligned} \quad (24)$$

and

$$\bar{k} = \tfrac{1}{6}k_1 + \tfrac{1}{3}(1 - \sqrt{\tfrac{1}{2}})k_2 + \tfrac{1}{3}(1 + \sqrt{\tfrac{1}{2}})k_3 + \tfrac{1}{6}k_4 \quad (25)$$

As given in (24), k_3 and k_4 each depend on quantities of two previous stages. In particular, k_3 is a function of k_2 and k_1; similarly, k_4 is a function of k_2 and k_3. By introducing auxiliary quantities (the q's defined below), which are linear combinations of the k's, a more efficient form equivalent to (24) and (25) can be expressed as follows:

$$\begin{aligned} k_1 &= hf(x_0, y_0) & y_1 &= y_0 + \tfrac{1}{2}k_1 \\ k_2 &= hf(x_0 + h/2, y_1) & y_2 &= y_1 + (1 - \sqrt{\tfrac{1}{2}}) \times (k_2 - q_1) \\ k_3 &= hf(x_0 + h/2, y_2) & y_3 &= y_2 + (1 + \sqrt{\tfrac{1}{2}}) \times (k_3 - q_2) \\ k_4 &= hf(x_0 + h, y_3) & y_4 &= y_3 + \tfrac{1}{6}(k_4 - 2q_3) \end{aligned}$$

$$\begin{aligned} q_1 &= k_1 \\ q_2 &= (2 - \sqrt{2})k_2 + (-2 + 3\sqrt{\tfrac{1}{2}})q_1 \\ q_3 &= (2 + \sqrt{2})k_3 + (-2 - 3\sqrt{\tfrac{1}{2}})q_2 \end{aligned} \quad (26)$$

As defined above, $y_4 = y_3 + \frac{1}{6}(k_4 - 2q_3) = y_0 + \bar{k} = \bar{y}(x_0 + h)$. At any stage in (26), the k, y, and q depend only on stored quantities of the previous stage and quantities computed during the current stage. Furthermore, as each quantity is computed it can be stored in the same register where the corresponding quantity of the previous stage was stored. The quantity replaced in storage is no longer required. Hence, the over-all process as given in (26) requires the storage of only three quantities at each stage and the same three registers can be used at each stage.

A further refinement of (26) can be made to yield greater accuracy by compensating for some of the round-off errors accumulated during each step. This refinement,* achieved without increasing the complexity of the procedure and with no increase in storage requirements, is obtained by introducing q_0 and q_4. The final form of the Gill procedure is then expressed as

$$\begin{aligned} k_1 &= hf(x_0, y_0) & y_1 &= y_0 + \tfrac{1}{2}(k_1 - 2q_0) \\ k_2 &= hf(x_0 + h/2, y_1) & y_2 &= y_1 + (1 - \sqrt{\tfrac{1}{2}})(k_2 - q_1) \\ k_3 &= hf(x_0 + h/2, y_2) & y_3 &= y_2 + (1 + \sqrt{\tfrac{1}{2}})(k_3 - q_2) \\ k_4 &= hf(x_0 + h, y_3) & y_4 &= y_3 + \tfrac{1}{6}(k_4 - 2q_3) \end{aligned}$$

$$\begin{aligned} q_1 &= q_0 + 3[\tfrac{1}{2}(k_1 - 2q_0)] - \tfrac{1}{2}k_1 \\ q_2 &= q_1 + 3[(1 - \sqrt{\tfrac{1}{2}})(k_2 - q_1)] - (1 - \sqrt{\tfrac{1}{2}})k_2 \\ q_3 &= q_2 + 3[(1 + \sqrt{\tfrac{1}{2}})(k_3 - q_2)] - (1 + \sqrt{\tfrac{1}{2}})k_3 \\ q_4 &= q_3 + 3[\tfrac{1}{6}(k_4 - 2q_3)] - \tfrac{1}{2}k_4 \end{aligned} \quad (27)$$

* Gill [8] chose this value of δ_2 since its use contributes a smaller term to the truncation error than would be the choice with the negative radical.

* The detailed derivation of this refinement is given in [8], pp. 102–106.

with q_0 initially zero. If the procedure were carried out with infinite precision, i.e., no round-off errors, the q_4 defined above would be zero. Such is not the case in practice and q_4 represents approximately three times the round-off error in y_4 accumulated during one step [8]. To compensate for this accumulated round-off, q_4 is used as the q_0 for the next step.

c. Error Analysis

Having expanded k and $\bar{k}$ through terms up to h^5, we can now obtain the general expression for the local truncation error; i.e., the truncation error committed at each step,

$$E = k - \bar{k}$$

The substitution from (12), (13), (14), and (15) into (10), and subtraction from (6), yields

$$\begin{aligned} E = h^5 \Bigg\{ & \left[\frac{1}{120} - \frac{(\mu_2\alpha^4 + \mu_3\alpha_1^4 + \mu_4\alpha_2^4)}{24}\right] D^4 f \\ & + \left[\frac{1}{20} - \frac{(\mu_3\gamma_1\alpha_2^2(\gamma_2\alpha + \delta_2\alpha_1))}{2}\right] D^2 f_y Df \\ & + \left[\frac{1}{30} - \frac{(\mu_3\gamma_1^2\alpha^2 + \mu_4(\gamma_2\alpha\alpha_2^2 + \delta_2\alpha_1^2))}{2}\right] Df_y D^2 f \\ & + \left[\frac{1}{120} - \frac{(\mu_4\delta_2\gamma_1\alpha^2)}{2}\right] f_y^2 D_f^2 \\ & + \left[\frac{1}{40} - \frac{(\mu_3\gamma_1^2\alpha^2 + \mu_4\delta_2\alpha_1(2\gamma_2\alpha + \delta_2\alpha_1))}{2}\right] \cdot f_{yy}(Df)^2 \\ & + \left[\frac{1}{120} - (\mu_3\gamma_1\alpha_3 + \mu_4(\delta_2\alpha_1^3 + \gamma_2\alpha^3))\right] f_y D^3 f \\ & + \left[\frac{7}{120} - (\mu_4\delta_2\gamma_1\alpha(\alpha_1 + \alpha_2))\right] f_y Df_y Df \\ & + \frac{f_y^3 Df}{120} - \frac{(\mu_4\gamma_2^2\alpha^2)}{2} f_{yy} D^2 f \Bigg\}_0 + \cdots \end{aligned} \tag{28}$$

Let $D^j f = T^j$; $D^j f_y = S^j$; and $(Df)^2 = P$. Then, corresponding to (20), the truncation error (28) becomes

$$\begin{aligned} E = \frac{h^5}{1440} \Bigg[& -\frac{T^4}{2} - 3(S^2T - ST^2 + f_y^2 T \\ & + 3f_{yy}P + 2f_y ST - 4f_y^3 T) + 2f_y T^3 \Bigg] + \cdots \end{aligned}$$

for which Lotkin [9] has obtained the bound,

$$|E| \le \frac{73}{720} ML^4 h^5 \tag{29}$$

where, in a certain region containing (x_0, y_0),

$$|f(x, y)| \le M$$

$$\left| \frac{\partial^{i+j} f}{\partial x^i \, \partial y^j} \right| \le \frac{L^{i+j}}{M^{j-1}}$$

where M and L are positive constants independent of (x, y), and $(i + j) \le 4$. A similar bound is given by Bieberbach [10]. A detailed analysis of the round-off errors in the Gill variation is included in [8].

With regard to the "stability" of the method, Carr [11] proved the following theorem.

THEOREM: Let f_y be continuous, negative, and bounded from above and below throughout a region D of the (x, y) plane, i.e., $-M_2 < f_y < -M_1$, $M_2 > M_1 > 0$; E be the absolute value of the maximum error (truncation, round-off, or both) introduced at each step, i.e., the bound of the local error; D^* be a region in which the solution of the difference equation approaches no closer than $Qh + |\epsilon_i|$ to the y boundary of D, where $Q \ge \max_{x,y \in D} |f(x, y)| \ge \max(|k_1/2|, |k_2/2|, |k_3|)$, and ϵ_i is the error in y_i at the ith step. Then, the Kutta fourth-order numerical integration procedure has a bound on the total error at the ith step of

$$|\epsilon_i| \le \frac{2E}{hM_1} \tag{30}$$

in the region D^*, when the step size

$$h < \min\left(\frac{M_1}{M_2^2}, \frac{4M_1^3}{M_2^4}\right)$$

If f_y is allowed to be zero or positive (the unstable case), but bounded throughout D, $0 \le f_y < M$, then the propagated error η_{i+1} in D^* at the $(i + 1)$st step is

$$|\eta_{i+1}| \le |\epsilon_i| \, e^{hM} \tag{31}$$

where

$$|\epsilon_i| \le E\left(\frac{e^{ihM} - 1}{e^{hM} - 1}\right)$$

and E and h are as given in the stable case $(f_y < 0)$.

The importance of (30) and (31) is that they relate the step size and propagated error. Hence, if a bound on the partial derivative is known, the step size h can be determined so that the propagated error remains less than a prescribed bound. Algorithms for determining the step size h required to make ϵ_i less than some bound can be found in [11] for both the stable and unstable cases.

A practical scheme for estimating the local truncation error, devised by Richardson [12], is based on the results of numerical integrations with steps h and $h/2$, respectively; i.e., the computation is performed a first time using step $h_1 = h$, then the computation is repeated using $h_2 = h/2$.

Let:

$C_1 h_1^{r+1}$ denote the truncation error in an rth-order process using step h_1;

$C_2 h_2^{r+1}$ denote the truncation error in an rth-order process using step h_2;

$y^{(1)}$ denote the value obtained at $x_0 + h$ when using step h_1;

$y^{(2)}$ denote the value obtained at $x_0 + h$ when using step h_2;

Y denote the true value of y at $x_0 + h$.

Then

$$Y - y^{(1)} = C_1 h_1^{r+1}$$
$$Y - y^{(2)} \approx 2C_2 h_2^{r+1}$$

For small h, $C_1 \approx C_2$. Therefore

$$Y - y^{(2)} \approx \frac{y^{(2)} - y^{(1)}}{2^r - 1} \tag{32}$$

The result, (32), can be shown to hold (for small enough h) for the accumulated truncation error ϵ propagated in the course of integration [13]. This procedure for estimating the error can be included in the integration procedure with the obvious increase in computing time and storage required, as shown in [14].

A "rule of thumb" for the Runge-Kutta fourth-order method, given by Collatz [2], indicates that when the ratio $|(k_2 - k_3)/(k_1 - k_2)|$ exceeds a "few hundredths" for a given step size h, then the step size should be decreased so as to yield a smaller truncation error.

3. SUMMARY OF THE CALCULATION PROCEDURE

Because of its applicability to digital computers, the calculation procedure outlined below for a system of first-order equations is the variation of the Runge-Kutta fourth-order process due to Gill [8].

a. Input

(1) The description of the system of $(n + 1)$ first-order equations,

$$y_i'(x) = f_i(y_0(x), y_1(x), \cdots, y_n(x)),$$
$$i = 0, 1, 2, \cdots, n$$

where $y_0'(x) = f_0 = 1$; i.e., $y_0(x) = x$ is used for convenience of notation and to simplify the form of the process.

(2) The initial conditions, $y_i(x_0) = y_{i0}$, $i = 0, 1, 2, \cdots, n$.

b. Order of the Calculation

(1) Let $j = 1$.

(2) Let $i = 0$.

(3) Compute:

$$y_{ij}' = k_{ij} = f_i(y_{0,j-1}, y_{1,j-1}, \cdots, y_{n,j-1})$$
$$= f_{i,j-1}$$

(4) Repeat step (3) for $i = 1, 2, \cdots, n$.

(5) Let $i = 0$.

(6) Compute:

$$y_{ij} = y_{i,j-1} + h[a_j(k_{ij} - b_j q_{i,j-1})]$$
$$q_{ij} = q_{i,j-1} + 3[a_j(k_{ij} - b_j q_{i,j-1})] - c_j k_{ij}$$

where

$$a_1 = \tfrac{1}{2} \qquad b_1 = 2 \qquad c_1 = \tfrac{1}{2}$$
$$a_2 = 1 - \sqrt{\tfrac{1}{2}} \qquad b_2 = 1 \qquad c_2 = 1 - \sqrt{\tfrac{1}{2}}$$
$$a_3 = 1 + \sqrt{\tfrac{1}{2}} \qquad b_3 = 1 \qquad c_3 = 1 + \sqrt{\tfrac{1}{2}}$$
$$a_4 = \tfrac{1}{6} \qquad b_4 = 2 \qquad c_4 = \tfrac{1}{2}$$

Initially, let $q_{i0}(x_0) = 0$ for all i; thereafter, in advancing the solution, let

$$q_{i0}(x_t) = q_{i4}(x_{t-1}), \qquad t = 1, 2, \cdots$$

To eliminate bias, q_{i1} and q_{i4} should be rounded in opposite directions, one rounded up and the other rounded down. Of fundamental importance in controlling the growth of round-off errors is that k_{ij} and q_{ij} be obtained to the same order of accuracy, with errors of the order of h times the error in y_{ij} [8]. In particular, for fixed point operations, omit the factor h in the definition of y_{ij} in step (6) and define $k_{ij} = hf_{i,j-1}$ in step (3).

(7) Repeat step (6) for $i = 1, 2, \cdots, n$.

(8) Repeat steps (2)–(7) for $j = 2$, 3, and 4.

c. Output

$$y_{i4} = \bar{y}_i(x_0 + h)$$

To advance the solution, repeat steps (1)–(8), letting the current y_{i4} be the initial values y_{i0} for the next step.

4. FLOW CHART

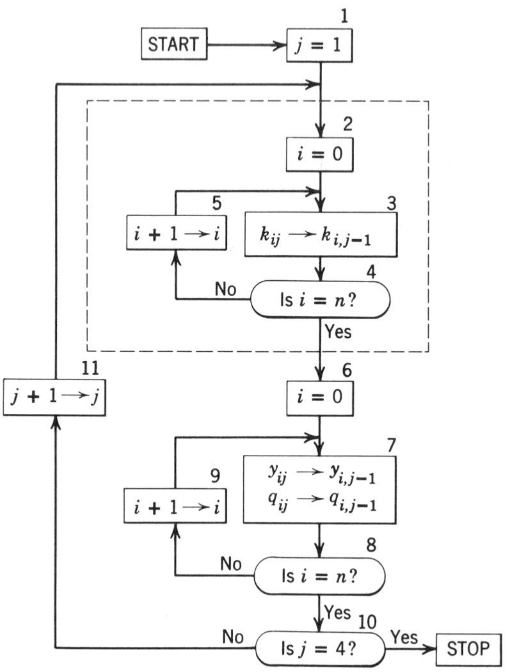

5. DESCRIPTION OF THE FLOW CHART

Box 1: The index j is set equal to one.

Box 2: The index i is set equal to zero.

Box 3: The quantity k_{ij} is computed and stored, say in K_i, for subsequent use in box 7.

Box 4: A decision is made to determine if k_{ij} has been computed and stored for all i and the fixed j.

Box 5: The index i is increased by one and control is directed to box 3.

Box 6: The index i is set equal to zero.

Box 7: The quantity y_{ij} is computed and stored, say in Y_i, for subsequent use in succeeding iterations.

Box 8: A decision is made to determine if y_{ij} and q_{ij} have been computed and stored for all i and the fixed j.

Box 9: The index i is increased by one and control is directed to box 7.

Box 10: A decision is made to determine if the calculation is complete.

Box 11: The index j is increased by one and control is directed to box 2.

The description of the flow chart, as given above, does not provide for printing the functional values at the end of any given step. The user must include with his program the instructions necessary to print the results as desired and to stop the computation when desired.

6. SUBROUTINES

The entire calculation procedure is generally classified as a subroutine, and as such the input to this subroutine (in addition to that listed under Section 3a) should consist of:

(a) The parameter, n, specifying the number of equations.

(b) Designation of the storage registers Y_i, K_i, and Q_i; in particular, specification of, say, Y_0, K_0, and Q_0, assuming these to be base addresses of three independent sets of $(n + 1)$ registers each.

(c) An exit address.

Furthermore, the coding corresponding to boxes 2, 3, 4, and 5 (enclosed by the broken lines on the flow chart) would have to be supplied by the user. As a part of the input to the subroutine, the address of the first instruction of box 2 should be specified by the user; it would then be the obligation of the subroutine to direct control to box 2 as required. Similarly, the user would be obligated to have the "yes" branch of box 4 direct control to box 6. If the procedure is used as a subroutine, an initial box should be appended to the flow chart to "set up" the base addresses, the parameter n, and the necessary control to box 2.

The output of the subroutine would be the same as that given in Section 3c, namely, the $\bar{y}_i(x_0 + h) = y_{i4}$, which would be available in registers Y_i, $i = 0, 1, 2, \cdots, n$.

7. SAMPLE PROBLEM

Given

$$\begin{aligned} y_1' &= 1/y_2 \qquad & y_1(0) &= 1 \\ y_2' &= -(1/y_1) \qquad & y_2(0) &= 1 \end{aligned} \tag{33}$$

Table 1

i	j	k_{ij}	y_{ij}	q_{ij}
0	1	$.10000000 \times 10^1$	$.50000000 \times 10^{-1}$	$.10000000 \times 10^1$
1		$.10000000 \times 10^1$	$.10500000 \times 10^1$	$.10000000 \times 10^1$
2		$-.10000000 \times 10^1$	.95000000	$-.10000000 \times 10^1$
0	2	$.10000000 \times 10^1$	$.50000000 \times 10^{-1}$	.70710678
1		$.10526316 \times 10^1$	$.10515415 \times 10^1$	.73793765
2		−.95238095	.95139473	−.67921219
0	3	$.10000000 \times 10^1$	$.99999999 \times 10^{-1}$	.50000000
1		$.10510884 \times 10^1$	$.11049997 \times 10^1$	.54736297
2		−.95098478	.90500025	−.44761411
0	4	$.10000000 \times 10^1$	$.99999999 \times 10^{-1}$	$-.18189894 \times 10^{-11}$
1		$.11049721 \times 10^1$	$.11051705 \times 10^1$	$-.36379788 \times 10^{-11}$
2		−.90497760	.90483776	$-.90947470 \times 10^{-12}$
0	1	$.10000000 \times 10^1$	.15000000	$.10000000 \times 10^1$
1		$.11051705 \times 10^1$	$.11604290 \times 10^1$	$.11051705 \times 10^1$
2		−.90483776	.85959587	−.90483776
0	2	$.10000000 \times 10^1$	.15000000	.70710678
1		$.11633374 \times 10^1$	$.11621327 \times 10^1$	.81554692
2		−.86175025	.86085787	−.61457684
0	3	$.10000000 \times 10^1$	.20000000	.50000000
1		$.11616319 \times 10^1$	$.12212131 \times 10^1$	.60492940
2		−.86048694	.81887839	−.40501815
0	4	$.10000000 \times 10^1$	.20000000	$-.18189894 \times 10^{-11}$
1		$.12211825 \times 10^1$	$.12214018 \times 10^1$	$-.36379788 \times 10^{-11}$
2		−.81885791	.81873137	$-.90949470 \times 10^{-12}$

determine $y_1(th)$ and $y_2(th)$ for $h = .1$ and $t = 1, 2, \cdots, 10$.

The results listed in Table 1 include the intermediate values for two complete steps in h. (This table is included primarily as an aid for code-checking.) The computations were performed on ORDVAC, a binary computer with a 40-bit register capacity (approximately 12-decimal-digit capacity); however, only 8 digits (rounded) are recorded as output.

Listed in Table 2 are the final values at each step, i.e., the $y_{i4}(x)$, and the corresponding errors ϵ_i, where

$$\epsilon_1 = e^x - y_{14}(x)$$

$$\epsilon_2 = e^{-x} - y_{24}(x)$$

Listed in Table 3 are the results of solving the equivalent system,

$$y_1' = y_1 \qquad y_1(0) = 1$$
$$y_2' = -y_2 \qquad y_2(0) = 1 \tag{34}$$

(Both systems have the solution: $y_1 = e^x$; $y_2 = e^{-x}$.)

The system, (34), differs from (33) in that here the y_1 and y_2 are independent of each other in the calculation procedure. These results exhibit explicitly the behavior of the propagated errors for the stable differential equation ($y_2' = -y_2$), and for the unstable equation ($y_1' = y_1$).

Table 2. $h = .1$

x	$y_1(x)$	$y_2(x)$	$10^7\epsilon_1$	$10^7\epsilon_2$
0	$.10000000 \times 10^1$	$.10000000 \times 10^1$	—	—
.1	$.11051705 \times 10^1$	.90483776	4	−3.4
.2	$.12214018 \times 10^1$	.81873137	10	−6.2
.3	$.13498573 \times 10^1$	.74081905	15	−8.3
.4	$.14918224 \times 10^1$	.67032105	23	−10.0
.5	$.16487181 \times 10^1$	.60653179	32	−11.3
.6	$.18221146 \times 10^1$	.54881286	42	−12.2
.7	$.20137474 \times 10^1$	.49658660	53	−13.0
.8	$.22255342 \times 10^1$	.44933030	67	−13.4
.9	$.24595947 \times 10^1$	.40657102	84	−13.6
1.0	$.27182715 \times 10^1$	.36788081	103	−13.7

Table 3. $h = .1$

x	$y_1(x)$	$y_2(x)$	$10^7\bar{\epsilon}_1$	$10^7\epsilon_2$
0	$.10000000 \times 10^1$	$.10000000 \times 10^1$	—	—
.1	$.11051708 \times 10^1$	.90483750	1	−0.8
.2	$.12214026 \times 10^1$	.81873090	2	−1.5
.3	$.13498585 \times 10^1$	.74081842	3	−2.0
.4	$.14918242 \times 10^1$	.67032029	5	−2.4
.5	$.16487206 \times 10^1$	.60653093	7	−2.7
.6	$.18221180 \times 10^1$	.54881193	8	−2.9
.7	$.20137416 \times 10^1$	.49658562	11	−3.2
.8	$.22255396 \times 10^1$	.44932929	13	−3.3
.9	$.24596014 \times 10^1$	.40656999	17	−3.3
1.0	$.27182797 \times 10^1$	.36787977	21	−3.3

To illustrate the use of (32) for obtaining estimates of the errors, computations were performed using the step size, $h = .2$, for the system (33). The values thus obtained for $x = 1$ are:

$$y_1^{(.2)}(1) = 2.7181164$$
$$y_2^{(.2)}(1) = 0.36790107$$

The substitution of these, and the corresponding values from Table 2, into (32) yields

$$\bar{\epsilon}_1 = Y_1 - y_1^{(.1)}(1) \approx \frac{y_1^{(.1)} - y_1^{(.2)}}{15} = \frac{(1551)10^{-7}}{15} = (103.4)10^{-7}$$

$$\bar{\epsilon}_2 = Y_2 - y_2^{(.1)}(1) \approx \frac{y_2^{(.1)} - y_2^{(.2)}}{15} = \frac{-(2026)10^{-8}}{15} = -(13.5)10^{-7}$$

That these estimates compare favorably with the true errors (the estimates are almost exact!) is due to the fact that the total errors represent primarily the effects of the propagated truncation errors. The round-off errors are negligible as compared to the truncation errors.

8. MEMORY REQUIREMENTS

Given below are the quantities which must be stored and the memory space required for each.

y_{ij}	$(n + 1)$ words
k_{ij}	$(n + 1)$ words
q_{ij}	$(n + 1)$ words

The program requires approximately 50 words; hence, the total memory required is approximately

$$3(n + 1) + 50 + B \text{ words}$$

where B represents the number of words required for boxes 2, 3, 4, and 5.

9. ESTIMATION OF THE RUNNING TIME

The procedure, not including boxes 2, 3, 4, and 5, requires approximately twenty additions and twenty multiplications per step h, per equation. Thus, the total calculation time per step h is

$$T \approx 4(n + 1)\,[T_a + 5(\mu + \nu)]$$

where T_a represents the time required for one pass through boxes 2, 3, 4, and 5; μ and ν are the multiply and add times respectively; and n is the number of equations in the system, not including the equation $y_0' = 1$.

10. REFERENCES

1. Z. Kopal, *Numerical Analysis*, John Wiley & Sons, New York, 1955, p. 195.
2. L. Collatz, *Numerische Behandlung von Differentialgleichungen*, Springer Verlag, Berlin, 1951, p. 34.
3. K. S. Kunz, *Numerical Analysis*, McGraw-Hill Book Co., New York, 1957.
4. C. Runge and H. Konig, *Vorlesungen über numerisches Rechnen*, Springer, Berlin, 1924.
5. E. L. Ince, *Ordinary Differential Equations*, Dover Publications, New York, 1944, p. 540.
6. C. Runge, Über die numerische Auflösung von Differentialgleichungen, *Math. Ann.*, vol. 46, 1895, pp. 167–178.
7. W. Kutta, Beitrag zur näherungsweisen Integration totaler Differentialgleichungen, *Z. Math. Phys.*, vol. 46, 1901, pp. 435–453.
8. S. Gill, A Process for the Step-by-Step Integration of Differential Equations in an Automatic Digital Computing Machine, *Proc. Cambridge Philos. Soc.*, vol. 47, 1951, pp. 96–108.
9. M. Lotkin, On the Accuracy of Runge-Kutta's Method, *MTAC*, vol. 5, Jan. 1951, p. 128.
10. L. Bieberbach, *Theorie der Differentialgleichungen*, Dover Publications, New York, 1944, p. 54.
11. J. W. Carr, III, Error Bounds for the Runge-Kutta Single-step Integration Process, *J. Assoc. Comp. Mach.*, vol. 5, Jan. 1958, p. 39.
12. L. F. Richardson and J. A. Gaunt, The Deferred Approach to the Limit, *Trans. Roy. Soc. London*, vol. 226A, 1927, p. 300.
13. B. Garfinkel, On the Choice of Mesh in the Integration of Ordinary Differential Equations, Ballistic Research Laboratories Report No. 907, Aberdeen Proving Ground, Md., 1954.
14. S. Gorn and R. Moore, Automatic Error Control—the Initial Value Problem in Ordinary Differential Equations, Ballistic Research Laboratories Report No. 893, Aberdeen Proving Ground, Md., 1954.
15. F. B. Hildebrand, *Introduction to Numerical Analysis*, McGraw-Hill Book Co., New York, 1956.
16. W. E. Milne, *Numerical Solution of Differential Equations*, John Wiley & Sons, New York, 1953.

The numerical solution of boundary value problems

10

Eugene L. Wachspress
Knolls Atomic Power Laboratory*

1. FUNCTION

This is a procedure for finding, numerically, the function $Y(x)$ which satisfies the differential equation

$$-\frac{d}{dx}[p(x)Y'] + q(x)Y = f(x) \qquad (1)$$

where p, q, and f are piecewise continuous in the interval $\alpha \leq x \leq \beta$; and Y and pY' are continuous in this interval and satisfy the boundary conditions

$$\begin{aligned} \alpha_1 Y(\alpha) - \alpha_2[p(\alpha)Y'(\alpha)] &= \alpha_3 \\ \beta_1 Y(\beta) + \beta_2[p(\beta)Y'(\beta)] &= \beta_3 \end{aligned} \qquad (2)$$

Equations (1) and (2) are expressed as simultaneous three-point difference equations which are then solved by Choleski's method [1]. This forward elimination, backward substitution technique for solving three-term linear systems is often referred to as line inversion, and is always numerically stable when p is positive and q is nonnegative. Alternative methods are available when line inversion is unstable [2].

* Operated for the U. S. Atomic Energy Commission by the General Electric Co. Contract No. W-31-109 Eng.-52.

2. MATHEMATICAL DISCUSSION

a. Symbols

i	Mesh point index ranging from 1 to n
h_i	Mesh spacing between i and $i+1$
a, b, c, d	Difference equation coefficients
R	Diagonal term of difference equation
s, e	Recursion parameters
ϕ, ψ, η	Round-off error

b. Derivation of the Method

DIFFERENCE EQUATIONS

Grid points may be placed at all discontinuities in p, q, and f, and at as many other points as necessary to describe Y without excessive truncation error. Average values replace variable coefficients between grid points:

$$p_i = \frac{\int_{x_i}^{x_{i+1}} p(x)\,dx}{h_i}; \qquad q_i = \frac{\int_{x_i}^{x_{i+1}} q(x)\,dx}{h_i};$$

$$f_i = \frac{\int_{x_i}^{x_{i+1}} f(x)\,dx}{h_i}$$

Often these quantities are material properties which change only at interfaces between different media so that the averaging, with its associated truncation error, is not required.

Equations relating Y_i to Y_{i+1} and Y_{i-1} may be derived by solving the differential equations with constant coefficients between mesh points and matching boundary conditions at i:

$$\begin{aligned} p_i Y'' + q_i Y &= f_i; \qquad x_i \le x \le x_{i+1} \\ p_{i-1} Y'' + q_{i-1} Y &= f_{i-1}; \qquad x_{i-1} \le x \le x_i \end{aligned} \tag{3}$$

Choosing x_i equal to zero for convenience, we obtain as the solution to the above equations:

$$z = \frac{z_i \sinh[\sqrt{q_i/p_i}(h_i - x)] + z_{i+1}\sinh(\sqrt{q_i/p_i}\,x)}{\sinh(\sqrt{q_i/p_i}\,h_i)} \qquad x_i \le x \le x_{i+1}$$

$$z = \frac{z_i \sinh[\sqrt{q_{i-1}/p_{i-1}}(h_{i-1} + x)] - z_{i-1}\sinh(\sqrt{q_{i-1}/p_{i-1}}\,x)}{\sinh(\sqrt{q_{i-1}/p_{i-1}}\,h_{i-1})} \qquad x_{i-1} \le x \le x_i \tag{4}$$

where $z = qY - f$ so that $pY' = (p/q)z'$. Applying the boundary condition of continuity of pY' at i we get

$$\begin{aligned} &[-\sqrt{p_i/q_i}\,z_i \coth(\sqrt{q_i/p_i}\,h_i) \\ &\quad + \sqrt{p_i/q_i}\,z_{i+1}\operatorname{csch}(\sqrt{q_i/p_i}\,h_i)]_{x=x_i+} \\ &= [\sqrt{p_{i-1}/q_{i-1}}\,z_i \coth(\sqrt{q_{i-1}/p_{i-1}}\,h_{i-1}) \\ &\quad - \sqrt{p_{i-1}/q_{i-1}}\,z_{i-1} \\ &\quad \times \operatorname{csch}(\sqrt{q_{i-1}/p_{i-1}}\,h_{i-1})]_{x=x_i-} \end{aligned} \tag{5}$$

Substituting $qY - f$ for z and applying the identity $\tanh(x/2) = \coth x - \operatorname{csch} x$, we obtain the three-point difference equation:

$$\begin{aligned} &\frac{-\sqrt{p_{i-1}q_{i-1}}\,Y_{i-1}}{\sinh(\sqrt{q_{i-1}/p_{i-1}}\,h_{i-1})} + \left\{\frac{\sqrt{p_{i-1}q_{i-1}}}{\sinh(\sqrt{q_{i-1}/p_{i-1}}\,h_{i-1})}\right. \\ &+ \frac{\sqrt{p_i q_i}}{\sinh(\sqrt{q_i/p_i}\,h_i)} + \sqrt{p_i q_i}\tanh[\sqrt{q_i/p_i}(h_i/2)] \\ &\left. + \sqrt{p_{i-1}q_{i-1}}\tanh[\sqrt{q_{i-1}/p_{i-1}}(h_{i-1}/2)]\right\} Y_i \\ &- \frac{\sqrt{p_i q_i}\,Y_{i+1}}{\sinh(\sqrt{q_i/p_i}\,h_i)} = \sqrt{p_{i-1}/q_{i-1}} \\ &\times \tanh[\sqrt{q_{i-1}/p_{i-1}}(h_{i-1}/2)]f_{i-1} + \sqrt{p_i/q_i} \\ &\times \tanh[\sqrt{q_i/p_i}(h_i/2)]f_i \end{aligned} \tag{6}$$

The mesh spacing is usually chosen small enough so that $\sqrt{q/p}\,h \ll 1$. The hyperbolic functions may then be replaced by their arguments to give the simplified difference equation:

$$\begin{aligned} &-\frac{p_{i-1}}{h_{i-1}}Y_{i-1} + \left(\frac{p_{i-1}}{h_{i-1}} + \frac{p_i}{h_i} + \frac{q_{i-1}h_{i-1}}{2} + \frac{q_i h_i}{2}\right)Y_i \\ &\qquad - \frac{p_i}{h_i}Y_{i+1} = \frac{h_{i-1}}{2}f_{i-1} + \frac{h_i}{2}f_i \end{aligned} \tag{7}$$

The boundary equations, derived in similar fashion, are:

$$\begin{aligned} &-\frac{\alpha_2 p_1}{h_1}Y_2 + \left(\frac{\alpha_2 p_1}{h_1} + \frac{\alpha_2 q_1 h_1}{2} + \alpha_1\right)Y_1 = \frac{\alpha_2 f_1 h_1}{2} + \alpha_3 \\ &-\frac{\beta_2 p_{n-1}}{h_{n-1}}Y_{n-1} + \left(\frac{\beta_2 p_{n-1}}{h_{n-1}} + \frac{\beta_2 q_{n-1}h_{n-1}}{2} + \beta_1\right)Y_n = \frac{\beta_2 f_{n-1}h_{n-1}}{2} + \beta_3 \end{aligned} \tag{8}$$

A boundary condition of specified value for Y at α or β is treated by setting α_2 or β_2 equal to zero and α_1 or β_1 equal to unity with α_3 or β_3 equal to the specified value.

The (fractional) error introduced by replacing the hyperbolic functions in (6) by their arguments is of order qh^2/p, and h is chosen so that this truncation error is negligible.

Line Inversion

The difference equation (6) at point i may be written as

$$-a_i Y_{i-1} + (a + b + c)_i Y_i - c_i Y_{i+1} = d_i$$

or

$$-a_i Y_{i-1} + R_i Y_i - c_i Y_{i+1} = d_i \tag{9}$$

In matrix form the n equations are

$$\begin{vmatrix} R_1 & -c_1 & & & & \\ -a_2 & R_2 & -c_2 & & 0 & \\ & -a_3 & R_3 & -c_3 & & \\ & \cdot & \cdot & \cdot & & \\ & & \cdot & \cdot & \cdot & \\ & & & \cdot & \cdot & \cdot \\ & 0 & & -a_{n-1} & R_{n-1} & c_{n-1} \\ & & & & -a_n & R_n \end{vmatrix} \begin{pmatrix} Y_1 \\ Y_2 \\ Y_3 \\ \cdot \\ \cdot \\ \cdot \\ Y_{n-1} \\ Y_n \end{pmatrix} = \begin{pmatrix} d_1 \\ d_2 \\ d_3 \\ \cdot \\ \cdot \\ \cdot \\ d_{n-1} \\ d_n \end{pmatrix} \tag{10}$$

The first step is to transform (10) into upper triangular form with unity on the diagonal, proceeding in succession from the first to the last row. This forward elimination transforms the matrix equation to:

$$\begin{vmatrix} 1 & -s_1 & & & \\ 0 & 1 & -s_2 & 0 & \\ & \cdot & \cdot & & \\ & & \cdot & \cdot & \\ & & & \cdot & \cdot \\ & 0 & & 1 & -s_{n-1} \\ & & & & 1 \end{vmatrix} \begin{pmatrix} Y_1 \\ Y_2 \\ \cdot \\ \cdot \\ \cdot \\ Y_{n-1} \\ Y_n \end{pmatrix} = \begin{pmatrix} e_1 \\ e_2 \\ \cdot \\ \cdot \\ \cdot \\ e_{n-1} \\ e_n \end{pmatrix} \quad (11)$$

The values for s_1 and e_1 are obtained by dividing the first equation in (10) by the diagonal element R_1:

$$s_1 = c_1/R_1; \qquad e_1 = d_1/R_1 \quad (12)$$

Recursion formulas for the remaining s_i and e_i may now be derived.

Consider row $i-1$ of (11) and row i of (10)

$$\begin{matrix} i-1 \text{ of } (11): \\ i \text{ of } (10): \end{matrix} \begin{vmatrix} \cdots & 1 & -s_{i-1} & 0\cdots \\ \cdots -a_i & R_i & -c_i\cdots & \end{vmatrix} \begin{pmatrix} Y_{i-1} \\ Y_i \end{pmatrix} = \begin{pmatrix} e_{i-1} \\ d_i \end{pmatrix}$$

Multiply row $i-1$ of (11) by a_i and add to row i of (10):

$$i \,|\cdots 0 \quad (R_i - a_i s_{i-1}) - c_i \cdots|(Y_i) = (d_i + a_i e_{i-1})$$

Divide by the diagonal element:

$$i \text{ of } (11) \left|\cdots 0 \quad 1 - \frac{c_i}{R_i - a_i s_{i-1}} \cdots\right|(Y_i) = \left(\frac{d_i + a_i e_{i-1}}{R_i - a_i s_{i-1}}\right)$$

Row i is now in the form required for (11), where

$$s_i = \frac{c_i}{R_i - a_i s_{i-1}}; \qquad e_i = \frac{d_i + a_i e_{i-1}}{R_i - a_i s_{i-1}} \quad (13)$$

Backward substitution in matrix equation (11) gives for the unknowns:

$$\begin{aligned} Y_n &= e_n \\ Y_i &= s_i Y_{i+1} + e_i \qquad i = 1, 2, \cdots, n-1 \end{aligned} \quad (14)$$

c. Stability

The calculation is stable when a small error introduced at any step in any calculated quantity does not lead to a large error in any quantity at any subsequent step. During the forward elimination process recursion parameters s and e are calculated at each point. Round-off errors are present. Let the errors in s and e at point $i-1$ be ϕ_{i-1} and ψ_{i-1} respectively. From (13), neglecting round-off at step i, we obtain the error in s_i resulting from the incorrect s_{i-1}:

$$s_i + \phi_i = \frac{c_i}{R_i - a_i(s_{i-1} + \phi_{i-1})}$$

$$s_i = \frac{c_i}{R_i - a_i s_{i-1}}$$

Subtracting the second equation from the first:

$$\phi_i = \frac{c_i}{c_i/s_i - a_i\phi_{i-1}} - s_i$$

or

$$\phi_i \frac{c_i}{s_i} - a_i\phi_i\phi_{i-1} = a_i\phi_{i-1}s_i$$

To first order in ϕ:

$$\phi_i = \frac{a_i}{c_i} s_i^2 \phi_{i-1}$$

For a_i and c_i the same order magnitude and $|s_i| \leq 1$, there is no significant error growth in the calculation of the s_i from (13). For b nonnegative with a and c positive, inspection of the recursion formula [(13) with $R = a + b + c$] shows that s_i is less than unity in magnitude. In this case the round-off error in s_i may be neglected.

The recursion formula for e_i gives similarly:

$$e_i + \psi_i = \frac{d_i + a_i(e_{i-1} + \psi_{i-1})}{R_i - a_i s_{i-1}}$$

$$e_i = \frac{d_i + a_i e_{i-1}}{R_i - a_i s_{i-1}}$$

Subtracting the second equation from the first:

$$\psi_i = \frac{a_i\psi_{i-1}}{R_i - a_i s_{i-1}} = \frac{a_i}{c_i} s_i \psi_{i-1}$$

Again stability depends only upon the magnitude of s_i for a and c the same order of magnitude.

The value of Y at each point is calculated during the backward substitution mesh sweep. Let the error at $i-1$ be η_{i-1} so that we obtain from (14):

$$Y_{i-1} + \eta_{i-1} = e_{i-1} - s_{i-1}(Y_i + \eta_i)$$

$$Y_{i-1} = e_{i-1} - s_{i-1}Y_i$$

Subtracting the second equation from the first gives $\eta_{i-1} = -s_{i-1}\eta_i$. There is no error

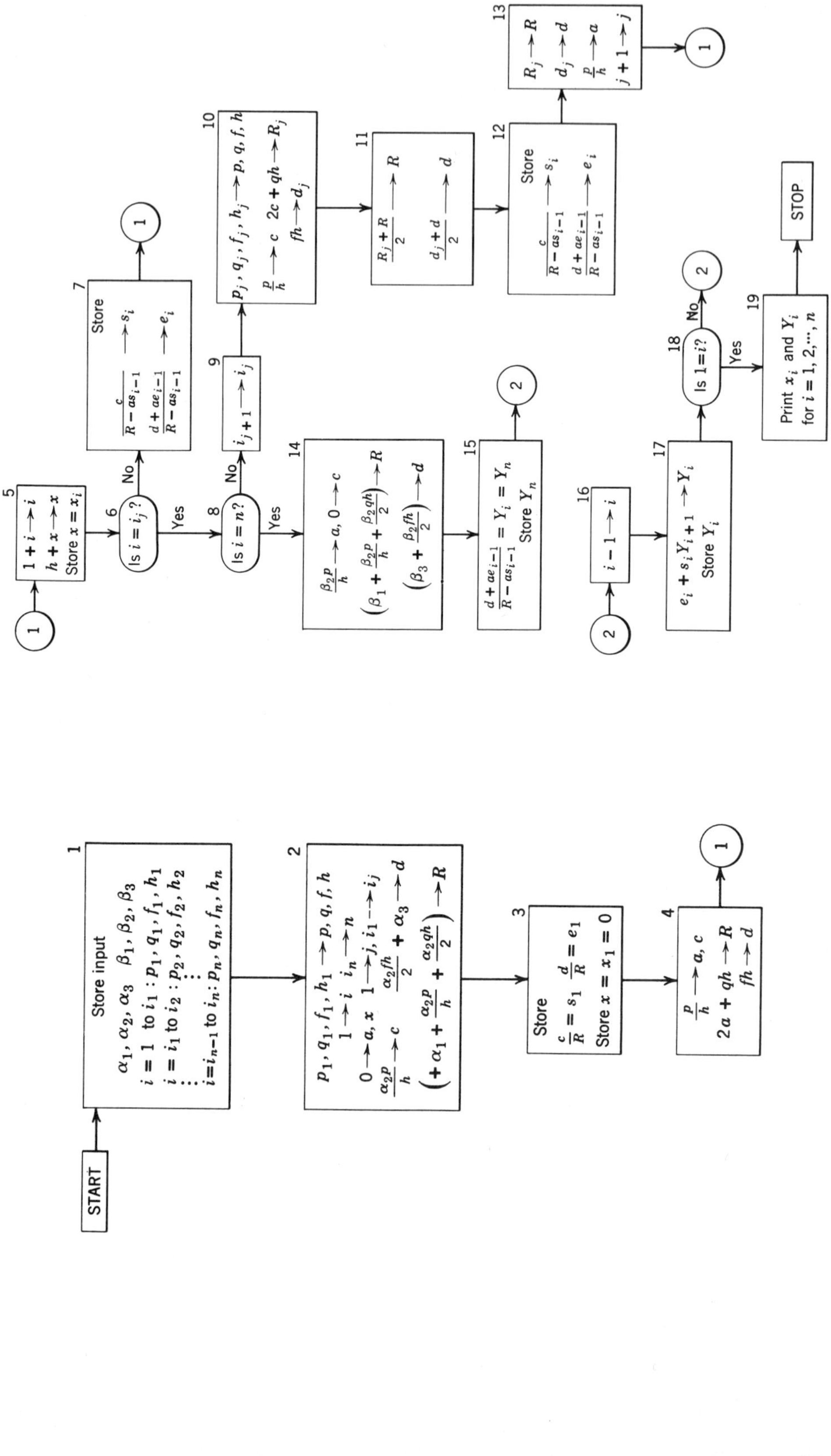
START
1
Store input
$\alpha_1, \alpha_2, \alpha_3 \quad \beta_1, \beta_2, \beta_3$
$i = 1$ to $i_1 : p_1, q_1, f_1, h_1$
$i = i_1$ to $i_2 : p_2, q_2, f_2, h_2$
$i = i_{n-1}$ to $i_n : p_n, q_n, f_n, h_n$
2
$p_1, q_1, f_1, h_1 \to p, q, f, h$
$1 \to i \quad i_n \to n$
$0 \to a, x \quad 1 \to j, i_1 \to i_j$
$\frac{\alpha_2 p}{h} \to c \quad \frac{\alpha_2 fh}{2} + \alpha_3 \to d$
$\left(+\alpha_1 + \frac{\alpha_2 p}{h} + \frac{\alpha_2 qh}{2}\right) \to R$
3
Store
$\frac{c}{R} = s_1 \quad \frac{d}{R} = e_1$
Store $x = x_1 = 0$
4
$\frac{p}{h} \to a, c$
$2a + qh \to R$
$fh \to d$
1
1
5
$1 + i \to i$
$h + x \to x$
Store $x = x_i$
6
Is $i = i_j$?
No
Yes
7
Store
$\frac{c}{R - as_{i-1}} \to s_i$
$\frac{d + ae_{i-1}}{R - as_{i-1}} \to e_i$
1
8
Is $i = n$?
No
Yes
9
$i_{j+1} \to i_j$
10
$p_j, q_j, f_j, h_j \to p, q, f, h$
$\frac{p}{h} \to c \quad 2c + qh \to R_j$
$fh \to d_j$
11
$\frac{R_j + R}{2} \to R$
$\frac{d_j + d}{2} \to d$
12
Store
$\frac{c}{R - as_{i-1}} \to s_i$
$\frac{d + ae_{i-1}}{R - as_{i-1}} \to e_i$
13
$R_j \to R$
$d_j \to d$
$\frac{p}{h} \to a$
$j + 1 \to j$
1
14
$\frac{\beta_2 p}{h} \to a, 0 \to c$
$\left(\beta_1 + \frac{\beta_2 p}{h} + \frac{\beta_2 qh}{2}\right) \to R$
$\left(\beta_3 + \frac{\beta_2 fh}{2}\right) \to d$
15
$\frac{d + ae_{i-1}}{R - as_{i-1}} = Y_i = Y_n$
Store Y_n
2
2
16
$i - 1 \to i$
17
$e_i + s_i Y_{i+1} \to Y_i$
Store Y_i
18
Is $1 = i$?
No
Yes
2
19
Print x_i and Y_i
for $i = 1, 2, \cdots, n$
STOP

growth in the Y calculation when $|s| \leq 1$. Hence, the entire calculation is stable when $|s| \leq 1$, which in turn is true when b is nonnegative while a and c are positive (or zero at $i = 1$ or n).

Let H be the solution to the homogeneous difference equations ($f_i = 0$ for all i) which satisfies only the initial boundary condition. Then $H_i = s_i H_{i+1}$, and the stability criterion is that the absolute value of H increase monotonically with i. An analogous mathematical development of the solution of second-order differential equations by the factorization method leads to the same conclusion. One considers the conditions under which the Wronskian of two auxiliary first-order differential equations, arising from the factorization, vanishes [3].

Although the initial boundary condition can cause difficulty in an otherwise stable system, a nonnegative b with positive a and c at all interior points ($i = 2$ to $n - 1$) is usually adequate for stability. Inspection of (7) shows that the stability requirement is satisfied when p is everywhere positive and q is nonnegative.

3. SUMMARY OF THE CALCULATION PROCEDURE

(a) Choose a mesh spacing for each region such that

(1) $qh^2/p \ll 1$, and

(2) average values may be substituted for p, q, and f between mesh points without excessive truncation error.

(b) Calculate successively the recursion parameters s and e.

(c) Calculate the Y_i by backward substitution, and list Y_i and x_i for $i = 1, 2, \cdots, n$.

4. FLOW CHART

The flow chart appears on page 124.

5. DESCRIPTION OF THE FLOW CHART

Box 1: Input quantities are stored. These include boundary conditions α and β, differential equation coefficients p, q, and f, and mesh increments h. One could calculate mesh increments such that qh^2/p is less than a predetermined truncation error bound. Also, one could include a differential equation coefficient averaging code here.

Box 2: Initial values are set and difference equation coefficients are calculated for point one.

Box 3: Recursion parameters s_1 and e_1 are calculated.

Box 4: Difference equation coefficients for region one are calculated.

Box 5: The i index is stepped up and the x value stored.

Box 6: Point i is tested to determine whether or not it lies on an interface where equation coefficients change.

Box 7: Recursion parameters s_i and e_i are calculated.

Box 8: Point i is compared with the last point, n.

Box 9: The index at the next interface (or boundary) is set.

Box 10: Difference equation coefficients for region j are calculated.

Box 11: Difference equation coefficients for interface point i_j are calculated.

Box 12: Recursion parameters s_i and e_i are calculated for i_j.

Box 13: Difference equation coefficients for region j are set, and the j index is stepped up.

Box 14: Boundary point β difference equation coefficients are calculated. This marks the end of the forward elimination.

Box 15: The value of Y_n is calculated.

Box 16: The i index is stepped down for the backward substitution.

Box 17: The value of Y_i is calculated.

Box 18: A test of i against unity is made to determine whether or not the calculation of the Y_i has been completed.

Box 19: Values of Y_i and x_i are printed.

6. SUBROUTINES

None.

7. SAMPLE PROBLEM

The following two-region problem is to be solved:

$x = 0$		$x = 2.5$		$x = 5.5$
	Region 1		Region 2	
$Y = 0$ at $x = 0$	$p_1 = 1.0$ $q_1 = .10$ $f_1 = 1.0$		$p_2 = .50$ $q_2 = .01$ $f_2 = 0$	$Y = 0$ at $x = 5.5$

The exact value of Y at $x = 2.5$ may be found by substituting in (6) with $Y(0) = Y(5.5) = 0$ as Y_{i-1} and Y_{i+1} respectively, and $Y(2.5) = Y_i$:

$$Y_i\left\{\frac{\sqrt{.10}}{\sinh[\sqrt{.10}(2.5)]} + \frac{\sqrt{.005}}{\sinh[\sqrt{.02}(3)]} + \sqrt{.10}\tanh\left[\sqrt{.10}\left(\frac{2.5}{2}\right)\right] + \sqrt{.005}\tanh\left[\sqrt{.02}\left(\frac{3}{2}\right)\right]\right\}$$

$$= \sqrt{10}\tanh\left[\sqrt{.10}\left(\frac{2.5}{2}\right)\right]$$

$$Y_i = 1.810 = Y(2.5)$$

Direct application of (7) gives

$$\left(\frac{1}{2.5} + \frac{.5}{3} + \frac{.1(2.5)}{2} + \frac{.01(3)}{2}\right)Y_i = \frac{1.0(2.5)}{2}$$

$$Y_i = 1.77$$

The percentage truncation error with no internal mesh points is

$$\frac{1.81 - 1.77}{1.81} \times 100 = 2.21\%$$

Since qh^2/p is this order of magnitude, an error of a few per cent is to be expected. To illustrate the line inversion technique, we choose one internal point in each region and proceed according to the diagram:

x	0	1.25	2.5	4.0	5.5
i	1	2	3	4	5

$\alpha_1 = 1.0$ $\quad p_1 = 1.0$ $\quad p_2 = .5$ $\quad \beta_1 = 1.0$
$\alpha_2 = \alpha_3 = 0$ $\quad q_1 = .10$ $\quad q_2 = .01$ $\quad \beta_2 = \beta_3 = 0$
$f_1 = 1.0$ $\quad f_2 = 0$
$h_1 = 1.25$ $\quad h_2 = 1.5$

$$c_1 = d_1 = 0, \qquad R_1 = \alpha_1 = 1.0$$

$$\frac{c_1}{R_1} = s_1 = 0, \qquad \frac{d_1}{R_1} = e_1 = 0, \qquad x_1 = 0$$

$$\frac{p_1}{h_1} = a_2 = c_2 = \frac{1.0}{1.25} = .80$$

$$2a_2 + q_1h_1 = R_2 = 1.6 + .125 = 1.725$$

$$f_1h_1 = d_2 = 1.25, \qquad x_2 = 1.25$$

$$\frac{c_2}{R_2 - a_2s_1} = \frac{.80}{1.725 - .8 \times 0} = .46377 = s_2$$

$$\frac{d_2 + a_2e_1}{R_2 - a_2s_1} = \frac{1.25}{1.725} = .72463 = e_2$$

As indicated in boxes 10–13 of the flow chart, the interface point equations require calculation of coefficients for the second region:

$$\frac{p_2}{h_2} = \frac{.5}{1.5} = .3333 = c_3 = a_4 = c_4$$

$$x_3 = 1.25 + 1.25 = 2.5$$

$$2c_3 + q_2h_2 = .6666 + .01(1.5) = .6816 = R_{j=2}$$

$$f_2h_2 = 0 = d_{j=2}$$

$$\frac{R_{j=2} + R_{j=1}}{2} = \frac{.6816 + 1.725}{2} = 1.2033 = R_{i=3}$$

$$\frac{d_{j=2} + d_{j=1}}{2} = \frac{0 + 1.25}{2} = .625 = d_{i=3}$$

$$\frac{c_3}{R_3 - a_3s_2} = \frac{.3333}{1.2033 - .80 \times .46377} = .40046 = s_3$$

$$\frac{d_3 + a_3e_2}{R_3 - a_3s_2} = \frac{.625 + .08 \times .72463}{1.2033 - .80 \times .46377} = 1.4474 = e_3$$

$$\frac{c_4}{R_4 - a_4s_3} = \frac{.3333}{.6816 - .3333 \times .40046} = .60806 = s_4$$

$$\frac{d_4 + a_4e_3}{R_4 - a_4s_3} = \frac{0 + .3333 \times 1.4474}{.6816 - .3333 \times .40046} = .88011 = e_4$$

$a_5 = 0$, $R_5 = 1.0$, $d_5 = 0$, $Y_5 = 0$ from box 15

From Box 17:

$x = 4$: $\quad Y_4 = e_4 + s_4Y_5 = .88011 + .60806 \times 0 = .88011$

$x = 2.5$: $\quad Y_3 = e_3 + s_3Y_4 = 1.4474 + .40046 \times .88011 = 1.7998$

$x = 1.25$: $\quad Y_2 = e_2 + s_2Y_3 = .72463 + .46377 \times 1.7998 = 1.5593$

$x = 0$: $\quad Y_1 = e_1 + s_1Y_2 = 0 + 0 \times 1.5593 = 0$

The error at $x = 2.5$ is

$$\frac{1.81 - 1.80}{1.81} \times 100 = 0.55\%$$

8. MEMORY REQUIREMENTS

Input data for J regions	$5J$
Recursion parameters for n mesh points	$2n$
x_i and Y_i for n points	$2n$
Program (approximately)	500

The total memory required is (approximately)

$$500 + 4n + 5J$$

One could reduce memory requirements by replacing the recursion parameter e by Y during the backward sweep. This decreases the required storage by n.

9. ESTIMATION OF THE RUNNING TIME

There are a minimum of five multiplications and three additions at each point. A conservative time estimate might include a factor of 2 to cover logical operations and coefficient calculation:

The running time estimate per point is $10\mu + 6\nu$, where μ and ν are the multiply and add times.

Although the solution of ordinary differential equations by this technique requires very little machine time, line inversion is sometimes used as a part of an iteration scheme for solving second-order boundary value problems. A considerable amount of machine time may be required for the repeated line inversions involved in such schemes [4].

10. REFERENCES

1. F. B. Hildebrand, *Methods of Applied Mathematics*, Prentice-Hall, New York, 1952.
2. L. Fox, *The Numerical Solution of Two Point Boundary Problems in Ordinary Differential Equations*, Oxford University Press, London, 1957.
3. E. C. Ridley, A Numerical Method of Solving Second-Order Linear Differential Equation with Two-Point Boundary Conditions, *Proc. Cambridge Phil. Soc.*, vol. 53, 1957, p. 442.
4. E. L. Wachspress and G. J. Habetler, An Alternating-Direction-Implicit Iteration Technique, *J. Soc. Indust. Appl. Math.* (to be published).
5. J. von Neumann and H. H. Goldstine, Numerical Inverting of Matrices of High Order, *Bull. Amer. Math. Soc.*, vol. 53, 1947, pp. 1021–1099.

The solution of ordinary differential equations with large time constants

11

J. Certaine
Nuclear Development Corporation of America

1. FUNCTION

The method described in this chapter was devised to integrate a system of first-order coupled differential equations where the derivatives take on a form which makes the system unsuitable for direct use of substitution processes such as that of Runge-Kutta.

2. MATHEMATICAL DISCUSSION

To see why special methods are required for differential equations having large time constants, consider the equation

$$\dot{y} = -50y + 1, \qquad y(0) = .04$$

It is easy to see that the solution of this equation is

$$y = .02(1 + e^{-50t})$$

The solution, therefore, is a slowly varying function of time, changing from .04 at the origin to .02 as $x \to \infty$. Since the solution is so slowly varying, one might be tempted to integrate such a system numerically with a conventional scheme such as the trapezoidal rule. If this is done, with a time spacing of 1 second, the values that result are shown in Table 1.

Table I

t	$y(t)_{\text{calc.}}$	$y(t)_{\text{exact}}$
0	.0400	.0400
1	.0370	.0200
2	.0043	.0200
3	.0035	.0200

We see that the process is wildly inaccurate even though the time step was chosen small in relation to the variation of the solution. The reason for this behavior is not far to seek. It results simply from the fact that in the trapezoidal rule, or indeed in using any conventional integration formula, the mesh size is governed not only by the behavior of the solution as a whole but also by that of rapidly varying transients which are of no importance to the solution. Since problems of this type occur frequently in practical applications, it is desirable to have a special technique which permits the use of a mesh size which is governed only by the rate of change of the over-all solution.

Let D represent a diagonal $N \times N$ matrix of large positive diagonal elements $\delta_1, \cdots, \delta_N$; $z(\tau)$ an $N \times 1$ vector with components $\zeta_1, \cdots, \zeta_N$; $z'(\tau)$ its derivative; and $s(z, \tau)$ an $N \times 1$ vector whose components are functions $\sigma_\nu(z, \tau) \equiv \sigma_\nu(\zeta_1, \cdots, \zeta_N; \tau)$. The differential equation system is assumed to have the form

$$z'(\tau) = -Dz(\tau) + s(z(\tau); \tau) \qquad (1)$$

It is assumed that the function $s(z(\tau); \tau)$ is a slowly varying function of τ.

The equations are rewritten by transposing $Dz(\tau)$, multiplying by $e^{D\tau}$, and integrating over the interval (τ_1, τ_2). We obtain

$$e^{D\tau_2} z_2 - e^{D\tau_1} z_1 = \int_{\tau_1}^{\tau_2} e^{D\tau'} s(\tau')\, d\tau'$$

and finally

$$z_2 = e^{-D\mu} z_1 + \int_{\tau_1}^{\tau_2} e^{D(\tau'-\tau_2)} s(\tau')\, d\tau' \qquad (2)$$

where $s(\tau') \equiv s(z(\tau'); \tau')$, $z_2 = z(\tau_2)$, $z_1 = z(\tau_1)$, and $\mu = \tau_2 - \tau_1$.

Equation (2) is rigorously equivalent to (1) but is in the form of an integrodifference equation. In this form it may be used to carry out a stepwise integration procedure. This process will be noniterative if $s(\tau')$ is extrapolated from previously calculated values of $z(\tau)$ and iterative otherwise. In any event, let $s(\tau')$ be approximated by a polynomial $a(\tau')$ in τ' of degree $\Lambda + 1$. For ease of integration, we express $a(\tau')$ as follows:

$$a(\tau') = \sum_{\lambda=0}^{\Lambda} \frac{(\tau' - \tau_1)^\lambda}{\lambda!\, \mu^\lambda} a_\lambda$$

Clearly,

$$\int_{\tau_1}^{\tau_2} e^{D(\tau'-\tau_2)} s(\tau')\, d\tau' \approx \sum_{\lambda=0}^{\Lambda} C_\lambda a_\lambda \qquad (3)$$

where

$$C_\lambda = \int_{\tau_1}^{\tau_2} e^{D(\tau'-\tau_2)} \frac{(\tau' - \tau_1)^\lambda}{\lambda!\mu^\lambda}\, d\tau'$$

It is immediately seen that

$$C_0 = D^{-1}(I - e^{-D\mu})$$

and we verify quite readily that

$$C_{\lambda+1} = D^{-1}\left(\frac{I}{(\lambda+1)!} - \mu^{-1}C_\lambda\right)$$

It may be remarked that all these matrices are diagonal.

Now the remaining quantities a_λ depend upon the $\Lambda + 1$ values of τ' which are chosen as interpolation points. We will restrict ourselves to $\Lambda \le 2$ and obtain two results, each of which will require an iterative process. We will take as our interpolation points the values $\tau' = \tau_2$, $\tau' = \tau_2 - \mu, \cdots, \tau' = \tau_2 - \Lambda\mu$, i.e., equally spaced intervals in τ'. This restriction allows us to use two previously computed points to carry forward the solution. The case $\Lambda = 1$ will be used for starting the calculation or when a value of μ is diminished in order to gain convergence or increased in order to gain speed in the calculation.

If we make the abbreviation $s_\lambda = s(\tau_\lambda)$, $\tau_\lambda = \tau_2 - (2 - \lambda)\mu$, we obtain for $\Lambda = 1$ the following:

$$s_1 = a_0$$
$$s_2 = a_0 + a_1$$

or

$$a_0 = s_1$$
$$a_1 = s_2 - s_1$$

Substituting these values in (3), and the ensuing result in (2), we get

$$z_2 = e^{-D\mu}z_1 + (C_0 - C_1)s_1 + C_1 s_2 \qquad (4)$$

For $\Lambda = 2$, we have

$$s_2 = a_0 + a_1 + \tfrac{1}{2}a_2$$
$$s_1 = a_0$$
$$s_0 = a_0 - a_1 + \tfrac{1}{2}a_2$$

or

$$a_0 = s_1$$
$$a_1 = \tfrac{1}{2}s_2 - \tfrac{1}{2}s_0$$
$$a_2 = s_0 - 2s_1 + s_2$$

Finally, we obtain

$$z_2 = e^{-D\mu}z_1 + (C_2 - \tfrac{1}{2}C_1)s_0 + (C_0 - 2C_2)s_1 + (\tfrac{1}{2}C_1 + C_2)s_2 \qquad (5)$$

A detailed error analysis is not available. It is clear, however, that if $s(y, \tau)$ is a quadratic in τ, the above process is exact within round-off errors. On the other hand, the error in a second-order Runge-Kutta process is given in [1] as

$$\mu^{-3}E = \left(\frac{1}{6} - \frac{\nu}{4}\right)D^2 r + \frac{1}{6} r_z D_r,$$
$$r = -Dz + s$$

where

$$D \equiv \frac{\partial}{\partial \tau} + r(\tau_1)\frac{\partial}{\partial z}$$

(We confine the discussion to the one-dimensional case so that z is a scalar.) We obtain

$$D_r = s'(\tau) + D(Dz_1 - s_1)$$

$$D_r^2 = s''(\tau)$$

$$r_z = -D$$

Hence

$$\mu^{-3}E = \left(\frac{1}{6} - \frac{\nu}{4}\right)s_1'' - \frac{D}{6}s_1' + \frac{D^2}{6}s_1 - \frac{D^3}{6}z_1$$

It is clear that for large values of D, small steps in τ will be required to obtain reasonable accuracy.

3. SUMMARY OF THE CALCULATION PROCEDURE

The detailed calculation procedure is most easily described by referring to the flow chart. In general terms, (4) is used to initiate a calculation. The value s_2 is assumed to equal s_1 and the first iterate is given by

$$z_2 = e^{-D\mu}z_1 + C_0 s_1$$

The value of z_2 is compared component by component with z_1 (or the previous iterate) and if satisfactory agreement results, z_2 is printed and the calculation proceeds to (5). Otherwise, s_2 is recalculated, and a new value z_2 is recalculated and tested for convergence. If more than, say five cycles are required for convergence the interval size is reduced by a factor α. Otherwise, we go to (5) since two consecutive values of z are available. Throughout this process, only those terms in (4) which depend on the iterated quantity z_2 are recalculated, the remaining terms being stored.

The process involving two known values is exactly as above, except that (5) is used. If this process fails because more than, say, five iterations are required to obtain convergence, the interval size is changed and the process carried forward by starting again with (4). This is done to avoid interpolation at points which were not in the grid, although the additional calculations required would not be very great.

It might be further noted that some variations are possible in the above scheme. The process could be accelerated by continuous replacement. That is, as each new component of z_2 is calculated, it is used to modify the next component of s_2. Also, a successive substitution process of Picard type could be used to calculate

$$\int_{\tau_1}^{\tau_2} e^{D(\tau' - \tau_2)}s(z(\tau');\ \tau')\,d\tau'$$

This, however, would require calculation of additional exponentials as well as additional values of s and might, depending upon the complexity of s, greatly add to the required machine time.

4. FLOW CHART

The flow chart appears on page 131.

5. DESCRIPTION OF THE FLOW CHART

Box 1: At the start (or after change of interval size as in box 13) the required constants are calculated.

Box 2: The values at $\tau + \mu$ are assumed to be equal to those in τ, to start off the iteration.

Box 3: The terms in (4) which are independent of the iteration are computed, stored, and control is transferred to box 7.

Box 4: Precomputed data is transferred to the proper location in memory.

Box 5: The values at $\tau + \mu$ are estimated by a linear extrapolation.

Box 6: The terms in (5) which are independent of the iteration are computed, stored, and control is transferred to box 7.

Box 7: The value of $s(\tau + \mu)$ is computed using the estimated values of $z(\tau + \mu)$ obtained in box 2 or box 5.

Box 8: A new approximation c to $z(\tau + \mu)$ is computed and stored. Ratios of each component are computed and stored for the convergence test (box 9).

Box 9: The absolute relative error is compared to the convergence vector e. If convergence is unsatisfactory, control is transferred to box 10. Otherwise control is transferred to box 11.

Box 10: If more than five iterations are required to satisfy the convergence criterion,

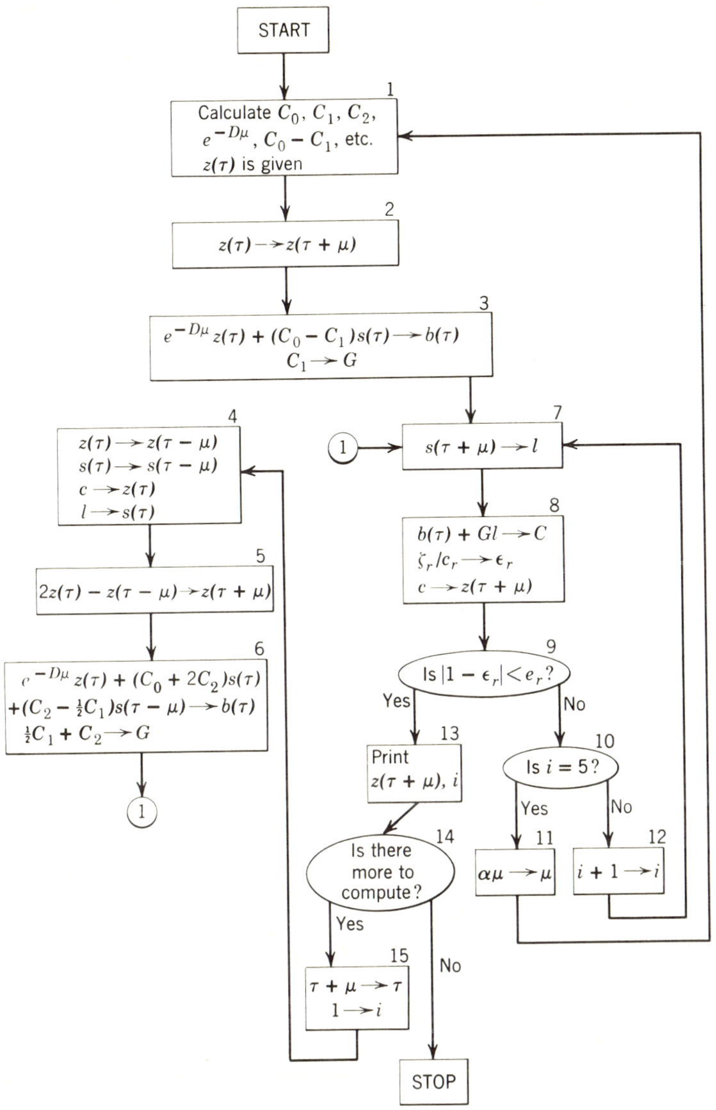

μ is reduced by the ratio α and control transferred to box 1 to restart. Otherwise, we go to box 11.

Box 11: If convergence is unsatisfactory, replace μ by $\alpha\mu$, where $0 < \alpha < 1$.

Box 12: Increase counter and iterate again.

Box 13: Convergence is satisfactory and results are printed along with the number of cycles required. If the number of cycles is small (say one or two), then a larger interval is indicated and manual insertion of a larger μ may be permitted.

Box 14: This is a test to see if the computation is complete or if more values of $z(\tau)$ are desired.

Box 15: Increase τ and reset counter.

6. SUBROUTINES

The following subroutines will be useful:

(a) Diagonal matrix times vector.
(b) Addition and subtraction of vectors.
(c) Exponential calculation.
(d) Subroutine for calculation of $s(y, \tau)$.

7. SAMPLE PROBLEM

Consider the one-dimensional problem:

$$z' = -z + \tau, \qquad z(0) = 10$$

The exact solution is given by

$$z = 11e^{-\tau} + \tau - 1$$

Let us now carry out the sequence of operations as given in the flow chart and compute the values $z(1)$ and $z(2)$ at intervals of $\mu = 1$. These should also be exact, since $s(z(\tau);\ \tau)$ is linear. We choose $e = 10^{-10}$ for the convergence criterion.

1. $C_0 = 1 - e^{-1}$
   $C_1 = e^{-1}$
   $C_2 = \frac{1}{2} - e^{-1}$
   $z(0) = 10$
2. $z(1) = 10$
3. $b(0) = 10e^{-1}\ [s(0) = 0]$
   $G = e^{-1}$, go to box 7.
7. $l = 1$
8. $C = 10e^{-1} + e^{-1} = 11e^{-1}$
   $$\epsilon = \frac{11e^{-1}}{10}$$
   $z(1) = 11e^{-1}$
9. $|1 - \epsilon| - 10^{-10} > 0$
10. $i \neq 5$, set $i = 2$, return to box 7.
7. $l = 1$
8. $C = 11e^{-1}$
   $\varepsilon = 1$
   $z(1) = 11e^{-1}$
9. $|1 - \varepsilon| - 10^{-10} \leq 0$, go to box 12.
12. Print $z(1)$, $i = 2$
13. Not finished, go to box 4, $\tau = 1$
4. $z(0) = 0$
   $s(0) = 0$
   $z(1) = 11e^{-1}$
   $s(1) = 1$
5. $z(2) = 22e^{-1} - 10$
6. $C_0 - 2C_2 = e^{-1}$
   $C_2 - \frac{1}{2}C_1 = \frac{1}{2} - \frac{3}{2}e^{-1}$
   $b(1) = 11e^{-2} + e^{-1}$
   $G = \frac{1}{2} - \frac{1}{2}e^{-1}$
7. $l = 2$
8. $C = 11e^{-2} + e^{-1} + 2(\frac{1}{2} - \frac{1}{2}e^{-1})$
   $= 11e^{-2} + 1$
   $$\varepsilon = \frac{11e^{-2} + 1}{22e^{-1} - 10} = -1.3$$
   $z(2) = 11e^{-2} + 1$
9. $|1 - \varepsilon| - 10^{-10} > 0$
10. $i \neq 5$, set $i = 2$, return to box 7.
7. $l = 2$
8. $C = 11e^{-2} + 1$
   $\varepsilon = 1$
   $z(2) = 11e^{-2} + 1$
9. $|1 - \varepsilon| - 10^{-10} \leq 0$, go to box 12
12. Print $z(1)$, $i = 2$
13. Finished. Stop.

8. MEMORY REQUIREMENTS

The storage required for calculation of $s(\tau)$ depends upon the complexity of this vector function. Approximately $20N$ words are required for the storage of vectors which arise during the calculation, but this may be reduced to perhaps $12N$ by careful allocation of temporary storage. The program (excluding the calculation of s) is approximately 600 words.

9. ESTIMATION OF THE RUNNING TIME

The calculation time per value of τ is given (approximately) by

$$40\mu + 120\nu + i(4\mu + 32\nu)$$

per equation, excluding source evaluation, for starting ($\Lambda = 1$), and

$$40\mu + 120\nu + i(7\mu + 63\nu)$$

for propagation ($\Lambda = 2$), where i is the number of iterations required and μ and ν are multiply and add times, respectively.

10. REFERENCE

1. Z. Kopal, *Numerical Analysis*, John Wiley & Sons, New York, 1955.

PART IV PARTIAL DIFFERENTIAL EQUATIONS

The numerical solution of parabolic partial differential equations

12

Herbert B. Keller
Institute of Mathematical Sciences
New York University

I. FUNCTION

In this chapter we study methods for the numerical solution of linear second-order parabolic partial differential equations of the form

$$L[u(x, t)] \equiv u_t - a(x, t)u_{xx} - 2b(x, t)u_x + c(x, t)u = d(x, t) \quad (1)$$

where

$$a(x, t) > 0 \quad (2)$$

A solution of (1) is uniquely determined in the semi-infinite strip

$$R:[0 \leq x \leq L; \quad t \geq 0] \quad (3)$$

by specifying appropriate initial and boundary conditions; say

$$\begin{aligned} u(x, 0) &= f(x), && 0 < x < L \\ u(0, t) &= g_0(t), && \\ u(L, t) &= g_1(t), && t > 0 \end{aligned} \quad (4)$$

It is assumed that the solution thus determined has as many continuous derivatives as required in the subsequent analysis.*

A brief summary of numerical methods for parabolic equations more general than (1) is contained in Section 2d.

2. MATHEMATICAL DISCUSSION

a. Difference Equations

On the strip R we place the net (also called grid or lattice)

$$R_{h,k}:[x_j = jh, j = 0, 1, \cdots, J + 1; \quad t_n = nk, n = 0, 1, \cdots] \quad (5)$$

where $h = L/(J + 1)$. At each point (x_j, t_n) of the net we seek a quantity $v(x_j, t_n)$ which approximates the solution $u(x_j, t_n)$ of (1) and (4). The approximating quantities are determined as the solution of a system of linear algebraic equations, the difference equations,

* This is insured by requiring the coefficients in (1) to be sufficiently differentiable and the initial data to be consistent with the boundary data [i.e., $f(0) = g_0(0)$, $(L) = g_1(0)$, and the derivatives of $f(t)$ and $g(x)$ satisfy (1) at the corners].

which in some sense approximate the parabolic equation (1) and boundary conditions (4). There is no unique procedure by which to derive such difference equations and indeed it cannot be determined *a priori* which of a class of equivalent* difference equations yields the best approximation to u. However, we can require that the difference equations have a solution which can be effectively computed and that this computed numerical solution, at least for sufficiently small h, k and round-off error, be close to the exact solution, u. The properties of consistency, convergence, and stability which guarantee this are defined in Section 2c and are shown to be satisfied by the difference equations of this section. The art of deriving such equations for any type of partial differential equation requires a knowledge of the theory of characteristics and their relation to difference methods as developed in the fundamental paper of Courant, Friedrichs, and Lewy [1].

To derive the difference equations we employ the usual notation

$$v(x_j, t_n) \equiv v_j^n, \qquad u(x_j, t_n) \equiv u_j^n \tag{6a}$$

for all quantities or functions defined at the points of $R_{h,k}$. In addition, for a quantity defined *only* at the points of the net, we use linear interpolation to define the quantity at intermediate values of t; i.e.,

$$v(x_j, t_n + \theta k) \equiv v_j^{n+\theta} \equiv \theta v_j^{n+1} + (1-\theta)v_j^n, \qquad 0 \le \theta \le 1 \tag{6b}$$

and similarly for other quantities. Interpolation in the x-variable is not needed as all spatial difference quotients are to be centered at some point x_j of the net. Then since v_j^n is to approximate $u(x_j, t_n)$ we may also require that the difference quotients of v approximate the derivatives of u as follows:

$$\begin{aligned} u_x(x_j, t_n + \theta k) &\approx \frac{1}{2h}(v_{j+1}^{n+\theta} - v_{j-1}^{n+\theta}) \\ u_{xx}(x_j, t_n + \theta k) &\approx \frac{1}{h^2}(v_{j+1}^{n+\theta} - 2v_j^{n+\theta} + v_{j-1}^{n+\theta}) \\ u_t(x_j, t_n + \theta k) &\approx \frac{1}{k}(v_j^{n+1} - v_j^n) \end{aligned} \tag{7}$$

* Two systems may be called equivalent if their truncation errors (see Section 2c) are of the same order in the net spacings, (h, k).

Letting $x = x_j$ and $t = t_n + \theta k$ in (1) and using the approximations (7) yields the difference equations (after multiplication by k)

$$\begin{aligned} kL_{h,k}[v(x_j, t_n + \theta k)] &\equiv v_j^{n+1} - v_j^n \\ &- \lambda a_j^{n+\theta}(v_{j+1}^{n+\theta} - 2v_j^{n+\theta} + v_{j-1}^{n+\theta}) \\ &- h\lambda b_j^{n+\theta}(v_{j+1}^{n+\theta} - v_{j-1}^{n+\theta}) + kc_j^{n+\theta}v_j^{n+\theta} \\ &= kd_j^{n+\theta} \end{aligned} \tag{8}$$

where we have introduced the mesh ratio $\lambda \equiv k/h^2$. The boundary and initial conditions (4) are replaced by

$$\begin{aligned} v_j^0 &= f(x_j), && 0 \le j \le J+1 \\ v_0^n &= g_0(t_n), && \\ v_{J+1}^n &= g_1(t_n), && n > 0 \end{aligned} \tag{9}$$

If $\theta = 0$, then, by (6*b*) in (8), the difference equations are immediately solved for v_j^{n+1}; the difference equations are called explicit in this case. However, if $\theta \ne 0$, the equations for v_j^{n+1} are a coupled linear system of the general form

$$\alpha_j v_{j-1}^{n+1} + \beta_j v_j^{n+1} + \gamma_j v_{j+1}^{n+1} = S_j^n$$

where S_j^n involves the inhomogeneous terms and the previously computed quantities v_j^n. The difference equations are now called implicit and somewhat more labor is required to solve them than in the explicit case. (A proof that these equations have a solution, under suitable conditions on the net spacing, is a simple consequence of the maximum principle established in Section 2b.) For the one space dimensional case considered here the above implicit equations are easily solved by well-known procedures (see Chapter 10 or [2], Part III, C).

A discussion of the relative merits of explicit versus implicit methods is deferred until Section 2d.

b. A Maximum Principle

In this section we establish a bound on the maximum absolute value of solutions of the difference equations (8) and (9). This bound immediately implies that the difference equations have a unique solution and it is used in Section 2c to prove stability and convergence.

In terms of the notation

$$V(t_n) \equiv \max_j |v_j^n| \,; \qquad F \equiv \max_j |f(x_j)| \,;$$

$$G(t_n) \equiv \max_{0 \le t \le t_n} (|g_0(t)|, |g_1(t)|) \tag{10}$$

$$B(t_n) \equiv \max [F, G(t_n)]; \quad C(t_n) \equiv \min_{\substack{j \\ 0 \le t \le t_n}} c(x_j, t);$$

$$D(t_n) \equiv \max_{\substack{j \\ 0 \le t \le t_n}} |d(x_j, t)|$$

we state the

Maximum Principle: On every net $R_{h,k}$ which satisfies:

$$1 + \theta k c(x, t) > 0 \tag{11a}$$

$$a(x, t) - h\,|b(x, t)| \ge 0 \tag{11b}$$

$$1 - (1 - \theta)[2\lambda a(x, t) + k c(x, t)] \ge 0 \tag{11c}$$

the solution v_j^n of the difference equations (8) and (9) is bounded by

$$V(t_n) \le \max [G(t_n);\; X^n(t_n)B(t_n)] + \frac{1 - X^n(t_n)}{C(t_n)} D(t_n) \tag{12}$$

where

$$X(t_n) \equiv \frac{1 - (1 - \theta)kC(t_n)}{1 + \theta k C(t_n)} \tag{13}$$

Proof: Using definition (6*b*) the difference equations (8) can be written as

$$\xi_j^{n+\theta}(\theta) v_j^{n+1} = \xi_j^{n+\theta}(\theta - 1) v_j^n + \lambda(a_j^{n+\theta} + h b_j^{n+\theta}) v_{j+1}^{n+\theta} + \lambda(a_j^{n+\theta} - h b_j^{n+\theta}) v_{j-1}^{n+\theta} + k d_j^{n+\theta}$$

where

$$\xi_j^{n+\theta}(\alpha) \equiv 1 + \alpha[2\lambda a_j^{n+\theta} + k c_j^{n+\theta}]$$

By (2) and (11) all coefficients of v in the above equation are positive. Then, taking absolute values and using (6*b*) and (10) yields

$$\xi_j^{n+\theta}(\theta)\,|v_j^{n+1}| \le [1 - (1 - \theta)k c_j^{n+\theta}]V(t_n) + \theta[2\lambda a_j^{n+\theta}]V(t_{n+1}) + kD(t_{n+1}) \tag{14}$$

If $\max_j |v_j^{n+1}|$ occurs at $j = 0$ or $j = J + 1$, i.e., on the boundary, we have by (10)

$$V(t_{n+1}) \le G(t_{n+1}) \tag{15}$$

Otherwise the maximum occurs at some interior point, say

$$V(t_{n+1}) = |v_m^{n+1}|$$

where $m \ne 0, J + 1$. Then taking (14) at the point $j = m$ and using the definition of $\xi_j^{n+\theta}(\theta)$ yields

$$[1 + \theta k c_m^{n+\theta}]V(t_{n+1}) \le [1 - (1 - \theta)k c_m^{n+\theta}]V(t_n) + kD(t_{n+1})$$

This inequality remains valid, and is in fact strengthened, when $c_m^{n+\theta}$ is replaced by $C(t_{n+1})$. By (11*a*), the coefficient of $V(t_{n+1})$ is still positive and we obtain by a division

$$V(t_{n+1}) \le X(t_{n+1})V(t_n) + \frac{kD(t_{n+1})}{[1 + \theta k C(t_{n+1})]}$$

From (15) and the above inequality we must have, reducing n by one for convenience,

$$V(t_n) \le \max [G(t_n), X(t_n)V(t_{n-1})] + \frac{kD(t_n)}{[1 + \theta k C(t_n)]} \tag{16}$$

By (13), (11), and (10) we see that $X(t_n) \ge X(t_{n-1})$ and a similar inequality is true of the last term in (16). Thus a recursive application of (16) yields

$$V(t_n) \le \max [G(t_n), X(t_n)G(t_n), X^2(t_n)V(t_{n-2})] + [1 + X(t_n)] \frac{kD(t_n)}{[1 + \theta k C(t_n)]}$$

Continuing this recursion n times, summing the geometric progression which arises, and using (13) yields the result (12), and the proof of the maximum principle is completed.

It should be observed from (10) that the sign of $C(t)$ is arbitrary and thus from (13)

$$X(t_n) \begin{cases} \le 1, & \text{if } C(t_n) \ge 0 \\ \ge 1, & \text{if } C(t_n) \le 0 \end{cases}$$

Also, on a sequence of nets for which $k \to 0$ while $t_n = nk$ is fixed,

$$\lim_{k \to 0} X^n(t_n) = e^{-C(t_n)t_n}$$

This exponential is an upper bound on X^n if $C \le 0$ and may be used in (12) to estimate the magnitude of the coefficients. However, the main feature concerning (13) is the fact that for all nets satisfying (11), $X^n(t_n)$ is finite (and in fact bounded by $e^{|C(t_n)|t_n}$).

If the difference equations (8) and (9) are homogeneous (i.e., $f \equiv g_0 \equiv g_1 \equiv d \equiv 0$), then by (10) $G \equiv B \equiv D \equiv 0$ and the maximum principle shows that the only solution for a net satisfying (11) is $v_j^n \equiv 0$. Thus, since the homogeneous system has only a zero solution, the fundamental theorem on linear systems implies that the inhomogeneous difference equations (8) and (9) have a unique solution.

c. Consistency, Convergence, and Stability

The difference scheme (8) is *consistent* with the differential equation (1) provided

$$\lim_{h,k\to 0} |L[w(x, t)] - L_{h,k}[w(x, t)]| = 0$$

for all sufficiently smooth functions $w(x, t)$ defined in R. This condition insures that the difference equations actually do "approximate" the differential equation. To verify consistency we apply Taylor's formula in a tedious but elementary calculation to obtain

$$w_x(x_j, t_n + \theta k) - \frac{1}{2h}(w_{j+1}^{n+\theta} - w_{j-1}^{n+\theta})$$
$$= \frac{h^2}{6}[\bar{w}_{xxx} + \frac{\theta(1-\theta)}{2}k^2\bar{w}_{xxxtt}] \equiv \tau^{(1)}$$
$$w_{xx}(x_j, t_n + \theta k) - \frac{1}{h^2}(w_{j+1}^{n+\theta} - 2w_j^{n+\theta} + w_{j-1}^{n+\theta})$$
$$= \frac{h^2}{12}[\bar{w}_{xxxx} + \frac{\theta(1-\theta)}{2}k^2\bar{w}_{xxxxtt}] \equiv \tau^{(2)} \quad (17)$$
$$w_t(x_j, t_n + \theta k) - \frac{1}{k}(w_j^{n+1} - w_j^n)$$
$$= \frac{1-2\theta}{2}k\bar{w}_{tt} + \frac{1 - 3\theta + 3\theta^2}{6}k^2\bar{w}_{ttt} \equiv \tau^{(3)}$$

Here the barred functions are to be evaluated at appropriate intermediate values. The right-hand sides of (17) are the truncation errors corresponding to the approximations (7). From (1) and (8) we have, using the above,

$$L[w] - L_{h,k}[w] = \tau^{(3)} - a\tau^{(2)} - 2b\tau^{(1)}$$
$$= 0(h^2) + 0(k) \quad (18)$$

Since the derivatives of w and the coefficients a and b are assumed bounded, the right side of (18) goes to zero with h and k and consistency is proved.

The right side of (18) is frequently called the *truncation error* of the difference scheme and we thus see that consistency merely implies that the truncation error vanishes with the net spacing. The faster it vanishes the more accurate is the difference approximation and the more likely is the numerical solution to be a good approximation to the exact solution. If $\theta = \frac{1}{2}$ we see from (17) that the truncation becomes $0(h^2) + 0(k^2)$ but the difference equations are implicit. This observation is of importance when comparing explicit and implicit schemes.

The difference equations are *convergent* if their solution satisfies

$$\lim_{h,k\to 0} |u(x_j, t_n) - v_j^n| = 0 \quad (19)$$

Convergence insures, at least for a sufficiently fine mesh, that the numerical solution is a "close" approximation to the exact solution of the differential equation. To demonstrate (19) we introduce the error e_j^n, defined by

$$e_j^n = v_j^n - u(x_j, t_n) \quad (20)$$

and show that it can be made arbitrarily small for a proper choice of net spacing.

Dividing (8) by k and subtracting (1) from it yields

$$0 = L_{h,k}[v] - L[u]$$
$$= L_{h,k}[v] - L_{h,k}[u] + L_{h,k}[u] - L[u]$$
$$= L_{h,k}[v - u] + (L_{h,k}[u] - L[u]) \quad (21)$$

Here we have used the linearity of the difference operators and have omitted the obvious arguments and subscripts. Using (20) and the definition of the truncation error, i.e.,

$$L[u(x_j, t_n + \theta k)] - L_{h,k}[u_j^{n+\theta}] \equiv \tau_j^{n+\theta} \quad (22)$$

in (21) we obtain

$$L_{h,k}[e_j^{n+\theta}] = \tau_j^{n+\theta} \quad (23)$$

Comparing this equation with (8) we find, as is usually the case for linear equations, that the error satisfies a system of difference equations formally identical to those of the numerical solution. They are identical if d is replaced by τ. Thus, given the proper boundary and initial conditions, the error satisfies the maximum principle of Section 2b. From (4), (9), and (20) the error vanishes initially and on the boundary. Then *if the net spacing satisfies* (11) we have by (12):

$$|e_j^n| \le \frac{1 - X^n(t_n)}{C(t_n)} \max_{j,n} |\tau_j^{n+\theta}| \quad (24)$$

Since the coefficient above is bounded, independent of the net spacing, (24) implies (19) and hence convergence, provided the truncation term approaches zero as the net is refined. However, by the proof of consistency we have, comparing (18) and (22),

$$\tau_j^{n+\theta} = 0(h^2) + 0(k)$$

which concludes the proof. Note, by (17), that if $\theta = \frac{1}{2}$,

$$\tau_j^{n+\theta} = 0(h^2) + 0(k^2)$$

and the error bound converges to zero faster than if $\theta \neq \frac{1}{2}$.

Since the maximum principle was used in the convergence proof, the conditions (11) are sufficient for convergence. However, on any sequence of nets for which $(h, k) \to 0$, conditions (11*a*) and (11*b*) are eventually satisfied and thus only (11*c*) may impose a restriction on the mesh ratio λ. If $\theta = 1$, this condition is automatically satisfied and λ is unrestricted.

The difference equations are *stable* if small errors introduced in the inhomogeneous term, at the boundary or initially (due, say, to inexact evaluation of the given functions), remain bounded as the computations progress. More precisely, using the notation (10), the difference equations (8) and (9) are stable if there exists a function $K(t)$ *independent of the mesh spacing*, (h, k), and bounded for any finite t such that

$$V(t_n) \leq K(t_n)[B(t_n) + D(t_n)]$$

The maximum principle immediately yields stability on all nets $R_{h,k}$ satisfying (11) since we may take, from (12) and (13),

$$K(t) = \begin{cases} 1 + C^{-1}(t), & C(t) > 0 \\ 1 + t, & C(t) = 0 \\ [1 - C^{-1}(t)]e^{-tC(t)}, & C(t) < 0 \end{cases}$$

From the above definition and proof of stability it is clear that stability is maintained on any sequence of nets with $(h, k) \to 0$ while satisfying (11). Applying this result to the difference equations (23) for the error, it follows that *stability implies convergence* on such a sequence of nets if only the difference equations are consistent. Indeed, it has been proved [3, 4], for a very large class of difference equations, that these two properties are equivalent. It should be noticed that stability is a property only of the difference equations. If stability is attained on *all* sequences of nets for which $(h, k) \to 0$, then the difference equations are said to be unconditionally stable (e.g., this is the case when $\theta = 1$).

The effects of round-off error throughout the calculation can be estimated. Assuming inexact calculations (say, in evaluating the coefficients and in solving the difference equations), the equations (23) for the error, now between the exact solution of the differential equation and the actual numerical solution obtained, must be modified by adding a term, $\rho_j^{n+\theta}$, to $\tau_j^{n+\theta}$. The stability proof now yields a bound on the actual error in which the D term is of the form

$$\max_{j,n} \left(|\rho_j^{n+\theta}| + |\tau_j^{n+\theta}| \right)$$

Thus we see that there is no great advantage to be gained in making the round-off terms much smaller than the truncation terms, as is frequently done by people inexperienced in computing.

d. Discussion and Generalizations

The explicit difference scheme, i.e., $\theta = 0$ in (8), has the advantage that the equations are immediately solved at each time step with a minimum of computations. However, the mesh ratio λ is then restricted by (11*c*) which implies

$$k \leq \frac{h^2}{2a(x, t) + h^2c(x, t)}$$

Thus a fine spatial net requires a "finer" time net and the number of steps to reach a time t is of the order t/h^2. The error bound for this scheme is by (17), (18), (22), (24), and the above:

$$|e_j^n| \leq 0(h^2) + 0(k) = 0(h^2)$$

The completely implicit difference scheme, i.e., $\theta = 1$ in (8), imposes no restriction on λ but requires somewhat more labor to solve the equations. The total computations required to reach a time t can be made considerably less than in the explicit case by choosing k sufficiently large. The error bound, however, is now of the form

$$|e_j^n| \leq 0(h^2) + 0(k)$$

and thus in the interest of accuracy k must not be too large.

The so-called Crank-Nicholson [3, 5] difference scheme, i.e., $\theta = \frac{1}{2}$ in (8), is implicit and as far as our analysis shows the restriction on λ implies

$$k \leq \frac{2h^2}{2a(x, t) + h^2c(x, t)}$$

The net spacings now are related as in the explicit scheme but the time step can be twice as large. Since the solutions of the implicit equations do not require twice the amount of computations used to solve the explicit equations, a saving may be made in the total work required to reach a time t. In addition, the error bound is now of the order

$$|e_j^n| \leq 0(h^2) + 0(k^2)$$

which is an improvement over the completely implicit scheme. As the above discussion indicates, the implicit methods yield solutions of a given accuracy with less computations and hence are to be preferred to the explicit methods; indeed, they are more popular in actual computations on high-speed digital computers.

The results of this chapter may be easily extended to parabolic equations in more than one space dimension. In such cases, however, the implicit equations cannot be solved in a simple manner and iterative methods are usually employed. The procedures of Chapter 13 are applicable for this purpose and if they, or similar procedures, are used the implicit schemes may still be preferable to the explicit ones.

For parabolic equations with constant coefficients (and similarly coefficients depending only on t) somewhat sharper results may be obtained. For such difference equations the von Neumann procedure [6] may be applied and frequently yields necessary and sufficient conditions for stability [4]. This analysis shows that the scheme corresponding to (8) is stable for arbitrary fixed λ if $\frac{1}{2} \leq \theta \leq 1$ (see [7], Chapter VI).

There is an extensive literature on difference methods for solving the classical diffusion or heat conduction equation,

$$u_t = \nabla^2 u$$

in one and two space dimensions. In addition to necessary and sufficient conditions for convergence and stability of schemes of the form (8), and even more accurate approximations [7, 8], detailed studies of the rate of convergence of the numerical to the exact solution have been given [9, 10].

Relatively few results have been published on the numerical solution of nonlinear parabolic equations. A thorough study of explicit difference methods for the pure initial value problem for quasi-linear equations is given in [11]. Implicit methods for such equations subject to initial and boundary conditions are discussed in [12, 13]. All of these papers also contain results on linear equations with variable coefficients. The more difficult case of "mildly nonlinear" parabolic equations, which necessarily lead to implicit schemes that must be solved by iterations, is studied in [13, 14].

3. SUMMARY OF THE CALCULATION PROCEDURE

To illustrate the calculation procedures required we consider the following initial, boundary value problem for the one-dimensional diffusion equation with constant coefficient $a > 0$:

$$u_t = au_{xx}; \qquad u(x, 0) = f(x), \qquad 0 \leq x \leq 1; \tag{25}$$

$$u(0, t) = 0, \quad u(1, t) = 0, \qquad t > 0$$

The numerical solution is sought in the rectangle $R_T:[0 \leq x \leq 1;\ 0 \leq t \leq T]$. Using the net points $x_j = jh$, $j = 0, 1, \cdots, J + 1$, $t_n = nk$, where $h \equiv 1/(J + 1)$ and $\lambda = k/h^2$ is an arbitrary fixed number, the difference procedure of Section 2a with $\theta = \frac{1}{2}$ yields, when applied to (25), the following numerical problem:

$$v_j^0 = f(x_j), \qquad 0 \leq j \leq J + 1 \tag{26a}$$

$$S_j^n \equiv -v_{j-1}^n + 2\left(1 - \frac{1}{a\lambda}\right)v_j^n - v_{j+1}^n, \qquad 1 \leq j \leq J,\ n \geq 0 \tag{26b}$$

$$v_{j-1}^{n+1} - 2\left(1 + \frac{1}{a\lambda}\right)v_j^{n+1} + v_{j+1}^{n+1} = S_j^n, \qquad 1 \leq j \leq J,\ n \geq 0 \tag{26c}$$

$$v_0^{n+1} = v_{J+1}^{n+1} = 0$$

The system (26*c*) can be solved by means of a well-known procedure (see Chapter 10) which, for the above constant coefficients, becomes:

$$P_1 = -2\left(1 + \frac{1}{a\lambda}\right); \quad P_j = -\left[2\left(1 + \frac{1}{a\lambda}\right) + \frac{1}{P_{j-1}}\right], \quad j = 2, 3, \cdots, J \quad (27a)$$

$$Z_1^{n+1} = S_1^n; \quad Z_j^{n+1} = \left[S_j^n - \frac{Z_{j-1}^{n+1}}{P_{j-1}}\right], \quad j = 2, 3, \cdots, J \quad (27b)$$

$$v_{J+1}^{n+1} = 0; \quad v_j^{n+1} = \frac{1}{P_j}[Z_j^{n+1} - v_{j+1}^{n+1}], \quad j = J, J-1, \cdots, 1 \quad (27c)$$

Thus, starting with the given values v_j^0 of (26*a*), the inhomogeneous term S_j^0 can be computed from (26*b*). Then, using these quantities in (27), the solution v_j^1 is obtained. Repeated use of (26*b*) and (27) in this manner yields the solution for any time step t_n. In applying (27) it should be noted that the P_j of (27*a*) are independent of t_n [for this problem, since the coefficients in (25) are time-independent] and hence need be computed only once.

4. FLOW CHART

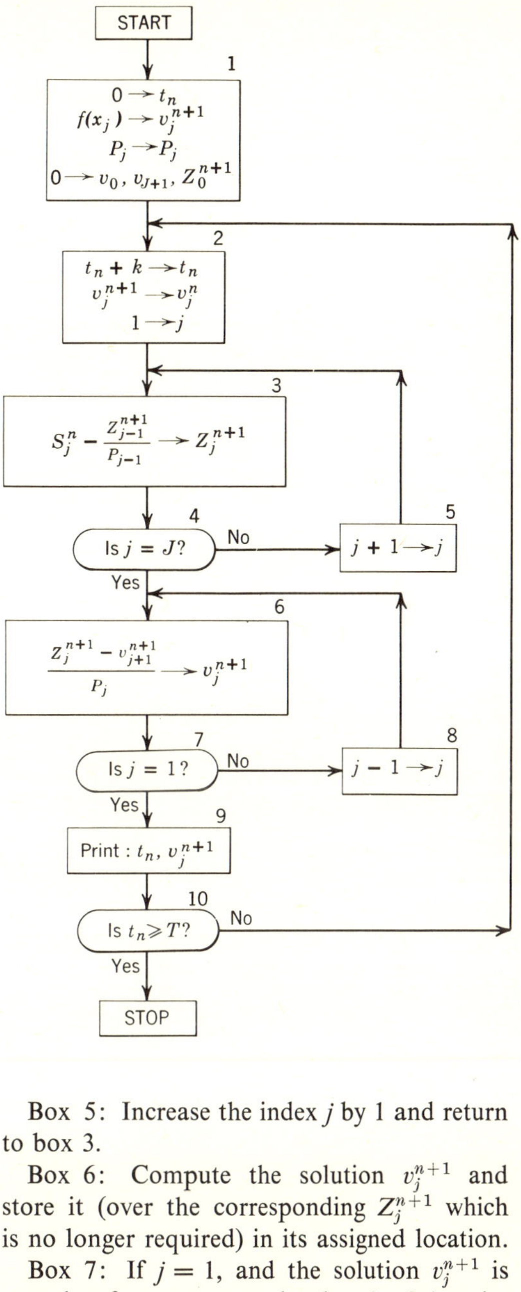

5. DESCRIPTION OF THE FLOW CHART

Box 1: The time is set to zero. The initial data (previously computed or computed by a subroutine) is stored in the v_j^{n+1} locations. The constant quantities P_j (previously computed or computed by a subroutine) are stored in their assigned locations. Zero is set into all locations where v_0 and v_{J+1} are stored (i.e., for t_n and t_{n+1}) as well as the location assigned to Z_0^{n+1}.

Box 2: The time is incremented by k. The quantities v_j^{n+1} are stored over the corresponding v_j^n. The counter j is set to one.

Box 3: The inhomogeneous term S_j^n [defined in (26*b*)] is evaluated and used to compute the Z_j^{n+1} [defined in (27*b*)] which may then be stored in the location assigned to v_j^{n+1}.

Box 4: If $j = J$, and all the Z_j^{n+1} have been computed, proceed to box 6. Otherwise go to box 5.

Box 5: Increase the index j by 1 and return to box 3.

Box 6: Compute the solution v_j^{n+1} and store it (over the corresponding Z_j^{n+1} which is no longer required) in its assigned location.

Box 7: If $j = 1$, and the solution v_j^{n+1} is complete for t_{n+1}, proceed to box 9. Otherwise go to box 8.

Box 8: Reduce the index j by one and return to box 6.

Box 9: Print the time t_n and solution v_j^{n+1} corresponding to it. (The printing should probably be done infrequently or in some problems only at the end.)

Box 10: If t_n is equal to or has exceeded the maximum time T for which the solution is desired, stop. Otherwise return to box 2.

6. SUBROUTINES

None.

7. SAMPLE PROBLEM

Take, in (25),

$$a \equiv \tfrac{1}{2}, \qquad f(x) \equiv x(1 - x)$$

and choose a net with

$$J = 3, \qquad \lambda = 2 \qquad (\text{i.e., } h = \tfrac{1}{4}, k = \tfrac{1}{8})$$

The computations yield, recalling that $v_0^n = v_4^n = 0$:

		$j = 1$	$j = 2$	$j = 3$	
Initial comp.	P_j	-4	$-\frac{15}{4}$	$-\frac{56}{15}$	$t = 0$
	v_j^0	$\frac{3}{16}$	$\frac{1}{4}$	$\frac{3}{16}$	
First time step	S_j^0	$-\frac{1}{4}$	$-\frac{3}{8}$	$-\frac{1}{4}$	$t = \frac{1}{8}$
	Z_j^1	$-\frac{1}{4}$	$-\frac{7}{16}$	$-\frac{11}{30}$	
	v_j^1	$\frac{11}{112}$	$\frac{1}{7}$	$\frac{11}{112}$	
Second time step	S_j^1	$-\frac{1}{7}$	$-\frac{11}{56}$	$-\frac{1}{7}$	$t = \frac{1}{4}$
	Z_j^2	$-\frac{1}{7}$	$-\frac{13}{56}$	$-\frac{43}{210}$	
	v_j^2	$\frac{43}{784}$	$\frac{15}{196}$	$\frac{43}{784}$	

8. MEMORY REQUIREMENTS

The storage allocations required for this program are, for the indicated quantities:

v_j^n	$J + 2$ words
P_j	J words
Z_j^{n+1}, v_j^{n+1}	$J + 2$ words
$(k, t_n, T, J, 1)$	5 words
Program	About 90 words (exclusive of print routine)

The total requirement is thus of the order of

$$[3J + 100] \text{ words}$$

The storage requirements can in fact be reduced to the order of $[2J + 100]$ words since all the v_j^n do not have to be retained when computing the S_j^n. This is accomplished by the standard technique of temporarily saving v_j^n in a fixed location and using it to compute S_{j+1}^n. The required modifications to the flow diagram are obvious but rather inconvenient to display.

9. ESTIMATION OF THE RUNNING TIME

The number of arithmetic operations used for one time step, i.e., from box 2 through box 9, is $2J$ divisions, J multiplications, and $4J$ additions. Thus, if ν is the addition time and μ is the multiplication and division time for the machine, one time step then requires about

$$(4\nu + 3\mu)J$$

units of machine time. For the total problem, exclusive of the generation of the initial data and P_j which is done only once, we require about

$$(4\nu + 3\mu)JT/k = (4\nu + 3\mu)J(J + 1)^2(T/\lambda)$$

units of machine time. These estimates should be doubled to take liberal account of the "bookkeeping" time required in the program.

10. REFERENCES

1. R. Courant, K. O. Friedrichs, and H. Lewy, Über die partiellen Differenzengleichungen der mathematischen Physik, *Math. Ann.*, vol. 100, 1928, pp. 32–74.
2. E. Bodewig, *Matrix Calculus*, Interscience Publishers, New York, 1956.
3. J. Douglas, On the Relation between Stability and Convergence in the Numerical Solution of Linear Parabolic and Hyperbolic Differential Equations, *J. Soc. Indust. Appl. Math.*, vol. 4, 1956, pp. 20–37.
4. P. D. Lax, and R. D. Richtmyer, Survey of the Stability of Linear Finite Difference Equations, *Comm. Pure Appl. Math.*, vol. 9, 1956, pp. 267–293.
5. J. Crank and P. Nicholson, A Practical Method for Numerical Evaluation of Solutions of Partial Differential Equations of the Heat-Conduction Type, *Proc. Cambridge Phil. Soc.*, vol. 43, 1947, pp. 50–67.
6. G. G. O'Brien, M. A. Hyman, and S. Kaplan, A Study of the Numerical Solution of Partial Differential Equations, *J. Math. Physics*, vol. 29, 1951, pp. 223–251.

7. R. D. Richtmyer, *Difference Methods for Initial-Value Problems*, Interscience Pub. Co., New York, 1958.
8. J. Douglas, The Solution of the Diffusion-Equation by a High Order Correct Difference Equation, *J. Math. Phys.*, vol. 35, 1956, pp. 145–151.
9. M. L. Juncosa and D. Young, On the Convergence of a Solution of a Difference Equation to a Solution of the Equation of Diffusion, *Proc. Amer. Math. Soc.*, vol. 5, 1954, pp. 168–174.
10. W. Wasow, On the Accuracy of Implicit Difference Approximations to the Equation of Heat Flow, U.S. Army Math. Research Center, Univ. Wisc., Tech. Summary Report No. 2, 1957.
11. F. John, On Integration of Parabolic Differential Equations by Difference Methods, *Comm. Pure Appl. Math.*, vol. 5, 1952, pp. 155–211.
12. J. Douglas, On the Numerical Integration of Quasi-Linear Parabolic Differential Equations, *Pacific J. Math.*, vol. 6, 1956, pp. 35–42.
13. M. Lees, Approximate Solutions of Parabolic Equations, *J. Soc. Indust. Appl. Math.*, vol. 7, 1959, pp. 167–183.
14. M. E. Rose, On the Integration of Non-Linear Parabolic Equations by Implicit Difference Methods, *Quart. Appl. Math.*, vol. 14, 1956, pp. 237–248.

Iterative methods for the solution of elliptic partial differential equations

13

J. W. Sheldon
Computer Usage Company, Inc.

I. FUNCTION

Iterative methods may be employed to solve systems of linear algebraic equations which arise when certain elliptic partial differential systems are replaced by systems of difference equations.

Suppose we are given a plane region G interior to a simple closed curve C. Let it be required to find the solution, z^*, of the elliptic partial differential system:

$$-\frac{\partial}{\partial x}\left(D\frac{\partial z^*}{\partial x}\right) - \frac{\partial}{\partial y}\left(D\frac{\partial z^*}{\partial y}\right) + az^* - S = 0 \text{ in } G \quad (1)$$

$$z^* - \sigma = 0 \text{ on } C \quad (2)$$

D, a, S, and σ are given functions of x, y. $D > 0$ and $a \geq 0$.

We may approximate the conditions of (1) and (2) by a system of algebraic equations. We overlay G with a rectangular x, y network choosing coordinates or "mesh lines" $x_0, x_1, \cdots, x_i, \cdots, x_I$; $y_0, y_1, \cdots, y_j, \cdots, y_J$. Let $z_{i,j} = z(x_i, y_j)$, $a_{i,j} = a(x_i, y_j)$, $S_{i,j} = S(x_i, y_j)$, $\sigma_{i,j} = \sigma(x_i, y_j)$. Let $D_{i+\frac{1}{2},j} = D[\frac{1}{2}(x_i + x_{i+1}, y_j]$ and $D_{i,j+\frac{1}{2}} = D[x_i, \frac{1}{2}(y_j + y_{j+1})]$. We approximate the derivatives occurring in (1) as follows:

$$\left(D\frac{\partial z}{\partial x}\right)_{i+\frac{1}{2},j} \doteq D_{i+\frac{1}{2},j}\frac{z_{i+1,j} - z_{i,j}}{x_{i+1} - x_i}$$

$$\left(D\frac{\partial z}{\partial y}\right)_{i,j+\frac{1}{2}} \doteq D_{i,j+\frac{1}{2}}\frac{z_{i,j+1} - z_{i,j}}{y_{j+1} - y_j}$$

$$\frac{\partial}{\partial x}\left(D\frac{\partial z}{\partial x}\right)_{i,j} \doteq \frac{2}{x_{i+1} - x_{i-1}}\left(D_{i+\frac{1}{2},j} \times \frac{z_{i+1,j} - z_{i,j}}{x_{i+1} - x_i} - D_{i-\frac{1}{2},j}\frac{z_{i,j} - z_{i-1,j}}{x_i - x_{i-1}}\right)$$

$$\frac{\partial}{\partial y}\left(D\frac{\partial z}{\partial y}\right)_{i,j} \doteq \frac{2}{y_{j+1} - y_{j-1}}\left(D_{i,j+\frac{1}{2}} \times \frac{z_{i,j+1} - z_{i,j}}{y_{j+1} - y_j} - D_{i,j-\frac{1}{2}}\frac{z_{i,j} - z_{i,j-1}}{y_j - y_{j-1}}\right)$$

If $p_{i,j}$ is a quantity associated with a mesh point interior to C, we write $p_{i,j} \in G$. If $p_{i,j}$ is a quantity associated with a mesh point on C, we write $p_{i,j} \in C$. We assume that C is a closed polygon composed of mesh lines. Then, replacing the derivatives in (1) and (2) by their approximations, we obtain the system of equations:

$$f_{i,j} = \frac{-2}{x_{i+1} - x_{i-1}}\left(D_{i+\frac{1}{2},j}\frac{z_{i+1,j} - z_{i,j}}{x_{i+1} - x_i} - D_{i-\frac{1}{2},j}\frac{z_{i,j} - z_{i-1,j}}{x_i - x_{i-1}}\right) - \frac{2}{y_{j+1} - y_{j-1}} \times \left(D_{i,j+\frac{1}{2}}\frac{z_{i,j+1} - z_{i,j}}{y_{j+1} - y_j} - D_{i,j-\frac{1}{2}}\frac{z_{i,j} - z_{i,j-1}}{y_j - y_{j-1}}\right) + a_{i,j}z_{i,j} - S_{i,j} = 0, \qquad f_{i,j} \in G \quad (3)$$

$$z_{i,j} - \sigma_{i,j} = 0, \qquad z_{i,j} \in C \quad (4)$$

$z_{i,j}$ is the solution of (3) and (4); $z^*_{i,j}$ the solution of (1) and (2) at $x = x_i$, $y = y_j$. Under appropriate assumptions on the regularity of D, a, S, σ, and C it can be established that $z_{i,j} \to z^*_{i,j}$, $(i,j) \in G$, when $x_{i+1} - x_i$, $y_{j+1} - y_j$ tend to zero for all $(i,j) \in G$. This question of *truncation error* is not discussed here (see Gerschgorin [1], for example). When C does not fall along mesh lines a special treatment is necessary (see [2], for example).

Using (4), $z_{i,j} \in C$ can be eliminated from (3), so that (3) becomes a system of N simultaneous equations for the N unknown values of $z_{i,j}$ interior to C. The matrix of coefficients of (3) becomes symmetric when (3) is multiplied by $(x_{i+1} - x_{i-1})(y_{j+1} - y_{j-1})$. Let the matrix of coefficients of these symmetric equations be A. Then the equations we wish to solve may be stated

$$Az - d = 0 \quad (5)$$

where A is a square symmetric matrix of order N, and d and z are column vectors.

In the present chapter our main concern is with iterative methods for solving (5), not with procedures for obtaining A such that A properly represents a differential operator. For the latter, see [1].

2. MATHEMATICAL DISCUSSION

Following Forsythe [3], we define *linear iterative processes of first degree*,

$$z^{(p)} = H_p z^{(p-1)} + M_p d, \qquad p = 1, 2, \cdots \quad (6)$$

where H_p and M_p are independent of d and where

$$H_p + M_p A = I \quad (7)$$

Equation (7) ensures that $z = A^{-1}d$ satisfies (6). To prove this, substitute $A^{-1}d$ for $z^{(p)}$ and $z^{(p-1)}$ in (6):

$$A^{-1}d \stackrel{?}{=} H_p A^{-1}d + M_p d \quad (8)$$

Since A, H_p, and M_p are independent of d,

$$A^{-1} \stackrel{?}{=} H_p A^{-1} + M_p \quad (9)$$

But the equality in (9) follows from (7) if we multiply (7) by A^{-1} on the right. Q.E.D.

Equation (6) is called a linear iterative process of *first degree* because $z^{(p)}$ is linearly related to $z^{(p-1)}$ and not to $z^{(p-1)}, z^{(p-2)}, \cdots$.

Let $e^{(p)} = z^{(p)} - z$. Then

$$e^{(p)} = H_p e^{(p-1)} \quad (10)$$

Equation (10) is obtained when we subtract the identity $z = H_p z + M_p d$ from (6). Let $K_p = \prod_{q=1}^{p} H_q$. Then from (10)

$$e^{(p)} = K_p e^{(0)} \quad (11)$$

As a measure of the *error* $e^{(p)}$, we use the Euclidean norm. If a is a column vector with real elements and a' the corresponding row vector, then the Euclidean norm of a, $\|a\|$, is

$$\|a\| = (a'a)^{\frac{1}{2}}$$

In the following we shall have occasion to make use of the *spectral norm* and the *spectral radius*. If A is a matrix the *spectral radius* of A, $\rho(A)$, is the magnitude of the characteristic roots of A of largest magnitude. The *spectral norm*, $\tau(A)$, is equal to the square root of the spectral radius of $A'A$, i.e., $\tau(A) = [\rho(A'A)]^{\frac{1}{2}}$, where A' is the transpose of A. The use of the terms spectral norm and spectral radius is not consistent in the literature. Here we are following Householder [4].

We shall use as a measure of the effectiveness of an iterative method

$$\max \frac{\|e^{(p)}\|}{\|e^{(0)}\|}$$

i.e., given p iterations in a particular iterative sequence (6), and given an arbitrary starting distribution, what is the maximum possible value of the ratio of the Euclidean norm of the error after p iterations to the Euclidean norm of the starting error?

THEOREM 1: Let A be a square matrix with real elements and b a vector with real elements. Then

$$\max \frac{\|Ab\|}{\|b\|} = \tau(A) \tag{12}$$

Proof: For any b

$$\frac{\|Ab\|^2}{\|b\|^2} - \tau^2(A) = \frac{b'A'Ab}{b'b} - \tau^2(A)$$
$$= \frac{b'[A'A - (\tau^2)I]b}{b'b} \leq 0$$

since the eigenvalues of $A'A - (\tau^2)I$ are nonpositive and, therefore, $A'A - (\tau^2)I$ is negative semidefinite. Since equality occurs when b is the eigenvector of $A'A$ corresponding to $\rho(A'A) = \tau^2(A)$, the theorem is proved.

Applying Theorem 1 to (11),

$$\max \frac{\|e^{(p)}\|}{\|e^{(0)}\|} = \tau(K_p)$$

Different iterative methods are characterized by different sequences of H_p. There will of course be an endless number of choices for H_p and M_p which satisfy (7). A popular subclass of iterative methods is *stationary* [3]. A *stationary* iterative process is one for which

$$H_1 = H_2 = \cdots = H \tag{13}$$

For stationary processes,

$$K_p = H^p \qquad \text{and} \qquad \tau(K_p) = \tau(H^p) \tag{14}$$

A second group of iterative methods consists in making *linear accelerations* to stationary iterative processes [3]. Let H, M be the iteration matrices for a stationary iteration process. Then

$$z^{(p)} = Hz^{(p-1)} + Md, \qquad p = 1, 2, \cdots$$
$$H + MA = I \tag{15}$$

Now consider the iterative process

$$\zeta^{(p)} = Hz^{(p-1)} + Md \tag{16}$$
$$z^{(p)} = \zeta^{(p)} + \omega_p(\zeta^{(p)} - z^{(p-1)}), \qquad p = 1, 2, \cdots \tag{17}$$

Equation (17) is called a *linear acceleration.* ω_p is an acceleration factor. Eliminating $\zeta^{(p)}$ from (16) and (17), we find that (16) and (17) are equivalent to a linear iterative process of first degree with

$$H_p = (1 + \omega_p)H - \omega_p I \tag{18}$$
$$M_p = (1 + \omega_p)M \tag{19}$$

Hence, from the definition of K_p,

$$K_p = \prod_{q=1}^{p} [(1 + \omega_q)H - \omega_q I]$$

Let

$$\omega_q = \frac{\Lambda_q}{1 - \Lambda_q} \tag{20}$$

Then

$$K_p = \prod_{q=1}^{p} \left(\frac{H - \Lambda_q I}{1 - \Lambda_q}\right) = Q_p \tag{21}$$

Equation (21) shows that when a stationary iterative process is subjected to a linear acceleration, the corresponding K_p becomes a polynomial $Q_p(H)$, of degree p. The roots of the polynomial $Q_p(x)$, where x is a scalar, are related to the acceleration factors by (20).

Note that from (21) $Q_p(x)$ must satisfy a "boundary condition"

$$Q_p(1) = 1 \tag{22}$$

In order to obtain most rapid convergence, we wish to choose the acceleration factors ω_p so as to minimize the spectral norm of K_p. It is shown, for H symmetric, in Theorem 3 below that

$$\tau(K_p) = \max_i |Q_p(\lambda_i)|$$

where the λ_i are the characteristic values of H. Let us assume that the λ_i are real and lie in a range $x_0 \leq \lambda_i \leq x_1 < 1$. Evidently we should choose the polynomial $Q_p(x)$, subject to (22), to have *minimum maximum magnitude* for $x_0 \leq x \leq x_1 < 1$. Then it is known ([5], [6], [7]) that

$$Q_p(x) = \frac{T_p(ax + b)}{T_p(a + b)} \tag{23}$$

where

$$a = \frac{2}{x_1 - x_0} \tag{24}$$

$$b = -\frac{x_1 + x_0}{x_1 - x_0} \tag{25}$$

and $T_p(\xi)$, where $\xi = ax + b$, is the Chebyshev polynomial of the first kind. $T_p(\xi)$ may be defined by the recursion formulas [8]

$$T_0(\xi) = 1$$
$$T_1(\xi) = \xi \tag{26}$$
$$T_p(\xi) = 2\xi T_{p-1}(\xi) - T_{p-2}(\xi), \qquad p > 1$$

It is also known ([5], [6], [7]) that with $Q_p(x)$ given by (23),

$$\max |Q_p(x)| = \frac{1}{T_p(a + b)} \qquad x_0 \leq x \leq x_1 < 1 \tag{27}$$

Now suppose we wish to construct an iterative process for which

$$K_p(H) = Q_p(H) = \frac{T_p(aH + bI)}{T_p(a + b)} \tag{28}$$

One way of doing this is to choose p and then compute the factors ω_q necessary for a linear acceleration process (16) and (17). The roots of the Chebyshev polynomial are known so that there is no special difficulty in doing this. The procedure is discussed by Shortley [6] and Young [7] as well as by others. When $a + b$ is close to one and p is large, some of the acceleration factors become large. This can make the iterative process become unstable because of rounding error. This difficulty can be obviated by replacing (17) by an acceleration of *second degree*. We make use of (26) to derive the second-degree acceleration equations.

Let

$$\alpha_p = \frac{T_{p-1}(a + b)}{T_p(a + b)} \tag{29}$$

$$\beta_p = \frac{T_{p-2}(a + b)}{T_p(a + b)} \tag{30}$$

Then, using (23) and (26),

$$Q_0 = \frac{T_0(aH + bI)}{T_0(a + b)} = I$$

$$\begin{aligned} Q_1 &= \frac{T_1(aH + b)}{T_1(a + b)} = \frac{aH}{a + b} + \frac{b}{a + b} I \\ &= a\frac{T_0(a + b)}{T_1(a + b)} H + b\frac{T_0(a + b)}{T_1(a + b)} I \\ &= a\alpha_1 H + b\alpha_1 I \end{aligned} \tag{31}$$

$$\begin{aligned} Q_p &= \frac{T_p(aH + bI)}{T_p(a + b)} \\ &= \frac{2(aH + bI)T_{p-1}(aH + bI)}{T_p(a + b)} \\ &\qquad - \frac{T_{p-2}(aH + bI)}{T_p(a + b)} \\ &= 2(aH + bI)\frac{T_{p-1}(a + b)}{T_p(a + b)} \\ &\quad \times \frac{T_{p-1}(aH + bI)}{T_{p-1}(a + b)} - \frac{T_{p-2}(a + b)}{T_p(a + b)} \\ &\quad \times \frac{T_{p-2}(aH + bI)}{T_{p-2}(a + b)} \\ &= 2(aH + bI)\alpha_p Q_{p-1} - \beta_p Q_{p-2}, \\ &\qquad\qquad p > 1 \end{aligned} \tag{32}$$

Equations (31) and (32) show us how to construct our iterative process of second degree. In place of (16) and (17) we have

$$\zeta^{(1)} = Hz^{(0)} + Md \tag{33}$$

$$z^{(1)} = a\alpha_1\zeta^{(1)} + b\alpha_1 z^{(0)} \tag{34}$$

$$\zeta^{(p)} = Hz^{(p-1)} + Md \tag{35}$$

$$z^{(p)} = 2a\alpha_p\zeta^{(p)} + 2b\alpha_p z^{(p-1)} - \beta_p z^{(p-2)}, \quad p > 1 \tag{36}$$

The iteration matrix K_p corresponding to the iterative sequence (33)–(36) is $Q_p(H)$ defined by (28). To prove this, consider the homogeneous system corresponding to (33)–(36). This is equivalent to setting $Md = 0$. Then eliminate $\zeta^{(1)}$ from (34) using (33). It follows that $z^{(1)} = Q_1 z^{(0)}$ from (31). Similarly, eliminate $\zeta^{(p)}$ from (36) using (35). It follows that

$$z^{(p)} = (2a\alpha_p H + 2b)\alpha_p z^{(p-1)} - \beta_p z^{(p-2)}$$

But if $z^{(p-1)} = Q_{p-1}z^{(0)}$ and $z^{(p-2)} = Q_{p-2}z^{(0)}$, then $z^{(p)} = Q_p z^{(0)}$ from (32). Thus the proof follows by induction if we note that $z^{(p-1)}$ does equal $Q_{p-1}z^{(0)}$ and $z^{(p-2)}$ does equal $Q_{p-2}z^0$ for $p = 2$.

The iterative sequence (33)–(36) has two important advantages over that of (16) and (17) for producing a "Chebyshev acceleration." The first is that the coefficients α_p and β_p use ratios of Chebyshev polynomials. These remain bounded and in fact approach constant values as $p \to \infty$. The second advantage is that $K_p = [T_p(aH + bI)]/[T_p(a + b)]$ for *every* p, using (33)–(36), whereas using (16) and (17) it is possible to make $K_p = T_p(aH + bI)/T_p(a + b)$ only for a *preassigned* value of p. In other words, if the Λ_q in (21) are roots of $T_{p^*}(ax + b)$, K_p will be the ratio of two Chebyshev polynomials given by (23) *only* when $p = p^*$. The α_p and β_p are defined by (29) and (30). They may be computed by generating $T_p(a + b)$ each iteration, i.e.,

$$\begin{aligned} T_0(a + b) &= 1 \\ T_1(a + b) &= a + b \\ T_p(a + b) &= 2(a + b)T_{p-1}(a + b) \\ &\qquad - T_{p-2}(a + b), \qquad p > 1 \end{aligned}$$

The iterative process (33)–(36) is a "best strategy" technique in the sense of Stiefel [8]. A disadvantage of the second-degree acceleration is that storage for an extra iterate $z^{(p-2)}$ is required.

In evaluating the effectiveness of stationary and linearly accelerated iterative sequences it is generally assumed that we know something about the characteristic roots of H. What is the connection between the spectral norm of K_p, $\tau(K_p)$, and the characteristic roots of H? There exists a *similarity transformation* which transforms H to its *Jordan canonical form* J. (The properties of similarity transformations and the Jordan form are treated by matrix theory. See [9], for example. One of the most important properties is, of course, that characteristic roots are preserved under a similarity transformation.) Let

$$\begin{aligned} J &= V^{-1}HV \\ H &= VJV^{-1} \end{aligned} \tag{37}$$

Since $H^2 = VJV^{-1}VJV^{-1} = VJ^2V^{-1}$, by induction it follows that $H^p = VJ^pV^{-1}$, and since $Q_p(H)$ is a sum of powers of H, $Q_p(H) = VQ_p(J)V^{-1}$. It is also an elementary property that $Q_p(H') = Q_p'(H)$. Then

$$\begin{aligned} K_p'K_p &= Q_p'(H)Q_p(H) \\ &= (VQ_p(J)V^{-1})'(VQ_p(J)V^{-1}) \\ &= V^{-1'}Q_p(J')V'VQ_p(J)V^{-1} \end{aligned} \tag{38}$$

Premultiply (38) by V' and postmultiply by V'^{-1}. This preserves the relationship of similarity [9]. Then note that $V^{-1}V'^{-1} = (V'V)^{-1}$. Then (38) becomes

$$K_p'K_p \cong Q_p(J')V'VQ_p(J)(V'V)^{-1}$$

The symbol $\cong$ stands for "is similar to."

THEOREM 2: If H is symmetric and the iterative process is stationary, then

$$\tau(K_p) = [\rho(H)]^p$$

Proof: If H is symmetric, V is an orthogonal matrix so that $V'V = I$ and J is diagonal [9]. Then

$$K_p'K_p \cong J^{2p}$$

The spectral radius of J^{2p} is $[\rho(H)]^{2p}$. Q.E.D.

THEOREM 3: If H is symmetric, and the iterative process is linearly accelerated, then $\tau(K_p) = \max_i |Q_p(\lambda_i)|$, where the λ_i are the characteristic roots of H.

Theorem 3 follows from the fact that $K_p'K_p \cong [Q_p(J)]^2$, and that the diagonal elements of the diagonal matrix $[Q_p(J)]^2$ are $[Q_p(\lambda_i)]^2$.

It does *not* follow *in general* that $\tau(K_p) = [\rho(H)]^p$ as $p \to \infty$ for all stationary iterative processes, as is sometime assumed. This proposition need not be true *even if* H has a single characteristic value of largest magnitude. Hence if H is not symmetric, it is important to estimate $\tau(K_p)$ as well as $\rho(H)$. Our success in doing this will usually depend on whether we can construct $V'V$ and $(V'V)^{-1}$. It is known [4] that if $\rho(H) < 1$ then $\lim_{p\to\infty} \tau(H^p) = 0$.

We now consider some specific iterative methods. We assume that the diagonal elements of A in (5) are positive. Let $B = I - D^{-r}AD^{-1+r}$, where D is a diagonal matrix of diagonal elements of A. Geiringer [10] has shown that the elements of the matrices belonging to algebraic approximations to elliptic partial differential equations often satisfy the condition

$$a_{ii} \geq \sum_{j\neq i} |a_{ij}| \tag{39}$$

with the *inequality* holding for at least one i. Let the exponent r in the definition of B be one. Then it follows from (39) that

$$\sum_j |b_{ij}| \leq 1 \tag{40}$$

where the b_{ij} are the elements of B. Equation (40) is sufficient to prove that $\rho(B) \leq 1$. This follows from Gerschgorin's theorem [11], which states that the spectral radius of a matrix is less than or equal to the largest magnitude absolute row sum of the matrix, where an absolute row sum is defined as the sum of the absolute values of the elements in a row. Actually if (39) is satisfied with the strict inequality for at least one i and if A is *irreducible*, as is usually the case, it can be shown that $\rho(B) < 1$. For the definition of irreducibility see Householder [4], for example.

Let $B_1 = I - D^{-r_1}AD^{-1+r_1}$ and $B_2 = I - D^{-r_2}AD^{-1+r_2}$. Let $r_2 = r_1 + \delta$. Then $B_2 = D^{-\delta}B_1D^{\delta}$, so that $B_2 \cong B_1$. This means that the roots of B are independent of the choice of r.

By a transformation, (5) may be transformed into a system in which the matrix of coefficients is symmetric and has ones on the diagonal. We let $A = D^{1/2}A^*D^{1/2}$, $z^* = D^{1/2}z$, and $d^* = D^{-1/2}d$. Then

$$A^*z^* = d^* \tag{41}$$

and A^* has ones on the diagonal.

Let $B^* = I - A^*$. $B^* = B$ when $r = \frac{1}{2}$, and so $\rho(B^*) = \rho(B) < 1$. Since $\rho(B^*) < 1$. B^* is a candidate for an iteration matrix, i.e., let $H \equiv B^*$. Then from (7), $B^* + M(I - B^*) = I$, from which it follows that $M = I$, so that (6) becomes

$$z^{*(p)} = B^* z^{*(p-1)} + d^*, \qquad p = 1, 2, \cdots \tag{42}$$

Alternatively, we may choose $r = 1$. Then from (15) $B + MD(I - B) = I$, from which it follows that $M = D^{-1}$, so that (6) becomes

$$z^{(p)} = Bz^{(p-1)} + D^{-1}d \tag{43}$$

Since B^* is symmetric while B is not, it is convenient to consider the iterative process in the form (42) rather than (43). However, (43) is usually more convenient for actual computing. The results we obtain for (42) will apply to (43) by transforming from the starred to the unstarred variables. In particular, we will have $e^* = D^{1/2}e$, whence

$$(e^{*\prime}e^*)^{1/2} = (e'De)^{1/2} \tag{44}$$

Equation (44) tells us that in measuring the error in the unstarred system we should weight the error components by the diagonal elements of A. Henceforth, we shall assume that we are in the starred system. For notational convenience we drop the stars, leaving them understood.

Equation (42) is usually called the Jacobi iteration process [12] or Richardson's method (unextrapolated) [13]. Since B is symmetric, Theorem 2 applies, and

$$\tau(K_p) = \tau(B^p) = [\rho(B)]^p \tag{45}$$

Equation (42) may be subjected to a linear acceleration process. The roots of B are real (because B is symmetric) and must lie between $\pm\rho(B)$. Then, using (24) and (25), $a = 1/[\rho(B)]$ and $b = 0$. The appropriate linearly accelerated process is that of (33)–(36) with these values of a and b and with $H \equiv B$, $M = I$. Then by Theorem 3

$$\tau(K_p) = \tau\left[\frac{T_p\left(\dfrac{B}{\rho(B)}\right)}{T_p\left(\dfrac{1}{\rho(B)}\right)}\right] = \max_i \frac{T_p\left(\dfrac{\lambda_i}{\rho(B)}\right)}{T_p\left(\dfrac{1}{\rho(B)}\right)} \leq \frac{1}{T_p\left(\dfrac{1}{\rho(B)}\right)}$$

Let $\rho(B) = 1 - \epsilon$, where ϵ is small. ϵ usually will be small if the order of A is large. [Cf. (68) following.] Then

$$\tau(B^p) = e^{p \log \rho(B)} \doteq e^{-p\epsilon}$$

We develop an approximate formula for

$$T_p(a + b) = T_p\left(\frac{1}{\rho(B)}\right) = T_p\left(\frac{1}{1 - \epsilon}\right) \doteq T_p(1 + \epsilon)$$

valid when ϵ is small and p is large. $T_p(\xi)$, $\xi > 1$, may be defined as [5]:

$$T_p(\xi) = \cosh\,(p \cosh^{-1} \xi)$$

From Dwight [14], formula 701,

$$\cosh^{-1} \xi = \log\,(\xi + \sqrt{\xi^2 - 1})$$

Then

$$\cosh^{-1}(1 + \epsilon) = \log\,(1 + \epsilon + \sqrt{2\epsilon - \epsilon^2}) \doteq \sqrt{2\epsilon}$$

Then

$$T_p(1 + \epsilon) \doteq \cosh p\sqrt{2\epsilon} \doteq \tfrac{1}{2}e^{p\sqrt{2\epsilon}} \tag{46}$$

Let $e^{-p_1\epsilon} = \frac{1}{2}e^{-p_2\sqrt{2\epsilon}}$. Then $p_1\epsilon \doteq p_2\sqrt{2\epsilon}$,

$$\frac{p_2}{p_1} \doteq \sqrt{\frac{\epsilon}{2}} \tag{47}$$

Equation (47) means that a factor of approximately $\sqrt{\epsilon/2}$ fewer iterations is required with the accelerated method. The advantage of the latter method in this case is obvious.

We now assume that B has a further property, namely that B has the form

$$B = \begin{pmatrix} 0 & R' \\ R & 0 \end{pmatrix} \tag{48}$$

where the zero entries are square matrices. A matrix which has the form (48) is said to be *cyclic of index* 2 [15]. A matrix which can be permuted by identical permutations of rows and columns to the form (48) is said by Young [16] to have property (A). The matrix still has property (A) if the zero entries are replaced diagonal matrices. Young [16] has shown that second-order elliptic self-adjoint partial differential equations may be replaced by difference equations in such a way that property (A) obtains.

We will demonstrate one way of writing equations (3) so that the corresponding B matrix has the form (48). Equations (3) are first symmetrized and normalized. Then we divide the equations $f_{i,j} = 0$ into two groups, those for which $i + j$ is even, followed by those for which $i + j$ is odd. This determines the row ordering of the matrix A. We then order the $z_{i,j}$ in exactly the same way as the $f_{i,j}$. This determines the column ordering of A. The resulting B matrix will have the form (48), as the reader may verify by considering an example. This ordering is discussed in more detail by Young [16] (who calls it the σ_1-ordering) and by Riley [17].

Let the components of B be b_{mn}. The successive overrelaxation sequence is [16]:

$$z_m^{(p)} = z_m^{(p-1)} + \omega\left(\sum_{n<m} b_{mn} z_n^{(p)} - z_m^{(p-1)} + \sum_{m>n} b_{mn} z_n^{(p-1)} + d_m\right) \tag{49}$$

ω is called the *overrelaxation factor*. If $\omega = 1$ we have the ordinary Gauss-Seidel method, and if $\omega > 1$ we are said to be applying "overrelaxation" [16].

Let E be the square matrix with elements equal to zero on and above the diagonal and equal to b_{mn} below the diagonal. Then (49) may be written

$$(I - \omega E)z^{(p)} = [(1 - \omega)I + \omega E']z^{(p-1)} + \omega d$$

Then

$$z^{(p)} = (I - \omega E)^{-1}[(1 - \omega)I + \omega E']z^{(p-1)} + (1 - \omega E)^{-1}\omega d \tag{50}$$

Hence the successive overrelaxation method with fixed ω is a stationary process with $H \equiv (I - \omega E)^{-1}[(1 - \omega)I + \omega E']$, $M = (I - \omega E)^{-1}\omega$.

Let us now assume that B has the form (48). Let the order of the zero matrix in the upper left corner of B be M and the order of zero matrix in the lower right corner be $\bar{M}$, where $M + \bar{M} = N$. Then

$$(I - \omega E)^{-1} = \begin{pmatrix} I_M & 0 \\ -\omega R & I_{\bar{M}} \end{pmatrix}^{-1} \tag{51}$$

$$(1 - \omega)I + \omega E' = \begin{pmatrix} (1 - \omega)I_m & \omega R' \\ 0 & (1 - \omega)I_{\bar{M}} \end{pmatrix} \tag{52}$$

I_M and $I_{\bar{M}}$ are unit matrices of order M and $\bar{M}$. Now

$$\begin{pmatrix} I_M & 0 \\ -\omega R & I_{\bar{M}} \end{pmatrix}^{-1} = \begin{pmatrix} I_M & 0 \\ \omega R & I_{\bar{M}} \end{pmatrix} \tag{53}$$

Equation (53) may be verified by multiplication. Then

$$H = \begin{pmatrix} I_M & 0 \\ \omega R & I_{\bar{M}} \end{pmatrix}\begin{pmatrix} (1 - \omega)I_M & \omega R' \\ 0 & (1 - \omega)I_{\bar{M}} \end{pmatrix}$$

$$= \begin{pmatrix} (1 - \omega)I_M & \omega R' \\ \omega(1 - \omega)R & (1 - \omega)I_{\bar{M}} + \omega^2 RR' \end{pmatrix} \tag{54}$$

Hereafter, H stands for the right-hand side of (54).

THEOREM 4: If μ is a characteristic value of B, then the two characteristic values λ_1, λ_2 of S, where S is the 2×2 matrix

$$S = \begin{pmatrix} 1 - \omega & \omega\mu \\ (1 - \omega)\omega\mu & \omega^2\mu^2 + 1 - \omega \end{pmatrix} \tag{55}$$

are characteristic values of H.

Proof: Let

$$Bh = \begin{pmatrix} 0 & R' \\ R & 0 \end{pmatrix}\begin{pmatrix} h_M \\ h_{\bar{M}} \end{pmatrix} = \mu h = \mu\begin{pmatrix} h_M \\ h_{\bar{M}} \end{pmatrix} \tag{56}$$

where h is a characteristic vector of B and where h_M and $h_{\bar{M}}$ are columns with M and $\bar{M}$ elements respectively. Equation (56) implies that

$$R'h_{\bar{M}} = \mu h_M$$
$$Rh_M = \mu h_{\bar{M}}$$
$$RR'h_{\bar{M}} = \mu^2 h_{\bar{M}}$$

Let a and b be scalars. Then

$$H\begin{pmatrix} ah_M \\ bh_{\bar{M}} \end{pmatrix} = \begin{pmatrix} (1 - \omega)I_M & \omega R' \\ (1 - \omega)\omega R & \omega^2 RR' + (1 - \omega)I_{\bar{M}} \end{pmatrix}\begin{pmatrix} ah_M \\ bh_{\bar{M}} \end{pmatrix}$$

$$= \begin{pmatrix} [(1 - \omega)a + \omega\mu b]h_M \\ [(1 - \omega)\omega\mu a + (\omega^2\mu^2 + 1 - \omega)b]h_{\bar{M}} \end{pmatrix}$$

$$= \begin{pmatrix} a'h_M \\ b'h_{\bar{M}} \end{pmatrix} \tag{57}$$

where

$$\begin{pmatrix} a' \\ b' \end{pmatrix} = S\begin{pmatrix} a \\ b \end{pmatrix}$$

Now let $\begin{pmatrix} ah_M \\ bh_{\bar{M}} \end{pmatrix}$ be a characteristic vector of H corresponding to a characteristic value λ. Then

$$\lambda\begin{pmatrix} a \\ b \end{pmatrix} = S\begin{pmatrix} a \\ b \end{pmatrix} \tag{58}$$

This establishes the theorem, since the determinant of coefficients of (58) must be zero to obtain a nontrivial solution for a, b.

THEOREM 5: If μ is a characteristic value of B and λ is a characteristic value of H, then $(\lambda + \omega - 1)^2 = \mu^2\omega^2\lambda$.

Proof: If λ is a root of S,

$$\begin{vmatrix} 1-\omega-\lambda & \omega\mu \\ (1-\omega)\omega\mu & \omega^2\mu^2+1-\omega-\lambda \end{vmatrix} = 0$$

hence

$$(1-\omega-\lambda)(\omega^2\mu^2+1-\omega-\lambda) - \omega^2\mu^2(1-\omega) = 0, \quad \text{or}$$

$$(\lambda+\omega-1)^2 - \omega^2\mu^2(\lambda+\omega-1) - \omega^2\mu^2(1-\omega) = 0, \quad \text{or}$$

$$(\lambda+\omega-1)^2 - \omega^2\mu^2\lambda = 0, \qquad \text{Q.E.D.}$$

COROLLARY: When $\omega = 1$, $\lambda = \mu^2$.

Theorem 5 was proved by Young [16] using a different and more general method which holds for all *consistent* orderings for the successive overrelaxation method. Young defines *consistent* orderings. Equation (48) is a special case. To display the dependence of H on the choice of ω we write $H(\omega)$. Then $\rho(H) = \rho[H(\omega)] = \rho(\omega)$. Young determined the value of ω, ω_b which minimizes $\rho(\omega)$ when the characteristic roots of B are real and $\rho(B) < 1$. Young obtained [16]

$$\omega_b = 1 + \left[\frac{\rho(B)}{1+[1-\rho^2(B)]^{1/2}}\right]^2 \tag{59}$$

$$\rho[H(\omega_b)] = \omega_b - 1 \tag{60}$$

For the ordering corresponding to (48) it is possible to obtain the spectral norm [18]. Let $\omega_b - 1 = r^2$.

THEOREM 6: Let l_+ be the larger root of

$$l^2 - \left[8p^2 + 4p^2\left(r^2 + \frac{1}{r^2}\right) + 2\right]l + 1 = 0$$

Then

$$\tau\{[H(\omega_b)]^p\} = l_+^{1/2}\{\rho[H(\omega_b)]\}^p$$

Since $r^2 + (1/r^2) \doteq 2$ when r is close to one, $l_+ \doteq 16p^2 + 2$. Hence for r close to one and p large, Theorem 6 tells us that the spectral norm of the successive overrelaxation method with $\omega = \omega_b$ goes to zero like

$$4p\{\rho[H(\omega_b)]\}^p$$

The corollary to Theorem 5 shows that the roots of $H(1)$ are real and nonnegative when the roots of B are real. Let the roots of $H(1)$ be λ_i. Then $0 \le \lambda_i \le \rho^2(B)$ from the corollary to Theorem 5. Then $x_0 = 0$, $x_1 = \rho^2(B)$, and from (24) and (25), $a = 2/\rho^2(B)$, $b = -1$. Equations (33)–(36) can be applied with these values of a and b, with $\omega = 1$, and with (33) and (35) representing (49).

THEOREM 7:

$$\tau(K_p) = \tau\left\{\frac{T_p\left[\dfrac{2H(1)}{\rho^2(B)} - I\right]}{T_p\left[\dfrac{2}{\rho^2(B)} - 1\right]}\right\} \doteq \frac{cp}{T_p\left(\dfrac{2}{\rho^2(B)} - I\right)}$$

where c is of order one and p is large.

Theorem 7 is proved by Sheldon [18]. Let $\rho(B) = 1 - \epsilon$ as before. Then $[2/\rho^2(B)] - 1 \doteq 1 + 4\epsilon$, so that, from (46),

$$\frac{1}{T_p\left(\dfrac{2}{\rho^2(B)} - 1\right)} \doteq 2e^{-2\sqrt{2\epsilon}p} \tag{61}$$

From (59) and (60),

$$\rho[H(\omega_b)] = \left[\frac{1-\epsilon}{1+[1-(1-\epsilon)^2]^{1/2}}\right]^2 \doteq 1 - 2\sqrt{2\epsilon}$$

so that

$$\{\rho[H(\omega_b)]\}^p \doteq e^{-2\sqrt{2\epsilon}p} \tag{62}$$

Comparing Theorem 6, Theorem 7, (61) and (62) we note that Chebyshev acceleration of $H(1)$ has a spectral norm not very different from a stationary process with $\omega = \omega_b$. Varga [12] has shown that no improvement can be obtained by applying a linear acceleration process to $H(\omega_b)$.

The acceleration of $H(1)$ may be improved by performing a single iteration without any acceleration before going into the Chebyshev acceleration process. In this case $Q_p(H)$ becomes

$$Q_p(H) = \frac{T_{p-1}\left[\dfrac{2H(1)}{\rho^2(B)} - 1\right]H(1)}{T_{p-1}\left[\dfrac{2}{\rho^2(B)} - 1\right]} \tag{63}$$

THEOREM 8: When Q_p is given by (63),

$$\tau[Q_p(H)] \doteq \frac{\sqrt{2}}{T_{p-1}\left(\dfrac{2}{\rho^2(B)} - 1\right)}$$

Theorem 8 is proved by Sheldon [18]. If $\rho(B) = 1 - \epsilon$, then, from (61),

$$T_p\left(\frac{2}{\rho^2(B)} - 1\right) \doteq \tfrac{1}{2}e^{2p\sqrt{2\epsilon}}$$

Hence by comparing with (46) we see that the iterative process described by (63) converges approximately twice as fast as that defined in connection with (36).

We have considered five iterative processes and have compared their spectral norms. The methods considered were:

(a) The Jacobi method; (42).

(b) The Chebyshev accelerated Jacobi method.

(c) The successive overrelaxation method, (49) with $\omega = \omega_b$, (59).*

(d) The Chebyshev accelerated successive overrelaxation method with $\omega = 1$.

(e) A modified Chebyshev accelerated successive overrelaxation method with $\omega = 1$. [Cf. (63)].

The actual use of the overrelaxation and recreational processes involves making an estimate of $\rho(B)$. This is often quite a difficult problem in itself. However, $\rho(B)$ is known for the Dirichlet problem in the rectangle with equal mesh spacing [13]. This problem is a special case of (3) and (4). We let $D = 1$, $a = 0$, $S = 0$, and C be a rectangle. Then we let $x_{i+1} - x_i = y_{j+1} - y_j = \Delta x$ for all $(i, j) \in G$. We assume that i goes from 1 to $M - 1$ in G, and j goes from 1 to $N - 1$. C falls on the mesh lines corresponding to $i = 0$, $i = M$; $j = 0$, $j = N$. Then (3) [after multiplying by $\frac{1}{4}(\Delta x^2)$] becomes

$$f_{i,j} = -\tfrac{1}{4}(z_{i+1,j} + z_{i,j+1} + z_{i-1,j} + z_{i,j-1}) + z_{i,j} = 0, \qquad f_{i,j} \in G \tag{64}$$

The matrix of (64) is symmetric and the diagonal elements are ones. Now we order the equations and $z_{i,j}$ values so that $i + j$ odd precedes $i + j$ even. Then the B matrix [(48)] is the matrix of coefficients of

$$\begin{aligned} &\tfrac{1}{4}(z_{i-1,j} + z_{i,j-1} + z_{i+1,j} + z_{i,j+1}), \quad (i + j) \text{ odd}, \ z_{i,j} \in G \\ &\tfrac{1}{4}(z_{i-1,j} + z_{i,j-1} + z_{i+1,j} + z_{i,j+1}), \quad (i + j) \text{ even}, \ z_{i,j} \in G \end{aligned} \tag{65}$$

$z_{i,j} \in C$ must be set to zero in (65), as we suppose $z_{i,j} \in C$ has been set to $\sigma_{i,j}$ in (64) and transferred to the *right-hand side* of (64) to form the inhomogeneous part. Then (56) becomes

$$\begin{aligned} &\tfrac{1}{4}(h_{i-1,j} + h_{i,j-1} + h_{i+1,j} + h_{i,j+1}) = \mu h_{i,j}, \quad (i + j) \text{ odd}, \ h_{i,j} \in G \\ &\tfrac{1}{4}(h_{i-1,j} + h_{i,j-1} + h_{i+1,j} + h_{i,j+1}) = \mu h_{i,j}, \quad (i + j) \text{ even}, \ h_{i,j} \in G \end{aligned} \tag{66}$$

In (66) $h_{i,j} = 0$ when $(i, j) \in C$. Now let

$$h_{i,j} = \sin\frac{m\pi i}{M}\sin\frac{n\pi j}{N} \tag{67}$$

We note that from (67) $h_{i,j} = 0$ when $(i, j) \in C$. Substituting (67) into (66) and making use of trigonometric addition formulas, we find that equations (66) are satisfied if

$$\mu = \frac{1}{2}\left(\cos\frac{m\pi}{M} + \cos\frac{n\pi}{N}\right) \tag{68}$$

Therefore these values of μ, for $m = 1, 2, \cdots, M - 1$; $n = 1, 2, \cdots, N - 1$, are the characteristic values of B. Then clearly

$$\rho(B) = \frac{1}{2}\left(\cos\frac{\pi}{M} + \cos\frac{\pi}{N}\right)$$

To get a feeling for the number of iterations required for a specific problem, let $M = N = 50$. Then

$$\rho(B) = \cos\frac{\pi}{50} \quad \text{and} \quad \epsilon \doteq \frac{\pi^2}{5000}$$

Suppose further that the Euclidean norm of the error is to be reduced by a factor of 10^{-3}. This

* Since writing this chapter, the author has shown that the spectral norm of the successive overrelaxation method with $\omega = \omega_b$ can be improved by performing a single iteration with $\omega = 1$ prior to iterating with $\omega = \omega_b$ [18]. The improvement is roughly the same as is obtained in going from method (d) to method (e). The proof assumes that the equations and unknowns have been ordered so that the matrix B has the form displayed in (48).

is adequate for most applications. Then we find for the unaccelerated Jacobi method, $p \leq 3500$; for the Chebyshev accelerated Jacobi method, $p \leq 120$; for successive overrelaxation with $\omega = \omega_b$, $p \leq 81$; for the modified Chebyshev accelerated successive overrelaxation with $\omega = 1, p \leq 63$.

We mention briefly some other results which have been obtained. Varga [15] has shown that if the B matrix is cyclic, then there is a generalization of Theorem 5 from which one may show that improvement in rate of convergence can be obtained by overrelaxation. By definition, if the B matrix is cyclic of index m it may be permuted by identical permutations of rows and columns to the form

$$\begin{bmatrix} 0 & L_1 & & & & \\ & 0 & L_2 & & & \\ & & \cdot & \cdot & & \\ & & & \cdot & \cdot & \\ & & & & \cdot & \cdot \\ & & & & & L_{M-1} \\ L_M & & & & & 0 \end{bmatrix}$$

where the zero entries are square matrices. The B matrix in (48) is cyclic of index 2. Finite difference approximations to (1) and (2) over a triangular network give B matrices which are cyclic of index 3 [15].

Varga [19] has studied the effect of choosing different orderings for the equations and unknowns in the successive overrelaxation scheme, (49). For large systems it appears to make little difference how the ordering is chosen, at least so far as *spectral radius* is concerned. Little is known about the effect of the ordering on the spectral norm, but Young [20] has observed empirically that different consistent orderings converge at approximately the same rate.

Equation (49) is called a "point relaxation" scheme because each $z_m^{(p)}$ can be obtained explicitly from a formula. In addition to point relaxation schemes, there are "block relaxation" schemes. In these, several unknowns are connected together in the iteration formula in such a way that the application of the formula involves solving simultaneous equations. Arms, Gates, and Zondek [21] have generalized the theory underlying successive point overrelaxation to apply to certain block relaxation schemes. "Line relaxation," in which all the $z_{i,j}$ across a row or column are treated as unknown and adjusted simultaneously, is a special block iterative method. Successive line overrelaxation has been shown to have a smaller spectral radius than the corresponding point overrelaxation scheme [21]. The degree of improvement, however, is often disappointingly small.

A method called the "alternating-direction-implicit" method has been studied by Wachspress and Sheldon and is described in [22], and has also been studied by Birkhoff and Varga [23]. The method is a line relaxation method in which the direction of the lines is alternated from "north-south" to "east-west" from iteration to iteration. The rate of convergence of this method has not been determined except for very specialized cases. However, the method has worked quite well on some specific problems [22].

3. SUMMARY OF THE CALCULATION PROCEDURE

As an example suppose we wish to solve the Dirichlet problem in the square using successive overrelaxation, with the equations ordered as indicated by (48). We summarize the calculation procedure as follows:

(a) We have an i, j network of storage locations, $0 \leq i \leq N$, $0 \leq j \leq N$. We start with the values of $\sigma_{i,j}$ (the boundary values) inserted in the corresponding i, j locations in the network. We start with an arbitrary set of values for $z_{i,j}$ at the i, j locations corresponding to interior points in the network. We set $\Sigma = 0$.

(b) We compute

$$(z'_{i,j} - z_{i,j}) = \omega[\tfrac{1}{4}(z_{i-1,j} + z_{i,j-1} + z_{i+1,j} + z_{i,j+1}) - z_{i,j}] \quad (69)$$

$$\Sigma' = \Sigma + (z'_{i,j} - z_{i,j})^2 \quad (70)$$

$$z'_{i,j} = z_{i,j} + (z'_{i,j} - z_{i,j}) \quad (71)$$

We compute (69), (70), and (71) first for points with $i + j$ odd and then for points with $i + j$ even. Σ' and $z'_{i,j}$ always replace Σ and $z_{i,j}$ as soon as computed.

(c) When (69), (70), and (71) have been

computed for all interior mesh points, we test to see if $\Sigma \leq \delta$, where δ is a preassigned constant. If so, we have converged. Otherwise we set Σ to zero and start step (b) over.

4. FLOW CHART

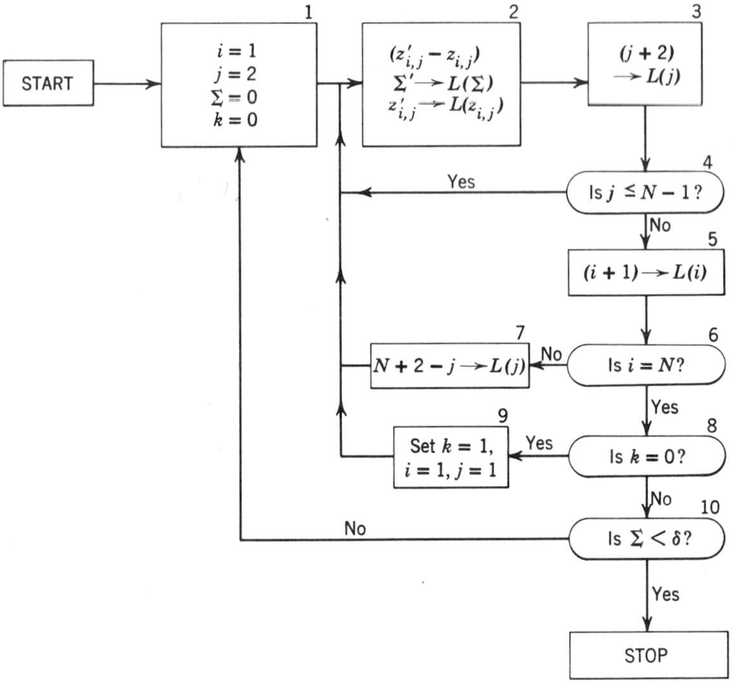

5. DESCRIPTION OF THE FLOW CHART

Box 1: Initialize "i-addresses" for $i = 1$. "j-addresses" for $j = 2$. Set Σ to zero and the index k to zero. k is an index to determine whether the sweep is for $i + j$ odd or $i + j$ even.

Box 2: Perform the computation described in step (b) of Section 3.

Box 3: Step up or index j-addresses by two. In boxes 2, 3, and 4 we are going across a row of the network computing $z'_{i,j}$ for $i + j$ even *or* odd.

Box 4: Have we completed a row?

Box 5: The row is complete so we step up i to go to the next row.

Box 6: Have we done the last row?

Box 7: We have not done the last row so we initialize j. The trick address computation ensures that we retain the evenness or oddness of $i + j$.

Box 8: We finished the last row but did we finish both sweeps?

Box 9: Initialize for the sweep with $i + j$ even. Set k equal to one.

Box 10: We have finished both sweeps. Test for convergence.

6. SAMPLE PROBLEM

Choose $N = 4$, $\sigma_{i,j} = 0$, and an arbitrary trial value $\omega = 1.96$. We choose the starting approximation as follows:

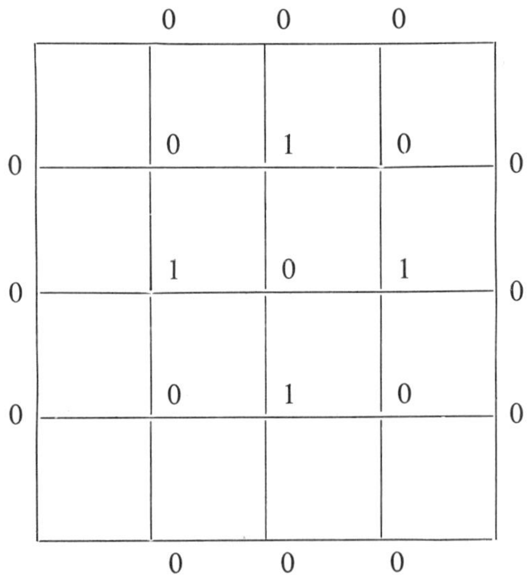

After the first iteration we obtain

	0	0	0	
0	−1.9208	−1.96	−1.9208	0
0	−1.96	−3.8916	−1.96	0
0	−1.9208	−1.96	−1.9208	0
	0	0	0	

We also compute $\|e^{(0)}\|$, $\|e^{(1)}\|$, although these quantities are not obtained in the computer. These quantities may be computed because we know that the solution for $z_{i,j}$ is $z_{i,j} = 0$.

$$\|e^{(0)}\| = 2 \qquad \text{and} \qquad \|e^{(1)}\| = 6.69941647$$

This illustrates that every iteration of the successive overrelaxation method does not necessarily reduce the error.

7. SUBROUTINES

None.

8. MEMORY REQUIREMENTS

$z_{i,j}$	$(N + 1)^2$ words
Σ, δ, ω	3 words

9. ESTIMATION OF THE RUNNING TIME

We have $3(N - 1)^2$ multiply/divide operations and $6(N - 1)^2$ additions. The total running time per iteration is then approximately

$$T = 3N^2(\mu + 2\nu)$$

where μ is the multiplication time and ν the addition time of the computer. If the computer has index registers, the time for logical overhead is practically nil.

The computation described in Sections (3)–(9) is a very simple case. If the boundary or the network is irregular, and if the coefficients must be computed from involved formulas, the running time may be much longer than given here and the program may be much more complicated.

10. REFERENCES

1. S. Gerschgorin, Fehlerabschatzung fur das Differenzenverfahren zur Losung partieller Differentialgleichungen, *Z. Angew. Math. Mech.*, vol. 10, 1930, pp. 373–382.
2. R. V. Viswanathan, Solution of Poisson's Equation by Relaxation Method–Normal Gradient Specified on Curved Boundaries, *MTAC*, vol. 11, 1957, pp. 67–78.
3. G. E. Forsythe, Solving Linear Algebraic Equations Can be Interesting, *Bull. Amer. Math. Soc.*, vol. 59, 1953, pp. 299–329.
4. A. S. Householder, The Approximate Solution of Matrix Problems, *J. Assoc. Comp. Mach.*, vol. 5, 1958, pp. 205–243.
5. D. Flanders and G. Shortley, Numerical Determination of Fundamental Modes, *J. Appl. Phys.*, vol. 21, 1950, pp. 1326–1332.
6. G. Shortley, Use of Tchebycheff Polynomial Operators in the Numerical Solution of Boundary Values Problems, *J. Appl. Phys.*, vol. 24, 1953, pp. 392–396.
7. D. Young, On Richardson's Method for Solving Linear Systems with Positive Definite Matrices, *J. Math. and Phys.*, vol. 32, 1954, pp. 243–255.
8. E. L. Stiefel, Kernel Polynomials in Linear Algebra and their Numerical Applications, *Nat. Bur. Standards Appl. Math. Series 49*, 1958, pp. 1–22.
9. S. Perlis, *Theory of Matrices*, Addison-Wesley, Cambridge, Mass., 1952, 237 pp.
10. H. Geiringer, On the Solution of Systems of Linear Equations by Certain Iteration Methods, *Reissner Anniversary Volume*, J. W. Edwards, Ann Arbor, Mich., 1949, pp. 365–393.
11. A. S. Householder, *Principles of Numerical Analysis*, McGraw-Hill Book Co., New York, 1953, p. 149.
12. R. S. Varga, A Comparison of the Successive Overrelaxation Method and Semi-Iterative Methods using Chebyshev Polynomials, *J. Soc. Indust. Appl. Math.*, vol. 5, 1957, pp. 39–45.
13. S. P. Frankel, Convergence Rates of Iterative Treatments of Partial Differential Equations, *MTAC*, vol. 4, 1950, pp. 65–75.
14. H. B. Dwight, *Tables of Integrals and Other Mathematical Data*, The Macmillan Co., New York, 1934.
15. R. S. Varga, *P*-cyclic Matrices: A Generalization of the Young-Frankel Successive Overrelaxation Method, Report WAPD-T-567, Bettis Plant, Westinghouse Electric Corp., Pittsburgh, Pa., Aug. 1957.

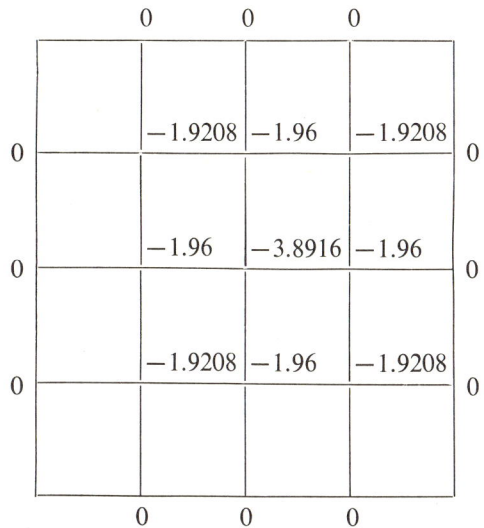

16. D. Young, Iterative Methods for Solving Partial Difference Equations of Elliptic Type, *Trans. Amer. Math. Soc.*, vol. 76, 1954, pp. 92–111.
17. J. D. Riley, Iteration Procedures for the Dirichlet Difference Problem, *MTAC*, vol. 8, 1954, pp. 125–131.
18. J. W. Sheldon, On the Spectral Norms of Several Iterative Processes, *J. Assoc. Comp. Mach.*, vol. 6, 1959, pp. 494–505.
19. R. S. Varga, Orderings of the Successive Over-relaxation Scheme, Report WAPD-T-788, Bettis Plant, Westinghouse Electric Corp., Pittsburgh, Pa., July 1958.
20. D. Young, Ordvac Solutions of the Dirichlet Problem, *J. Assoc. Comp. Mach.*, vol. 2, 1955, pp. 137–161.
21. R. J. Arms, L. D. Gates, Jr., and B. Zondek, A Method of Block Iteration, *J. Soc. Indust. Appl. Math.*, vol. 4, 1956, pp. 220–229.
22. E. L. Wachspress, CURE: A Generalized Two-Space-Dimension Multi-Group Coding for the IBM-704, Report KAPL-1724, Knolls Atomic Power Lab., General Electric Co., Schenectady, N.Y., April 30, 1957.
23. G. Birkhoff and R. S. Varga, Implicit Alternating Direction Methods, Report WAPD-T-650, Bettis Plant, Westinghouse Electric Corp., Pittsburgh, Pa., Oct. 1957.

A Monte Carlo method for the solution of elliptic partial differential equations

14

Carl N. Klahr
Technical Research Group

1. FUNCTION

The purpose of this chapter is to describe the use of the Monte Carlo method of statistical (or numerical) experimentation in solving elliptic partial differential equations. In particular, we consider the solution of Poisson's equation

$$\nabla^2 V(r) = F(r) \tag{1}$$

for the potential V as a function of the position vector r. The function $F(r)$ is assumed given within a closed boundary curve C (in two dimensions) or within a closed surface S (in three dimensions) which is specified. V assumes a given set of values $\phi(r)$ on the boundary and it is desired to find the values of $V(r)$ within the boundary (Dirichlet's problem).

The present discussion is restricted to two dimensions but the techniques presented are equally applicable to a higher dimensional space. Furthermore, one can solve the more general elliptic partial differential equation in two dimensions

$$\beta_{11}\frac{\partial^2 V}{\partial x^2} + 2\beta_{12}\frac{\partial^2 V}{\partial x\,\partial y} + \beta_{22}\frac{\partial^2 V}{\partial y^2} + 2\alpha_1\frac{\partial V}{\partial x} + 2\alpha_2\frac{\partial V}{\partial y} = F(x, y) \tag{2}$$

using the same procedures, in many cases.

The technique described here enables one to obtain the solution $V(P)$ at a single point P without obtaining the solution at any other points. In a context where the solution at one point or a few points only is required, e.g., finding the temperature at suspected hot spots, the Monte Carlo method may be particularly economical.

2. MATHEMATICAL DISCUSSION AND DERIVATION

The steps in the solution of (1) by the Monte Carlo method are the following:

(a) The region enclosed within C is replaced by a rectangular mesh of lattice points. The bounding curve C is replaced by a set of lattice points C_h which are nearest neighbors to mesh points within C, or are diagonally across from mesh points within C.

(b) Poisson's equation is rewritten as a difference equation. $\phi(r_h)$ is assumed given for the set C_h of lattice points. The difference equation relates values of the potential at adjacent points of the lattice. The coefficients in the difference equation are interpreted as transition probabilities from one point to its neighbors.

(c) To find the solution at any point P within C, construct a set of random walks of fictitious particles, the walks starting at P and terminating at the boundary points C_h. Steps in the walk from any point to an adjacent lattice point are taken by random sampling techniques using the transition probabilities derived from the difference equation. A tally is made for each walk, from which the solution $V(P)$ is estimated. The tally depends on the transition probabilities between successive points visited, the value of $F(x, y)$ at the points visited in the walk, and the values $\phi(r_h)$ at the boundary point terminating the walk.

These three steps are now considered in detail.

a. The Lattice Mesh

The replacement of the area bounded by C with a point lattice proceeds as follows: A rectangular lattice of points of the (x, y) plane is selected, consisting of vertices of squares of side h formed by the system of lines

$$x = x_0 + jh$$
$$y = y_0 + jh \qquad j = 0, \pm 1, \pm 2, \cdots$$

x_0 and y_0 are chosen for convenience and h is the mesh width. By choosing h sufficiently small any point in the area can be brought arbitrarily close to a lattice point. Note that a sequence of lattice points connecting nearest neighbors can always be found to proceed from any one point to any other. Each lattice point has four nearest neighbors.

The boundary C is replaced by a set of lattice points C_h that includes the nearest exterior neighbors of all points within C, and in addition, for all lattice points (x, y) within C, the diagonal exterior points $(x \pm h, y \pm h)$.

b. The Difference Equation

Poisson's equation can be approximated in difference equation form for this lattice, for example as follows:

$$\Delta_{xx}V + \Delta_{yy}V = F(x, y) \qquad (3)$$

where the differences $\Delta_{xx}V$ and $\Delta_{yy}V$ are symbols defined as

$$\Delta_{xx}V = \frac{V(x + h, y) + V(x - h, y) - 2V(x, y)}{h^2}$$

$$\Delta_{yy}V = \frac{V(x, y + h) + V(x, y - h) - 2V(x, y)}{h^2}$$

In writing the general elliptic equation (2) in difference form we shall follow the development of [1], where the following additional replacements of derivatives by differences are made:

$$\frac{\partial V}{\partial x} = \frac{V(x + h, y) - V(x, y)}{h}$$

$$\frac{\partial V}{\partial y} = \frac{V(x, y + h) - V(x, y)}{h}$$

$$\frac{\partial^2 V}{\partial x\, \partial y} = \frac{V(x + h, y + h) - V(x + h, y) - V(x, y + h) + V(x, y)}{h^2}$$

The point (x, y) is denoted by P. (Boundary points are denoted by Q.) The neighbors of P in counterclockwise order starting from $(x + h, y)$ are denoted by subscripts 1, 2, 3, 4, and the point $(x + h, y + h)$ by the subscript 5. Equation (3) can be rewritten in this notation as

$$V(P) = \tfrac{1}{4}\sum_{i=1}^{4} V(P_i) - \tfrac{1}{4}h^2[F(P)] \qquad (4)$$

Thus the transition probabilities are $\frac{1}{4}$ for a step to each of the nearest neighbors.

The general elliptic equation (2) in difference form reads as follows in this notation:

$$V(P) = \sum_{i=1}^{5} p_i(P)V(P_i) - \frac{h^2F(P)}{D(P)} \qquad (5)$$

where $p_i(P)$ are the transition probabilities at P

which are defined in terms of the coefficients of (2).

$$p_1(P) = \frac{1}{D}(\beta_{11} - 2\beta_{12} + 2h\alpha_1)$$

$$p_2(P) = \frac{1}{D}(\beta_{22} - 2\beta_{12} + 2h\alpha_2)$$

$$p_3(P) = \beta_{11}/D$$

$$p_4(P) = \beta_{22}/D$$

$$p_5(P) = 2\beta_{12}/D$$

$$D(P) = 2\beta_{11} + 2\beta_{22} - 2\beta_{12} + 2h(\alpha_1 + \alpha_2)$$

It is assumed that the $p_i(P)$ are all positive. Sufficient conditions for all the $p_i(P)$ to be nonnegative for an h sufficiently small are:

(i) $\beta_{12} = 0$.

(ii) $\beta_{11}\beta_{22} > 0$ (condition for elliptic equation).

The sum $\sum_{i=1}^{5} p_i(P) \equiv 1$ independent of the β_{ij} or the α_i.

In the following development the more general equation (5) will be considered since (4) is a special case of it, and the same procedures apply.

It can be shown ([1], Section 13) that as the mesh width h approaches zero the solution to the difference equation approaches the solution of the differential equation.

c. Random Walk Procedure

The development of the random walk procedure for solving (4) or (5) depends on the following theorem:

THEOREM: Let a particle start at a point P_0 of the lattice and perform a random walk on the lattice. The conditions of the walk are that if the particle is momentarily at the arbitrary point P, the probabilities of stepping to P_1, P_2, P_3, P_4, P_5 are $p_1(P)$, $p_2(P)$, $p_3(P)$, $p_4(P)$, $p_5(P)$, respectively. When the particle reaches a boundary point Q_i, the walk is terminated. Then the solution of (4) is the expected value of the tally Z_i (for the ith such walk)

$$Z_i = -\sum_j \tfrac{1}{4}h^2[F(P_j)] + \phi(Q_i) \qquad (6)$$

That is, $V(P_0)$ = average Z_i over many such walks. The summation on j is for all stopping points (including P_0 but not including Q_i) of the ith walk. For (5) the tally is

$$Z_i = -\sum_j \frac{h^2F(P_j)}{D(P_j)} + \phi(Q_i) \qquad (7)$$

In proving this theorem we may assume that all the $F(P)$'s are of the same sign, since particular solutions of linear differential equations are additive, and one can write $F(P)$ as the sum of a positive and a negative part. For convenience assume $-F(P)$ to be positive. Correspondingly, one can assume that the boundary values $\phi(Q)$ are all positive. Since increasing all the boundary values by a constant C merely adds C to the solution of (4) at each point, this assumption is also not restrictive.

To carry out a proof for (7) we first assume the walk to be limited to m steps at most. Let $W_m(P)$ be the mean value of

$$Z^*_{mi} = -\sum_{j=1}^{m} \frac{h^2F(P_j)}{D(P_j)} + \phi_m(Q_i) \qquad (7a)$$

where $\phi_m(Q_i) = \phi(Q_i)$ if the walk reaches the boundary within m steps
$= 0$ if the walk does not reach the boundary in m steps

The function $W_m(P)$ satisfies the following equation for each P in the lattice:

$$W_m(P) = \sum_{i=1}^{5} p_i(P)W_{m-1}(P_i) - \frac{h^2}{D(P)}F(P) \qquad (8)$$

The last term is added to $W_m(P)$ at every step, since there is a contribution from P according to the definition of $W_m(P)$ as the mean value of (7a). The first five terms represent the transitions possible in one step; i.e., the probability of moving from P to P_i is $p_i(P)$. The mean value of Z^*_{m-1} from P_i is $W_{m-1}(P_i)$. Thus the mean value of Z^*_m starting from P is $p_i(P)W_{m-1}(P_i)$, assuming the first move is to P_i. Since these five possible transitions on the first move are mutually exclusive and collectively exhaustive, one sums them up.

An equation of the form (8) is called a Chapman-Kolmogorov [2] equation. It has the property that if $W_{m+1} \geq W_m$ for all lattice points, the inequality will be true for all greater values of m, since

$$W_{m+1}(P) - W_m(P) = \sum_{i=1}^{5} p_i(P)[W_m(P_i) - W_{m-1}(P_i)] \quad (9)$$

by simple subtraction.

The sum (7*a*) implies that $W_1(P) \geq W_0(P)$. Hence the $W_m(P)$'s form a monotonic sequence in m. An upper bound on $W_m(P)$ is given from (7*a*) by

$$\phi_{\max} + \left[-\frac{h^2}{D}F\right]_{\max} \cdot \bar{k}_m$$

where $\phi_{\max}$ = maximum boundary value

$\left[-\frac{h^2}{D}F\right]_{\max}$ = maximum value of the bracketed quantity within the area bounded by C

k_n = number of steps to the termination of a walk with a maximum of n steps

$\bar{k}_n$ = mean value of k_n over all walks

It can be shown ([1], Sections 11 and 12) that $\bar{k}_n$ is uniformly bounded independent of n as long as the bounded region is finite in area. Hence the monotonic sequence $W_m(P)$ is bounded, and therefore it must have a limit. The limit is obtained by letting $m \to \infty$. In the limit, (8) is equivalent to (5) and the mean value of (7*a*) is equal to the mean value of (7). Since $W_m(P)$ is known to satisfy (8), it must satisfy (5) in the limit when (5) and (8) become identical. Hence the mean value of (7) is the solution to (5).

The solution $V(P)$ can be estimated from the tallies Z_i of N random walks by $\bar{V}$

$$\bar{V} = \frac{1}{N}\sum_{i=1}^{N} Z_i \quad (10)$$

The error of estimate is determined by the variance $\sigma^2(\bar{V})$

$$\sigma^2(\bar{V}) = \frac{1}{N}\sigma^2(Z_i) \quad (11)$$

The probability that $V(P)$ differs from V by a given amount can be roughly estimated from a normal distribution with variance $\sigma^2(\bar{V})$.

$\sigma^2(\bar{V})$ can be estimated from

$$\sigma^2(\bar{V}) = \frac{1}{N}\sum_{i=1}^{N} Z_i^2 - \bar{V}^2 \quad (12)$$

An a priori formula for $\sigma^2(Z)$ is

$$\sigma^2(Z) = -V^2(P) + \sum_{Q_i}\sum_{\substack{\text{paths}\\ \text{to } Q_i}} Z_i^2 \prod_j p(j \to j+1) \quad (13)$$

where the summation over Q_i is over all possible terminal boundary points for a walk. The path summation is over all possible paths from P to Q_i, and $p(j \to j+1)$ is the transition probability on the jth step of the path.

An a priori estimate of $\sigma^2(Z)$ is often useful to estimate the number of walks that will be necessary for given accuracy *before* the decision is made to do the calculation. One can show, for example, that

$$\sigma^2(Z_i) \leq \min_M \max_{Q \epsilon C} [\phi(Q) - M]^2$$

When the $F(P_i)$'s are not zero and the transition probabilities are all equal, one may estimate $\sigma^2(Z)$ by an intuitive procedure of the following sort:

1. Divide the boundary into a small number of sectors. Estimate the probability of reaching any sector from P from the angle subtended $L/2\pi R$. L is the arc subtended at the boundary. R is the "average" distance.

2. Estimate the average number of steps required by $n \approx \frac{R^2}{2h^2}$ (see [3]).

3. Estimate $\sum_j h^2 \frac{F(P_j)}{D(P_j)}$ as nh^2 times the average value of F/D on a pie-shaped slice of the lattice going to this boundary sector. Estimate Z_i by $-\frac{n}{D}\left(\frac{F}{D}\right)_{\text{average}} + \phi(Q_i)_{\text{average}}$ for this section.

4. Sum $Z_i^2 \frac{L_i}{R_i}$ over all sectors of the boundary.

This is an estimate of the second term on the right-hand side of (13). If one has an estimate of $V(P)$ one can then roughly estimate the variance from (13).

Statistical Estimation

Alternative sampling procedures (sometimes called statistical estimation or quota sampling) can be used which may greatly increase the accuracy of the estimate for a given number of walks.

Instead of p_1, p_2, p_3, p_4, p_5, a new set of nonnegative transition probabilities $p_1^*, p_2^*, p_3^*,$

p_4^*, p_5^* is chosen at each point P of the lattice, such that none are zero and their sum is unity. The difference equation (5) can be written

$$V(P) = \sum_{i=1}^{5} \left[p_i^*(P) \frac{p_i(P)}{p_i^*(P)} V(P_i) \right] - \frac{h^2 F(P)}{D(P)} \tag{14}$$

One interprets this as follows: The transition probabilities are $p_j^*(P)$ but the particle gets a weight

$$\omega_i(P) = \frac{p_i(P)}{p_i^*(P)} \tag{15}$$

after each step, i.e., one particle becomes $\omega_j(P)$ particles after each step. Weights are multiplied from step to step in a walk and the total weight at the kth step is

$$\omega(k) = \prod_{j=1}^{k} \omega_{i_j}(P_j) \tag{16}$$

where i_j is the neighbor to which the particle moved on the jth step. One can show that the estimate of $V(P_0)$ now becomes the mean value of the tally

$$Z_m = -\sum_{j=1}^{m} \frac{h^2 F(P_j)}{D(P_j)} \omega(j) + \omega(k_m)\phi(Q_m) \tag{17}$$

where k_m is the number of steps needed to reach the boundary on the mth walk ending on Q_m.

Suppose a guess of the solution $V(P)$ at all points of the lattice is available. A good choice for the transition probabilities $p_i^*(P)$ is

$$p_i^*(P) = p_i(P) \cdot \frac{V(P_i)}{V(P)} \qquad i = 1, 2, \cdots, 5 \tag{18}$$

One can show that if $V(P)$ is the exact solution, the a priori variance $\sigma^2(Z)$ with this choice of transition probabilities is zero. Hence the use of statistical estimation with a choice of transition probabilities and weights based on information available concerning the solution will tend to increase the accuracy. In addition, one may revise the p^*'s during the calculation by the use of (18), according to the latest estimate of the solution obtained from the random walks that have been completed.

3. SUMMARY OF THE CALCULATION PROCEDURE

Consider the solution of Poisson's equation in two dimensions. Statistical estimation is used. It is assumed that the region within C is convex.

Input Quantities

1. Lattice points and boundary points: The point P_0 at which the solution is to be obtained is the origin. Let

$n_+ \equiv$ maximum y coordinate (positive number)

$n_- \equiv$ minimum y coordinate (negative number)

$(i, j) \equiv$ lattice point designation

$m_R(j) \equiv$ maximum x coordinate for a given j on a horizontal slice

$m_L(j) \equiv$ minimum x coordinate for a given j on a horizontal slice

The lattice points (L, j) then are bounded as follows:

$$n_- \le j \le n_+$$
$$m_L(j) \le i(j) \le m_R(j)$$

To fix boundaries one must therefore specify $n_+, n_-, m_R(j), m_L(j)$, a total of $2(n_+ + |n_-| + 2)$ values.

2. Boundary values: $\phi(m, n)$ for each of the $2(n_+ + |n_-| + 1)$ boundary points.

3. The quantities $-\frac{1}{4}h^2[F(i, j)]$ for all lattice points. [Note that $F(i, j) \equiv 0$ on the boundary.]

4. Number of walks N to be made.

b. Calculation Procedure

1. Clear and set to zero all the walk registers.
2. Add $-\frac{1}{4}h^2[F(0, 0)]$ to tally register Z.
3. Generate two random numbers.*
4. Choose $\Delta i = \pm 1$ or 0 and $\Delta j = \pm 1$ or 0, on the basis of the two random numbers. The point $(\Delta i, \Delta j) = (0, 0)$ is excluded.
5. Register the new current point in the walk in the (i, j) register.
6. Add $-\frac{1}{4}h^2[F(i, j)]$ to the tally register Z.
7. Check whether current point is on the boundary.

   (a) If not, go back to step 3.

   (b) If it is, add $\phi(m, n)$ to the tally register $(m = i, n = j)$.

Go on to the next walk. Keep a running sum of walk tables and of sum of squares of walk tables.

* Reference [4] gives methods for the generation of random numbers on computers.

8. Have all walks been completed?

(a) If not, go back to step 1.

(b) If so, calculate from the sum of walk tables and the sum of squares of walk tables the estimated solution and its variance.

c. Output Quantities

1. The estimated solution $\bar{V}(P_0)$.
2. The variance $\sigma^2(\bar{V})$ of the estimate.

4. FLOW CHART

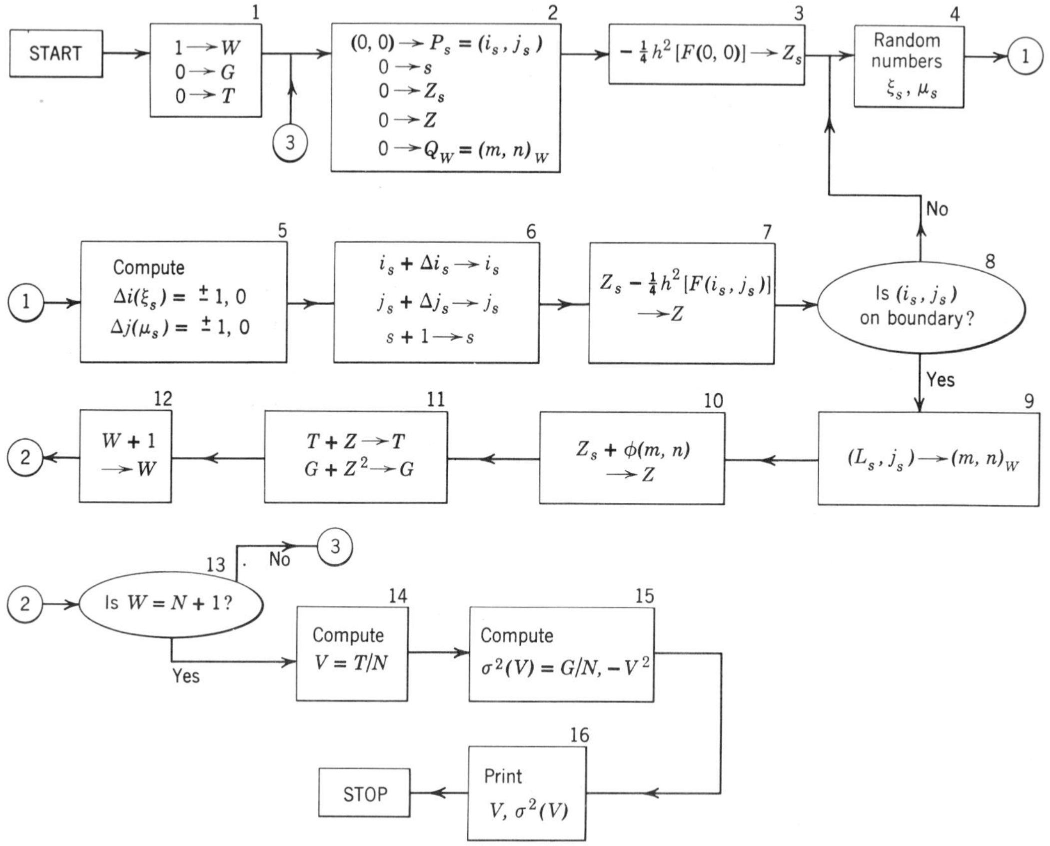

is set to zero. The boundary point register (end of the walk) denoted $Q_W = (m, n)_W$ is set to zero.

Box 3: The weighted charge density, $-\frac{1}{4}h^2[F(0, 0)]$ at the solution point (0, 0) is added to the tally register Z_s.

Box 4: Random numbers ξ_s and μ_s are generated for the sth step.

Box 5: The step changes in the abscissa Δi_s and in the ordinate Δj_s are each computed as zero or ± 1 depending on the two random numbers ξ_s and μ_s. Let ξ_s and μ_s be integers

5. DETAILED DESCRIPTION OF THE FLOW CHART

Box 1: The counter W (indexing the walks) is set to one. The sum of squares register G of the tallies Z^2 is set to zero. The sum register T of the tallies Z is set to zero.

Box 2: The current point counter $P_s = (i_s, j_s)$ is set to the solution point (0, 0) (beginning of the walk). The step counter s is set to zero. The tally register Z_s during the walk is set to zero. The end of walk tally register Z

from 0 to 9. Then Table 1 gives Δi_s and Δj_s.

Box 6: The now current point (i, j) is computed. The step counter is advanced.

Table I

ξ_s	μ_s	Δi_s	Δj_s
≥ 5	≥ 5	1	0
<5	≥ 5	0	1
≥ 5	<5	0	-1
<5	<5	-1	0

Box 7: The weighted source density $-\frac{1}{4}h^2[F(i_s, j_s)]$ is obtained from the memory and added to the tally register.

Box 8: The new current point is compared with the boundary points to answer the question: Is the new current point on the boundary? If not, one proceeds to generate new random numbers (box 4) for the next step. If the new current point is on the boundary, one proceeds to box 9.

Box 9: The current point is transferred to the boundary point register $(m, n)_W$ for the Wth walk.

Box 10: The boundary value $\phi(m, n)$ corresponding to the boundary point reached is added to the tally register contents and the sum transferred to the end of walk tally register Z.

Box 11: The T register is increased by the contents of the Z register. The G register is increased by the square of the quantity in the Z register.

Box 12: The W register is increased by one.

Box 13: If the W register is $N + 1$, all the walks have been completed and one proceeds to box 14. If the W register is less than $N + 1$, one proceeds to box 2 for the next walk.

Box 14: The estimate of the solution is computed from the contents of the T register.

Box 15: The variance of the estimated solution is computed.

Box 16: The solution estimate and its variance are printed out.

6. SUBROUTINES

SR1—Generation of two random numbers. Input is first location of this subroutine. Output is two random numbers. Used in box 4.

7. SAMPLE PROBLEM

A simple problem suitable for debugging is the following: Consider the solution of

$$\nabla^2 V = -4\pi \sum_{j=1}^{J} A_j \frac{\delta(r - a_j)}{2\pi a_j}$$

within the circle $0 \leq r \leq b$, with the boundary condition that $V = V_0$ on the circle. The sources are circular delta function rings of strength A_j at radii a_j. The solution is given by

$$V = V_0 + 2 \sum_{j=1}^{J} A_j \ln \frac{b}{a_j} - 2 \sum_{j=1}^{J} A_j H(r - a_j) \ln \frac{r}{a_j}$$

where $H(x)$ is the Heaviside step function

$$H(x) \begin{matrix} = 1 & \text{for} & x > 0 \\ = 0 & \text{for} & x < 0 \end{matrix}$$

Thus the solution at the center of the circle is

$$V = V_0 + 2 \sum_{j=1}^{J} A_j \ln \frac{b}{a_j}$$

A convenient choice of constants for a simple debugging problem is

$$V_0 = 1 \qquad J = 2 \qquad a_1 \doteq \tfrac{1}{3} \qquad a_2 = \tfrac{2}{3}$$
$$b = 1 \qquad A_1 = A_2 = \tfrac{1}{2}$$

Then

$$V = 1 + \ln 2 + \ln 3 = 2.79176$$

One step of the calculation, starting from the center point, is illustrated below. Generation of the random numbers is not shown; they are assumed to be integers from 0 to 9 inclusive. The criterion for a step is given in Table 1. It is assumed that $h = 0.05$.

$(0, 0) \rightarrow (L_0, j_0)$	box 2
$0 \rightarrow Z_0$	box 3
$\xi_0 = 4,\ \mu_0 = 7$	box 4
$\Delta i_0 = 0,\ \Delta j_0 = 1$	box 5
$(0, 1) \rightarrow (i_1, j_1)$	box 6
$-\frac{1}{4}h^2[F(0, 1)] = 0 \rightarrow Z$	box 7
Is (0, 1) on the boundary? No	box 8

Return to box 4 and choose two new random numbers.

8. MEMORY REQUIREMENTS

Given below is a table of the quantities to be stored in permanent memory locations and the space required for each, corresponding to the calculation symbolized by the flow chart.

INPUT

Boundary points: (m, n) where $n_- \leq n \leq n_+$

$m_R(n)$	$n_+ + \lvert n_- \rvert + 1$ words
$m_L(n)$	$n_+ + \lvert n_- \rvert + 1$ words

Boundary values:

$\phi[m_R(n), n]$	$n_+ + \lvert n_- \rvert + 1$ words
$\phi[m_L(n), n]$	$n_+ + \lvert n_- \rvert + 1$ words

Sources and sinks:

$-\frac{1}{4}h^2[F(i,j)]$ where $n_- \le j \le n_+$

$$m_L(j) \le i(j) \le m_R(j)$$

$$\text{words required} = \sum_{j=n_-}^{n_+} [m_R(j) + |m_L(j)| + 2]$$

Number of walks N: 1 word

COUNTERS

W counter	1 word
s counter	1 word
Z_s counter	1 word
Z_W counter	1 word
T_W counter	1 word
G_W counter	1 word
P counter	2 words
Q counter	2 words

Total number of memory locations required =

$$4(n_+ + |n_-| + 1) + 10 + \sum_{j=n_-}^{n_+} [m_R(j) + |m_L(j)| + 2]$$

If the transition probabilities were not constant and equal, a maximum of $5 \sum_{j=n_-}^{n_+} [m_R(j) + |m_L(j)| + 2]$ additional locations would be required. If statistical estimation were used, an extra $5 \sum_{j=n_-}^{n_+} [m_R(j) + |m_L(j)| + 2]$ locations would be required.

9. ESTIMATION OF THE RUNNING TIME

Let R be the average distance of the point at which the solution is desired from all boundary points. The mean number of steps required to end a walk is approximately

$$S \sim \frac{R^2}{2h^2}$$

Assume that the generation of two random numbers requires 8μ, where μ is the multiply time. Computation of the new current point requires about 8ν, where ν is the add time. Assume that the comparison time to test the equality of two numbers is about equal to the add time. Then each walk requires approximately $8S(\mu + \nu)$ in boxes 4, 5, and 6, $S \cdot 2\nu[m_R + |m_L| + 2]_{\text{average}}$ for boxes 7 and 8, and $5\nu + \mu$ additional. The total problem requires N times this for N walks. Thus the total calculation time required is approximately

$$T \sim [8NS(\mu + \nu) + NS] \cdot 2\nu[m_R + |m_L| + 2]_{\text{average}} + N(5\nu + \mu)$$

10. REFERENCES

1. J. H. Curtiss, Sampling Methods Applied to Differential and Difference Equations, *Proceedings of Seminar on Scientific Computation*, IBM Corp., Nov. 1949, Sections 9, 11, 12, 13, 14, and 16.
2. J. L. Doob, *Stochastic Processes*, John Wiley & Sons, New York, 1953.
3. S. Chandrasekhar, Stochastic Problems in Physics and Astronomy, *Rev. Modern Phys.*, vol. 15, no. 1, Jan. 1943, Chap. 1.
4. O. Taussky, and J. Todd, Generation of Pseudo-Random Numbers, in *Symposium on Monte Carlo Methods*, edited by H. A. Meyer, John Wiley & Sons, New York, 1956, pp. 15–28.

The numerical solution of hyperbolic partial differential equations by the method of characteristics

15

Mary Lister
The Pennsylvania State University

I. FUNCTION

The techniques described in this chapter can be used to find the numerical solution to systems of partial differential equations of the hyperbolic type. The discussion is primarily related to two simultaneous quasi-linear partial differential equations for the case of two dependent and two independent variables. Here, linear combinations of the differential equations are sought which contain derivatives of the two unknown functions in one direction only. For this type of equation there are two such directions, called characteristic directions. The finite difference approximations to the two partial differential equations are derived using the properties of these directions. Both first- and second-order approximations are obtained and extrapolation procedures, which increase the accuracy of the solution, are given. In order to give details of the sequence of the operations in a calculation, an example is taken from the field of fluid mechanics.

Courant and Friedrichs [1] show how the notion of characteristic directions can be extended to n quasi-linear partial differential equations in two independent variables. Again, characteristic directions are sought. For n equations, there are n characteristic directions through each point. Some may be coincident. With this extension, the size of the computation is greatly increased and not too many problems have been solved for $n > 2$. Further generalizations can be made when more than two independent variables are involved. Courant and Friedrichs [1] have again outlined possible procedures. Here, the characteristic curves become characteristic surfaces and many more computational difficulties are introduced. Some are described by Roberts [2]. The basic techniques can be related to the case of two equations in two unknowns; hence this case has been used to illustrate the computational procedures.

2. MATHEMATICAL DISCUSSION

a. Definitions and Derivation of the Method

The general form of a quasi-linear system of equations for the case of two independent variables x, y and two dependent variables u, v can be written in the form:

$$L_1 = A_1 u_x + B_1 u_y + C_1 v_x + D_1 v_y + E_1 = 0 \qquad (1)$$

$$L_2 = A_2 u_x + B_2 u_y + C_2 v_x + D_2 v_y + E_2 = 0 \qquad (2)$$

where $A_1, A_2, \cdots, E_2$ are known functions of x, y, u, v, and $u_x = \partial u/\partial x$, $v_y = \partial v/\partial y, \cdots$. (Subscript notation will be used for partial derivatives throughout the discussion.)

In the following considerations, it is assumed that all the functions introduced above are continuous and possess as many continuous derivatives as may be required. It is also assumed that nowhere does the following condition exist:

$$A_1{:}A_2 = B_1{:}B_2 = C_1{:}C_2 = D_1{:}D_2 \qquad (3)$$

Consider a linear combination of L_1 and L_2:

$$\begin{aligned} L &= \lambda_1 L_1 + \lambda_2 L_2 \\ &= (\lambda_1 A_1 + \lambda_2 A_2)u_x + (\lambda_1 B_1 + \lambda_2 B_2)u_y \\ &\quad + (\lambda_1 C_1 + \lambda_2 C_2)v_x + (\lambda_1 D_1 + \lambda_2 D_2)v_y \\ &\quad + (\lambda_1 E_1 + \lambda_2 E_2) \end{aligned} \qquad (4)$$

Now, if $y = y(x)$ is the equation of a curve, then dy/dx is the slope of the tangent at any point of this curve and, if $u = u(x, y)$ and $v = v(x, y)$ are solutions to (1) and (2), then

$$du = \frac{\partial u}{\partial x} dx + \frac{\partial u}{\partial y} dy, \qquad dv = \frac{\partial v}{\partial x} dx + \frac{\partial v}{\partial y} dy \qquad (5)$$

Now, the differential expression L can be written in the form

$$\begin{aligned} dxL = (\lambda_1 A_1 + \lambda_2 A_2)\, du + (\lambda_1 C_1 + \lambda_2 C_2)\, dv \\ + (\lambda_1 E_1 + \lambda_2 E_2)\, dx \end{aligned} \qquad (6)$$

if the constants λ_1 and λ_2 are chosen so that

$$\frac{dx}{dy} = \frac{\lambda_1 A_1 + \lambda_2 A_2}{\lambda_1 B_1 + \lambda_2 B_2} = \frac{\lambda_1 C_1 + \lambda_2 C_2}{\lambda_1 D_1 + \lambda_2 D_2} \qquad (7)$$

In this case, in the differential expression L, the derivatives of u and those of v are combined so that their derivatives are in the same direction, namely, dy/dx. This direction is called a *characteristic direction.*

From (7), the ratio λ_1/λ_2 can be obtained:

$$-\frac{\lambda_1}{\lambda_2} = \frac{A_2\, dy - B_2\, dx}{A_1\, dy - B_1\, dx} = \frac{C_2\, dy - D_2\, dx}{C_1\, dy - D_1\, dx} \qquad (8)$$

hence

$$a(dy)^2 - 2b\, dx\, dy + c(dx)^2 = 0 \qquad (9)$$

Here

$$a = A_1 C_2 - A_2 C_1, \qquad 2b = A_1 D_2 - A_2 D_1, \qquad c = B_1 D_2 - B_2 D_1 \qquad (10)$$

For the case of hyperbolic partial differential equations, two distinct roots of the quadratic equation (9) exist. Therefore

$$b^2 - ac > 0 \qquad (11)$$

This excludes the exceptional case of all three coefficients vanishing. Moreover, it is assumed for convenience that

$$a \neq 0 \qquad (12)$$

The latter condition can always be satisfied, if necessary, by introducing new coordinates instead of x and y. Consequently, $dx \neq 0$ for a characteristic direction (dx, dy) as seen from (9); thus the slope

$$\zeta = \frac{dy}{dx} \qquad (13)$$

can be introduced, and ζ satisfies the equation:

$$a\zeta^2 - 2b\zeta + c = 0 \qquad (14)$$

This equation has two different real solutions ζ_+ and ζ_-,

$$\zeta_+ \neq \zeta_- \qquad (15)$$

Thus, at the point (x, y), the two different characteristic directions are given by:

$$\frac{dy}{dx} = \zeta_+, \qquad \frac{dy}{dx} = \zeta_- \qquad (16)$$

Since a, b, and c are in general functions of u, v, x, and y, ζ_+ and ζ_- will also be functions of these quantities:

$$\frac{dy}{dx} = \zeta_+(x, y, u, v), \qquad \frac{dy}{dx} = \zeta_-(x, y, u, v) \qquad (17)$$

Once a solution $u(x, y)$, $v(x, y)$ of (1) and (2) has been obtained, equations (17) become two separate ordinary differential equations of the first order. These define two one-parameter families of characteristic curves (often abbreviated to characteristics), C_+ and C_- in the (x, y) plane, belonging to this solution $u(x, y)$, $v(x, y)$. These two families form a curvilinear coordinate net.

If ζ_+ and ζ_- are functions of x and y only, then

$$\frac{dy}{dx} = \zeta_+(x, y), \qquad \frac{dy}{dx} = \zeta_-(x, y)$$

and it is not necessary to find a solution to (1) and (2) in order to find the equations of the characteristics; hence the problem is simplified.

Substituting the solutions (16) into the expressions for λ_1/λ_2 given in (8) yields

$$\frac{\lambda_1}{\lambda_2} = -\frac{A_2\zeta_+ - B_2}{A_1\zeta_+ - B_1}, \qquad \frac{\lambda_1}{\lambda_2} = -\frac{A_2\zeta_- - B_2}{A_1\zeta_- - B_1} \tag{18}$$

Finally, combining (18) and (6) gives

$$F\,du + (a\zeta_+ - G)\,dv + (K\zeta_+ - H)\,dx = 0 \tag{19}$$

$$F\,du + (a\zeta_- - G)\,dv + (K\zeta_- - H)\,dx = 0 \tag{20}$$

where

$$F = A_1B_2 - A_2B_1, \qquad G = B_1C_2 - B_2C_1 \tag{21}$$
$$K = A_1E_2 - A_2E_1, \qquad H = B_1E_2 - B_2E_1$$

Thus, the following four characteristic equations have been obtained:

$$\left.\begin{aligned} dy - \zeta_+\,dx &= 0 \quad (22)\\ F\,du + (a\zeta_+ - G)\,dv &\\ + (K\zeta_+ - H)\,dx &= 0 \quad (23)\end{aligned}\right\}\text{ along } C_+$$

$$\left.\begin{aligned} dy - \zeta_-\,dx &= 0 \quad (24)\\ F\,du + (a\zeta_- - G)\,dv &\\ + (K\zeta_- - H)\,dx &= 0 \quad (25)\end{aligned}\right\}\text{ along } C_-$$

Equations (22)–(25) are of a particularly simple form, inasmuch as each equation contains only total derivatives of all the variables.

According to the derivation, every solution of the original system (1) and (2) satisfies the system (22)–(25). Courant and Friedrichs [1] show that the converse is also true.

b. Characteristic Equations for Two Specific Problems

The equations which govern the motion in the case of one-dimensional isentropic flow are

$$u_t + uu_x + \frac{c^2}{\rho}\rho_x = 0 \tag{26}$$

$$\rho u_x + \rho_t + u\rho_x = 0 \tag{27}$$

where u is the velocity of the fluid, ρ the density, c the velocity of sound, x the space coordinate, and t the time variable. If, in addition, it is assumed that the flow is adiabatic, so that $c^2 = \rho^{\gamma-1}$, where γ is the adiabatic exponent, (26) and (27) become

$$u_t + uu_x + \frac{2cc_x}{\gamma - 1} = 0 \tag{28}$$

$$cu_x + \frac{2}{\gamma - 1}(c_t + uc_x) = 0 \tag{29}$$

Combining (28) and (29) linearly, it is found that

$$u_t + (u + \lambda c)u_x + \frac{2}{\gamma - 1}[(c + \lambda u)c_x + \lambda c_t] = 0 \tag{30}$$

The conditions for (dt, dx) are, therefore,

$$dx = (u + \lambda c)\,dt, \qquad dx = (\lambda u + c)\,dt$$

from which

$$\lambda^2 = 1 \tag{31}$$

Hence, the two characteristic directions are given by

$$dx = (u + c)\,dt, \qquad dx = (u - c)\,dt \tag{32}$$

and the characteristic equations for u and c become

$$du + \frac{2}{\gamma - 1}dc = 0, \qquad -du + \frac{2}{\gamma - 1}dc = 0 \tag{33}$$

The differential equations for three-dimensional isentropic spherical flow are:

$$u_t + uu_x + \frac{2cc_x}{\gamma - 1} = 0 \tag{34}$$

$$cu_x + \frac{2}{\gamma - 1}(c_t + uc_x) + \frac{2uc}{x} = 0 \tag{35}$$

Equations (34) and (35) differ from (28) and (29) by the term $2uc/x$ in (35). This does not involve a derivative and hence the two characteristic directions are given by (32). The characteristic equations for u and c differ from equations (33); they are

$$\begin{aligned} du + \frac{2}{\gamma - 1}\,dc + \frac{2uc}{x}\,dt = 0 \\ -du + \frac{2}{\gamma - 1}\,dc + \frac{2uc}{x}\,dt = 0 \end{aligned} \qquad (36)$$

Further examples are given in Courant and Friedrichs [1].

c. Finite Difference Approximations

Two methods of approximating (22)–(25) will be described here. One finite difference approximation is expressed by the formula:

$$\int_{x_0}^{x_1} f(x)\,dx \approx f(x_0)(x_1 - x_0) \qquad (37)$$

and will be called a first-order or linear approximation.

The second finite difference approximation is expressed by the trapezoidal rule formula:

$$\int_{x_0}^{x_1} f(x)\,dx \approx \tfrac{1}{2}[f(x_0) + f(x_1)](x_1 - x_0) \qquad (38)$$

and will be called a second-order approximation.

Let P be the point of intersection of the C_+ characteristic through A and the C_- characteristic through B,* as shown in Fig. 1. It is assumed that x_A, y_A, u_A, v_A, x_B, y_B, u_B, and v_B are known, and that x_P, y_P, u_P, and v_P are to be found. x_A means value of x at A, etc.

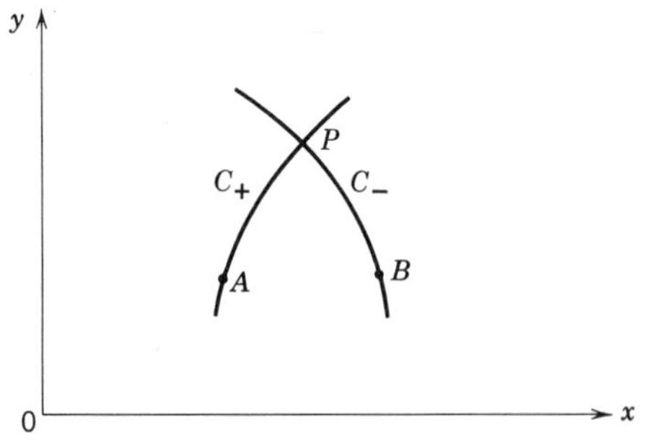

Fig. 1

* Such a point of intersection exists if A and B are sufficiently close together, since $\zeta_+ \neq \zeta_-$.

Using (37), the linear approximation to (22)–(25) is

$$y_P - y_A - (\zeta_+)_A(x_P - x_A) = 0 \qquad (39)$$

$$F_A(u_P - u_A) + (a\zeta_+ - G)_A(v_P - v_A) + (K\zeta_+ - H)_A(x_P - x_A) = 0 \qquad (40)$$

$$y_P - y_B - (\zeta_-)_B(x_P - x_B) = 0 \qquad (41)$$

$$F_B(u_P - u_B) + (a\zeta_- - G)_B(v_P - v_B) + (K\zeta_- - H)_B(x_P - x_B) = 0 \qquad (42)$$

Now, ζ_+, ζ_-, F, G, K, H, and a are functions of x, y, u, and v. They are evaluated at the points A and B and hence are known. Therefore, the unknowns y_P and x_P can be obtained by solving simultaneously (39) and (41). The solutions of (39) and (41) can then be substituted into (40) and (42). These can then be solved simultaneously to give values for u_P and v_P.

A second order approximation to (22)–(25) using (38) is

$$y_P - y_A + \tfrac{1}{2}[(\zeta_+)_P + (\zeta_-)_A](x_P - x_A) = 0 \qquad (43)$$

$$\tfrac{1}{2}(F_A + F_P)(u_P - u_A) + \tfrac{1}{2}[(a\zeta_+ - G)_A + (a\zeta_+ - G)_P](v_P - v_A) + \tfrac{1}{2}[(K\zeta_+ - H)_A + (K\zeta_+ - H)_P(x_P - x_A)] = 0 \qquad (44)$$

$$y_P - y_B - \tfrac{1}{2}[(\zeta_-)_B + (\zeta_-)_P](x_P - x_B) = 0 \qquad (45)$$

$$\tfrac{1}{2}(F_B + F_P)(u_P - u_B) + \tfrac{1}{2}[(a\zeta_- - G)_B + (a\zeta_- - G)_P](v_P - v_B) + \tfrac{1}{2}[(K\zeta_- - H)_B + (K\zeta_- - H)_P](x_P - x_B) = 0 \qquad (46)$$

Now, unless ζ_+, ζ_-, F, G, K, H, and a are independent of x, y, u, and v, then (43)–(46) are no longer linear equations in the unknowns x_P, y_P, u_P, and v_P and some iterative method will have to be used, in general, to find a solution. More details of this type of solution will be given later when a specific set of equations is solved.

There are several ways of using either the set of equations (39)–(42) or the set (43)–(46) to obtain an approximate numerical solution to the original set of partial differential equations. One is to use a grid of characteristics. This is particularly simple if ζ_+ and ζ_- depend on x and y only. Equations (39) and (41) or (43) and (45) can be integrated immediately and the grid of characteristics found before the calculation for u and v is started. Another way is to use specified intervals in the y-direction, for example, and relate the values of u and v at the beginning of the interval with those at the end by means of either relations (39)–(42) or (43)–(46). Both methods are described by Hartree [3], in relation to second-order procedures.

d. Grid of Characteristics

In this method, for each pair of points A and B located on the grid of characteristics, as shown in Fig. 2, (39)–(42) or (43)–(46) are regarded as equations for the four unknowns x_P, y_P, u_P, and v_P. They can be used in the following way to obtain an approximate numerical solution:

1. A solution of (39)–(42) will yield directly a set of values for x_P, y_P, u_P, and v_P.
2. These values can be used as the initial estimates for x_P, u_P, y_P, and v_P when the solution of (43)–(46) is to be obtained iteratively.

The calculation can proceed in the following manner: I is a curve in the (x, y) plane, on which the values of x, y, u, and v are given. That is, x, y, u, v are known at A_1, A_2, A_3, A_4, A_5 in Fig. 3. From these values, u, v, x, y can be calculated at the points B_1, B_2, B_3, B_4 using steps 1 and 2 described above. Proceeding in the same way, u, v, x, y can be found at C_1, C_2, C_3, then D_1, D_2, and finally E_1. It should be noted that at each stage there is one point less at which u, v, x, y can be found. This illustrates the "domain of dependence" described by Courant and Friedrichs [1]. It should also be noted that the values of x, y, u, v affect the values of x, y, u, v at succeeding stages. For the point A_3, these points are

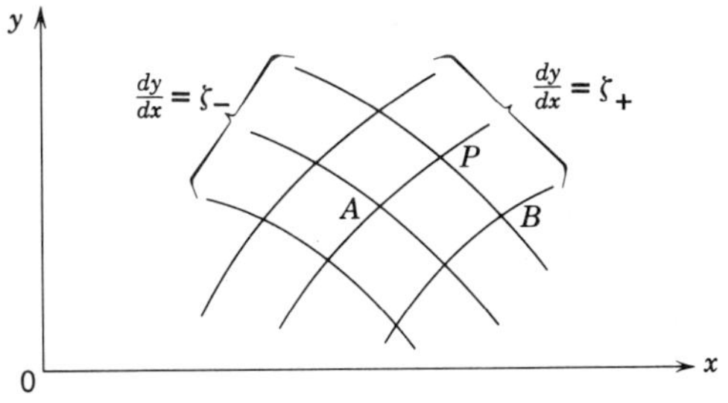

Fig. 2

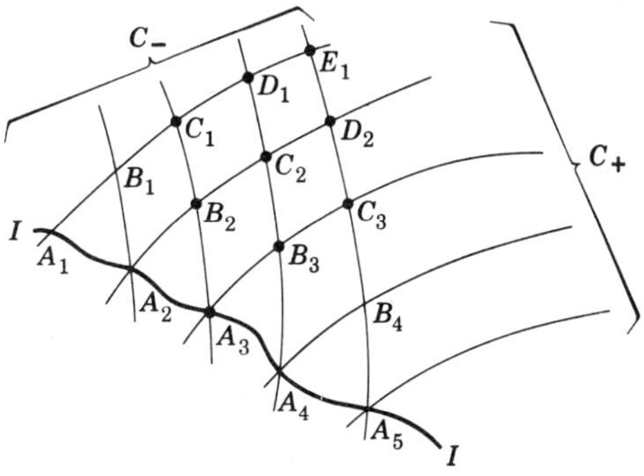

Fig. 3

marked by a circle. This region is bounded by the characteristics through A_3 and is known as the "range of influence" of the point A_3.

e. Boundaries

Consider the situation as shown in Fig. 4. $A_1A_2A_3A_4A_5 \cdots$ is the boundary of the region on which x, y, u, v are known. $A_1C_0E_{-1} \cdots$ is a second boundary of the region. The problem is to find the unknowns x, y, u, v on the boundary. Values of x, y, u, v have been obtained at B_1, B_2, B_3, B_4, C_1, $\cdots$, using the method given in Section 2d.

The characteristic A_2B_1 is extended until it intersects the boundary curve at C_0. To find the values of x, y, u, v at C_0, the following relations are known

$$y_{C_0} - y_{B_1} = (\zeta_-)_{B_1}(x_{C_0} - x_{B_1}) \tag{47}$$

$$F_{B_1}(u_{C_0} - u_{B_1}) + (a\zeta_- - F)_{B_1}(v_{C_0} - v_{B_1}) + (K\zeta_- - H)_{B_1}(x_{C_0} - x_{B_1}) = 0 \tag{48}$$

from the linear approximations, and

$$y_{C_0} - y_{B_1} = \tfrac{1}{2}[(\zeta_-)_{B_1} + (\zeta_-)_{C_0}](x_{C_0} - x_{B_1}) \tag{49}$$

$$\tfrac{1}{2}[F_{B_1} + F_{C_0}](u_{C_0} - u_{B_1}) + \tfrac{1}{2}[(a\zeta_- - G)_{C_0} + (a\zeta_- - G)_{B_1}](v_{C_0} - v_{B_1}) + \tfrac{1}{2}[(K\zeta_- - H)_{B_1} + (K\zeta_- - H)_{C_0}](x_{C_0} - x_{B_1}) = 0 \tag{50}$$

from the second-order approximation to the equations along the C_- characteristic. Thus, there are two relations connecting four unknowns in either approximation. Two more relations are therefore needed along the boundary $A_1C_0E_{-1} \cdots$ in order to determine the four unknowns. Hence, either

$$\begin{array}{l}\text{the equation of the curve } x = f(y) \\ \text{and the dependent variable } u = \phi(x)\end{array} \tag{51}$$

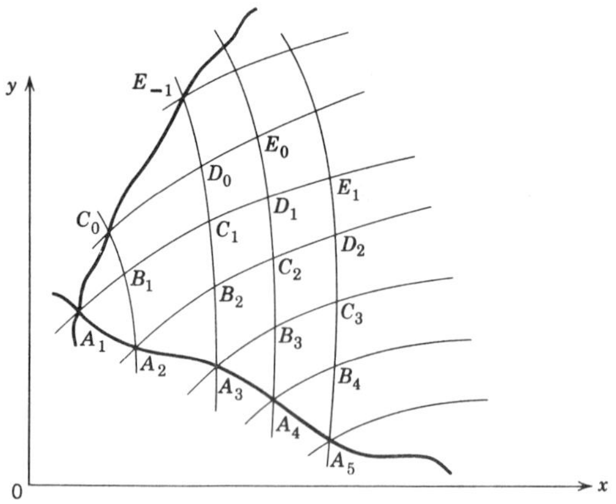

Fig. 4

or another combination such as

$$y = g(x) \quad \text{and} \quad v = \psi(y) \tag{52}$$

must be given. If $y = g(x)$ and $v = \psi(y)$ are given, then the calculation is as follows:

1. Equation (47) gives

$$g(x_{C_0}) - y_{B_1} = (\zeta_-)_{B_1}(x_{C_0} - x_{B_1}) \tag{53}$$

and this can be solved for x_{C_0}.

2. y_{C_0} can then be calculated from $y_{C_0} = g(x_{C_0})$. (54)

3. v_{C_0} can then be calculated from $v_{C_0} = \psi(y_{C_0})$. (55)

4. Equation (48) contains u_{C_0} as the only unknown; hence u_{C_0} can be found.

Thus, initial estimates of x, y, u, v are obtained at C_0.

Equations (49) and (50) can be solved iteratively as follows:

5. $$g(x_{C_0}^{(k+1)}) - y_{B_1} = \tfrac{1}{2}[(\zeta_-)_{B_1} + (\zeta_-)_{C_0}^{(k)}] \times [(x)_{C_0}^{(k+1)} - x_B]. \tag{56}$$

6. $$y_{C_0}^{(k+1)} = g(x_{C_0}^{(k+1)}). \tag{57}$$

7. $$v_{C_0}^{(k+1)} = \psi(y_{C_0}^{(k+1)}). \tag{58}$$

8. $$\tfrac{1}{2}(F_{B_1}^{(k)} + F_{C_0}^{(k)})(u_{C_0}^{(k+1)} - u_{B_1}) = -\tfrac{1}{2}[(a\zeta_- - G)_{C_0}^{(k)} + (a\zeta_- - G)_{B_1}] \times (v_{C_0}^{(k+1)} - v_{B_1}) - \tfrac{1}{2}[(K\zeta_- - H)_{B_1} + (K\zeta_- - H)_{C_0}^{(k)}](x_{C_0}^{(k+1)} - x_{B_1}) \tag{59}$$

and can be solved for $u_{C_0}^{(k+1)}$, where the superscript $(k + 1)$ indicates the value of the $(k + 1)$st iterate in this scheme.

f. Specified Time (i.e., y-Direction) Intervals

In this process u and v are considered as known functions of x at distance y, either as given initial conditions or as the results of a previous stage of the calculations. For example, it is assumed that u and v are known on the line $y = Y$ and are to be found on the line $y = Y + \Delta y$. Let P be a typical point on the line $y = Y + \Delta y$ and A, B, C be three adjacent points on the line $y = Y$, as shown in Fig. 5. Let the C_+ characteristic at P intersect ACB at R and the C_- characteristic at P intersect ACB at S.

Since x_P and y_P are known, u_P and v_P are to be found. The computation proceeds as follows:

1. The equations

$$y_P - y_R = (\zeta_+)_C(x_P - x_R) \tag{60}$$

$$y_P - y_S = (\zeta_-)_C(x_P - x_S) \tag{61}$$

give the x-coordinates of R and S respectively.

2. Using a formula for linear interpolation, u_R, u_S, v_R, v_S can be found from:

$$u_R = u_C[1 - (\zeta_+)_C^{-1}\theta] + u_A\theta(\zeta_+)_C^{-1} \tag{62}$$

$$v_R = v_C[1 - (\zeta_+)_C^{-1}\theta] + v_A\theta(\zeta_+)_C^{-1} \tag{63}$$

$$u_S = u_C[1 - (\zeta_-)_C^{-1}\theta] + u_B\theta(\zeta_-)_C^{-1} \tag{64}$$

$$v_S = v_C[1 - (\zeta_-)_C^{-1}\theta] + v_B\theta(\zeta_-)_C^{-1} \tag{65}$$

where $\theta = \Delta y/\Delta x$.

It has been assumed that Δy is sufficiently small that the parts of the characteristics between P and R and between P and S are straight lines and that the slope of PR at P is $(\zeta_+)_C$ and the slope of PS at P is $(\zeta_-)_C$.

3. u_P and v_P can now be obtained by

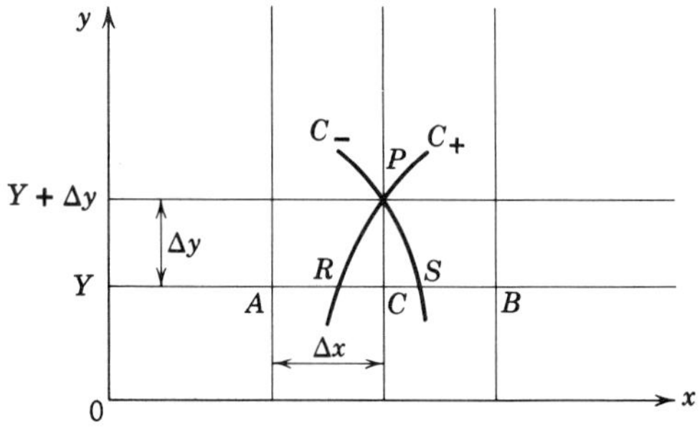

Fig. 5

solving simultaneously the following two equations:

$$F_C(u_P - u_R) + (a\zeta_+ - G)_C(v_P - v_R) + (K\zeta_+ - H)_C(x_P - x_R) = 0 \quad (66)$$

$$F_C(u_P - u_S) + (a\zeta_- - G)_C(v_P - v_S) + (K\zeta_- - H)_C(x_P - x_S) = 0 \quad (67)$$

Equations (60)–(67) form a process with first-order accuracy. If a higher degree of accuracy is required, the values of u_P and v_P obtained from steps 1, 2, and 3 can be used as initial estimates for the second-order process, which is as follows:

4. The equations

$$y_P - y_R = \tfrac{1}{2}[(\zeta_+)_P^{(k)} + (\zeta_+)_R^{(k)}](x_P - x_R^{(k+1)}) \quad (68)$$

$$y_P - y_S = \tfrac{1}{2}[(\zeta_-)_P^{(k)} + (\zeta_-)_S^{(k)}](x_P - x_S^{(k+1)}) \quad (69)$$

give the x-coordinates of R and S at the $(k + 1)$st iteration.

5. Using a formula for quadratic interpolation, $u_R^{(k+1)}$, $u_S^{(k+1)}$, $v_R^{(k+1)}$, and $v_S^{(k+1)}$ can be calculated:

$$u_R^{(k+1)} = u_C - \frac{1}{2\Delta x}(u_A - u_B)(x_R^{(k+1)} - x_C) + \frac{1}{2(\Delta x)^2}(u_A + u_B - 2u_C)(x_R^{(k+1)} - x_C)^2 \quad (70)$$

$$v_R^{(k+1)} = v_C - \frac{1}{2\Delta x}(v_A - v_B)(x_R^{(k+1)} - x_C) + \frac{1}{2(\Delta x)^2}(v_A + v_B - 2v_C)(x_R^{(k+1)} - x_C)^2 \quad (71)$$

$$u_S^{(k+1)} = u_C + \frac{1}{2\Delta x}(u_A - u_B)(x_S^{(k+1)} - x_C) + \frac{1}{2(\Delta x)^2}(u_A + u_B - 2u_C)(x_S^{(k+1)} - x_C)^2 \quad (72)$$

$$v_S^{(k+1)} = v_C + \frac{1}{2\Delta x}(v_A - v_B)(x_S^{(k+1)} - x_C) + \frac{1}{2(\Delta x)^2}(v_A + v_B - 2v_C)(x_S^{(k+1)} - x_C)^2 \quad (73)$$

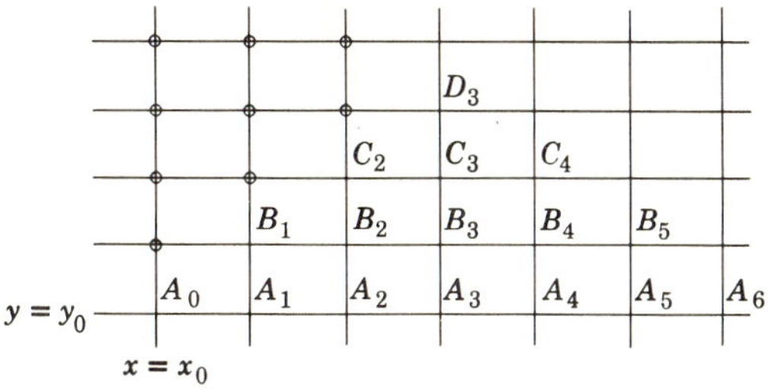

Fig. 6

6. $u_P^{(k+1)}$ and $v_P^{(k+1)}$ can then be obtained by solving simultaneously the equations:

$$\tfrac{1}{2}(F_R^{(k+1)} + F_P^{(k)})(u_P^{(k+1)} - u_R^{(k+1)}) + \tfrac{1}{2}[(a\zeta_+ - G)_R^{(k+1)} + (a\zeta_+ - G)_P^{(k)}] \times (v_P^{(k+1)} - v_R^{(k+1)}) + \tfrac{1}{2}[(K\zeta_+ - H)_R^{(k+1)} + (K\zeta_+ - H)_P^{(k)}](x_P - x_R^{(k+1)}) = 0 \quad (74)$$

$$\tfrac{1}{2}(F_S^{(k+1)} + F_P^{(k)})(u_P^{(k+1)} - u_S^{(k+1)}) + \tfrac{1}{2}[(a\zeta_- - G)_S^{(k+1)} + (a\zeta_- - G)_P^{(k)}] \times (v_P^{(k+1)} - v_S^{(k+1)}) + \tfrac{1}{2}[(K\zeta_- - H)_S^{(k+1)} + (K\zeta_- - H)_P^{(k)}](x_P - x_S^{(k+1)}) = 0 \quad (75)$$

Hence the computation can proceed as follows: If u and v are given at $A_0, A_1 \cdots A_6$ as shown in Fig. 6, then u and v can be computed at $B_1 \cdots B_5$ using either (60)–(67) or (60)–(75). In a similar manner, u and v can be calculated at C_2, C_3, C_4, then finally D_3. It will depend upon the number of points taken along $y = y_0$ as to how far the computation will proceed. If further information is given on $x = x_0$, then values of u and v can be obtained at the points marked by a circle.

g. Boundary Points and Points Near the Boundary

Again, it is assumed that the equation of the boundary is known and either u or v is known:

$$y = f(x) \quad (76)$$

$$\text{and} \quad u = \phi(x) \quad \text{or} \quad v = \psi(y) \quad (77)$$

where f, ϕ, and ψ are known functions.

Boundary Points

The values of u and v at N, A, B, and C in Fig. 7 are assumed known. The following are the steps in the computation for u_M if v_M is given, or for v_M if u_M is known.

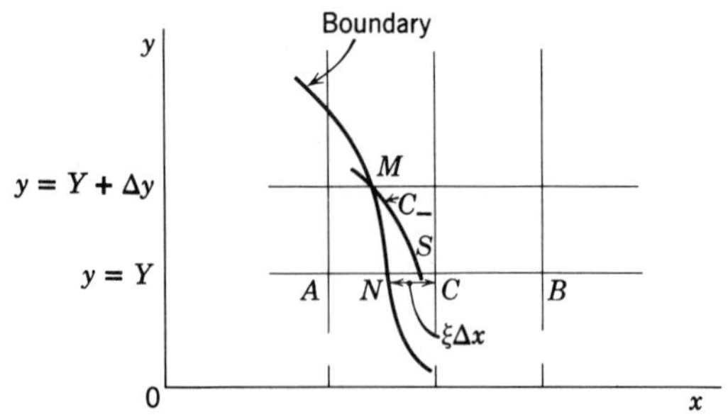

Fig. 7

1. x_M can be obtained from (76).
2. x_S can be obtained from the relation

$$x_S = x_M - (\zeta_-)_N^{-1}(y_M - y_S) \qquad (78)$$

3. u_S and v_S are obtained using linear interpolation:

$$u_S = u_N + \frac{u_C - u_N}{\xi\,\Delta x}(x_S - x_N) \qquad (79)$$

$$v_S = v_N + \frac{v_C - v_N}{\xi\,\Delta x}(x_S - x_N) \qquad (80)$$

where $NC = \xi\,\Delta x, 0 < \xi < 1$,

4. The relationship between u_M and v_M is the following:

$$F_N(u_M - u_S) + (a\zeta_- - G)_N(v_M - v_S) + (K\zeta_- - H)_N(x_M - x_S) = 0 \qquad (81)$$

and can be used to find which of u_M or v_M is unknown. Equations (78)–(81) are the linear approximations to the original partial differential equations.

The formulas for the second-order process are as follows:

5. $x_S^{(k+1)}$ can be calculated from the equation

$$y_M - y_S = \tfrac{1}{2}[(\zeta_-)_M^{(k)} + (\zeta_-)_S^{(k)}](x_M - x_S^{(k+1)}) \qquad (82)$$

6. $u_S^{(k+1)}$ and $v_S^{(k+1)}$ can be found using the following quadratic interpolation formulas:

$$u_S^{(k+1)} = u_C - \frac{\xi^2 u_B - u_N + (1 - \xi^2)u_C}{\xi(\xi + 1)\,\Delta x} \times (x_C - x_S^{(k+1)}) + \frac{\xi u_B + u_N - u_C(1 + \xi)}{\xi(\xi + 1)(\Delta x)^2} \times (x_C - x_S^{(k+1)})^2 \qquad (83)$$

$$v_S^{(k+1)} = v_C - \frac{\xi^2 v_B - v_N + (1 - \xi^2)v_C}{\xi(\xi + 1)\,\Delta x} \times (x_C - x_S^{(k+1)}) + \frac{\xi v_B + v_N - v_C(1 + \xi)}{\xi(\xi + 1)(\Delta x)^2} \times (x_C - x_S^{(k+1)})^2 \qquad (84)$$

7. The relationship between $u_M^{(k+1)}$ and $v_M^{(k+1)}$ is the following:

$$\tfrac{1}{2}(F_S^{(k+1)} + F_M^{(k)})(u_M^{(k+1)} - u_S^{(k+1)}) + \tfrac{1}{2}[(a\zeta_- - G)_M^{(k)} + (a\zeta_- - G)_S^{(k+1)}](v_M^{(k+1)} - v_S^{(k+1)}) + \tfrac{1}{2}[(K\zeta_- - H)_M^{(k)} + (K\zeta_- - H)_S^{(k+1)}] \times (x_M - x_S^{(k+1)}) = 0 \qquad (85)$$

and can be used to find either $u_M^{(k+1)}$ or $v_M^{(k+1)}$.

Points Near the Boundary

The formulas for the first-order process are as follows:

1. To find the x- and y-coordinates of W, the point of intersection of the C_+ characteristic curve at P and the boundary in Fig. 8, the following equations have to be solved, simultaneously:

$$y_W = f(x_W) \qquad (86)$$

$$y_P - y_W = (\zeta_+)_C(x_P - x_W) \qquad (87)$$

2. The x-coordinate of T, the point of intersection of the C_- characteristic through W and the line $y = Y$, is found from:

$$y_W - y_T = (\zeta_-)_N(x_W - x_T) \qquad (88)$$

3. u_T and v_T are found from linear interpolation formulas:

$$u_T = u_N + \frac{u_C - u_N}{\xi\,\Delta x}(x_T - x_N) \qquad (89)$$

$$v_T = v_N + \frac{v_C - v_N}{\xi\,\Delta x}(x_T - x_N) \qquad (90)$$

4. The relationship between u_T and u_W is as follows:

$$F_N(u_W - u_T) + (a\zeta_- - G)_N(v_W - v_T) + (K\zeta_- - H)_N(x_W - x_T) = 0 \qquad (91)$$

Again, this equation can be solved either for v_W in terms of u_W or vice versa.

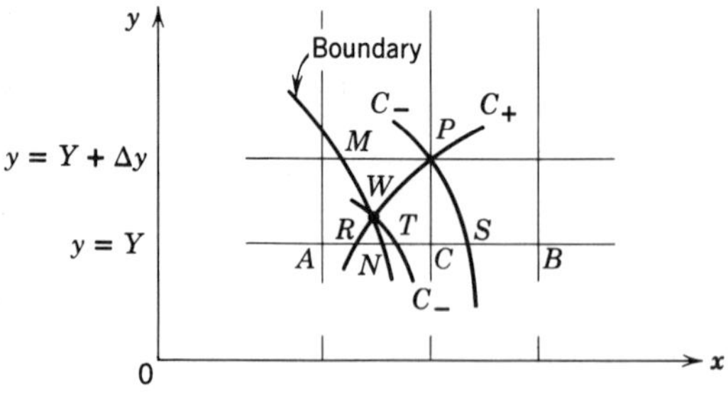

Fig. 8

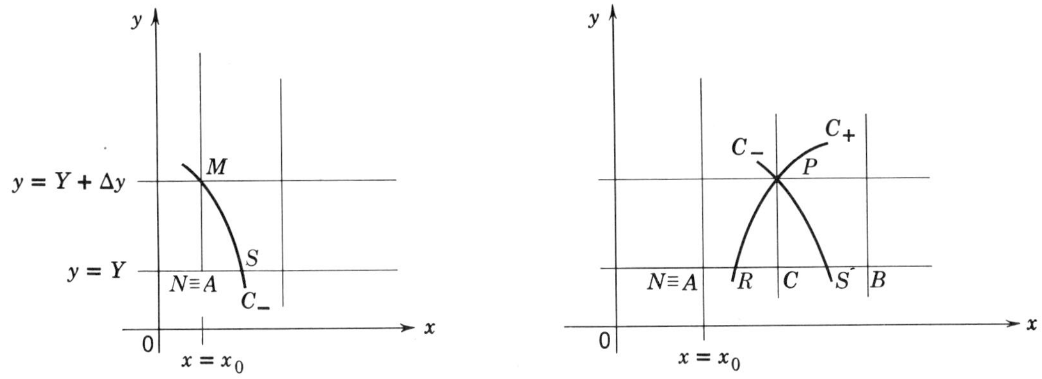

Fig. 9

The formulas for the second-order process are as follows:

5. The equation for finding the x-coordinate of W is:

$$y_P - f(x_W^{(k+1)}) = \tfrac{1}{2}[(\zeta_+)_P^{(k)} + (\zeta_+)_W^{(k)}](x_P - x_W^{(k+1)}) \tag{92}$$

and will be solved directly or iteratively depending on the complexity of $f(x)$.

6. $$y_W^{(k+1)} = f(x_W^{(k+1)}). \tag{93}$$

7. The $x^{(k+1)}$ coordinate of T is found from the following equation:

$$y_W^{(k+1)} - y_T = \tfrac{1}{2}[(\zeta_+)_W^{(k)} + (\zeta_+)_T^{(k)}](x_W^{(k+1)} - x_T^{(k+1)}) \tag{94}$$

8. $u_T^{(k+1)}$ and $v_T^{(k+1)}$ can be computed from second-order interpolation formulas:

$$u_T^{(k+1)} = u_C - \frac{\xi^2 u_B - u_N + (1 - \xi^2)u_C}{\xi(\xi + 1)\,\Delta x} \times (x_C - x_T^{(k+1)}) + \frac{\xi u_B + u_N - u_C(1 + \xi)}{\xi(\xi + 1)(\Delta x)^2} \times (x_C - x_T^{(k+1)})^2 \tag{95}$$

$$v_T^{(k+1)} = v_C - \frac{\xi^2 v_B - v_N + (1 - \xi^2)v_C}{\xi(\xi + 1)\,\Delta x} \times (x_C - x_T^{(k+1)}) + \frac{\xi v_B + v_N - v_C(1 + \xi)}{\xi(\xi + 1)(\Delta x)^2} \times (x_C - x_T^{(k+1)})^2 \tag{96}$$

9. $v_W^{(k+1)}$ and $u_W^{(k+1)}$ are related as follows:

$$\tfrac{1}{2}(F_T^{(k+1)} + F_W^{(k)})(u_W^{(k+1)} - u_T^{(k+1)}) + \tfrac{1}{2}[(a\zeta_- - G)_W^{(k)} + (a\zeta_- - G)_T^{(k+1)}] \times (v_W^{(k+1)} - v_T^{(k+1)}) + \tfrac{1}{2}[(K\zeta_- - H)_W^{(k)} + (K\zeta_- - H)_T^{(k+1)}](x_W^{(k+1)} - x_T^{(k+1)}) = 0 \tag{97}$$

Alternatively, after u_M and v_M have been computed, and x_W and y_W found, u_W and v_W could be calculated using suitable interpolation formulas.

10. $u_S^{(k+1)}$ and $v_S^{(k+1)}$ can be calculated using (69), (72), and (73).

11. Finally, the following two equations can be solved for $u_P^{(k+1)}$ and $v_P^{(k+1)}$:

$$\tfrac{1}{2}(F_P^{(k)} + F_W^{(k+1)})(u_P^{(k+1)} - u_W^{(k+1)}) + \tfrac{1}{2}[(a\zeta_+ - G)_P^{(k)} + (a\zeta_+ - G)_W^{(k+1)}] \times (v_P^{(k+1)} - v_W^{(k+1)}) + \tfrac{1}{2}[(K\zeta_+ - H)_P^{(k)} + (K\zeta_+ - H)_W^{(k+1)}](x_P - x_W^{(k+1)}) = 0 \tag{98}$$

$$\tfrac{1}{2}(F_P^{(k)} + F_S^{(k+1)})(u_P^{(k+1)} - u_S^{(k+1)}) + \tfrac{1}{2}[(a\zeta_- - G)_P^{(k)} + (a\zeta_- - G_S^{(k+1)})] \times (v_P - v_S) + \tfrac{1}{2}[(K\zeta_- - H)_P^{(k)} + (K\zeta_- - H)_S^{(k+1)}](x_P - x_S^{(k+1)}) = 0 \tag{99}$$

If the boundary is the line $x = x_0$, as in Fig. 9, then the greater part of this computation is eliminated. In this case, (76)–(85) are used, with $A \equiv N$ and $\xi = 1$. The scheme developed in (86)–(99) is not required. Variations on this procedure are given by Hartree [3].

h. Extrapolation Procedures

When specified time intervals are used, it is possible to employ extrapolation procedures to increase the accuracy of the computation.

Richardson and Gaunt [4] discussed this method, and it has been used by several other authors. This description is given in a paper by Roberts [5].

Consider a function $f(x, y)$ which is to be determined at $y = 2n\,\Delta y$, in terms of its known value at $y = 0$. This may be accomplished in n steps of $2\Delta y$ by repeating a linear process at a constant value of x. Let this value of $f(x, y)$ be denoted by $f^2(2n\,\Delta y)$. Alternatively, $2n$ steps of length Δy can be used. Let $f^1(2n\,\Delta y)$ denote this value of $f(x, y)$. Then, if $\bar{f}(2n\,\Delta y)$ is computed, where

$$\bar{f}(2n\,\Delta y) = 2f^1(2n\,\Delta y) - f^2(2n\,\Delta y) \quad (100)$$

$\bar{f}(2n\,\Delta y)$ and the true value $f(2n\,\Delta y)$ agree when terms of the order $(\Delta y)^2$ are neglected. Further, it can be shown that, after n steps if $n\,\Delta y = 0(1)$, then the error is of $0(\Delta y)^2$ [6].

The linear procedure given by (60)–(67) is of this form. Thus, if u and v are computed for two different step sizes ($2\Delta y$ and Δy), and combined using (100), then the error in u and v after n steps is of the order $(\Delta y)^2$. This is the accuracy achieved when the procedure given by (68)–(75) is used for the smaller step, in the y-direction.

If it is desirable to eliminate higher order errors, then one of the following methods can be used:

(i) Compute $f(2n\,\Delta y)$ for the three steps, $2\Delta y$, Δy, $0.5\Delta y$. If $f^{1/2}(2n\,\Delta y)$ denotes the value of $f(2n\,\Delta y)$ computed by the linear process in $4n$ steps of length $0.5\Delta y$, then $\bar{f}(2n\,\Delta y)$ given by

$$\bar{f}(2n\,\Delta y) = \tfrac{8}{3}f^{1/2}(2n\,\Delta y) - 2f'(2n\,\Delta y) + \tfrac{1}{3}f^2(2n\,\Delta y) \quad (101)$$

and the true value $f(2n\,\Delta y)$ are related by

$$\bar{f}(2n\,\Delta y) = f(2n\,\Delta y) + 0(\Delta y)^3 \quad (102)$$

(ii) Compute $f(2n\,\Delta y)$ for two steps, $2\Delta y$ and Δy, using the second-order process given by (60)–(75). Then, $\bar{f}(2n\,\Delta y)$ computed from the relation

$$\bar{f}(2n\,\Delta y) = \tfrac{1}{3}[4f^1(2n\,\Delta y) - f^2(2n\,\Delta y)] \quad (103)$$

is again related to $f(2n\,\Delta y)$ by (102).

Further details for method (i) can be found in [5] and for method (ii) in [3].

i. Comparison of the Two Methods

[Method of characteristics is referred to as method (i). Method of fixed time intervals is referred to as method (ii).]

One reason why method (ii) may be preferred over method (i) is that x_P and y_P are known exactly, and only the two quantities u_P and v_P have to be determined at each point. Further, with method (ii) it is possible to use extrapolation procedures to increase the accuracy of the calculation. For method (i) it would be difficult to arrange the computation so that the points of intersection of the characteristics occurred at values of x and y on a rectangular mesh. Hence, interpolation formulas would have to be used before extrapolating to "zero mesh size."

Further, for the calculation of unsteady flow method (ii) has the advantage that it produces results directly in the form most likely to be needed, namely the velocity distribution in space at different times. In addition, the values of Δx, Δy are under the control of the individual organizing the computation; hence at any time the mesh widths can be halved or doubled depending on the behavior of the dependent variables.

On the other hand, the grid of characteristics is useful in determining physical characteristics. An example of this is described in a paper by Fox and Ralston [7]. Here a numerical solution of the equations for spherical waves of finite amplitude is given. In this problem a shock wave could form. This phenomenon would be shown by the convergence of the characteristics.

From the point of view of programming, the ordering of the computation would appear to be easier for method (ii) as opposed to (i), since the values of u and v are known at predetermined points and the additional information of their x- and y-coordinates does not have to be recorded. If an extrapolation procedure is used with method (ii), then the results of two separate computations have to be stored within the machine.

3. SUMMARY OF THE CALCULATION PROCEDURE

We consider the case of isentropic spherically symmetric flow. The equations which govern this flow are (32) and (36). An approximate numerical solution to these equations has been obtained using both a characteristic grid [7] and specified time intervals [5], [6]. Roberts [5] gives graphically the results of both computations and there is agreement between the two sets of values. For his approximation Roberts did not use the characteristic network but rather a finite difference approximation to the governing equations. Lister and Roberts [6] show that the two approximations are equivalent.

The boundary conditions for this problem are:

$$x = 0 : u = 0, \qquad \frac{\partial c}{\partial x} = 0 \tag{104}$$

and the initial conditions are of the form:

$$t = 0 : u = 0,$$

$$c = f(x) = [1 + d \exp(-4x^2)]^{1/5} \tag{105}$$

where d = constant.

We will consider the calculation using specified time intervals and a linear process with extrapolation. For this problem (60)–(67) take the form

$$x_R = x_C - (u + c)_C(t_P - t_C) \tag{106}$$

$$x_S = x_C - (u - c)_C(t_P - t_C) \tag{107}$$

$$c_R = c_C[1 - \theta(u_C + c_C)] + c_A\theta(u_C + c_C) \tag{108}$$

$$c_S = c_C[1 + \theta(u_C - c_C)] - c_B\theta(u_C - c_C) \tag{109}$$

$$u_R = u_C[1 - \theta(u_C + c_C)] + u_A\theta(u_C + c_C) \tag{110}$$

$$u_S = u_C[1 + \theta(u_C - c_C)] - u_B\theta(u_C - c_C) \tag{111}$$

$$c_P = .5(c_R + c_S) + .1(u_R - u_S) - .4\left(\frac{uc}{x}\right)_C (t_P - t_C) \tag{112}$$

$$u_P = .5(u_R + u_S) + 2.5(c_R - c_S) \tag{113}$$

where $\theta = \Delta t/\Delta x$. (See Fig. 5.)

Equations (106)–(113) can be used to obtain values of u and c at all points of the region except on the line $x = 0$. Here, the term uc/x takes the form 0/0. Fox and Ralston [7] developed a power series solution in the neighborhood of $x = 0$. This showed that, near $x = 0$,

$$\frac{u}{x} \sim k_0 + k_2x^2 \tag{114}$$

where k_0 and k_2 are constants.

Using (114), it is found that

$$\left(\frac{u}{x}\right)_{x=0} = \frac{1}{3\Delta x}(4u_C - .5u_B) \tag{115}$$

hence, (78)–(81) take the following form:

$$x_S = c_A(t_M - t_S) \tag{116}$$

$$c_S = c_A + \frac{(c_C - c_A)}{\Delta x} x_S \tag{117}$$

$$u_S = \frac{u_C x_S}{\Delta x} \tag{118}$$

$$c_M = c_S - .2u_S - \frac{c_A(1.6u_C - .2u_B)(t_M - t_S)}{3\Delta x} \tag{119}$$

For the initial conditions (105) formulas (106)–(113) do not yield a linear process in the neighborhood of $t = 0$. This is because the assumption of zero initial velocity causes the error in the first step to be of the order of $(\Delta t)^2$ instead of Δt. Hence, the extrapolation procedure (100) could not be used. The computation was, therefore, started $t = .1$. u and c were computed at $t = .1$, at intervals $x = .05$, using a power series approximation [with d in (105) equal to 5].

The input is then the values of u and c on the line $t = .1$ at intervals $\Delta x = .05$. If the storage is available it is convenient to store the values of u and c at intervals $\Delta x = .1$ in one set of storage locations and the values of u and c at intervals $\Delta x = .05$ in a second set of storage locations.

The steps in the computation are:

1. Calculate u and c for $2\Delta t = .05$ using (106)–(113) and (116)–(119).
2. Calculate u and c for $\Delta t = .025$ using the same equations.
3. Using the results of step 2 calculate u and c for $\Delta t = .025$.
4. Use the results of steps 1 and 3 to calculate extrapolated values of u and c using (100).
5. Repeat steps 1–4 as long as desired. The output is the results of step 4.

The same block of instructions can be used for the computation of u and c. This means keeping a record of the row where the computation was being done. Here, a location can be used to store a number m. If $m = 1$, then $\Delta x = .1$ and values of u and c have been computed at $t = T + 2\Delta t$ from values of u and c at $t = T$, $2\Delta t = .05$. If $m = 2$, then $\Delta x = .05$ and values of u and c have been computed at $t = T + \Delta t$ from values of u and c at $t = T$, $\Delta t = .025$. If $m = 4$, then $\Delta x = .05$ and values of u and c have been computed at $t = T + 2\Delta t$ from values of u and c at $t = T + \Delta t$, $\Delta t = .025$.

4. FLOW CHART

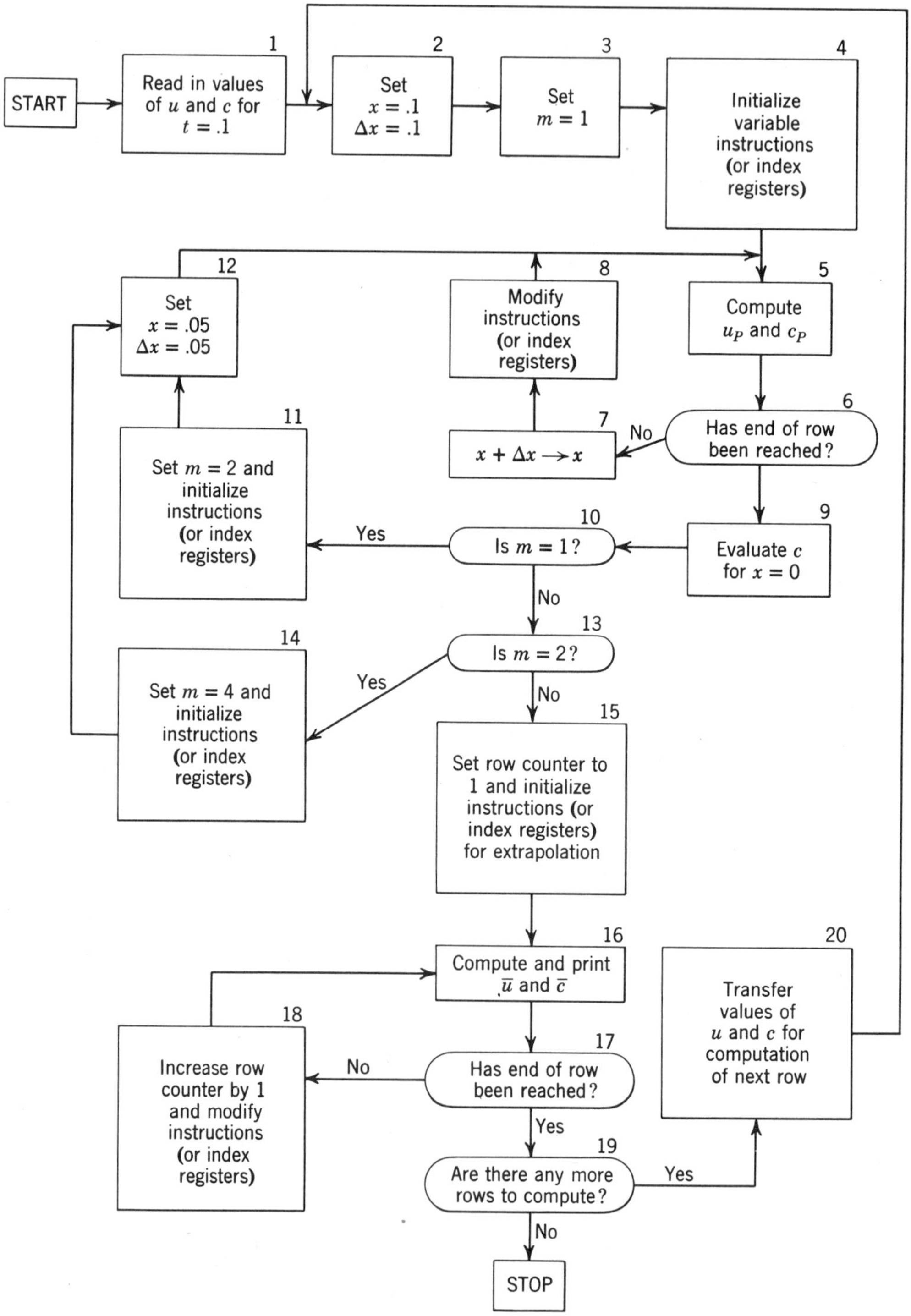

5. DESCRIPTION OF THE FLOW CHART

Box 1: Values of u and c are stored in memory.

Box 2: The x-coordinate is set at .1, and the increment Δx at .1.

Box 3: The parameter for indicating the computation being performed is set at its initial value.

Box 4: The instructions (or index registers), which are modified in the computation in boxes 5 and 9, are set at their initial value, for $x = .1$.

Box 5: Values of u_P and c_P are computed using (106)–(113).

Box 6: If the end of the row has been reached go to box 9, otherwise go to box 7. Note that each successive row contains one less point.

Box 7: The value of x is increased by Δx and stored in the location for x.

Box 8: Variable instructions (or index registers) in the calculation of u_P and c_P are modified.

Box 9: The value of c_M is computed using (116)–(119).

Box 10: If $m = 1$, all the values of u and c have been computed for $t = T + 2\Delta t$ in one step, and the next procedure is taken from box 11. Otherwise, the procedure is taken from box 13.

Box 11: The parameter for indicating the progress of the computation is changed and the instructions (or index registers) which are modified in the computation in boxes 5 and 9 are set at their initial values for the first step when $\Delta x = .05, x = .05$.

Box 12:. The x-coordinate and Δx are set at .05, and the computation in box 5 is repeated.

Box 13: If $m = 2$, all the values of u and c have been calculated for $t = T + \Delta t$ and the procedure in box 14 is followed. Otherwise, the procedure in box 15 is followed.

Box 14: The parameter for indicating the progress of the computation is changed and the instructions (or index registers) which are modified in the computation in boxes 5 and 9 are set at their initial values for the second step when $\Delta x = .05$, $x = .05$. The procedure in box 12 is then followed.

Box 15: The counter used in the extrapolation and the instructions (or index registers) which are modified in the computation in box 16 are initialized.

Box 16: $\bar{u}$ and $\bar{c}$ are calculated using (100).

Box 17: If all the extrapolated values of u and c on $t = T + 2\Delta t$ have not been computed, the procedure in box 18 is followed. Otherwise the procedure in box 19 is followed.

Box 18: The counter is increased and variable instructions (or index registers) are modified.

Box 19: According to a criterion set up by the programmer, the computation is stopped or proceeds to the next row.

Box 20: Values of u and c corresponding to the last computed row are put in the locations of the initial values.

6. SUBROUTINES

None.

7. SAMPLE PROBLEM FOR SPECIFIED TIME INTERVALS: LINEAR PROCESS WITH EXTRAPOLATION

The values of u and c given in Fig. 10, with c above the line and u below, are taken from the power series solution to the problem and the numbers needed in this calculation are inserted at each node point. Taking $x = .1$, $\dot{\theta} = .5$ and substituting into (106)–(113), we obtain

$$c_P = 1.3759132$$
$$u_P = .0977342$$

In this way all the values of u and c can be found for $t = .15$ except at $x = 0$. At $x = 0$ we use (116)–(119) and obtain

$$x_S = .0703465$$
$$c_S = 1.4030710$$
$$u_S = .0326688$$
$$c_M = 1.3832675$$

Then with $\Delta t = .025$ and $\Delta x = .05$, values of u and c are computed at $t = .125$ and $t = .15$. The results are shown in Fig. 11 (with u below the line and c above). Now, using the results

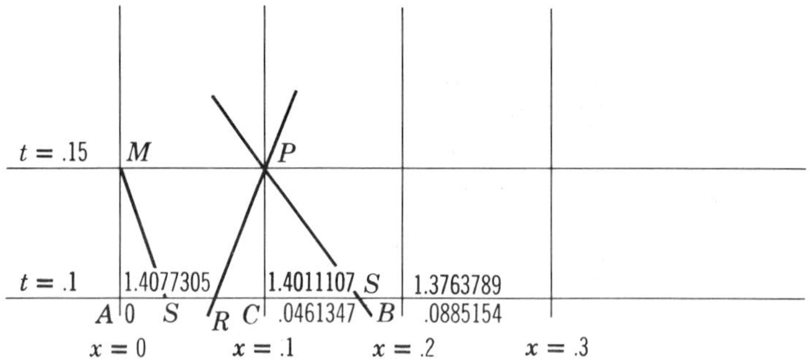

Fig. 10

	x = 0	x = .05	x = .1	x = .15
t = .15	1.3819145 0	1.3796244 .0484929	1.3733367 .0970458	
t = .125	1.3975813 0	1.3954768 .0344839	1.3902287 .0745921	
t = .1	1.4077305 0	1.4074342 .0234779	1.4011107 .0464134	

Fig. 11

of the two computations and applying (100), we obtain the extrapolated value of c at $t = .15, x = 0$ as

$$\bar{c}_M = 2(1.3819145) - 1.3832675 = 1.3805615$$

and the extrapolated value of c at $t = .15$, $x = .1$ as

$$\bar{c}_P = 2(1.3733367) - 1.3759132 = 1.3707602$$

8. MEMORY REQUIREMENTS

If there are α points on the initial line ($t = .1$ in the example) for the step size $2\Delta t$ ($2\Delta t = .05$ in the example), then 2α locations are required for the initial data on u and c for the step size $2\Delta t$ and 4α locations for the initial data for the step size Δt. If the calculated values are, for convenience, not immediately written over the values on the previous line, then 6α locations are needed for the calculated values. Finally, 2α locations are needed for the extrapolated values so that the total is 14α locations. This can be reduced to about 8α if the results of one line are immediately written over the results of the previous line, although in this case care must be taken not to erase needed data.

In general the program itself depends on the equations being solved. In the example considered here about 120 registers are needed for the program to evaluate u_P and c_P and for the bookkeeping operations.

9. ESTIMATION OF THE RUNNING TIME

The number of multiplications involved in the calculation of u_P and c_P as given in sample calculation is 18, and the number of additions is 13. Hence, if α is the number of points on the initial line for the step $2\Delta t$, then the total number of multiplications will be 90α and additions will be 65α. The extrapolation procedure involves M multiplications and M additions.

Finally, if the calculation is taken from $t = 0$ to $t = 2N\,\Delta t$, then (neglecting the fact that there is one less point to compute on each successive row) the number of multiplications involved is $91N\alpha$ and the number of additions is $66N\alpha$ so that the running time is approximately

$$N\alpha(91\mu + 66\nu)$$

where μ is the multiplication time of the computer and ν the addition time.

10. REFERENCES

1. R. Courant and K. O. Friedrichs, *Supersonic Flow and Shock-Waves*, Interscience Publishers, New York, 1948.
2. R. C. Roberts, Some Difficulties Encountered in Using the Method of Characteristics in Three Dimensions, NAVORD Report 2491, Aug. 1952.
3. D. R. Hartree, Some Practical Methods of Using Characteristics in the Calculation of Non-Steady Compressible Flows, Los Alamos Report LA-HU-1, 1952.
4. L. F. Richardson and J. A. Gaunt, The Deferred Approach to the Limit, *Trans. Roy. Soc. London*, vol. 226, 1927, p. 300.
5. L. Roberts, On the Numerical Solution of the Equations for Spherical Waves of Finite Amplitude, II, *J. Math. and Phys.*, vol. XXXVI, no. 4, Jan. 1958.
6. M. Lister and L. Roberts, On the Numerical Solution of Spherical Waves of Finite Amplitude. Technical Report on the Project of Machine Methods of Computation, MIT-Proj. DIC 6915, June 1956.
7. P. Fox and A. Ralston, On the Numerical Solution of the Equations for Spherical Waves of Finite Amplitude, I. *J. Math. and Phys.*, vol. XXXVI, no. 4, Jan. 1958.

The solution of hyperbolic partial differential equations by difference methods

16

P. Fox
Massachusetts Institute of Technology*

1. FUNCTION

In the previous chapter the numerical solution of hyperbolic partial differential equations using a characteristic mesh was described. In this chapter we will discuss other methods of integration based on differences taken in the directions of the original independent variables. The latter method has the advantage that the values of the solution are found at evenly spaced points in the physical plane, but we shall find that certain restrictions must be placed on the ratio of mesh spacings in various directions in order for a correct solution to be found.

2. MATHEMATICAL DISCUSSION

a. Hyperbolic Partial Differential Equations

Higher order systems of partial differential equations can always be reduced to systems of first order, and in this chapter we shall consider hyperbolic systems expressed in the form

$$\frac{\partial u}{\partial t} = \sum_{s=1}^{n} A^s \frac{\partial u}{\partial x^s} + Bu \tag{1}$$

where u is a vector with m components, and the x^s ($s = 1, \cdots, n$) denote the n independent variables other than t. The A^s and B are matrices depending on x, t, and u. Such a system is called hyperbolic if any linear combination (with real coefficients) of the A^s has a complete set of eigenvectors. An equivalent and more familiar way of expressing the criterion is to state that the system has a complete set of real characteristics.† In this case the initial value problem, i.e., the problem of determining the solution at later times given the prescribed values at $t = 0$,

$$u(x, 0) = f(x) \tag{2}$$

is properly posed and yields a unique solution (see [3]).

* Formerly with the Institute of Mathematical Sciences, New York University.

† For more precise discussions of the condition of hyperbolicity see, for example, [1] and [2].

For future reference we note that the value of the vector u at some point $\bar{x}$ at later times is determined entirely by the data on the segment of the initial line lying between the outermost characteristics drawn backward from the point P to the initial line. This segment is called the domain of dependence of the point P. The situation is illustrated in Fig. 1 for the one-dimensional case.

It is also appropriate to discuss the *mixed initial value–boundary value* problem for hyperbolic systems. For the mixed problem one prescribes:

the initial condition:

$$u(x, 0) = f(x) \qquad \text{for} \qquad 0 \leq x \leq X \tag{3}$$

and the boundary condition:

$$u(x, t) = g(x, t) \qquad \text{on the curve } \Gamma$$

where of course $f(0)$ must equal $g(0, 0)$. Then (see Fig. 2) it is known that the solution in the region A, enclosed by the characteristics, is determined by the initial condition and that the extension of the solution into region B is determined by the boundary condition (see [4], for instance). More generally, the number of unknown components of u that can properly be prescribed on a curve Γ is equal to the number of characteristics entering region B from Γ. (See, for example, [5], p. 180.)

b. Finite Difference Aspects

Leading up to later general discussions of difference methods for solving hyperbolic systems we consider the familiar wave equation and the initial value (Cauchy) problem:

$$\frac{\partial^2 y}{\partial t^2} = c^2 \frac{\partial^2 y}{\partial x^2}, \qquad c^2 = \text{constant} \tag{4}$$

$$y(x, 0) = f_1(x) \tag{5a}$$

$$\frac{\partial y}{\partial t}(x, 0) = f_2(x) \tag{5b}$$

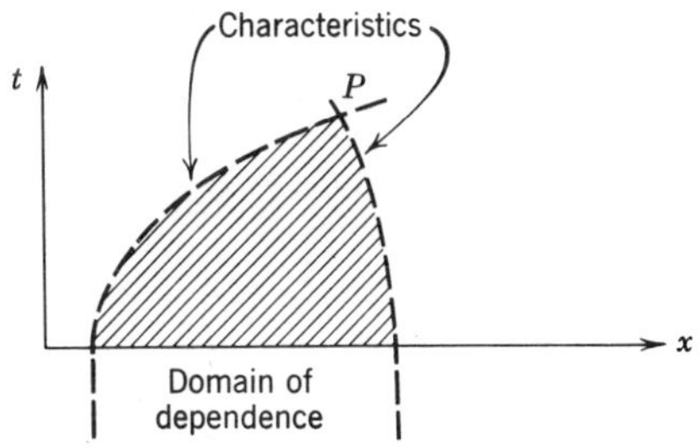

Fig. 1

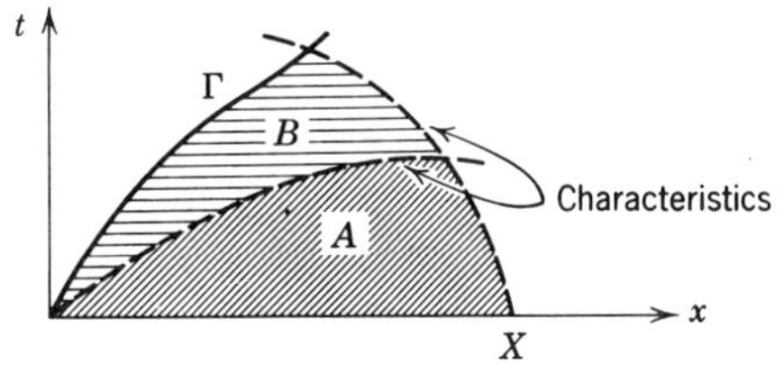

Fig. 2

By the usual methods ([6], pp. 10 ff.), (4) may be reduced to the equivalent first-order system

$$\frac{\partial y}{\partial t} = c\frac{\partial w}{\partial x} \tag{6a}$$

$$\frac{\partial w}{\partial t} = c\frac{\partial y}{\partial x} \tag{6b}$$

$$y(x, 0) = f_1(x) \tag{7a}$$

$$w(x, 0) = \frac{1}{c}\int_0^x f_2(x)\, dx + \text{arbitrary constant} \tag{7b}$$

where (7*b*) is the analog of (5*b*) under the condition that (6*a*) be satisfied at $t = 0$. The vector u of (1) has components y and w for this system, and the matrix A is simply the matrix $\begin{pmatrix} 0 & c \\ c & 0 \end{pmatrix}$. The problem is properly posed.

Various types of difference schemes are available for solving numerically the problem (6), (7). The following scheme, used here for purposes of illustration, will later be shown to be satisfactory. Let the usual $(\Delta x, \Delta t)$ mesh be imposed on the region of solution, and let, for integral values n and j, the following notation be used for the solution of the difference equation:

$$y(j\,\Delta x, n\,\Delta t) \equiv y_j^n$$

so that, for example,

$$y[(j - \tfrac{1}{2})\,\Delta x, (n + 1)\,\Delta t] = y_{j-\frac{1}{2}}^{n+1}$$

Then we can approximate (6) by

$$\frac{y_j^{n+1} - y_j^n}{\Delta t} = c\,\frac{w_{j+\frac{1}{2}}^n - w_{j-\frac{1}{2}}^n}{\Delta x} \tag{8a}$$

$$\frac{w_{j-\frac{1}{2}}^{n+1} - w_{j-\frac{1}{2}}^n}{\Delta t} = c\,\frac{y_j^{n+1} - y_{j-1}^{n+1}}{\Delta x} \tag{8b}$$

Here the half-interval points have been used to "center" the formulas. The advantage in centering the formulas lies in the fact that the truncation error is reduced. This can be shown by expanding the solution in a Taylor series and noting that the two terms involving $\Delta x/2$ cancel out. Equations (8) are an appropriate approximation to the differential equation in this sense, and also from a general point of view discussed below.

Using (8) as a typical example, we will consider explicit finite difference schemes for differential equation systems of type (1) from a somewhat generalized point of view. The general expression for a one-level difference equation (in which the value of u at one time step depends only on its values at the previous time step) may be written in the form*

$$u(x, t + \Delta t) = \sum_p C^p u(x + \Delta_p x, t) \tag{9}$$

where the $\Delta_p x$ are vectors with n components. The notation denotes that C^p multiplies the function u evaluated at some mesh point a vector distance $\Delta_p x$ from x. For example, if u is evaluated at $x + h$ and at $y - h$, where $x = x^{(1)}$ and $y = x^{(2)}$, then $\Delta_p x$ for this term has $+h$ and $-h$ as its first two components and the rest of the components are zero. The number of points p at which u is evaluated in (9) depends on the particular difference scheme chosen. The components of $\Delta_p x$ are assumed to be related to Δt linearly,

$$\frac{\Delta_p x}{\Delta t} = r_p \tag{10}$$

The u is a vector with m components, of course, as in (1), and the C^p are coefficient matrices.

A finite difference approximation (9) is satisfactory if and only if it satisfies the two following criteria of convergence and stability (see [7], [8]):

Convergence: A difference scheme is called convergent if the solution of the difference equation tends to that of the differential equation as Δt tends to zero.

Stability: A difference scheme is called stable if solutions of the difference equation are uniformly bounded functions of the initial data for all sufficiently small Δt and all $n\ \Delta t$ less than some finite value T.

* The development and notation of this section is based on the work of Hahn [7], Lax [3], and Richtmyer [8], [9].

Convergence relates the solution of the difference equation to the solution of the differential equation, and clearly convergence will be possible only if the solution of the *differential* equation u satisfies the *difference* equation (9) at least through terms of first order, so that from (9)

$$\begin{aligned} u(x, t + \Delta t) &= u(x, t) + \frac{\partial u}{\partial t}(x, t)\,\Delta t + 0\,(\Delta t^2) \\ &= \sum_p C^p u(x + \Delta_p x, t) \\ &= \sum_p C^p \left[u(x, t) + \sum_{s=1}^{n} \frac{\partial u}{\partial x^s} \cdot \Delta_p x^s + 0(\Delta t^2) \right] \end{aligned}$$

Using (1) and (10), this is found to yield the so-called *consistency conditions* on the matrices C^p,

$$\sum_p C^p = I \tag{11}$$

where I is the identity matrix, and

$$\sum_p C^p r_p^s = A^s \tag{12}$$

In view of the definitions for convergence and stability it is clear that convergence implies stability. Lax and Richtmyer in [9] have succeeded in proving that the converse is also true; stability gives convergence.

If the coefficients C^p in (9) are constants, the stability of the difference equation can be explored by making use of the fact that an initial exponential function remains exponential. Taking advantage of this fact, we assume an initial value for u in the form

$$u(x, 0) = u_0 e^{ik\cdot x} \tag{13}$$

Then the solution of (9) at $t = \Delta t$ will be

$$\begin{aligned} u(x, \Delta t) &= \sum_p C^p e^{ik\cdot\Delta_p x} e^{ik\cdot x} u_0 \\ &= \left(\sum_p C^p e^{ik\cdot r_p \Delta t}\right) u(x, 0) \end{aligned} \tag{14a}$$

and at $t = n\ \Delta t$ it will be

$$u(x, t) = \left(\sum_p C^p e^{ik\cdot r_p \Delta t}\right)^n u(x, 0) \tag{14b}$$

The matrix $\sum_p C^p e^{ik\cdot r_p \Delta t}$ is called the *amplification matrix*. The amplification matrix is said to satisfy *von Neumann's condition* if its eigenvalues do not exceed one in absolute value for any real value of $k \cdot \Delta t$. This condition is discussed in [3], [7], [8], [9], [10], and elsewhere, and in its necessary aspects is equivalent to the condition given by Courant, Friedrichs, and Lewy in their 1928 paper [11]. Roughly speaking, the condition arises from expressing the bound on the solution in terms of the bound on the amplification matrix since in a sense this matrix is a measure of rate of increase of the solution from one step to the next. Then, in turn, the bound on the matrix is related to the largest eigenvalue of the matrix (see in particular Chapter IV of [8]).

Since the nth power of a quantity greater than one grows without bound as n increases, the von Neumann condition is clearly necessary for convergence. Somewhat refined versions of the condition are also sufficient for cases of constant coefficients. (An excellent and detailed treatment of these matters is given by Richtmyer in his book [8].) When *variable* coefficients are encountered the sufficiency of the von Neumann condition no longer holds except in some special cases, and for nonlinear cases practically nothing is known. In practice the condition is usually applied as a local criterion of sufficiency even in these latter cases and in fact usually seems to work experimentally.

A sufficiency condition applicable to the case of variable coefficients has been given by Friedrichs [12], described by Lax and Richtmyer [9], and lately extended to the nonsymmetric case by Lax [3]. Friedrichs' condition states that if the matrices C^p in (9) satisfy (11), are symmetric, and are Lipschitz continuous,* then the difference scheme is stable provided the C^p are nonnegative. This condition is often useful in practice.

We return now to the example given earlier, and use it to illustrate some of the ideas discussed above. As a first step in analyzing the finite difference scheme the amplification matrix for the scheme given by (8) is obtained as follows.

Assume an initial condition in the form of the exponential,

$$u(x, 0) = u_0 e^{ikx} \tag{15}$$

where in this one-dimensional case x is no longer a vector, but u has two components so that

$$u_0 = \begin{pmatrix} y_0 \\ w_0 \end{pmatrix} \tag{16}$$

Then from (8) one obtains

$$y(x, \Delta t) = y_0 e^{ikx} + c\frac{\Delta t}{\Delta x} w_0[e^{ik[x+\frac{1}{2}(\Delta x)]} - e^{ik[x-\frac{1}{2}(\Delta x)]}] \tag{17a}$$

$$w[x - \tfrac{1}{2}(\Delta x), \Delta t] - c\frac{\Delta t}{\Delta x}[y(x, \Delta t) - y(x - \Delta x, \Delta t)] = w_0 e^{ik[x-\frac{1}{2}(\Delta x)]} \tag{17b}$$

Then using (17*a*) in (17*b*) to obtain an equation of the form (14*a*), one finds

$$w[x - \tfrac{1}{2}(\Delta x), \Delta t] = c\frac{\Delta t}{\Delta x}\Big\{y_0 e^{ikx} + c\frac{\Delta t}{\Delta x} w_0 \times [e^{ik[x+\frac{1}{2}(\Delta x)]} - e^{ik[x-\frac{1}{2}(\Delta x)]}] - y_0 e^{ik(x-\Delta x)} - c\frac{\Delta t}{\Delta x} w_0[e^{ik[x-\frac{1}{2}(\Delta x)]} - e^{ik\{x-3[\frac{1}{2}(\Delta x)]\}}]\Big\} + w_0 e^{ik[x-\frac{1}{2}(\Delta x)]}$$

Let

$$c\frac{\Delta t}{\Delta x}\{2 \sin [\tfrac{1}{2}k(\Delta x)]\} = a$$

Then

$$y(x, \Delta t) = y(x, 0) + iaw(x, 0) \tag{18a}$$

and since, from (14*a*)

$$w(x, \Delta t) = e^{ik[\frac{1}{2}(\Delta x)]} w[x - \tfrac{1}{2}(\Delta x), \Delta t]$$

it follows that

$$w(x, \Delta t) = iay(x, 0) + w(x, 0) \times \left[1 + \left(c\frac{\Delta t}{\Delta x}\right)^2 (e^{ik\Delta x} - 2 + e^{-ik\Delta x})\right]$$

or

$$w(x, \Delta t) = iay(x, 0) + (1 - a^2)w(x, 0) \tag{18b}$$

In matrix form (18) is equivalent to

$$\begin{pmatrix} y(x, \Delta t) \\ w(x, \Delta t) \end{pmatrix} = \begin{pmatrix} 1 & ia \\ ia & 1 - a^2 \end{pmatrix} \begin{pmatrix} y(x, 0) \\ w(x, 0) \end{pmatrix} \tag{19}$$

* A matrix $C(x)$ is Lipschitz continuous at $\tilde{x}$ if for a given $\delta > 0$ there exists a constant M such that $\|C(x) - C(\tilde{x})\| < M|x - \tilde{x}|$ for $|x - \tilde{x}| < \delta$, where the norms defined for the matrix $C(x)$ and the vector x must be related in some reasonable way so that $\|C(x) \cdot x\| \le \|C(x)\| \, \|x\|$. For example, if a norm $\|x\|$ is defined for vectors, the matrix norm can be taken as $\|C(x)\| = \max_{\|x\|=1} \|C(x) \cdot x\|$.

with amplification matrix

$$\begin{pmatrix} 1 & ia \\ ia & 1 - a^2 \end{pmatrix}$$

The von Neumann condition states that the eigenvalues of this matrix must not exceed one in absolute value. The eigenvalues λ satisfy the equation

$$\lambda^2 - \lambda(2 - a^2) + 1 = 0 \tag{20}$$

and they will not exceed one in absolute value if and only if $a^2 \leq 4$. This means that

$$\left(c \frac{\Delta t}{\Delta x} \{2 \sin [\tfrac{1}{2}k(\Delta x)]\}\right)^2 \leq 4$$

or, since this inequality must hold for any k,

$$c \frac{\Delta t}{\Delta x} \leq 1 \tag{21}$$

Since the characteristics of (4) may be shown to be $x \pm ct = \text{constant}$, (21) prescribes that the ratio of mesh widths must be such as to force the domain of dependence of the differential equation to fall *within* the domain of dependence of the difference equation. Of course, one would expect this to be a necessary condition. In the case of this particular problem with constant coefficients, Courant, Friedrichs, and Lewy have shown (Section, 3 Part II of [11]) that the condition (21) is also a sufficient condition for stability.

We remark that if the values of y at y^n instead of at y^{n+1} had been used in (8*b*), the resultant scheme would have had the amplification matrix

$$\begin{pmatrix} 1 & ia \\ ia & 1 \end{pmatrix}$$

The eigenvalues μ of this matrix are

$$\mu = 1 \pm ia$$

and since a is real the eigenvalues are greater than one in absolute value and the scheme is not stable. (Actually the scheme will be stable if $c(\Delta t/\Delta x)$ is of lower order, i.e., if $c[\Delta t/(\Delta x)^2]$ is bounded as $\Delta t, \Delta x \rightarrow 0$ [8, p. 168], but this is an impractical restriction on Δt.)

Other difference schemes can be used for this equation. Lax [5] has suggested the so-called "staggered" scheme wherein the space derivatives are replaced by symmetric difference quotients so that in (6*a*), for example,

$$\frac{\partial w}{\partial x} \sim \frac{w(x + \Delta x, \Delta t) - w(x - \Delta x, t)}{2\Delta x} \tag{22}$$

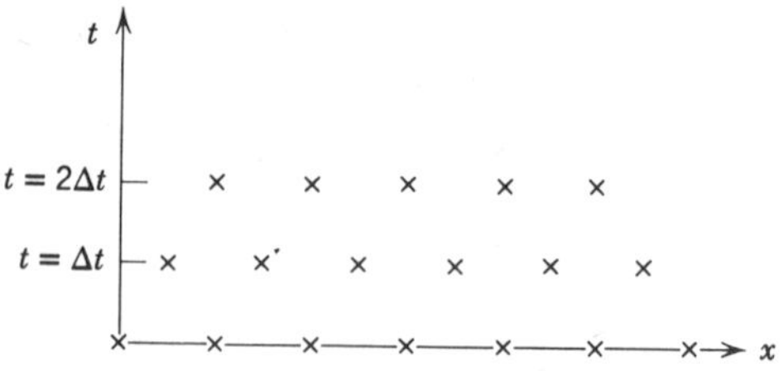

Fig. 3

and the time derivatives are replaced by forward difference quotients of the form

$$\frac{\partial w}{\partial t} \sim \frac{w(x, t + \Delta t) - \tfrac{1}{2}[w(x + \Delta x, t) + w(x - \Delta x, t)]}{\Delta t} \tag{23}$$

This method of differencing automatically centers the equations. Values of the unknowns are computed on a mesh of points as shown in Fig. 3. The results of various computations using this scheme of differencing are given by Lax in [5].

Still a different category of difference schemes is available under the heading of "implicit" schemes. In schemes of this category the values of the unknown at *several* mesh points on the advance time line appear in the equations. The equations must then be solved for this set of unknowns. For example, one such scheme [8] for equations (6) is

$$\begin{aligned} \frac{y_j^{n+1} - y_j^n}{\Delta t} &= c\, \frac{w_{j+\frac{1}{2}}^n - w_{j-\frac{1}{2}}^n + w_{j+\frac{1}{2}}^{n+1} - w_{j-\frac{1}{2}}^{n+1}}{2\Delta x} \\ \frac{w_{j-\frac{1}{2}}^{n+1} - w_{j-\frac{1}{2}}^n}{\Delta t} &= c\, \frac{y_j^{n+1} - y_{j-1}^{n+1} + y_j^n - y_{j-1}^n}{2\Delta x} \end{aligned} \tag{24}$$

This scheme has the advantage of being stable no matter what mesh ratio is used. [In proving that this is so the reader will find that instead of a vector difference equation of the form

$$u(x, \Delta t) = Mu(x, 0)$$

where M is an amplification matrix—as was obtained, for example, in (19)—he will arrive at an equation of the form

$$Nu(x, \Delta t) = Mu(x, 0)$$

since the values at the advanced time step are implicit. However, the latter system cannot be solved at all unless the matrix N has an inverse, and so in this case the matrix $N^{-1}M$ becomes the amplification matrix.]

The complication introduced by the requirement of solving an implicit set of equations can be simplified by the following useful algorithm [8].

Assume that along a line of constant t a three-term recursion formula is given for y [this will be the case if w is eliminated between the two equations (24)]. Assume further that the end values y_0 and y_J are known. Let the recursion have the form

$$A_j y_{j+1} + B_j y_j + C_j y_{j-1} = D_j \qquad (a)$$

for known A_j, B_j, and C_j. Then assume the following form for y_j:

$$y_j = E_j y_{j+1} + F_j \qquad (b)$$

or

$$y_{j-1} = E_{j-1} y_j + F_{j-1} \qquad (c)$$

Substituting (*c*) into (*a*) we find

$$y_j = \frac{-A_j}{B_j + C_j E_{j-1}} y_{j+1} + \frac{D_j - C_j F_{j-1}}{B_j + C_j E_{j-1}} \qquad (d)$$

Comparing (*b*) and (*d*) we see that we must have

$$E_j = \frac{-A_j}{B_j + C_j E_{j-1}} \qquad (e)$$

$$F_j = \frac{D_j - C_j F_{j-1}}{B_j + C_j E_{j-1}} \qquad (f)$$

Equations (*e*) and (*f*) are a recursion formula for the unknowns E_j and F_j. For $j = 0$ one takes $E_0 = 0$ and $F_0 = y_0$ to satisfy (*b*) at the left-hand end point. Then (*e*) and (*f*) are used to find successive E's and F's up to E_J and F_J. Finally, using y_J, the right-hand value, in (*c*), the y values can be computed back to y_1 using the known values for E_{j-1} and F_{j-1}, and the line has been computed.

So-called multilevel difference schemes which use more than two time levels can also be evolved and analyzed, but the author knows of no case where they have proved particularly advantageous for hyperbolic equations. Also, of course, the storage requirements for machine solution are increased by adding the requirement of storing another line of computation.

In passing, we remark that a scheme using a characteristic grid together with uniform time intervals has been proposed by Hartree [13]. This method was discussed in detail in the previous chapter.

There is not space within the confines of this chapter to discuss adequately certain very interesting phenomena which arise in systems of *nonlinear* hyperbolic partial differential equations. The most notable of these enters in the case of the equations of compressible flow in fluid dynamics, and is of course the appearance of shock waves within the field of flow. Various numerical methods have been developed for treating shocks, but the science, especially for flow in more than one dimension, is quite young. A field of flow containing a shock is difficult to calculate because one must locate the shock, give it the right velocity, and satisfy the Rankine-Hugoniot shock transition conditions relating the state of the flow on the two sides of the shock (cf. [14])—all while computing the variables of the flow field. Two main methods of calculation have been used. The first of these is called "shock-fitting" and endeavors to locate the shock properly, adjust the field, relocate the shock, etc., in an iterative fashion. An example of such a computation is given by Goldstine and von Neumann [15]. The second method is to introduce an additional "pseudo-viscosity" term into the equations. Such a method was suggested by von Neumann and Richtmyer [16] and has been studied from a different point of view by Lax [5]. The viscosity terms have little effect in shock-free regions, but as shock waves develop the terms both locate the shocks and "smear" them out somewhat so that the calculation can be continued even into regions containing shocks.

Not only for the nonlinear case, but also sometimes even in the linear case certain difficulties can arise. For example, the system of partial differential equations can change type in various regions of a calculation. The very simple equation

$$y \frac{\partial^2 u}{\partial x^2} + \frac{\partial^2 u}{\partial y^2} = 0 \qquad (25)$$

is elliptic in the region $y > 0$ and hyperbolic in the region $y < 0$. Since quite opposing finite difference schemes must be used for the two types of equations (see Chapter 13), the computation is very difficult. For the nonlinear case where the coefficient which changes sign depends on the solution itself the problem becomes even harder. Various methods have been developed for the numerical computation of such equations (see in particular [17] and [18] and the references given therein), but again the matter, especially for the nonlinear supersonic-subsonic equations of fluid mechanics, is not at all resolved.

3. SUMMARY OF THE CALCULATION PROCEDURE

Returning to more explored fields we will describe briefly a computer solution for the example problem given by (6) and (8). From equations (8)

$$y_j^{n+1} = y_j^n + c\frac{\Delta t}{\Delta x}(w_{j+1/2}^n - w_{j-1/2}^n) \qquad (26a)$$

$$w_{j-1/2}^{n+1} = w_{j-1/2}^n + c\frac{\Delta t}{\Delta x}(y_j^{n+1} - y_{j-1}^{n+1}) \qquad (26b)$$

The mixed initial value–boundary value problem will be considered with initial conditions for the differential equation as stated in (7) and with the simple boundary condition

$$y(0, t) = g(t)$$

These conditions must be given at the mesh points and we will assume that $f_1(x)$, $f_2(x)$ can be evaluated at the mesh points along the x-axis for $0 \leq x \leq X$, where X is some constant, and that $g(t)$ is defined similarly along the t-axis. Further, we assume that an integration formula is available for computing (7b) (see Chapter 22). Of course, the numerical error of integration introduced in (7b) must not be greater than the over-all error of the approximations (26).

From stability considerations we found in (21) that Δt can be taken no larger than $(1/c)\,\Delta x$, so in order to advance the calculation as fast as possible in time we will take Δt equal to $(1/c)\,\Delta x$.* The region of calculation is shown in Fig. 4; the quantity y is computed at the

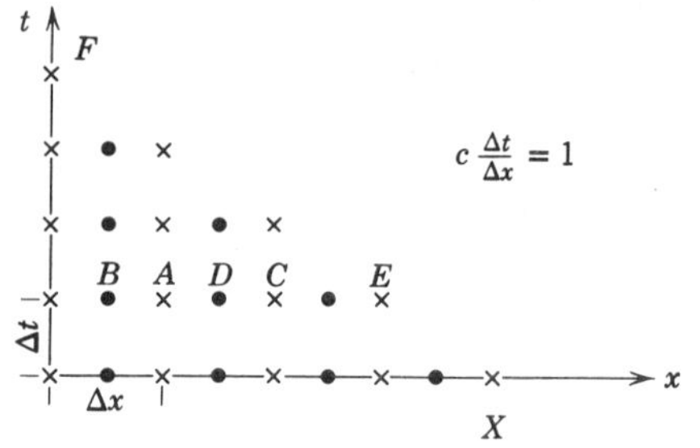

Fig. 4

points defined by small crosses and w at the points defined by small circles. Initially both y and w are known on the x-axis for $0 \leq x \leq X$, where the mesh points are $x = j\,\Delta x$, $j = 0, 1, 2, \cdots, J$, with $J\,\Delta x = X$. The values of $g(t)$ are assumed available for all t.

The computation proceeds as follows. The initial values for y_j^0 and $w_{j+1/2}^0$ are read in. To compute the next line ($t = \Delta t$) the value of $y(0, t) = g(t)$ is first determined. Then (26a) is used to find y_1^1 at point A, and (26b) to find $w_{1/2}^1$ at point B. Then (26a) gives y_2^1 at point C, and (26b) $w_{1+1/2}^1$ at point D. The computation continues across the line to point E. Any output of results for this line is then made and the computation is set up for the next line ($t = 2\Delta t$). Lines are computed successively in this fashion over the entire range of definition of the results, i.e., up to $n = N$, where

$$cN\,\Delta t = J\,\Delta x$$

In Fig. 4 the time axis is the curve Γ discussed with reference to Fig. 2. We noted that the region enclosed by Γ and the characteristic launched from the point X was completely determined; above we have just shown from a finite difference point of view that the rule is indeed true for this example.

The programming of the computation for a computer is quite simple in this case. We will assume that the same storage locations can be used for the initial data and for each line calculated thereafter. Any intermediate results desired as output can be obtained after each line is calculated. The flow chart for the program is given here.

* Since (21) is a sufficient as well as a necessary condition for stability, it is better in practice to take $\Delta t = (1/c)\,\Delta x$, rather than some smaller value, since error growth is not exponential for this value and since a smaller Δt would make for additional calculation and round-off error. Also from other considerations (e.g., the fact that the numerical spreading out of the shock wave depends on the time step [5]) the largest possible Δt is to be preferred.

4. FLOW CHART

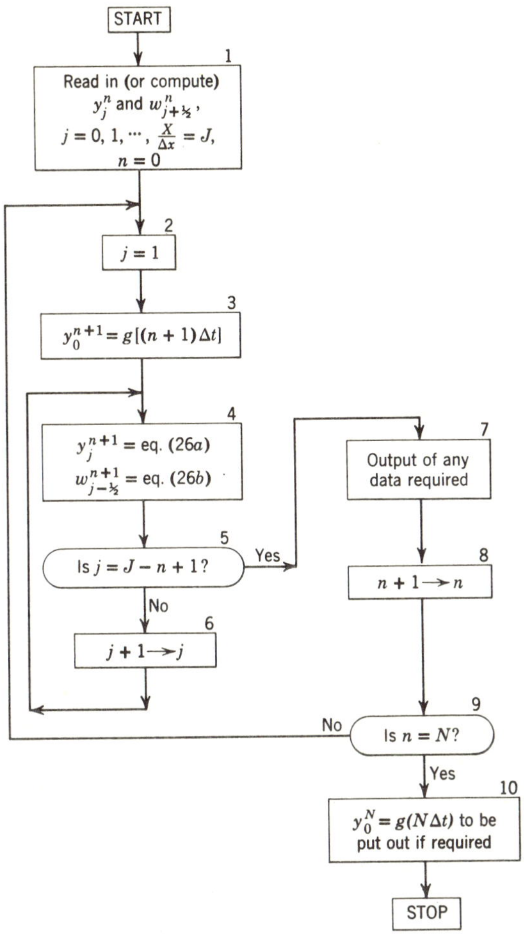

5. DESCRIPTION OF THE FLOW CHART

Box 1: Initial values are read in (or computed) and the time is set to zero.

Box 2: The computation is set to the beginning of a line.

Box 3: The value of y on the t-axis is found for this line.

Box 4: The value of y is computed at one mesh step away from the axis on this line, and then the value of w at one-half mesh step is evaluated.

Box 5: If the line has been completed go to box 7.

Box 6: Otherwise advance the j (x-direction) counter and return to box 4.

Box 7: Put out any data required from this line (it will be written over by the next lines computed).

Box 8: Advance the time step.

Box 9: If the computation has been finished up to the final point go to box 10. Otherwise return to box 2.

Box 10: Compute the final point on the time axis and record it.

6. SUBROUTINES

Besides the input and output routines required, the only subroutines used are:

(a) The evaluation of the equations (26).

(b) The evaluation of $g(n\,\Delta t)$ given n and Δt.

7. SAMPLE PROBLEM

The initial steps of a calculation for the problem described in Section 3 are given in Table 1. The conditions here are taken to be:

$$y(x, 0) = f_1(x) = \sin x$$

$$\frac{\partial y}{\partial x}(x, 0) = f_2(x) = -\cos x$$

$$y(0, t) = g(t) = \sin t$$

and the constant c is taken as 1. The solution in this case is $y = \sin(x + t)$. We assume that $w(x, 0)$ has been obtained from (7b) although in this simple case w clearly can be taken equal to $\sin(x + t)$. We set $\Delta x = \Delta t = .024$.

Table I

x	$y(x, 0)$	$w(x, 0)$	$y(x, \Delta t)$	$w(x, \Delta t)$	$y(x, 2\Delta t)$
0	0		.02399		.04798
.012		.01199		.03599	
.024	.02399		.04799		.07195
.036		.03599		.05995	
.048	.04798		.07195		.09588
.060		.05996		.08388	
.072	.07194		.09587		
.084		.08390			
.096	.09585				

From the analytic solution we note that y should remain constant on the characteristic $x + t$ = constant. This is approximately true here—the error is that of the truncation error and is of order of $(\Delta x)^2$ as one would expect. The computation pattern shows that the domain of dependence of the value of $y(2\Delta x, 2\Delta t)$ is the portion $0 \leq x \leq .096$ of the x-axis.

8. MEMORY REQUIREMENTS

For the data, storage locations are required for the values only along one line in the one-dimensional case. This is true because as new data are computed they can be stored over the previous ones without destroying any required values. In Section 7, for example, the values for $y(x, \Delta t)$ can be placed in the locations of $y(x, 0)$, and the values for $w(x, \Delta t)$ in the locations of $w(x, 0)$.

The program, exclusive of input-output, and the evaluation of the functions $f_1(x)$, $f_2(x)$, and $g(t)$ can be written for the simple sample problem in less than thirty instructions.

For more general systems, such as finite difference equivalents of (1), the storage for data will be the number of mesh points on the plane $t = 0$ multiplied by m, the number of components of the vector u. The program will depend, of course, on the complexity of the system and the finite difference equations.

9. ESTIMATION OF THE RUNNING TIME

For the problem of Section 7 the running time, based on J points along the x-axis, for multiplication and addition times of μ and ν respectively, would be in the neighborhood of

$$J^2(6\nu + \mu) + J(4\nu + g)$$

where g is the time required for evaluating $g(t)$. The estimate omits time for input-output and initial value calculations. In general, no estimate can be made without knowing the form of the equations and the form and amount of data to be read in and of results to be taken out.

10. REFERENCES

1. O. A. Ladyzhenskaia, The Method of Finite Differences in the Theory of Partial Differential Equations, *Uspekhi Mathematicheskikh Nauk*, vol. 12, 1957, pp. 123–148. (Translation at the Institute of Mathematical Sciences, New York University.)
2. P. D. Lax, Partial Differential Equations, Lecture Notes, New York University, 1950.
3. P. D. Lax, Differential Equations, Difference Equations and Matrix Theory, *Communs. Pure Appl. Math.*, vol. 11, 1958, pp. 175–194.
4. Stephen H. Crandall, *Engineering Analysis, a Survey of Numerical Procedures*, McGraw-Hill Book Co., New York, 1956.
5. P. D. Lax, Weak Solutions of Nonlinear Hyperbolic Equations and Their Numerical Computation, *Communs. Pure Appl. Math.*, vol. 7, 1954, pp. 159–192.
6. R. Courant and D. Hilbert, *Methoden der Mathematischen Physik*, vol. 2, Springer, Berlin, 1937.
7. S. G. Hahn, Stability Criteria for Difference Schemes, *Communs. Pure Appl. Math.*, vol. 11, 1958, pp. 243–255.
8. R. D. Richtmyer, *Difference Methods for Initial-Value Problems*, Interscience Publishers, New York, 1957.
9. P. D. Lax and R. D. Richtmyer, Survey of the Stability of Linear Finite Difference Equations, *Communs. Pure Appl. Math.*, vol. 9, 1956, pp. 267–293.
10. G. G. O'Brien, M. A. Hyman, and S. Kaplan, A Study of the Numerical Solution of Partial Differential Equations, *J. Math. and Phys.*, vol. 29, 1951, pp. 223–251.
11. R. Courant, K. O. Friedrichs, and H. Lewy, Über die Partiellen Differenzengleichungen der Mathematischen Physik, *Math. Ann.*, vol. 100, 1928, pp. 32–74.
12. K. O. Friedrichs, Symmetric Hyperbolic Linear Differential Equations, *Communs. Pure Appl. Math.*, vol. 7, 1954, pp. 345–392.
13. D. R. Hartree, Some Practical Methods of Using Characteristics in the Calculation of Non-Steady Compressible Flows, Los Alamos Report LA-HU-1, 1952.
14. R. Courant and K. O. Friedrichs, *Supersonic Flow and Shock Waves*, Interscience Publishers, New York, 1948.
15. H. H. Goldstine and J. von Neumann, Blast Wave Calculation, *Communs. Pure Appl. Math.*, vol. 8, 1955, pp. 327–353.
16. J. von Neumann and R. D. Richtmyer, A Method for the Numerical Calculation of Hydrodynamic Shocks, *J. Appl. Phys.*, vol. 21, 1950, pp. 232–237.
17. Dorothy Levy, A Numerical Scheme for Solving a Boundary Value Problem for the Tricomi Equation, Report NYO-7979, AEC Computing and Applied Mathematics Center, Institute of Mathematical Sciences, New York University, 1958.
18. A. F. Filipov, On the Difference Method for Solving the Tricomi Problem (in Russian), *Prikladnaia Matematika i Mekhanika*, vol. 21, 1957, pp. 73–88.
19. R. Courant, E. Isaacson, and M. Rees, On the Solution of Nonlinear Hyperbolic Differential Equations by Finite Differences, *Communs. Pure Appl. Math.*, vol. 5, 1952, pp. 243–255.

PART V STATISTICS

Multiple regression analysis

17

M. A. Efroymson*
Esso Research and Engineering Company

I. FUNCTION

Multiple regression is used in data analysis to obtain the best fit of a set of observations of independent and dependent variables by an equation of the form:

$$y = b_0 + b_1x_1 + b_2x_2 + \cdots + b_nx_n \quad (1)$$

where y is the dependent variable; $x_1, x_2, \cdots$ are the independent variables; and $b_0, b_1, \cdots$ are the coefficients to be determined.

A multiple regression solution gives the least squares "best" value of these coefficients for a particular sample of observations. The solution also gives a measure of the reliability of each of the coefficients so that inferences can be made regarding the parameters of the population from which the sample of observations was taken.

Multiple regression can also be used to fit nonlinear equations of the form:

$$y = b_0 + b_1z_1 + b_2z_1^2 + b_3(z_1z_2) + \cdots + b_nf_n(z_1, z_2, \cdots, z_m) \quad (2)$$

* Acknowledgement is made of the assistance of Dr. John W. Tukey of Princeton University, who was consulted in the development of the method. Also, acknowledgement is made of the assistance of O. Schricker and F. H. Blanding and other personnel of the Esso Research and Engineering Company.

Such an equation is made equivalent to (1) by the substitutions

$$\begin{aligned} x_1 &= z_1 \\ x_2 &= z_1^2 \\ x_3 &= z_1z_2 \\ &\vdots \\ x_n &= f_n(z_1, z_2, \cdots, z_m) \end{aligned}$$

For a large number of variables, any method of regression analysis requires a large number of calculations. Most methods of regression analysis given in the literature [1], [2] are based on techniques particularly adaptable to a desk calculator where a minimum transcription of intermediate answers is desirable.

In the stepwise procedure, to be described below, intermediate results, which are not even recorded by normal calculation methods, are used to give valuable statistical information at each step in the calculation. These intermediate answers are also used to control the method of calculation. Essentially, without adding greatly to the number of arithmetic steps, a number of intermediate regression equations are obtained, as well as the complete multiple regression equation. These equations are obtained by adding one variable at a time and thus give the following intermediate equations:

$$y = b_0 + b_1x_1 \quad (3)$$

$$y = b_0' + b_1'x_1 + b_2'x_2 \quad (4)$$

$$y = b_0'' + b_1''x_1 + b_2''x_2 + b_3''x_3 \quad (5)$$

$$\cdot$$
$$\cdot$$
$$\cdot$$

The variable added is that one which makes the greatest improvement in "goodness of fit." The coefficients represent the best values when the equation is fitted by the specific variables included in the equation.

An important property of the stepwise procedure is based on the facts that (a) a variable may be indicated to be significant in any early stage and thus enter the equation, and (b) after several other variables are added to the regression equation, the initial variable may be indicated to be insignificant. The insignificant variable will be removed from the regression equation before adding an additional variable. Therefore, only significant variables are included in the final regression.

2. MATHEMATICAL DISCUSSION

a. Symbols Used

n = number of independent + dependent variables

m = number of sets of observations

x_{it} = tth observation of ith variable

$x_{nt} = y_t$ = tth observation of dependent variable

w_t = weighting factor of tth observation

$\bar{x}_i$ = weighted mean of ith variable

s_{ij} = weighted residual sum of squares and cross products of ith and jth variables

$\sigma_i = \sqrt{s_{ii}}$

r_{ij} = simple correlation coefficient of ith and jth variables

$N_{\min}, N_{\max}$ = subscripts of selected independent variables

$V_{\min}$ = variance increase by deleting a variable $N_{\min}$

$V_{\max}$ = variance reduction by adding a variable $N_{\max}$

β_i = "true" value of coefficient of ith variable

b_i = estimated coefficient of ith variable

s_y = standard error of dependent variable

s_{bi} = standard error of coefficient of ith variable

$\hat{y}_t$ = predicted value of dependent variable for tth observation

Dev. = deviation between observed and predicted value of tth observation of dependent variable $(y_t - \hat{y}_t)$

c_{ij} = element of inverse matrix of r_{ij}

F_1 = F value for entering variable

F_2 = F value for removing variable

b. Derivation of the Method

Let us assume that y is to be estimated by the equation

$$y_t - \bar{y} = \sum_{i=1}^{n-1} \beta_i(x_{it} - \bar{x}_i) \qquad (t = 1, \cdots, m)$$

The error of the estimate of the tth observed value of y is

$$e_t = (y_t - \bar{y}) - \sum_{i=1}^{n-1} \beta_i(x_{it} - \bar{x}_i) \qquad (t = 1, \cdots, m)$$

The purpose of regression analysis is to determine the β_i in such a way that the length of the vector $e = (e_t)$ is minimized. But

$$\|e\|^2 = (e, e) = \sum_{t=1}^{m} \{(y_t - \bar{y}) - \sum_{i=1}^{n-1} \beta_i(x_{it} - \bar{x}_i)\}^2$$

Taking partial derivatives with respect to one of the β_i and equating the result to zero, we get

$$\sum_{j=1}^{n-1} \left\{ \sum_{t=1}^{m} (x_{it} - \bar{x}_i)(x_{jt} - \bar{x}_j) \right\} \beta_j = \sum_{t=1}^{m} (x_{it} - \bar{x}_i)(y_t - \bar{y}) \quad (6)$$

These are the normal equations. They are a set of $n - 1$ simultaneous linear algebraic equations in the β_j and can be solved by any method. The technique that we will adopt here is, in essence, precisely that described by Orden in Chapter 2, i.e., the Gaussian elimination method. However, we note an extremely interesting feature of the method as it applies to the regression problem: not only is the final solution of interest, but at each stage in the elimination we have a "partial regression equation," where the variables which have already been eliminated are in the equation, and the others are not. We shall exploit this observation by making a decision at each stage as to what variable shall next be included in the regression.

The procedure followed is to apply linear transformations to the partitioned matrix

$$\begin{pmatrix} S & T' & I \\ T & Z & D \\ -I & B & C \end{pmatrix} \tag{7}$$

where S, C, and I are $(n-1) \times (n-1)$, T and D are $1 \times (n-1)$, B is $(n-1) \times 1$, and Z is a scalar. Specifically,

$$(S)_{ij} = s_{ij} = \Sigma(x_{it} - \bar{x}_i)(x_{jt} - \bar{x}_j)$$
$$(T)_{1j} = t_{1j} = \Sigma(x_{jt} - \bar{x}_j)(y_t - \bar{y})$$
$$Z = \Sigma(y_t - \bar{y})(y_t - \bar{y}) \quad \text{or}$$
$$\Sigma(x_{nt} - \bar{x}_n)(x_{nt} - \bar{x}_n)$$
$$\text{where} \quad x_{nt} = y_t$$
$$(T')_{i1} = (T)_{1i}$$
$$B = C = D = O \text{ (initially)}$$
$$(I)_{ij} = \delta_{ij}$$

The applications of linear transformations will, of course, cause nonzero elements to enter the B, C, and D positions.

Each successive row elimination of the S matrix results in a regression equation with one *more* variable in the regression equation. The same algorithm applied to eliminate a row in the C matrix results in a regression equation with one *less* variable in the regression equation.

At every step, the B matrix contains the regression coefficients and the C matrix contains the inverse of the *partitioned* part of the S matrix corresponding to the variables in the regression at that time.

The criterion used to select the x_i variable to add or remove from the regression is as follows:

1. If the variance contribution of a variable in the regression is insignificant at a specified F level, this variable is removed from the regression. If no variable is to be removed, then the following criterion is used.

2. If the variance reduction obtained by adding a variable to the regression is significant at a specified F level, this variable is entered into the regression.

The form given in (7) can be used directly. However, in the program outlined in the flow chart, the S, T, T', and Z matrices are normalized to obtain unity in the diagonal elements; these four matrices together form the R matrix. The elements are therefore transformed to simple correlation coefficients by

$$r_{ij} = \frac{\Sigma(x_{it} - \bar{x}_i)(x_{jt} - \bar{x}_j)}{\sqrt{\Sigma(x_{it} - \bar{x}_i)^2 \Sigma(x_{jt} - \bar{x}_j)^2}} \tag{8}$$

or, if $\sigma_i = \sqrt{\Sigma(x_{it} - \bar{x}_i)^2}$, by

$$r_{ij} = \frac{\Sigma(x_{it} - \bar{x}_i)(x_{jt} - \bar{x}_j)}{\sigma_i \sigma_j} \tag{9}$$

3. SUMMARY OF THE CALCULATION PROCEDURE

a. The Square Matrix

The full matrix discussed in Section 2b is a square matrix of order $2n - 1$. The matrix, however, can be handled in storage as an $(n \times n)$ square matrix, since all other elements are zero or unity and can be generated as needed. The basis for operating on this $(n \times n)$ square matrix is discussed in the following paragraphs.

b. Basis of Matrix Operations

The general algorithm or algebraic rules used in generating a succeeding matrix are as follows:

Given: a_{kk} is a diagonal element, k corresponding to one of the independent variables.

Take:

$$\text{The new } a_{ij} = \begin{cases} \dfrac{a_{ij}a_{kk} - a_{ik}a_{kj}}{a_{kk}} & \text{if } i \neq k; j \neq k \quad (10) \\[2ex] \dfrac{a_{kj}}{a_{kk}} & \text{if } i = k; j \neq k \quad (11) \\[2ex] -\dfrac{a_{ik}}{a_{kk}} & \text{if } i \neq k; j = k \quad (12) \\[2ex] \dfrac{1}{a_{kk}} & \text{if } i = j = k \quad (13) \end{cases}$$

c. Definition of Matrix Elements

The general algorithm applies whether an element has an *r*, *b*, *d*, or *c* designation. These designations are made by the following rules:

(a) $a_{ij} = r_{ij}$ when *both* x_i and x_j are *not* in the regression at this step. In applying this rule, the dependent variable is *never* considered to be in the regression.

(b) $a_{ij} = b_{ij}$ when x_i is in the regression and x_j is *not* in the regression. b_{ij} is a normalized regression coefficient of x_i on x_j adjusted for any other variable that is in the regression.

(c) $a_{ij} = d_{ij}$ when x_i is *not* in the regression and x_j is in the regression.

$$d_{ij} = -b_{ji}$$

(d) $a_{ij} = c_{ij}$ when *both* x_i and x_j are in the regression. The matrix with elements c_{ij} is the inverse of the r_{ij} matrix for all *i*'s and *j*'s that are in the regression at this step.

The specific application of the general algorithm to generate the elements of a new matrix based on the *r*, *b*, *d*, and *c* designations of elements is summarized in Table 1.

d. Selection of the Key Element

The logic used in selecting the key element, i.e., the a_{kk} used in generating a new matrix, is shown in the flow chart, to be discussed in detail later. V_i terms are calculated as $(a_{in}a_{ni})/a_{ii}$ for each $a_{ii} > \text{Tol}$.

The control on the size of a_{ii} reduces the possibility of degeneracy when an "independent" variable is approximately a linear combination of other independent variables. If the multiple correlation coefficient between a number of so-called independent variables is so large that most of the variability in one "independent" variable is related to the other independent variables, this variable will not be placed in the regression.

If V_i is negative, then x_i is in the regression at this stage. If V_i is positive, then x_i is not in the regression at this time. The criteria used to select the x_i variable to add or remove from the regression are as follows:

(a) x_i corresponding to the minimum $|V_i|$ for negative V_i's is removed from the regression if the variance contribution of x_i is insignificant at this step. This rule takes priority over adding a variable. If no variable is to be removed, then the following rule is applied:

(b) x_i corresponding to the maximum V_i for positive values of V_i is added to the regression, if the variance reduction due to adding x_i is significant.

Applying either rule above, the a_{ii} corresponding to the variable selected is used as the key element. The general algorithm given

Table I

	General Algorithm	Condition of Variables at Start of Step: *i* in regression, *j* in regression	*i* in regression, *j* not in regression	*i* not in regression, *j* in regression	*i* not in regression, *j* not in regression
$i \neq k$, $j \neq k$	$a_{ij} - \dfrac{a_{ik}a_{kj}}{a_{kk}} = a'_{ij}$	$c_{ij} - \dfrac{b_{ik}d_{kj}}{r_{kk}} = c'_{ij}$	$b_{ij} - \dfrac{b_{ik}r_{kj}}{r_{kk}} = b'_{ij}$	$d_{ij} - \dfrac{r_{ik}d_{kj}}{r_{kk}} = d'_{ij}$	$r_{ij} - \dfrac{r_{ik}r_{rj}}{r_{kk}} = r'_{ij}$
		$c_{ij} - \dfrac{c_{ik}c_{kj}}{c_{kk}} = c'_{ij}$	$b_{ij} - \dfrac{c_{ik}b_{kj}}{c_{kk}} = b'_{ik}$	$d_{ij} - \dfrac{d_{ik}c_{kj}}{c_{kk}} = d'_{ij}$	$r_{ij} - \dfrac{d_{ik}b_{kj}}{c_{kk}} = r'_{ij}$
$i = k$, $j \neq k$	$a_{kj}/a_{kk} = a'_{kj}$	—	—	$d_{kj}/r_{kk} = c_{kj}$	$r_{kj}/r_{kk} = b_{kj}$
		$c_{kj}/c_{kk} = d_{kj}$	$b_{kj}/c_{kk} = r_{kj}$	—	—
$i \neq k$, $j = k$	$-a_{ik}/a_{kk} = a'_{ik}$	—	$-b_{ik}/r_{kk} = c_{ik}$	—	$-r_{ik}/r_{kk} = d_{ik}$
		$-c_{ik}/c_{kk} = b_{ik}$	—	$-d_{ik}/c_{kk} = r_{ik}$	—
$i = j = k$	$1/a_{kk} = a'_{kk}$	—	—	—	$1/r_{kk} = c_{kk}$
		$1/c_{kk} = r_{kk}$	—	—	—

earlier is then used to calculate the elements of the next matrix.

e. Coefficients and Standard Deviation

In addition to the matrix elements a_{ij}, which are calculated at each step, there are also stored in the computer the "degrees of freedom" ϕ, the mean of each variable x_i, and the σ_i used to obtain the correlation coefficients. From these data in storage, the standard error of y, the coefficients for the regression equation, and the standard errors of these coefficients are calculated at the end of each stage.

Standard Error of Dependent Variable

The standard error of y at each step is equal to

$$s_y = \sigma_n \sqrt{r_{nn}/\phi} \tag{14}$$

where x_n = dependent variable.

Calculation of Regression Coefficients

The regression coefficients at each step are calculated as follows:

$$b_i = b_{in} \frac{\sigma_n}{\sigma_i} \tag{15}$$

where x_n = dependent variable

x_i = a variable in regression at this stage

The constant in the regression equation at each step is calculated as $b_0 = \bar{y} - \Sigma b_i \bar{x}_i$.

Standard Errors of Regression Coefficients

The standard error of each regression coefficient is calculated as follows:

$$s_{bi} = \frac{s_y}{\sigma_i} \sqrt{c_{ii}} \tag{16}$$

f. Calculation of Predicted Values for the Dependent Variable and Deviations Between Actual and Predicted Values

The final calculation in the stepwise procedure predicts the value of the dependent variable for each set of observations, based on the final regression equation. The deviation between the actual and predicted value of the dependent variable for each set of operations is also calculated.

4. FLOW CHART

The flow chart appears on pages 196-199.

5. DESCRIPTION OF THE FLOW CHART

Box 1: The observed data for the m sets of independent and dependent variables x_{it} are entered together with the weighting factor w_t for each set of observations. When a regression is made on a homogeneous set of data, each set of observations has equal weight and a weighting factor of one is used. There are instances, however, where the reliability of certain sets is better than that of others, and weighting factors can be applied in relation to their estimated reliabilities. It is generally desirable to normalize the weighting factors in advance so that their average is unity. Unless the sum of weighting factors is equal to the number of observations, the standard errors of the regression coefficients and the criterion of significance will be distorted.

F_1 and F_2, respectively, are F values for entering and removing variables from the regression; $F_1 \geq F_2$. Typical F values are ≈ 4.

Box 2: For matrix purposes consider X as an $[m \times (n + 1)]$ matrix where the first column is a column of one's and the remaining columns are the independent and dependent variables. The W matrix is an $m \times m$ diagonal matrix with w_t's as the diagonal elements. The values in box 2 can be calculated by the matrix multiplication $X'WX$. If all w's are equal to one, then the elements can be calculated as $X'X$. In each case a symmetric matrix is generated and only the upper triangle of the result need be retained.

Box 3: The weighted means and weighted residual sums of squares and cross products are calculated from the quantities obtained from box 2. These calculations also give the upper triangular part of a symmetric matrix.

Box 4: The residual sums of squares and cross products are normalized to a simple correlation coefficient matrix. The upper triangular matrix is expanded into a square matrix.

Box 5: Initializes the procedure for the determination of the most significant variable to be added to the regression.

Box 6: The control on the size of a_{ii} reduces the possibility of degeneracy when an "independent" variable is approximately a linear

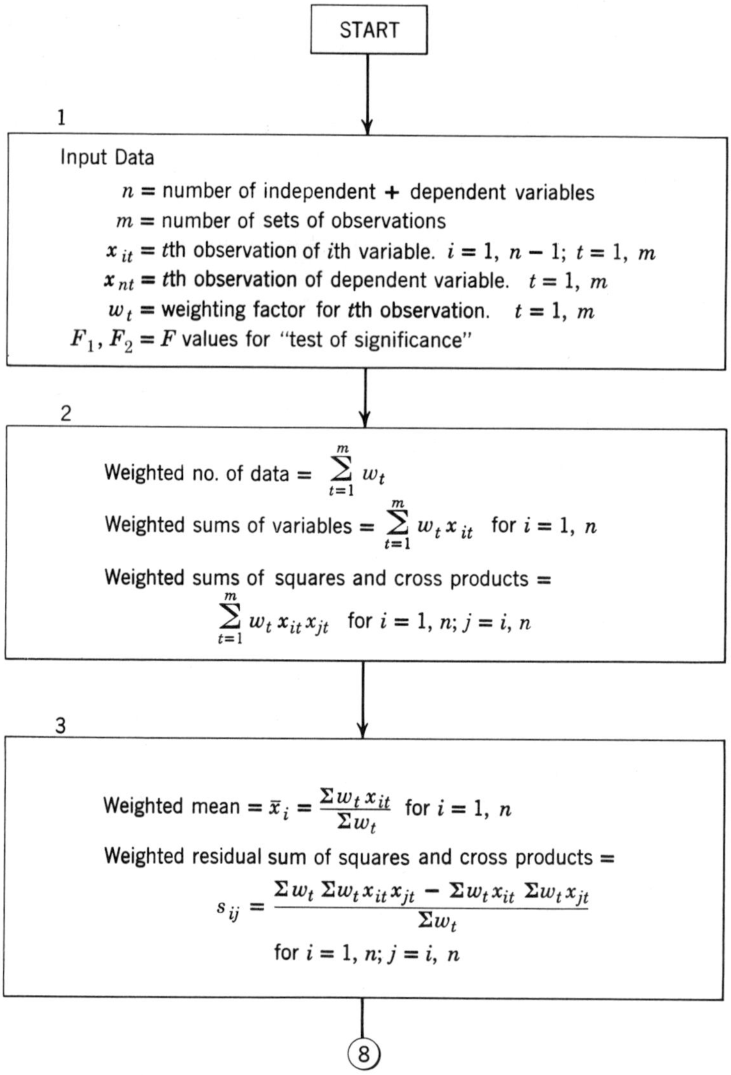
START
1
Input Data
n = number of independent + dependent variables
m = number of sets of observations
x_{it} = tth observation of ith variable. $i = 1, n - 1;\ t = 1, m$
x_{nt} = tth observation of dependent variable. $t = 1, m$
w_t = weighting factor for tth observation. $t = 1, m$
$F_1, F_2 = F$ values for "test of significance"
2
Weighted no. of data = $\sum_{t=1}^{m} w_t$
Weighted sums of variables = $\sum_{t=1}^{m} w_t x_{it}$ for $i = 1, n$
Weighted sums of squares and cross products =
$\sum_{t=1}^{m} w_t x_{it} x_{jt}$ for $i = 1, n;\ j = i, n$
3
Weighted mean = $\bar{x}_i = \frac{\Sigma w_t x_{it}}{\Sigma w_t}$ for $i = 1, n$
Weighted residual sum of squares and cross products =
$s_{ij} = \frac{\Sigma w_t\ \Sigma w_t x_{it} x_{jt} - \Sigma w_t x_{it}\ \Sigma w_t x_{jt}}{\Sigma w_t}$
for $i = 1, n;\ j = i, n$
8

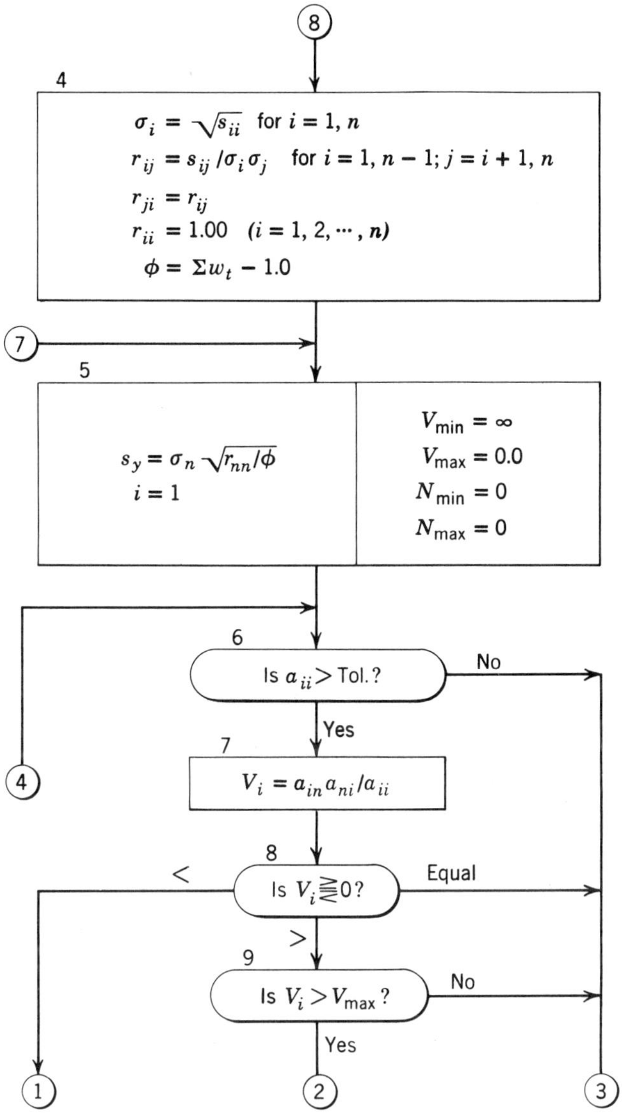
8
4
$\sigma_i = \sqrt{s_{ii}}$ for $i = 1, n$
$r_{ij} = s_{ij}/\sigma_i\sigma_j$ for $i = 1, n-1; j = i+1, n$
$r_{ji} = r_{ij}$
$r_{ii} = 1.00$ $(i = 1, 2, \cdots, n)$
$\phi = \Sigma w_t - 1.0$
7
5
$s_y = \sigma_n \sqrt{r_{nn}/\phi}$
$i = 1$
$V_{min} = \infty$
$V_{max} = 0.0$
$N_{min} = 0$
$N_{max} = 0$
6
Is $a_{ii} >$ Tol.?
No
Yes
7
4
$V_i = a_{in}a_{ni}/a_{ii}$
8
<
Is $V_i \gtreqless 0$?
Equal
>
9
Is $V_i > V_{max}$?
No
Yes
1
2
3

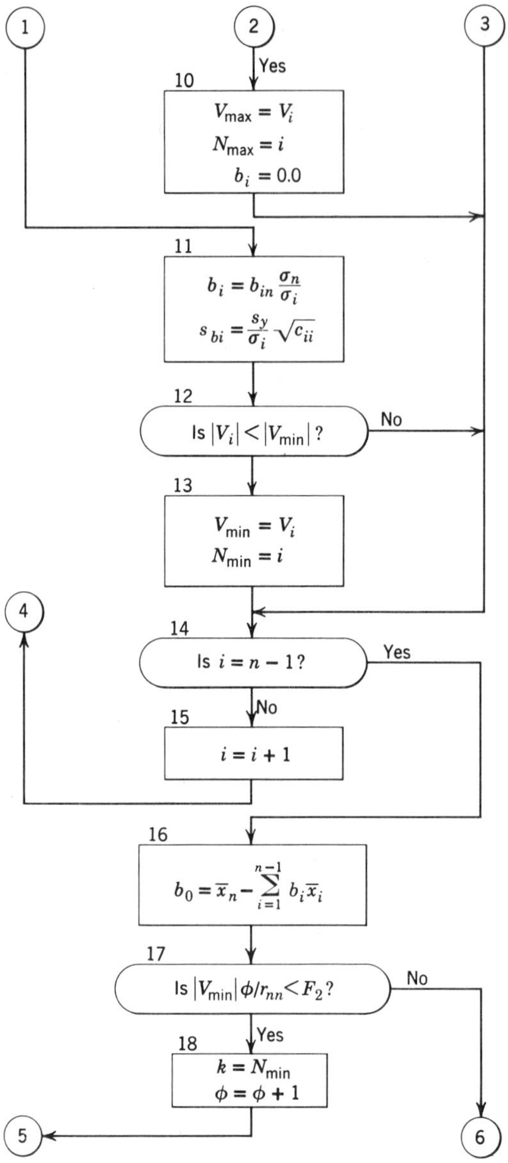
1
2
3
Yes
10
$V_{max} = V_i$
$N_{max} = i$
$b_i = 0.0$
11
$b_i = b_{in} \frac{\sigma_n}{\sigma_i}$
$s_{bi} = \frac{s_y}{\sigma_i} \sqrt{c_{ii}}$
12
Is $|V_i| < |V_{min}|$?
No
13
$V_{min} = V_i$
$N_{min} = i$
4
14
Is $i = n - 1$?
Yes
No
15
$i = i + 1$
16
$b_0 = \bar{x}_n - \sum_{i=1}^{n-1} b_i \bar{x}_i$
17
Is $|V_{min}| \phi / r_{nn} < F_2$?
No
Yes
18
$k = N_{min}$
$\phi = \phi + 1$
5
6

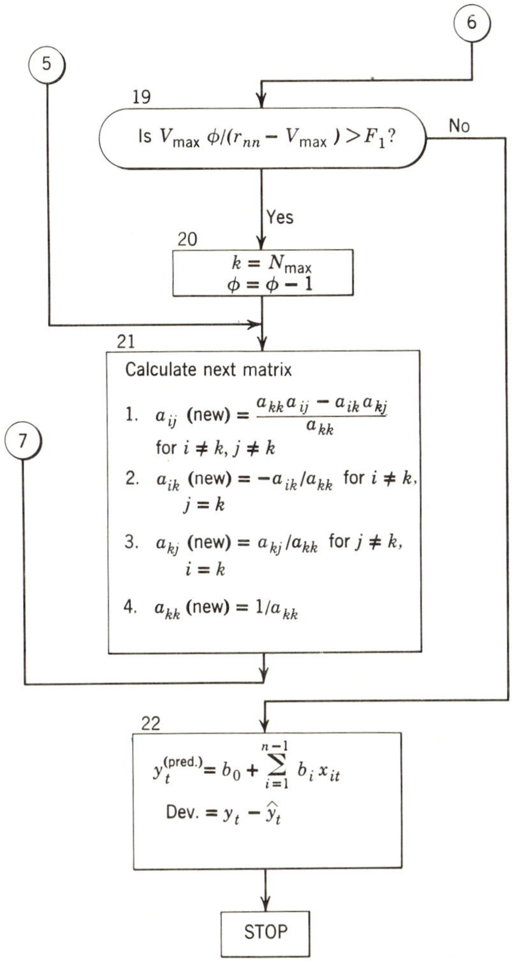
6
5
19
Is $V_{max}\,\phi/(r_{nn} - V_{max}) > F_1$?
No
Yes
20
$k = N_{max}$
$\phi = \phi - 1$
21
Calculate next matrix
1. a_{ij} (new) $= \dfrac{a_{kk}a_{ij} - a_{ik}a_{kj}}{a_{kk}}$
for $i \neq k, j \neq k$
2. a_{ik} (new) $= -a_{ik}/a_{kk}$ for $i \neq k$, $j = k$
3. a_{kj} (new) $= a_{kj}/a_{kk}$ for $j \neq k$, $i = k$
4. a_{kk} (new) $= 1/a_{kk}$
7
22
$y_t^{(pred.)} = b_0 + \sum_{i=1}^{n-1} b_i x_{it}$
Dev. $= y_t - \hat{y}_t$
STOP

```
PROBLEM NO     1

NO OF DATA =   15

NO OF VARIABLES =  4

WEIGHTED DEGREES OF FREEDOM =  15.00

F LEVEL TO ENTER VARIABLE =  2.500

F LEVEL TO REMOVE VARIABLE =  2.500


                              SUM OF VARIABLES

   SUM X( 1) =   343.0000     SUM X( 2) =   530.0000     SUM X( 3) =   692.0000
   SUM   Y   =   270.0000

                   RAW SUM OF SQUARES AND CROSS PRODUCTS

   X( 1) VS X( 1) =   9817.000000   X( 1) VS X( 2) =  12793.000000   X( 1) VS X( 3) =  16561.000000
   X( 2) VS X( 2) =  19456.000000   X( 2) VS X( 3) =  24935.000000   X( 3) VS X( 3) =  34936.000000
   X( 1) VS   Y   =   6511.000000   X( 2) VS   Y   =   9657.000000   X( 3) VS   Y   =  13080.000000
   Y     VS   Y   =   5088.000000

                        AVERAGE VALUE OF VARIABLES

   SUM X( 1) =    22.8667     SUM X( 2) =    35.3333     SUM X( 3) =    46.1333
   SUM   Y   =    18.0000

                 RESIDUAL SUMS OF SQUARES AND CROSS PRODUCTS

   X( 1) VS X( 1) =   1973.733398   X( 1) VS X( 2) =    673.666748   X( 1) VS X( 3) =    737.266846
   X( 2) VS X( 2) =    729.333740   X( 2) VS X( 3) =    484.333740   X( 3) VS X( 3) =   3011.733643
   X( 1) VS   Y   =    337.000061   X( 2) VS   Y   =    117.000122   X( 3) VS   Y   =    624.000122
   Y     VS   Y   =    228.000000
```

```
                              CORRELATION COEFFICIENTS

   X( 1) VS X( 2)  =      0.561484     X( 1) VS X( 3)  =      0.302393
   X( 2) VS X( 3)  =      0.326793
   X( 1) VS   Y    =      0.502364     X( 2) VS   Y    =      0.286917     X( 3) VS   Y    =      0.753024

 STANDARD ERROR OF Y =    4.035556

STEP NO.     1
  VARIABLE ENTERING        3
  F LEVEL       17.0263
  STANDARD ERROR OF Y =    2.7556
  CONSTANT      8.44165
           VARIABLE    COEFFICIENT     STD ERROR OF COEF

            X- 3         0.20719            0.05021

STEP NO.     2
  VARIABLE ENTERING        1
  F LEVEL        2.8472
  STANDARD ERROR OF Y =    2.5785
  CONSTANT      7.25256

           VARIABLE       COEFFICIENT     STD ERROR OF COEF

            X- 1            0.10274            0.06089
            X- 3            0.18204            0.04929

   DIAGONAL ELEMENTS

VAR.NO.        VALUE

   1          1.100645
   2          0.657604
   3          1.100645
```

PREDICTED VS. ACTUAL RESULTS

RUN NO.	ACTUAL	PREDICTED	DEVIATION
1.	15.00000	20.37043	-5.37043
2.	16.00000	14.41007	1.58993
3.	14.00000	13.02194	0.97806
4.	22.00000	20.13593	1.86407
5.	24.00000	21.87363	2.12637
6.	19.00000	19.36644	-0.36644
7.	13.00000	14.48380	-1.48380
8.	15.00000	19.31954	-4.31954
9.	23.00000	21.31301	1.68699
10.	12.00000	14.80098	-2.80098
11.	25.00000	24.25464	0.74536
12.	17.00000	15.82286	1.17714
13.	18.00000	16.57446	1.42554
14.	19.00000	17.74260	1.25740
15.	18.00000	16.50967	1.49033

INPUT DATA

RUN NO.	X(1)	X(2)	X(3)	Y
1.	32.	48.	54.	15.
2.	36.	33.	19.	16.
3.	3.	28.	30.	14.
4.	12.	33.	64.	22.
5.	36.	34.	60.	24.
6.	24.	36.	53.	19.
7.	19.	42.	29.	13.
8.	20.	33.	55.	15.
9.	27.	36.	62.	23.
10.	15.	22.	33.	12.
11.	45.	46.	68.	25.
12.	9.	28.	42.	17.
13.	11.	32.	45.	18.
14.	33.	34.	39.	19.
15.	21.	45.	39.	18.

combination of other independent variables. If the multiple correlation coefficient between a number of so-called independent variables is so large that practically all the variability in one "independent" variable is related to the other independent variables, this variable will not be placed in the regression. The value of Tol. (tolerance) is arbitrary but has generally been used between .001 and .00001.

Box 8: If V_i is negative, then x_i is in the regression at this stage. If V_i is positive, then x_i is not in the regression at this time.

Boxes 9 and 10: x_i corresponding to the maximum V_i is the variable which will cause the greatest variance reduction by introducing it into the regression.

Box 11: The regression coefficient and its standard error are calculated for each variable in the regression at this stage.

Boxes 12 and 13: x_i corresponding to the minimum $|V_i|$ for negative V_i's will cause the least variance increase by deleting this variable from the regression.

Box 17: If the variance contribution of a variable in regression is insignificant, this variable is to be removed before adding another variable.

Box 19: If the variance reduction due to adding x_i is significant, this variable is to be added to the regression.

Box 21: The algorithm for deleting or adding a variable as given in box 21 is identical for either deleting or adding a variable. At the end of these calculations, the control is always transferred to box 5.

Box 22: The program is terminated by calculating the predicted values of the dependent variables and the deviation between actual and predicted values. The entrance to this terminating box is always from box 19.

6. SUBROUTINES

(a) Square root.

(b) Although not necessary, general-purpose matrix routines can simplify programming.

7. SAMPLE PROBLEM FOR A STEPWISE REGRESSION

On pages 200–202 are shown the input data, intermediate results, and output quantities for a problem with four variables and fifteen observations of each. The data were taken, as shown, from an actual computer printout. The identifications are, it is hoped, self-explanatory.

8. MEMORY REQUIREMENTS

Table 2 lists the memory space requirements.

Table 2

	Total Words Required	(Desirable in High-Speed Memory)
x_{it}	$N \times M$	N or $2M$*
w_t	M	1 or M*
A†	$(N + 1)(N + 1)$	
σ_i	N	
b_i	N	
s_y	1	
s_{bi}	N	
y	M	
Dev.	M	
Program	1500–3500‡	

* The program can be coded so that either only one row or two columns of an $N \times M$ matrix of observations are required in memory.

† The A matrix space has multiple use as (a) a sum of squares and cross products, (b) residual sums of squares and cross products, (c) simple correlation coefficients, (d) partial correlation coefficients, and (e) an inverse matrix.

‡ The size of the program will depend on the amount of output desired and on options for alternate output forms.

9. ESTIMATION OF THE RUNNING TIME

There are approximately $(n + 2)^2(m + n)$ additions and multiplications and $(n + 2)^2 \times (n)/2$ divisions. Allowing double the amount of time for bookkeeping, input, and output gives for the running time

$$2(n + 2)^2[m(\nu + \mu) + n(\nu + \mu + \omega/2)]$$

where ν is the addition time, μ the multiplication time, and ω the division time.

10. REFERENCES

1. C. A. Bennett and N. L. Franklin, *Statistical Analysis in Chemistry and the Chemical Industry*, John Wiley & Sons, New York, 1954.
2. A. Hald, *Statistical Theory with Engineering Applications*, John Wiley & Sons, New York, 1952.

18 Factor analysis

Harry H. Harman
System Development Corporation

1. FUNCTION

In the relatively short period since its introduction by Spearman [1] in 1904, "factor analysis" has had an interesting and colorful history. The method came into being specifically to provide mathematical models for the explanation of psychological theories of human ability as proposed by Spearman [1], Kelley [2], Thurstone [3], Thomson [4], and Burt [5]. However, the mathematical techniques inherent in factor analysis certainly are not limited to psychological applications. For a detailed description of the concepts, theory, and techniques of factor analysis, rather than the development of "psychological theories," the reader is referred to Harman [6], upon which the present chapter is based.

Factor analysis in psychology and multivariate analysis in statistics have developed more or less independently, although they have much in common and could very profitably benefit from cross-fertilization. One problem which represents the common interests is the determination of the dimensionality of a multivariate normal distribution. As viewed by factor analysts, the question is the extent to which a set of variables can be described in terms of a smaller number of categories or factors. While formal tests of significance have been developed by Hotelling [7], Lawley [8], Bartlett [9], and Rao [10], factor analysts more often are content to select a minimum number of factors that have "practical" significance.

2. MATHEMATICAL DISCUSSION

The simplest mathematical model—a linear one—is employed in factor analysis to represent a variable in terms of the factors. The factors are usually put in two classes: *common factors*, being those involved in more than one variable of a set; and *unique factors*, appearing exclusively in individual variables of the set. If the set of n variables is designated by the vector z, the m common factors by the vector f, and the n unique factors by the vector u, the factor model for the composition of variables is:

$$z = Af + Cu \tag{1}$$

where the elements of the $n \times m$ matrix A are to be determined, and the diagonal matrix C merely contains the complements needed to bring the variance of each variable to unity. One of the basic problems of factor analysis is to determine the coefficients of the common factors, employing the observed correlations among the n variables.

Thus, we will consider the basic purpose to be that of determining a factor matrix A, given the off-diagonal elements of a correlation matrix R, such that

$$AA^T = R \tag{2}$$

where R is assumed to be positive semidefinite and symmetric (referred to as Gramian in the psychological literature). From this statement of the problem it follows that the values in the principal diagonal of the correlation matrix determine what portions of the unit variances are accounted for by the common factors. The variances associated with the composition of a variable (1) are referred to as common-factor variance, or *communality*, and *uniqueness*. Hence, if the communalities are inserted in the principal diagonal of R and a factor matrix A is obtained, it will in fact reproduce the communalities and the factor solution will be of the form (1). On the other hand, if unities are in the principal diagonal of R, the factor solution will be entirely in terms of common factors (all elements of uniqueness being part of the common variance). In general, factor analysts accept the composition of variables in the form (1) and expect to analyze some quantity less than the total variance into common-factor variance. The communality of a variable j is designated by h_j^2, where

$$h_j^2 = \sum_{p=1}^{m} a_{jp}^2 \qquad (j = 1, 2, \cdots, m) \tag{3}$$

While many ingenious, but subjective, schemes for estimating communalities have been proposed [6, Chapter 5], [11], none has proved wholly satisfactory. It has been proved by Dwyer [12] that a lower bound for the communality is given by

$$\rho_j^2 \leqq h_j^2 \tag{4}$$

where ρ_j^2 is the squared multiple correlation for variable j with the $(n - 1)$ remaining variables of the set. Furthermore, the squared multiple correlations measure only common variance of the particular set of variables. For this reason it is recommended [11] that in place of some vague estimate of communality its objective lower bound (4) be employed.

When communalities are inserted in the principal diagonal of R it is referred to as the "reduced correlation matrix." If such values are found which preserve the positive semidefinite property of R, then the rank m of R will be less than or equal to its order n, and will correspond to the number of common factors in (1). For a reduced correlation matrix R, the popular approach to a solution A of (2) has been the centroid method [6, Chapter 10]. More recently, with the advent of high-speed digital computers, this method gave way to the mathematically sounder principal-factor solution. The foundation for "the method of principal axes" was laid at the turn of the century by Karl Pearson [13]. However, it was not until the 1930's that Hotelling [7, 14] presented the theory and method for calculating a principal-factor solution.

The analysis is begun by selecting a factor F_1 whose contribution to the communalities of the variables has as great a total as possible. The effect of this factor is removed, and from the residual correlations a second factor F_2, independent of F_1, is selected with a maximum contribution to the residual communality. This process is continued until the total communality is analyzed.

a. Symbols Used

$R = (r_{jk})$ represents a given symmetric nth-order matrix of correlation coefficients, which is assumed to be positive semidefinite. The symbol R will be employed either for the complete correlation matrix, or for the "reduced correlation matrix," i.e., the matrix with r_{jj} replaced by h_j^2, the context making the distinction perfectly clear where necessary. The $n \times m$ $(m \leqq n)$ matrix $A = (a_{jp})$ of common-factor coefficients is to be determined. Certain conventional designations from the specialized usage in factor analysis are retained, e.g., N for the number of observations, n for the number of variables, m for the number of common factors, F_p for the factors, and indices like i, j, k, p, q. Otherwise, unless specifically stated to the contrary, matrices are denoted by capital English letters, vectors by lower case English letters, and scalars by lower case Greek letters. The transpose of a matrix, as well as a vector, is denoted by a T superscript.

b. Derivation of the Principal-Factor Method

The first stage of the principal-factor method involves the selection of the first factor coefficients a_{j1} so as to make the sum of the contributions of that factor to the total communality a maximum; i.e.,

$$S_1 = \sum_{j=1}^{n} a_{j1}^2 \tag{5}$$

must be maximized under the conditions

$$r_{jk} = \sum_{p=1}^{m} a_{jp} a_{kp} \qquad (j, k = 1, 2, \cdots, n) \tag{6}$$

in which $r_{jj} = h_j^2$. Employing the method of Lagrange multipliers, let

$$2T = S_1 - \sum_{j,k=1}^{n} \mu_{jk} r_{jk} = S_1 - \sum_{j,k=1}^{n} \sum_{p=1}^{m} \mu_{jk} a_{jp} a_{kp} \tag{7}$$

where $\mu_{jk}\ (= \mu_{kj})$ are the Lagrange multipliers and the last equality follows from the conditions (6). Then, differentiating the function T with respect to any one of the nm variables a_{jp}, and setting the result equal to zero, produces:

$$\frac{\partial T}{\partial a_{jp}} = \delta_{1p} a_{j1} - \sum_{k=1}^{n} \mu_{jk} a_{kp} = 0$$

$$(j = 1, 2, \cdots, n;\ p = 1, 2, \cdots, m) \tag{8}$$

where δ_{1p} is the Kronecker delta (zero when $p \neq 1$ and unity when $p = 1$).

Multiply (8) by a_{j1} and sum with respect to j, obtaining

$$\delta_{1p} \sum_{j=1}^{n} a_{j1}^2 - \sum_{j=1}^{n} \sum_{k=1}^{n} \mu_{jk} a_{j1} a_{kp} = 0 \tag{9}$$

The last term can be simplified, since $\sum_{j=1}^{n} \mu_{jk} a_{j1} = a_{k1}$ according to (8). Then, setting $\sum_{j=1}^{n} a_{j1}^2 = \lambda_1$, (9) becomes:

$$\delta_{1p} \lambda_1 - \sum_{k=1}^{n} a_{k1} a_{kp} = 0 \tag{10}$$

Upon multiplying (10) by a_{jp} and summing for p, this equation becomes

$$a_{j1}\lambda_1 - \sum_{k=1}^{n} a_{k1} \left(\sum_{p=1}^{m} a_{jp} a_{kp} \right) = 0$$

or, upon applying the conditions (6),

$$\sum_{k=1}^{n} r_{jk} a_{k1} - \lambda_1 a_{j1} = 0 \qquad (j = 1, 2, \cdots, n) \tag{11}$$

Thus, it has been shown that the maximization of (5) under the conditions (6) leads to the system of n equations (11) for the determination of the n unknowns a_{j1}. A necessary and sufficient condition for this system of n homogeneous equations to have a nontrivial solution is the vanishing of the determinant of coefficients of the a_{j1}. This condition may be written:

$$\begin{vmatrix} (h_1^2 - \lambda) & r_{12} & \cdots & r_{1n} \\ r_{21} & (h_2^2 - \lambda) & \cdots & r_{2n} \\ \cdots & \cdots & \cdots & \cdots \\ r_{n1} & r_{n2} & \cdots & (h_n^2 - \lambda) \end{vmatrix} = 0 \tag{12}$$

where, it will be recalled, $r_{jj} = h_j^2$ and the parameter of (11) is designated by λ without a subscript. Of course, it will immediately be recognized that the determinantal equation (12) is an nth order polynomial in λ, known as the characteristic equation of the system (11). It is desired to determine values of λ which admit nontrivial solutions of (11).

It has been shown that the first step in solving the factor analysis problem as stated in (2) by the principal-factor method leads to a classical problem that has had very extensive treatment in the mathematical literature (e.g., [15]). The resulting problem—solution of (11)—might be restated in the form: Find a number λ and an n-dimensional vector $x \neq 0$ such that

$$Rx = \lambda x \tag{13}$$

A number λ_p is called an *eigenvalue* of R and its associated vector $x_p = \{\alpha_{1p}, \alpha_{2p}, \cdots, \alpha_{np}\}$ is called an *eigenvector* of R.

When a simple root of the characteristic equation (12) is substituted for λ in (11), the set of equations becomes redundant, yielding a family of solutions, all of which are proportional to one particular solution. From the assumed normalization of (10), the factor of proportionality is taken to be

$$\lambda_1 = \sum_{j=1}^{n} a_{j1}^2$$

Thus, the quantity to be maximized, S_1, is equal to one of the eigenvalues, namely, the largest one; and the problem of finding the coefficients a_{j1} of the first factor F_1 is solved. The largest root of (12) is substituted in (11) and any solution $\{\alpha_{11}, \alpha_{21}, \cdots, \alpha_{n1}\}$ is obtained. Then this arbitrary solution is converted to the desired coefficients of F_1 in (1) as follows:

$$a_{j1} = \alpha_{j1}\sqrt{\lambda_1} \Big/ \sum_{j=1}^{n} \alpha_{j1}^2 \tag{14}$$

Having determined the coefficients of the first factor, the next problem is to find a factor which will account for a maximum of the remaining communality. This is accomplished in a manner similar to the foregoing procedure, but operating on a correlation matrix with the effect of the first factor removed. Such a residual matrix is given by

$$R_1 = R - a_1 a_1^T \tag{15}$$

where a_1 is the vector $\{a_{11}, a_{21}, \cdots, a_{n1}\}$ of first-factor coefficients. The elements $({}_1r_{jk})$ of the new matrix are called residual correlations, including the residual communalities $({}_1h_j^2)$ in the principal diagonal.

In determining the coefficients a_{j2} of the second factor F_2, it is necessary to maximize the quantity

$$S_2 = \sum_{j=1}^{n} a_{j2}^2 \tag{16}$$

which is the sum of the contributions of F_2 to the residual communality. This maximization is subject to the conditions

$${}_1r_{jk} = r_{jk} - a_{j1}a_{k1} = \sum_{p=2}^{m} a_{jp}a_{kp} \tag{17}$$

analogous to the restrictions (6) in the case of the first factor. Instead of carrying through the analysis as before, it will be shown that the eigenvectors of R_1 are identical with those of R, and that they have corresponding eigenvalues except that corresponding to the eigenvector a_1 in R_1 is a zero eigenvalue in place of the λ_1 in R.

Employing the notation a_p for the m eigenvectors of R (properly scaled), and inquiring if they are also eigenvectors of R_1, we proceed as follows:

$$R_1 a_p = (R - a_1 a_1^T)a_p = Ra_p - a_1 a_1^T a_p$$

and from relation (13) for any eigenvector of R, the right-hand side simplifies to yield

$$R_1 a_p = \lambda_p a_p - a_1 a_1^T a_p \tag{18}$$

Now consider the two cases: $p = 1$ and $p \neq 1$. (a) When $p = 1$, $a_1^T a_1 = \lambda_1$ according to (10), so that the above expression reduces to

$$R_1 a_1 = 0 \tag{19}$$

In other words, the eigenvector corresponding to the largest eigenvalue λ_1 of R is also an eigenvector of R_1 but its associated eigenvalue in R_1 is zero. (b) When $p \neq 1$, $a_1^T a_p = 0$ according to (10), and expression (18) becomes:

$$R_1 a_p = \lambda_p a_p - a_1 \cdot 0 = \lambda_p a_p \quad (p \neq 1) \tag{20}$$

which says that, except for λ_1, the eigenvalues of R_1 are identical with those of R and their associated eigenvectors are also identical. The expressions (19) and (20) prove the statement made above.

From the foregoing it is clear that the λ_2 of R is the largest eigenvalue of R_1. In other words, to obtain the largest eigenvalue of the residual matrix R_1, it suffices to extract the second largest eigenvalue of the original matrix R. By the same type of argument, the successive eigenvalues and their associated eigenvectors are obtained directly from the original correlation matrix R, until m factors have been extracted.

When unities are placed in the principal diagonal of R then, in practice, one usually finds $m = n$. If some numbers less than unities (estimates of communalities or squared multiple correlations) are placed in the diagonal, and the positive semidefinite property of R is preserved, then m will usually be less than n, and all eigenvalues will be real and nonnegative. When negative eigenvalues are obtained in the course of the computation, it is evidence that the requirement of positive semidefiniteness has been violated.

Another property of the principal-factor solution should be noted. The m eigenvectors a_p of R satisfy the following orthogonality relations:

$$a_p^T a_q = \delta_{pq}\lambda_p, \qquad (p, q = 1, 2, \cdots, m) \tag{21}$$

which are a generalization of the property exhibited in (10) between the first eigenvector and all others.

3. SUMMARY OF THE CALCULATION PROCEDURE

While the theory in the foregoing section can be reduced to an efficient iterative procedure appropriate for desk calculators [6, Sections 9.3 and 9.4], it is not the most expeditious method for programming a high-speed digital computer. The primary purpose of the preceding discussion was to bring the specially phrased problem of factor analysis into the scope of well-established mathematical procedures. Thus, when it is seen that the problem posed in (2) reduces to that of finding the eigenvalues and associated eigenvectors of the reduced correlation matrix R, it is not necessary that these values be calculated by the same routine as their theoretical determination.

In recent years extensive literature has been building up on numerical methods for the solution of systems of linear equations, with the associated problems of matrix inversion and characteristic equations. In the present volume, Chapters 2–7 are devoted to these topics, and Chapter 7, in particular, has a direct bearing on the factor analysis problem. Such a modified Jacobi method has been found very effective in programming the JOHNNIAC [16], the ILLIAC [17], the IBM 704 [18], and other electronic computers. This method diagonalizes the matrix R by performing a sequence of orthogonal transformations on it, designed to reduce *one* off-diagonal element to zero at each stage. Each orthogonal transformation is of the form $B_{jk}RB_{jk}^T$, where

$$B_{jk} = \begin{pmatrix} 1 & & & & & & \\ & \ddots & & & & & \\ & & 1 & & & & \\ & & & \cos\theta_{jk} & & -\sin\theta_{jk} & \\ & & & & 1 & & \\ & & & & & \ddots & \\ & & & & & & 1 \\ & & & \sin\theta_{jk} & & \cos\theta_{jk} & \\ & & & & & & 1 \\ & & & & & & & \ddots \\ & & & & & & & & 1 \end{pmatrix} \tag{22}$$

with the four elements in the intersections of rows and columns j and k as indicated, all other diagonal elements unity, and all other elements zero. The angle of rotation θ_{jk} is chosen so as to transform the element r_{jk} into zero, and is defined by

$$\tan 2\theta_{jk} = \frac{2r_{jk}}{h_j^2 - h_k^2} \tag{23}$$

When an element is reduced to zero it does not, in general, remain at zero during subsequent transformations. However, the sum of squares of off-diagonal elements is decreased each time by an amount equal to $2r_{jk}^2$, and thus Jacobi's method guarantees the convergence of the off-diagonal elements of R to zeros (to a designated number of decimal places) with a sufficient number of iterations.

In the process of reducing the original matrix R to a diagonal matrix D, the off-diagonal elements are considered systematically. Since the original correlations r_{jk} are altered by the transformations (22), it would seem more appropriate to designate a general off-diagonal element by d_{jk} and the intermediate derived matrices by D's with suitable subscripts indicated below. Thus, at any stage of a specific transformation directed at reducing one off-diagonal element d_{jk} to zero, the following notation will be employed:

$$_{\nu}D = B_{jk\ (\nu-1)}DB_{jk}^T \tag{24}$$

where the sequencing of the transformations is derived from:

$$\nu = n(j-1) - \frac{j(j+1)}{2} + k \tag{25}$$

in which $j = 1, 2, \cdots, n-1$ and $k = j+1, j+2, \cdots, n$, producing the $\binom{n}{2}$ possible pairs (j, k).

At any stage the product of the ν individual transformations may be designated

$$_{\nu}B = B_{jk} \cdots B_{13}B_{12} \tag{26}$$

where ν is given by (25); and when *all* combinations of j and k have been tried, the product of these transformations in the ith iteration is designated

$$B_i = \Pi(B_{jk}; j < k = 1, 2, \cdots, n) \qquad (i = 1, 2, \cdots, s) \tag{27}$$

The diagonalizing effect of the ith iteration is given by:

$$D_i = B_iD_{i-1}B_i^T \qquad (i = 1, 2, \cdots, s) \tag{28}$$

If the product (in the indicated order) of the transformation matrices through the ith iteration is defined by

$$P_i = B_i B_{i-1} \cdots B_2 B_1 \quad (29)$$

then (28) can be expressed in terms of the original matrix R as follows:

$$D_i = P_i R P_i^T \quad (30)$$

After a specified number of iterations s, the solution to the problem will be in the form:

$$D_s = B_s D_{s-1} B_s^T = BRB^T = D = (\delta_{pq}\lambda_p) \quad (31)$$

where

$$B = P_s = B_s B_{s-1} \cdots B_2 B_1 \quad (32)$$

The diagonal elements λ_p of D are the eigenvalues of R, and the rows of the final transformation matrix B contain the corresponding eigenvectors $a_p^* = (\alpha_{1p}, \alpha_{2p}, \cdots, \alpha_{np})$ of R. Since each individual transformation matrix is orthogonal (and hence the product of such matrices is orthogonal), the resulting final matrix B is orthogonal and therefore the eigenvectors are normalized. Then the coefficients of each factor F_p are obtained simply by multiplying the square root of the eigenvalue by its associated eigenvector, namely:

$$a_p^T = \sqrt{\lambda_p}\, a_p^* \quad (33)$$

In general, when the rank m of R is equal to its order n, there will be n real nonnegative eigenvalues, and coefficients for n common factors will ensue. If, however, the reduced correlation matrix R is positive semidefinite and of rank $m < n$, then there will be $n - m$ zero eigenvalues and there will be only m common factors with nonzero coefficients.

4. FLOW CHART

The flow chart appears on page 210.

5. DESCRIPTION OF THE FLOW CHART

Box 1: Store the reduced correlation matrix R [of course, only $n(n + 1)/2$ elements of this symmetric matrix are distinct, so that the detailed programming would normally capitalize on this]. Also stored are the trace $T(R)$, the order of the matrix n, and the maximum number of iterations s (it may be desired to stop the machine after a predetermined number of iterations regardless of the degree of convergence). The matrix R also is put in the working locations ${}_\nu D$ as the initial ${}_0 D$.

Box 2: The identity matrix is put in the ${}_\nu B$ locations as the initial ${}_0 B$. The counter j is set to 1 and k to 2.

Box 3: The d_{jk} element of ${}_\nu D$ (initially, r_{jk} of R) is tested to see if it is zero to the specified (ϵ_1) degree of accuracy. In order to make the program more flexible, the maximum allowable error ϵ_1 (as well as ϵ_2 and ϵ_3 required later) could be predetermined and stored as parameters in box 1. If d_{jk} is zero to the degree ϵ_1 proceed to box 7, otherwise to box 4.

Box 4: Compute B_{jk} as indicated in (22), using (23) and well-known trigonometric identities to get the values of the sine and cosine of the transformation angle (employing only algebraic functions of the argument).

Box 5: Compute ${}_\nu D$ according to (24), remembering that ${}_0 D = R$.

Box 6: Compute ${}_\nu B$ according to (26), which is equivalent to ${}_\nu B = B_{jk\,(\nu-1)}B$, and again, ${}_0 B = I$.

Box 7: If k has reached n, then all columns have been considered for the particular choice of the row j and the test of box 9 is made. Otherwise proceed to box 8.

Box 8: The index k is increased to continue with the next individual transformation in the particular iteration.

Box 9: If j has not reached $n - 1$, proceed to box 10. When j has reached $n - 1$, then the ith iteration has been completed, i.e., the B_i of (27) are the last computed ${}_\nu B$ when all combinations of j and k have been considered in this iteration, and the result of the iteration is D_i. The test of box 11 is then made.

Box 10: The index j is increased to $j + 1$ (with the associated k as $j + 2$) to continue with the next series of individual transformations, involving the elements of the new row, in the particular iteration i.

Box 11: As a check on the preceding calculations, the trace of the resulting matrix D_i is determined to see that it does not vary from the original value, according to the property:

$$\sum_{j=1}^{n} h_j^2 = T(R) = T(D_i) = \sum_{j=1}^{n} \lambda_j$$

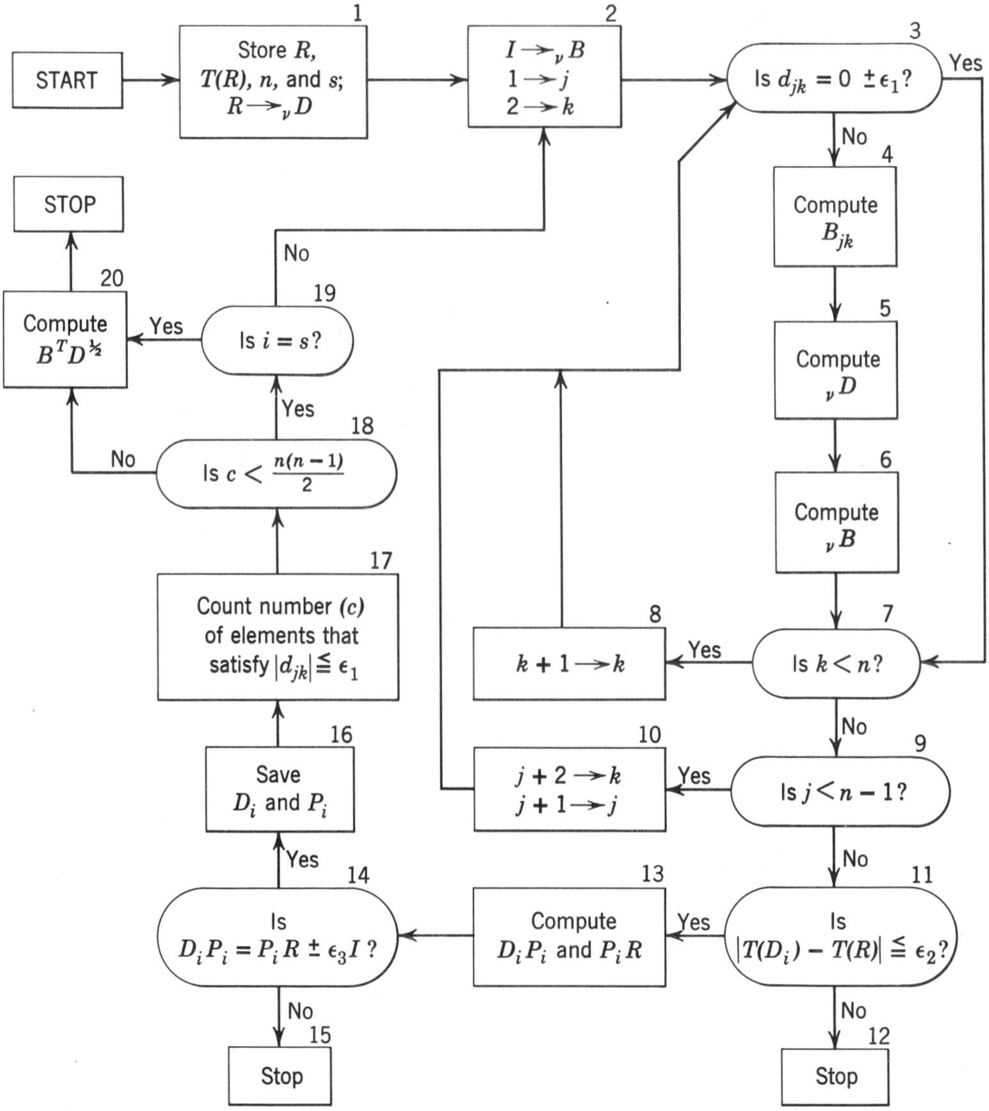

If the test is satisfied go to box 13. Otherwise proceed to box 12.

Box 12: This is a conditional stop. The magnitude of the allowable error ϵ_2 is to be determined by the user of the program. Instructions to the computer operator should include what to do in the event the stop occurs. The instructions may call for certain memory locations to be printed out; or if the user desires, they may even include a continuation of the process after a record is made of the stop.

Box 13: Compute $D_i P_i$ and $P_i R$, employing the P_i of (29) for the product of all the transformation matrices through the ith iteration.

Box 14: Another check is provided by postmultiplying both sides of (30) by P_i, noting that $P_i^T P_i = I$ from the orthogonality property of P_i. If each element of $D_i P_i$ agrees with the corresponding element of $P_i R$ within the specified (ϵ_3) degree of accuracy, proceed to box 16. Otherwise go to box 15.

Box 15: This conditional stop is treated the same way as box 12.

Box 16: The D_i and P_i are retained in temporary storage, being changed as a result of each successive complete iteration.

Box 17: Count the number (c) of off-diagonal elements d_{jk} that are zero within the specified (ϵ_1) degree of accuracy.

Box 18: If any off-diagonal element is not zero to the degree ϵ_1, go to box 19. Otherwise the process has converged, so proceed to box 20.

Box 19: If the number of iterations i has not reached the maximum number s, go back to box 2 to initiate the next iteration. Otherwise go to box 20.

Box 20: Since the eigenvectors appear in the rows of B, and since the factor matrix A conventionally contains the coefficients of the respective factors in columns, the conversion (33) for all the common factors is given by:

$$A = B^T D^{1/2}$$

This completes the problem.

6. SUBROUTINES

Employing the criterion of "normally available" subroutines, only two major instances are listed. However, many other parts of the computational procedures are amenable to subroutine treatment, e.g., differences of elements of two matrices. The specific subroutines follow:

SR1. Square root. Used in box 4.

SR2. Matrix multiplication. Input is initial location of the two matrices, dimensions of each of the matrices, and the desired initial location of the product matrix. Output is the product of the matrices. Used in boxes 5, 6, 13, 20.

7. SAMPLE PROBLEM

In order to illustrate the preceding theory, a simple example [6, Section 5.11] of six hypothetical variables will be employed. The reduced correlation matrix is

$$R = \begin{pmatrix} .74 & .72 & .75 & .49 & .42 & .28 \\ .72 & .72 & .78 & .42 & .36 & .24 \\ .75 & .78 & .89 & .35 & .30 & .20 \\ .49 & .42 & .35 & .49 & .42 & .28 \\ .42 & .36 & .30 & .42 & .36 & .24 \\ .28 & .24 & .20 & .28 & .24 & .16 \end{pmatrix}$$

in which the communalities in the diagonal were obtained by exact algebraic solution [6, Section 5.5] since there were no sampling or rounding errors in the data. The problem was designed for R to be of rank $m = 2$, so that four of the eigenvalues should be zero. By actual computation on an IBM 704, the eigenvalues were found to be:

$$\lambda_1 = 2.87037808, \qquad \lambda_2 = 0.48962192$$

with two of the remaining eigenvalues being zero to eight places and two containing unity in the eighth decimal place. The complete transformation matrix B of eigenvectors will not be listed here, but the two rows associated with the two nonzero eigenvalues were converted to factor coefficients according to the last step of the description of the flow chart. The resulting factor solution is shown in Table 1.

Table 1. COEFFICIENTS OF COMMON FACTORS

Variable	F_1	F_2
1	.8600	−.0169
2	.8341	−.1559
3	.8646	−.3773
4	.5776	.3956
5	.4950	.3391
6	.3300	.2261

8. MEMORY REQUIREMENTS

The bulk of the memory space required is for the storage of the elements of the original matrix R and the intermediate values in the course of arriving at the diagonal matrix D. Under the assumption of infinite rapid-access memory, all matrices will be shown completely rather than economizing on any symmetry properties that may exist. Then each of the following matrices requires space for n^2 words:

R, which is the primary input;
D_i, which ultimately becomes D;
P_i, which ultimately becomes B;
${}_vD$, which is computed for each j and k, and leads to the resulting D_i for the ith iteration;
${}_vB$, which is computed for each j and k, and leads to B_i when all (j, k) pairs have been considered in the ith iteration.

In addition, n^2 memory locations are required for intermediate matrix results. The storage space needed for various constants is so trivial by comparison that it will be ignored. Then, excluding the space for the program instructions (approximately 600 words), the total memory requirements is about $6n^2$ words.

9. ESTIMATION OF THE RUNNING TIME

In developing the flow chart it was tacitly assumed that the digital computer has an infinite rapid-access memory. If one would desire to program the factor analysis problem for a particular machine, it is presumed he would find the material helpful in making decisions about using secondary storage. Of course, the use of such supplementary storage and the input-output characteristics of a particular computer make significant impact on the actual time for getting a solution. However, the following estimate is limited to the running time of the program only.

Actual experience indicates that reasonable convergence will be obtained in six to ten iterations for problems with $n < 20$, and that for larger n's convergence may be expected in about $\frac{1}{4}n$ iterations. Thus, it is relatively safe to set $s = 10$ for problems with $n < 20$ and $s = \frac{1}{4}n$ for larger problems. In each iteration there will be $n(n - 1)/2$ individual orthogonal transformations involving one addition and two multiplications for the determination of each of the $n(n + 1)/2$ distinct elements. Hence, after allowing 50% of the arithmetic time for "bookkeeping" operations, and throwing away inconsequential factors, an estimate of the total calculation time required is given by:

$$T = \frac{n^5}{10}(\alpha + 2\beta)$$

where α and β are the add and multiply times, respectively, on the particular computer.

10. REFERENCES

1. C. Spearman, General Intelligence, Objectively Determined and Measured, *Am. J. Psych.*, vol. 15, 1904, pp. 201–293.
2. T. L. Kelley, *Crossroads in the Mind of Man: A Study of Differentiable Mental Abilities*, Stanford University Press, 1928.
3. L. L. Thurstone, *Multiple Factor Analysis*, University of Chicago Press, 1947.
4. G. H. Thomson, *The Factorial Analysis of Human Ability*, 5th ed., Houghton Mifflin Co., New York, 1951.
5. C. Burt, Group Factor Analysis, *Brit. J. Psych. Stat. Sec.*, vol. 3,1950, pp. 40–75.
6. H. H. Harman, *Modern Factor Analysis*, University of Chicago Press, 1960.
7. H. Hotelling, Analysis of a Complex of Statistical Variables into Principal Components, *J. Ed Psych.*, vol. 24, Sept. and Oct. 1933, pp. 417–441 and 498–520.
8. D. N. Lawley, The Estimation of Factor Loadings by the Method of Maximum Likelihood, *Proc. Roy. Soc. Edinburgh*, vol. 60, 1940, pp. 64–82.
9. M. S. Bartlett, Tests of Significance in Factor Analysis, *Brit. J. Psych. Stat. Sec.*, vol. 3, 1950, pp. 77–85; vol. 4, 1950, pp. 1–2.
10. C. R. Rao, Estimation and Tests of Significance in Factor Analysis, *Psychometrika*, vol. 20, 1955, pp. 93–111.
11. C. Wrigley, The Distinction between Common and Specific Variance in Factor Theory, *Brit. J. Stat. Psych.*, vol. 10, 1957, pp. 81–98.
12. P. S. Dwyer, The Contribution of an Orthogonal Multiple Factor Solution to Multiple Correlation, *Psychometrika*, vol. 4, 1939, pp. 163–171.
13. K. Pearson, On Lines and Planes of Closest Fit to Systems of Points in Space, *Phil. Mag.*, vol. 6, Ser. II, 1901, pp. 559–572.
14. H. Hotelling, Simplified Calculation of Principal Components, *Psychometrika*, vol. 1, 1936, pp. 27–35.
15. *Nat. Bur. Standards Appl. Math. Series 29, 39, 49* (on the solution of linear equations and the determination of eigenvalues), 1953–1958.
16. G. H. Golub, Eigenvalues and Eigenvectors of a Real Symmetric Matrix, Computing Program J124 and J125 for the JOHNNIAC at The RAND Corp., Santa Monica, Calif., 1955.
17. Programs M3-117, M0-141, and M7-150 for eigenvalues and eigenvectors of a symmetric matrix and for principal axes in factor analysis, University of Illinois Digital Computer Laboratory, Urbana, Ill.
18. Y. Sawanobori, Characteristic Roots and Factors, Computing Program N.Y. CRV3 for the IBM 704 at SBC N.Y. Data Processing Center, International Business Machines Corp., Jan. 1957.

Autocorrelation and spectral analysis

19

Raymond W. Southworth
Yale University

1. FUNCTION

The calculation of autocorrelation coefficients and the determination of the power spectrum are methods used in studying cyclical components in a time series such as a set of economic data, the output of electrical communications devices, or the yearly index of sunspots. A thorough discussion of the measurement and calculation of power spectra is given in [1]–[4]. In these references both continuous and discrete functions of time are considered, as well as stationary and nonstationary series, and methods of computation are derived. Also discussed are the reliability of estimates and inherent problems in sampling and computation. The presentation here will be limited to a consideration of data which are taken at equal intervals of time and which form a stationary time series, i.e., one that oscillates or fluctuates about a constant mean. The distributional properties or general structure of the series should not change with time. If the series does contain a trend or secular component, it is necessary first to adjust the data, for example, by means of a moving average method.

The output of the code described herein includes the autocorrelation coefficients, the power spectrum, and secondary and intermediate results such as means and lagged products.

2. MATHEMATICAL DISCUSSION

a. Symbols Used

C_p = lagged product
f = frequency = $\omega/2\pi$
L_p = raw estimate of spectral density
M = maximum lag
N = total number of observations
$P(f)$ = spectral distribution function, also called power spectrum or spectral density
p = lag between observations
q = an index used in several summations
r_{yx} = correlation coefficient between variates x and y
r_p = autocorrelation coefficient
S_x = sample standard deviation
t = time
T = total time over which observations are taken
U_h = smoothed spectral density
W = autocovariance
x_i = individual observation
$\bar{x}$ = sample mean
ω = angular frequency
θ = angle
T_p, F_p, S_p, G_p = special sums, defined in Section 3

b. Derivation of the Method

The simple correlation coefficient between two variates x and y is a measure of the fit of an equation of the form $y_c = a + bx$, where the constants a and b are determined by the method of least squares. It may be calculated from

$$r_{yx} = \frac{N\Sigma x_i y_i - (\Sigma x_i)(\Sigma y_i)}{\sqrt{N\Sigma(x_i^2) - (\Sigma x_i)^2}\,\sqrt{N\Sigma(y_i^2) - (\Sigma y_i)^2}} \tag{1}$$

where the summations all extend from $i = 1$ to $i = N$. The range of r_{yx} is $(-1, 1)$. A value of $r_{yx} = 0$ corresponds to the absence of correlation, whereas $r_{yx} = +1.0$ corresponds to perfect positive linear correlation and $r_{yx} = -1.0$ to perfect negative linear correlation, the sign indicating whether y increases or decreases with an increase in x. It is assumed that the values of the dependent variate are normally distributed around the least squares line used to fit the data and that the observations, as well as their deviations from the line, are independent of one another. Mathematically, so far as r_{yx} is concerned, the variates x and y are interchangeable.

The square of r_{yx} is known as the coefficient of determination. Physically, it represents the fraction of the variation in y that is associated with a given variation in x.

If, on the other hand, we are concerned with a set of time-series data, we may apply (1) to the calculation of an *auto*correlation coefficient between values of the variate x_i and the same variate at a constant interval or lag of time x_{i+p}. The autocorrelation coefficient for a particular lag p is then given by

$$r_p = \frac{(N-p)\Sigma x_i x_{i+p} - (\Sigma x_i)(\Sigma x_{i+p})}{\sqrt{(N-p)\Sigma(x_i^2) - (\Sigma x_i)^2}} \times \left(\sqrt{(N-p)\Sigma(x_{i+p}^2) - (\Sigma x_{i+p})^2}\right)^{-1} \tag{2}$$

which is seen to be the same as the equation for the simple correlation coefficient between two variates x and y with the value x_{i+p} being used in place of y_i. The summations extend over the range $i = 1$ to $i = N - p$, or, as is sometimes done, the upper limit is fixed at $N - M$, where M is the maximum value assigned to the lag. The factor $N - p$ is correspondingly changed to $N - M$. The difference is usually small, since M is ordinarily chosen as not over 10% of N.

An examination of r calculated as a function of p indicates those lags or periods over which the data of the time series seem to be correlated. It must be remembered, however, that the assumption of a normal distribution of the fluctuations around the trend line may not be justified. Extreme deviations will markedly affect the value of r_p.

If the data are first normalized by subtracting off the mean and dividing by the standard deviation, then the expression for r_p becomes

$$r_p = \frac{\Sigma x_i x_{i+p}}{\Sigma x_i^2} \qquad i = 1 \text{ to } N - p \tag{3}$$

where we have assumed a constant standard deviation, independent of the range of the summation.

Also, if we consider a time series with a mean value of zero and if we define an autocovariance function $W(p)$ as

$$W(p) = \frac{1}{N-p}\sum_{i=1}^{N-p} x_i x_{i+p} \tag{4}$$

it follows that

$$r_p = \frac{W(p)}{S_x^2} \tag{5}$$

For a continuous function $x(t)$ the autocovariance function $W(p)$ is given by

$$W(p) = \lim_{T\to\infty} \frac{1}{T}\int_{-T/2}^{T/2} x(t)x(t+p)\,dt \tag{6}$$

As shown by Blackman and Tukey [1], it may also be written as the Fourier transform of a distribution function $P(f)$

$$W(p) = \int_{-\infty}^{\infty} e^{i2\pi fp} P(f)\,df \tag{7}$$

where

$$P(f) = \lim_{T\to\infty} \frac{1}{T}\left[\int_{-T/2}^{T/2} x(t)e^{-i2\pi ft}\,dt\right]^2 \tag{8}$$

The function $P(f)$ is fundamental in the harmonic analysis of $x(t)$ and is called the spectral distribution function for the stationary process $x(t)$. It is also said to describe the power spectrum of the process, since $P(f)$ may be shown to represent the contribution to the variance of $x(t)$ from frequencies between f and $f + df$. If we think of $x(t)$ as the voltage across a pure resistance of 1 ohm, the average power dissipated in the resistance will be proportional to the variance (average square) of $x(t)$, hence the use of the adjective "power." $P(f)$ may also be referred to as the spectral density, the covariance spectrum, or the second-moment spectrum.

The relation (7) giving the autocovariance function $W(p)$ as the Fourier transform of the power spectrum $P(f)$ may be inverted to give $P(f)$ as the transform of $W(p)$. Thus,

$$P(f) = \int_{-\infty}^{\infty} W(p)e^{-i2\pi fp}\, dp \tag{9}$$

Since the autocovariance function and the power spectrum are Fourier transforms of each other, either might be used in the analysis of $x(t)$. The power spectrum is more frequently used, however, and many studies have been made with respect to it on the effects of sampling, length of series, and method of computation.

If $x(t)$ is real, $W(p)$ is real and symmetric around $p = 0$. Hence $W(p)$ and $P(f)$ may be expressed more simply as two-sided cosine transforms:

$$W(p) = \int_{-\infty}^{\infty} P(f) \cos 2\pi fp\, df \tag{10}$$

$$P(f) = \int_{-\infty}^{\infty} W(p) \cos 2\pi fp\, dp \tag{11}$$

or even more simply as

$$W(p) = 2\int_{0}^{\infty} P(f) \cos 2\pi fp\, df \tag{12}$$

$$P(f) = 2\int_{0}^{\infty} W(p) \cos 2\pi fp\, dp \tag{13}$$

It should be noted that this definition of the power spectrum differs from that which associates the power spectrum only with positive frequencies, which would be equal to twice the $P(f)$ given here.

In dealing with a discrete set $x_i(t)$, it is necessary to use a finite Fourier series transformation rather than the infinite integral transformation. With a spectrum involving angular frequencies no greater than π, *raw estimates* of the spectral density are given by

$$L_p = W_0 + 2\sum_{q=1}^{M-1} W_q \cos\frac{qp\pi}{M} + W_M \cos p\pi \tag{14}$$

where the $W_0, W_1, \cdots, W_M$ are calculated by (4), and M is the maximum value of the lag p.

The problem of smoothing the raw values has been discussed by Bartlett [5], Blackman and Tukey [1], and Tukey and Hamming [4]. The last-named recommend the use of

$$U_p = 0.23L_{p-1} + 0.54L_p + 0.23L_{p+1} \tag{15}$$

where

$$L_{-1} = L_1 \qquad \text{and} \qquad L_{M+1} = L_{M-1} \tag{16}$$

and the U values represent *corrected* estimates of the smoothed power density. These may be thought of as giving the power in the frequency interval $\left(\frac{\pi p}{M} - \frac{\pi}{2M}, \frac{\pi p}{M} + \frac{\pi}{2M}\right)$.

Tukey and Hamming [4] also show that the autocorrelation coefficients r_p may be treated as though they were $W(p)$ values, giving

$$L_p = 1 + 2\sum_{q=1}^{M-1} r_q \cos\frac{pq\pi}{M} + r_M \cos p\pi \tag{17}$$

$$U_p = 0.23L_{p-1} + 0.54L_p + 0.23L_{p+1} \tag{18}$$

The effect of using r_p instead of $W(p)$ will depend principally on the ratio of p to N. If this is small, the main change is to reduce L_0.

Among the numerous variations on the above method, one that has been widely used involves the calculation of *circular correlations.* The data are assumed to be periodic of period N, and the circular autocovariance is defined as

$$W(p) = \frac{1}{N}\left(\sum_{i=1}^{N-p} x_i x_{i+p} + \sum_{i=N-p+1}^{N} x_i x_{i+p-N}\right) \tag{19}$$

According to Tukey and Hamming [4], however, an examination of the average values of the power spectrum estimates obtained from circular autocovariances shows them to be inferior to values obtained by using the autocovariance previously described.

c. Errors

The main source of error in the calculations is due to small differences of large sums and sums of squares or products. It is therefore desirable to carry as many digits in the sums as possible. Some of the difficulty can be avoided by first normalizing the data, thus reducing the mean to zero.

3. SUMMARY OF THE CALCULATION PROCEDURE

(a) If the data are to be normalized, calculate the mean and the standard deviation. Replace the raw data by the normalized data, and determine the frequency distribution if desired.

(b) Calculate the sums, sums of squares, and lagged products over the range from $i = 1$ to $i = N - p$ (or to $N - M$ if desired) for $p = 0, 1, \cdots, M$. The sums that are needed are

$$T_p = \Sigma x_{i+p} \quad (20) \qquad G_p = \Sigma x_i^2 \quad (23)$$

$$F_p = \Sigma x_i \quad (21) \qquad C_p = \Sigma x_i x_{i+p} \quad (24)$$

$$S_p = \Sigma x_{i+p}^2 \quad (22) \qquad W(p) = \frac{C_p}{N - p} \quad (25)$$

It is most efficient in calculating T_p, F_p, S_p, and G_p first to get the sum from $i = 1$ to $i = N$, corresponding to $p = 0$, and then to use

$$T_p = T_{p-1} - x_p \tag{26}$$

$$F_p = F_{p-1} - x_{N-p+1} \tag{27}$$

$$S_p = S_{p-1} - x_p^2 \tag{28}$$

$$G_p = G_{p-1} - x_{N-p+1}^2 \tag{29}$$

(If the range $i = 1$ to $i = N - M$ is chosen, the recurrence formulas will have to be modified.)

(c) Substitute these sums into the expression for the autocorrelation coefficient.

(d) In calculating the power spectrum it might seem most efficient first to set up a matrix of the values of $\cos pq\pi/M$, for $q = 0, 1, \cdots, M$ and $p = 0, 1, \cdots, M$, and then to use a subroutine to multiply the C_p vector by this matrix to obtain the L_p vector. If M is at all large, however, the matrix will take up too many storage locations. It is better to calculate only the $M + 1$ distinct values of the cosine that occur for $pq\pi/M = 0, \pi/M, 2\pi/M, \cdots, M\pi/M$ and then to reduce the argument to this range. It is still more efficient to calculate only the cosines in the first quadrant and then to obtain those in the second quadrant by the relationship $\cos(\pi - \theta) = -\cos\theta$.

(e) The raw power spectrum may be smoothed to give the U_p values.

4. FLOW CHART

The flow chart appears on pages 218-219.

5. DESCRIPTION OF THE FLOW CHART

Box 1: Read control information, probably from a punched card. This is needed to initialize or specialize the program for particular values of N and M.

Box 2: Read the data either from punched cards or from magnetic tape.

Box 3: If it is desired to normalize the data, go to box 4. Otherwise go to box 15.

Box 4: Calculate the mean by summing the data and dividing by N. The details of this loop are omitted.

Box 5: Set index i equal to one, and store zero in the location used in summing for the standard deviation.

Box 6: Replace the original x_i by $x_i - \bar{x}$.

Box 7: Square x_i (the newly stored value), and add to the partial sum for S_x.

Box 8: If the index i has reached N, go to box 10. Otherwise go to box 9.

Box 9: Increase i, and return to box 7.

Box 10: Complete the calculation of the standard deviation.

Box 11: Reset index i to one.

Box 12: Finish calculating the normalized data by dividing by the standard deviation. Store over the old values.

Box 13: If the index i has reached N, go to box 15. Otherwise go to box 14.

Box 14: Increase i, and return to box 12.

Box 15: Set the index p equal to zero, the index i to one, and clear the locations used in forming the T_p and S_p sums.

Box 16: Calculate x_i^2, add it to the S_p partial sum, and add x_i to the T_p partial sum.

Box 17: If the index i has reached N, go to box 19. Otherwise go to box 18.

Box 18: Increase i, and return to box 16.

Box 19: Set F_p (still for $p = 0$) equal to T_p; set G_p equal to S_p. Now set p equal to one.

Box 20: Using the formulas shown on the flow chart, calculate T_p, F_p, S_p, and G_p.

Box 21: If the index p has reached M, go to box 23. Otherwise go to box 22.

Box 22: Increase p, and return to box 20.

Box 23: Reset the index p to zero.

Box 24: Set the index i equal to one, and clear the C_p location.

Box 25: Add the lagged product $x_i x_{i+p}$ into the partial sum for C_p.

Box 26: If the index i has reached $N - p$, go to box 28. Otherwise go to box 27.

Box 27: Increase i, and return to box 25.

Box 28: Compute the value of $W(p)$.

Box 29: Calculate the autocorrelation coefficient by substituting the various sums into (2).

Box 30: If the index p has reached M, go to box 32. Otherwise go to box 31.

Box 31: Increase p, and return to box 24.

Box 32: Print the set of r_p values and any other desired intermediate sums.

Box 33: Set the index p equal to zero.

Box 34: Set the index q equal to one, and clear the location for the L_p summation.

Box 35: Add the product of W_q and $\cos \pi pq/M$ into the L_p partial sum.

Box 36: If the index q has reached $M - 1$, go to box 38. Otherwise go to box 37.

Box 37: Increase q and return to box 35.

Box 38: Finish the calculation of L_p.

Box 39: If the index p has reached M, go to box 41. Otherwise go to box 40.

Box 40: Increase p, and return to box 34.

Box 41: Reset the index p to zero.

Box 42: Calculate the smoothed spectral density U_p. The complete details are not shown.

Box 43: If the index p has reached M, go to box 45. Otherwise go to box 44.

Box 44: Increase p, and return to box 42.

Box 45: Print the L_p and U_p values.

6. SUBROUTINES

(a) SR1—cosine. Input is argument in radians. Used in box 38.

(b) SR2—square root. Used in box 29 in calculating r_p.

7. SAMPLE PROBLEM

Table 1 shows a set of data with $N = 50$, $M = 5$, which was synthesized for this example. For this data, $\bar{x} = 9.7$ and $S_x = 1.92$.

Table I

Observation No.	x	Observation No.	x	Observation No.	x
1	10	18	8	35	14
2	11	19	9	36	12
3	12	20	12	37	10
4	11	21	10	38	7
5	10	22	8	39	8
6	9	23	7	40	12
7	8	24	10	41	13
8	10	25	13	42	10
9	12	26	11	43	8
10	13	27	8	44	7
11	10	28	7	45	8
12	7	29	10	46	12
13	8	30	12	47	10
14	9	31	10	48	8
15	10	32	8	49	7
16	12	33	7	50	8
17	9	34	10		

The results of the calculation are given in Table 2. The upper limit used in forming the sums was $N - p$.

Table 2

p	T_p	F_p	S_p	G_p	C_p	r_p
0	485	485	4889	4889	4889	1.00
1	475	477	4789	4825	4695	0.40
2	464	470	4668	4778	4451	−0.52
3	452	462	4524	4712	4319	−0.71
4	441	452	4403	4612	4328	−0.03
5	431	440	4303	4468	4330	+0.68

The results for the calculation of the power spectrum are given in Table 3. Little significance can be attached to the L_p and U_p values here because of the small size of N and M.

Table 3

p	$W(p)$	p	L_p	U_p
0	97.8	0	950	517
1	95.8	1	8.0	227
2	92.7	2	20.4	12.6
3	91.9	3	−1.2	4.5
4	94.1	4	1.6	0.6
5	96.2	5	0.0	0.7

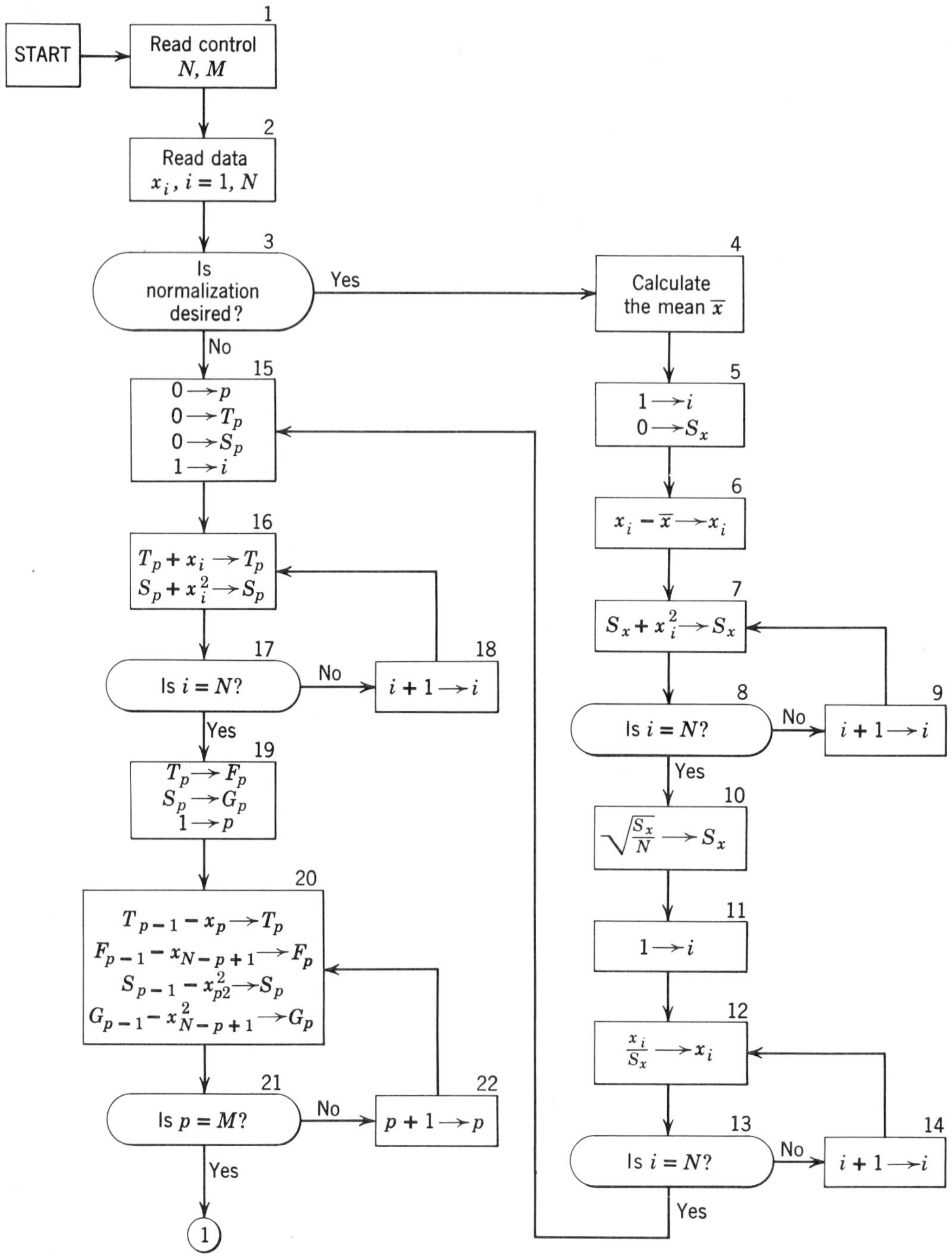
START
Read control N, M
Read data $x_i, i = 1, N$
Is normalization desired?
Yes
No
Calculate the mean $\overline{x}$
$1 \to i$
$0 \to S_x$
$x_i - \overline{x} \to x_i$
$S_x + x_i^2 \to S_x$
Is $i = N$?
$i + 1 \to i$
$\sqrt{\frac{S_x}{N}} \to S_x$
$\frac{x_i}{S_x} \to x_i$
$0 \to p$
$0 \to T_p$
$0 \to S_p$
$T_p + x_i \to T_p$
$S_p + x_i^2 \to S_p$
$T_p \to F_p$
$S_p \to G_p$
$1 \to p$
$T_{p-1} - x_p \to T_p$
$F_{p-1} - x_{N-p+1} \to F_p$
$S_{p-1} - x_{p2}^2 \to S_p$
$G_{p-1} - x_{N-p+1}^2 \to G_p$
Is $p = M$?
$p + 1 \to p$
1

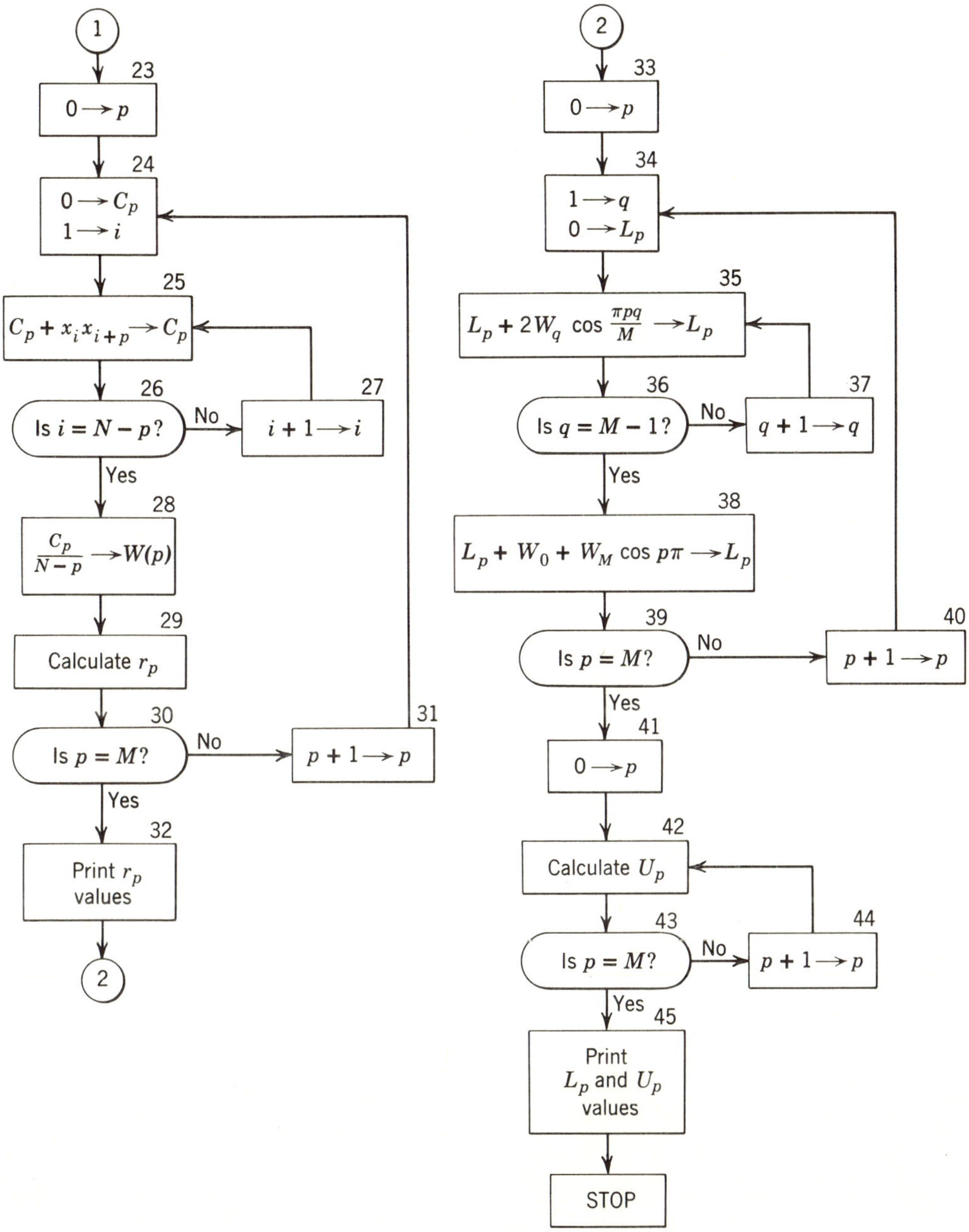
1
23
0 → p
24
0 → C_p
1 → i
25
C_p + x_i x_{i+p} → C_p
26
Is i = N − p?
No
27
i + 1 → i
Yes
28
C_p/(N − p) → W(p)
29
Calculate r_p
30
Is p = M?
No
31
p + 1 → p
Yes
32
Print r_p values
2
2
33
0 → p
34
1 → q
0 → L_p
35
L_p + 2W_q cos πpq/M → L_p
36
Is q = M − 1?
No
37
q + 1 → q
Yes
38
L_p + W_0 + W_M cos pπ → L_p
39
Is p = M?
No
40
p + 1 → p
Yes
41
0 → p
42
Calculate U_p
43
Is p = M?
No
44
p + 1 → p
Yes
45
Print L_p and U_p values
STOP

8. MEMORY REQUIREMENTS

The quantities that must be stored are:

x_i	N words
N	1 word
M	1 word
T_p	$M + 1$ words
F_p	$M + 1$ words
S_p	$M + 1$ words
G_p	$M + 1$ words
C_p	$M + 1$ words
r_p	$M + 1$ words
L_p	$M + 1$ words
U_p	$M + 1$ words

The program requires about 1000 locations, not including input and output subroutines, which will vary considerably from one machine to another. It is possible to store the r_p, L_p, and U_p values in the original data area. The minimum total memory requirement is, therefore, about $N + 5M + 1000$.

9. ESTIMATION OF THE RUNNING TIME

The greatest amount of work comes in calculating the C_p lagged product values. Here there are about $N \times (M + 1)$ additions and the same number of multiplications. T_p and S_p require about N additions each; S_p also takes N multiplications. The F_p and G_p sums are obtained with little additional work from the other sums. Assuming that bookkeeping and other operations take about the same length of time as the above, then the total calculation time is given by

$$T \sim (3N + NM)\nu + (2N + NM)\mu$$

where μ and ν are the multiply and add times for the machine being used.

10. REFERENCES

1. R. B. Blackman and J. W. Tukey, The Measurement of Power Spectra from the Point of View of Communication Engineering, *Bell System Tech. J.* vol. 37, 1958, pp. 185–282, 485–569.
2. U. Grenander and M. Rosenblatt, *Statistical Analysis of Stationary Time Series*, John Wiley & Sons, New York, 1956.
3. S. O. Rice, Mathematical Analysis of Random Noise, *Bell System Tech. J.*, vol. 23, 1944, pp. 283–332; and vol. 24, 1945, pp. 46–156.
4. J. W. Tukey and R. W. Hamming, Measuring Noise Color, an unpublished manuscript.
5. M. S. Bartlett and J. Mehdi, On the Efficiency of Procedures for Smoothing Periodograms from Time Series with Continuous Spectra, *Biometrika*, vol 42, 1955, pp. 143–150.
6. W. Z. Hirsch, *Introduction to Modern Statistics*, The Macmillan Co., New York, 1957.
7. R. Mosher and R. W. Southworth, Autocorrelation and Power Spectrum Analysis, an IBM 704 program, SHARE distribution No. 296, 1957.
8. H. Wold, *Stationary Time Series*, Almqvist and Wiksell, Stockholm, 1953

20 Analysis of variance

H. O. Hartley
Iowa State University

I. FUNCTION

From a numerical point of view, the analysis of variance can be regarded as an algebraic decomposition of a measure of variation into additive components. More specifically, it is concerned with data x (usually readings from a designed experiment) and their variation as measured by the sum of squares of deviations from the arithmetic mean $\bar{x}$, i.e., the sum of squares $\Sigma(x - \bar{x})^2$.* The decomposition of this sum of squares takes account of various criteria of classification into which the data x can be grouped. For example, if the yield x of a chemical process is measured for twelve combinations of three different additives, each tried at four different temperatures, the above sum can be shown to contain components measuring respectively yield variation due to temperature and yield variation due to additives, also computed in the form of sum of squares (of deviations) (see Section 2a).

Although a considerable volume of computational work is expended on analysis of variance at numerous research centers, most of this work is still carried out on desk computers and this is even true of some centers at which the services of a high-speed computer are available. The reason for this is undoubtedly the great variety of experimental designs, each of which gives rise to a different type of analysis of variance each applied to a small body of data. There is no difficulty in setting up and testing suitable programs every time data from a new design are ready for analysis, but insofar as the time and effort of doing this are usually much greater than the effort of completing the analysis of variance on a desk computer, there is clearly no point in enlisting the high-speed machines.† The question, however, is whether it is not possible to so adapt the analyses of variance for the diverse designs that they can all be covered by a standard computing program. Such a standard program would be set up and tested once and for all, and would then be available for the analysis of variance of data from any design. Because of its generality such a standard program will, by necessity, be slower than "tailor-made" programs specifically prepared for the rapid analysis of data from a large number of

* In this chapter all symbols denote scalars or operators.

† The comparative clerical labor of preparing the data for input into the respective machines, although by no means negligible, is not discussed here as this depends on details of the organization of the computing center.

experiments all of virtually identical design, a situation which some centers encounter. Some very excellent tailor-made programs have been written (see, for example, Yates, Healy, and Lipton [1]; Rowell [2]), but as their "tailoring" has taken account of the particular features of the high-speed computer for which they were written (and, indeed, sometimes of the organizational requirements of the research center) it is not suitable to give an account of them in this context.

In confining ourselves, then, to "standard" analysis of variance programs we further distinguish two main situations:

(i) The analysis of data from equal number or equal weight designs. In this category fall most of the common orthogonal designs such as the randomized block experiment, split plots, Latin squares, etc., but also certain nonorthogonal but balanced designs such as balanced incomplete blocks.

(ii) The analysis of variance of data when the representation of the groups into which the data fall is unbalanced, as indeed it usually will be when data which do not come from designed experiments are subjected to an analysis of variance.

While the analysis under (ii) is best treated by reduction to regression analysis as indicated briefly in Section 2c, that under (i) is here treated by developing a general program. This is specifically based on what is known as the analysis of a factorial experiment, considered in the next section.

2. MATHEMATICAL DISCUSSION

a. The Calculus of the General k-Factor Experiment

We now introduce the notation and calculus for the general k-factor experiment. Although the description is presented in terms of $k = 3$ factors, generalizations to k factors are obvious. Let x_{abc} denote the experimental reading from the ath level of factor A, the bth level of factor B, and the cth level of factor C. The symbols A, B, and C will also denote the number of levels for each factor so that $a = 1, 2, \cdots, A$; $b = 1, 2, \cdots, B$; and $c = 1, 2, \cdots, C$. In the example shown in Table 1 we have $A = 3$, $B = 3$, and $C = 2$ and the $3 \times 3 \times 2 = 18$ values x_{abc} are entered on the left-hand side. For later reference we require the following symbols for certain group totals of the data, namely:

$$X_{\cdot bc} = \sum_{a=1}^{A} x_{abc} \tag{1}$$

and similarly for $X_{a\cdot c}$, $X_{ab\cdot}$.

For example, in Table 1 we have entered $X_{\cdot 11} = 11$, $X_{\cdot 12} = 13$; $X_{\cdot 21} = 12$, $X_{\cdot 22} = 5$; $X_{\cdot 31} = 8$, $X_{\cdot 32} = 11$.

Table I. DATA x_{abc} IN A $3 \times 3 \times 2$ EXPERIMENT AND RESULTS FOR OPERATORS $\sum_a$ AND Δ_a

		x_{abc}		$\sum_a$ and Δ_a		
	a	$c = 1$	$c = 2$	$c = 1$	$c = 2$	
$b = 1$	1	4	6	1	5	
	2	5	3	4	-4	Δ_a
	3	2	4	-5	-1	
				11	13	$\sum_a$
$b = 2$	1	2	2	-6	1	
	2	6	1	6	-2	Δ_a
	3	4	2	0	1	
				12	5	$\sum_a$
$b = 3$	1	3	4	1	1	
	2	2	2	-2	-5	Δ_a
	3	3	5	1	4	
				8	11	$\sum_a$

Also we define

$$X_{\cdot\cdot c} = \sum_{a=1}^{A} \sum_{b=1}^{B} x_{abc} \tag{2}$$

and similarly for $X_{a\cdot\cdot}$, $X_{\cdot b\cdot}$.

And finally we define

$$X = \sum_{a=1}^{A} \sum_{b=1}^{B} \sum_{c=1}^{C} x_{abc} \tag{3}$$

Corresponding arithmetic means are denoted by $\bar{x}_{\cdot bc}$, etc.

While these and similar symbols are used in the classical treatment of analysis of variance computations, we employ here an operational calculus based on the following operators, which will be defined with regard to the

Table 2. SCHEDULE OF OPERATIONS FOR 3-FACTOR ANALYSIS OF VARIANCE

Run	Line No.	Operator	Applied to Values in Lines	Will Form Totals or Deviates	Deviates Used for Analysis of Variance Components
Read in	1	Input		x_{abc}	
A-run	2	$\sum_a$	1	$X_{\cdot bc}$	
	3	Δ_a	1 and 2	$Ax_{abc} - X_{\cdot bc}$	
B-run	4	$\sum_b$	2	$X_{\cdot\cdot c}$	
	5		3	$AX_{a\cdot c} - X_{\cdot\cdot c}$	
	6	Δ_b	2 and 4	$BX_{\cdot bc} - X_{\cdot\cdot c}$	
	7		3 and 5	$ABx_{abc} - BX_{\cdot bc} - AX_{a\cdot c} + X_{\cdot\cdot c}$	
C-run	8	$\sum_c$	4	X	
	9		5	$AX_{a\cdot\cdot} - X$	A
	10		6	$BX_{\cdot b\cdot} - X$	B
	11		7	$ABX_{ab\cdot} - BX_{\cdot b\cdot} - AX_{a\cdot\cdot} + X$	$A \cdot B$
	12	Δ_c	4 and 8	$CX_{\cdot\cdot c} - X$	C
	13		5 and 9	$ACX_{a\cdot c} - CX_{\cdot\cdot c} - AX_{a\cdot\cdot} + X$	$A \cdot C$
	14		6 and 10	$BCX_{\cdot bc} - CX_{\cdot\cdot c} - BX_{\cdot b\cdot} + X$	$B \cdot C$
	15		7 and 11	$ABCx_{abc} - BCX_{\cdot bc} - ACX_{a\cdot c} + CX_{\cdot\cdot c}$ $-ABX_{ab\cdot} + BX_{\cdot b\cdot} + AX_{a\cdot\cdot} - X$	$A \cdot B \cdot C$

first factor A but will be applied sequentially with regard to all factors A, B, C.

Operator $\sum_a \equiv$ sum over all levels $a = 1, \cdots, A$, holding the other subscripts at constant levels (4)

Operator $\Delta_a \equiv$ multiply all items by A and subtract the result of $\sum_a$ from all items (5)

Operator $(\;\bar{}\;)^2 \equiv$ the mean square operator. Form the sum of the squares of the items inside the parentheses and divide by the number of items (6)

For example, if we apply the first two operators to the original set of results x_{abc}, we have

$$\sum_a x_{abc} \equiv X_{\cdot bc} \equiv \text{the total for the } b, c \text{ combination of factors } B \text{ and } C \quad (7)$$

$$\Delta_a x_{abc} \equiv Ax_{abc} - X_{\cdot bc} \equiv \text{the deviate of } x_{abc} \text{ from the } b, c \text{ mean, multiplied by } A$$
$$\equiv A(x_{abc} - \bar{x}_{\cdot bc}) \quad (8)$$

Numerical values obtained from these operators are shown on the right-hand side of Table 1.

These simple operations (which may be regarded as the analysis of variance operators of integration and differentiation) represent the first two lines in the schedule of operations shown in Table 2, which gives complete formulas for the totals and deviates resulting from the sequence of operations $\sum_a \Delta_a \sum_b \Delta_b \sum_c \Delta_c$ applied to the data x_{abc} in the sequence shown. The $2^3 - 1 = 7$ sets of deviates finally reached in lines 9 through 15 are then subjected to the mean squares operator $(\;)^2$ and the results are the sums of squares of deviations (all multiplied by ABC) for the respective analysis of variance components shown in the last column of Table 2. It can be shown by elementary algebra that the total of these components is the sum of squares for total times ABC, i.e.

$$ABC \sum_{abc} (x_{abc} - \bar{x})^2 = ABC \sum_{abc} x_{abc}^2 - X^2 \quad (9)$$

Table 3. Reduction of analysis of variance of certain designs to the standard factorial program

Description of Design	Single Classification: A groups, B replicates	2-way Classification with Cell Repetition	Randomized Block with Two Factor Treatments	Split Plot: A main treatments, B sub treatments, C blocks	Split-Split Plot	3-factor Experiment in Randomized Blocks
Factor no. 1 2 3 4	Groups $= A$ Reps. $= B$	Rows $= A$ Columns $= B$ Reps. $= C$	Factor 1 $= A$ Factor 2 $= B$ Blocks $= C$	Main T. $= A$ Sub T. $= B$ Blocks $= C$	Main T. $= A$ Sub T. $= B$ Sub-sub T. $= D$ Blocks $= C$	Factors $\begin{cases} A \\ B \\ D \end{cases}$ Blocks C
Summary instruction	Error $= B + (A \cdot B)$	Error = within cells $\begin{cases} C \\ +(A \cdot C) \\ +(C \cdot B) \\ +(A \cdot B \cdot C) \end{cases}$	Error $\begin{cases} (A \cdot C) \\ +(B \cdot C) \\ +(A \cdot B \cdot C) \end{cases}$	Error (b) $\begin{cases} (B \cdot C) \\ +(A \cdot B \cdot C) \end{cases}$	Error (b) $\begin{cases} (B \cdot C) \\ +(A \cdot B \cdot C) \end{cases}$ Error (c) $\begin{cases} (D \cdot C) \\ +(D \cdot C \cdot B) \\ +(D \cdot C \cdot A) \\ +(D \cdot C \cdot A \cdot B) \end{cases}$	Error = Sum of all interactions with C
Analysis of variance	Groups A Error	Rows A Columns B Inter. $A \cdot B$ Error	Factor 1 A Factor 2 B Inter. $A \cdot B$ Blocks C Error	Main T. A Blocks C Error (a) $A \cdot C$ Sub T. B Inter. $A \cdot B$ Error (b)	Main T. A Blocks C Error (a) $A \cdot C$ ——— Sub T. B Main X sub $A \cdot B$ Error (b) ——— Sub-sub T. D Main X sub-sub $A \cdot D$ Sub X sub-sub $B \cdot D$ 2nd-order inter. $(A \cdot B \cdot D)$ Error (c)	Blocks C Main effects $\begin{cases} A \\ B \\ D \end{cases}$ 1st-order inter. $\begin{cases} A \cdot B \\ A \cdot D \\ C \cdot D \end{cases}$ 2nd order inter. $\{(A \cdot C \cdot D)$ Error

The numerical results of all of these operations applied to the data in Table 1 are shown in Tables 4 and 7.

b. The Reduction of Other Designs to the Factorial Program

The k-factor experiment will result in an analysis of variance in the general case consisting of $2^k - 1$ components of variance. While such designs are not uncommon, they are certainly not predominant and in the majority of cases the design employed will not be of this kind. For example, the simplest of all analysis of variance situations is an experiment in which A experimental groups are each replicated B times, and x_{ab} denotes the result from the bth replicate in the Ath treatment group. Since the B replicates in each treatment group are completely unrelated they do not constitute a factor. Nevertheless, we may analyze the data x_{ab} formally as if they came from a 2-factor experiment with factors A and B and thus obtain from the standard computing procedure the sums of squares (multiplied by AB) for the components A, B, and $A \cdot B$. To complete the proper analysis of variance for this experiment the program should contain a routine for a summary of components which would be appropriate for this particular design and would read as follows:

Summary: $B + (A \cdot B) =$ Error (within treatments)

Again if we have a split plot design with A main treatments and B subtreatments arranged in C blocks, we can obtain the appropriate analysis of variance by performing first the standard factorial analysis and then adding the summary instruction:

Summary: Error $(b) = (B \cdot C) + (A \cdot B \cdot C)$

The general principle of this procedure is therefore to perform first a formal factorial analysis and then pool certain components in accordance with summary instructions which specifically apply to the particular design. This would certainly be a wasteful procedure for a desk computer but is convenient for a high-speed computer. In Table 3 we list examples of such experiments which can be dealt with by this method. We give for each design:

(a) The factors which would be associated with the standard factors of the factorial analysis.

(b) The summary instructions for the design.

(c) The final analysis of variance in terms of the main effects and interactions of the factorial analysis.

It will be seen that a considerable diversity of designs can be covered in this manner.

Designs not covered by the above principle of reduction are those involving a higher degree of balance, i.e., a higher degree of restriction in the randomization, such as the Latin square, Youden square, lattices, and incomplete randomized blocks. The analysis of variance of these can be reduced to the factorial program (see Hartley [3]) by introducing a further operation called "rearrangement." We refer also to this paper and to Healy and Westmacott [4] for the treatment of missing values in the data.

c. Unequal Number and Weighted Analysis of Variance

The well-known method of obtaining the appropriate least square estimates of the components of variance in this case is by reduction to regression analysis. The principle of this reduction is to introduce effect variables $u_a(A)$, $u_b(B)$, $u_c(C)$ for each level of each factor. The effect variable $u_1(A_1)$, for example, is defined by

$$u_1(A) = \begin{cases} 1 & \text{if} \quad a = 1 \\ 0 & \text{if} \quad a \neq 1 \end{cases}$$

Most of the analysis of variance test criteria can then be defined as particular situations in the linear hypothesis model of the form

$$x_{abc} = \sum_a \beta_a(A)u_a(A) + \sum_b \beta_b(B)u_b(B) + \sum_c \beta_c(C)u_c(C) + z_{abc}$$

where z is an error variable.

A good account of this approach is given by Tocher [5].

3. SUMMARY OF THE CALCULATION PROCEDURE

a. Input

The input consists of:

(a) The characteristic numbers for the factorial design:

k = number of factors (3 in example)
A = number of levels for 1st factor (3 in example)
B = number of levels for 2nd factor (10) (3 in example)
C = number of levels for 3rd factor (2 in example)

(b) The data x_{abc} along with factor levels a, b, c. x_{abc} is stored in the store numbered

$$S(a, b, c) = a + (A + 1)(b - 1) + (A + 1)(B + 1)(c - 1) \quad (11)$$

(For the example these are shown in Table 4.)

b. Initialization

The following auxiliary quantities are required for the main computations:

For factor	Factor index	Limit	Step $s(f)$	
A	$f = 1$	A	$s(1) = 1$	
B	$f = 2$	B	$s(2) = A + 1$	(12)
C	$f = 3$	C	$s(3) = (A + 1) \cdot (B + 1)$	

The last store S^* is

$$S^* = (A + 1)(B + 1)(C + 1) \quad (13)$$

The form of these quantities for more than three factors is clear from their formulas for $f = 2$ and $f = 3$.

c. The Main Calculation

This consists of $k + 1$ runs (four runs in example), one run for each factor, i.e., the A-run, B-run, C-run, and the mean square run. The A-run performs the operations $\sum_a$ and Δ_a, the B-run the operations $\sum_b$ and Δ_b, and the C-run $\sum_c$ and Δ_c. For the example, the numerical results of these operations are shown in Table 4, while the symbols for the respective store contents are shown in Table 5. Both tables are more fully explained by flow chart 1. Flow chart 2 explains the mean square run and the output.

4. THE Σ AND Δ OPERATIONS, DESCRIPTION OF FLOW CHART I

This section should be followed in terms of the example in Table 4. At the beginning of the A-run the following temporary storage locations must be initialized:

Counter	$n = 0$,	Limit	$N = A$ (i.e., N for $f = 1$)	
Stores	$S, S' = 1$,	Step	$s = s(1) = 1$	(14)
Sum	$\Sigma x = 0$,	Factor count	$f = 1$	
Enter at "Start"				

EXPLANATION OF BOXES

Box 1: Control for end of operation Σ.

Box 2: Addition of current datum, deviate, or total denoted by x to progressive sum Σx.

Box 3: Step up of current store number S.

Box 4: Step up of item count n.

Box 5: Store current total Σx in store numbered S.

Box 6: Reset count n to zero.

Here the Σ operation is completed for a group of items.

Box 7: Control for end of operation Δ.

Box 8: Formation of current deviate and storing of same in same store (S') where datum $x(S')$ came from. $x(S')$ denotes the contents of store S' which may itself be a datum, deviate, or total.

Box 9: Step up of current store number S'.

Box 10: Step up of item counter n.

Box 11: Reset item count to zero, reset current Σx store to zero.

Box 12: Control for end of (factor) run. It should be noted that all runs stop when the final store $S^* = (A + 1)(B + 1)(C + 1)$ is

Table 4. NUMERICAL RESULTS FOR OPERATIONS $\sum_a \Delta_a, \sum_b \Delta_b, \sum_c \Delta_c$ AND LOCATION OF THESE IN STORES OF MEMORY

		$c = 1$					$c = 2$					$c = 3$				
		Store No.	Data Read In	Cont. after a-run	Cont. after b-run	Cont. after c-run	Store No.	Data Read In	Cont. after a-run	Cont. after b-run	Cont. after c-run	Store No.	Data Read In	Cont. after a-run	Cont. after b-run	Cont. after c-run
	$a = 1$	1	4	1	7	−1	17	6	5	8	1	33				15
	2	2	5	4	4	5	18	3	−4	−1	−5	34				3
$b = 1$	3	3	2	−5	−11	−4	19	4	−1	−7	4	35				−18
	4	4		11	2	−8	20		13	10	8	36				12
	$a = 1$	5	2	−6	−14	−10	21	2	1	−4	10	37				−18
	2	6	6	6	10	5	22	1	−2	5	−5	38				15
$b = 2$	3	7	4	0	4	5	23	2	1	−1	−5	39				3
	4	8		12	5	19	24		5	−14	−19	40				−9
	$a = 1$	9	3	1	7	11	25	4	1	−4	−11	41				3
	2	10	2	−2	−14	−10	26	2	−5	−4	10	42				−18
$b = 3$	3	11	3	1	7	−1	27	5	4	8	1	43				15
	4	12		8	−7	−11	28		11	4	11	44				−3
	$a = 1$	13			−4	−11	29			7	11	45				3
	2	14			8	19	30			−11	−19	46				−3
$b = 4$	3	15			−4	−8	31			4	8	47				0
	4	16			31	2	32			29	−2	48				60

Table 5. CONTENTS OF STORES OF MEMORY AT THE END OF EACH RUN IN SYMBOLIC FORM

		$c = 1$					$c = 2$					$c = 3$				
		Store No.	Data Read In	Cont. after a-run	Cont. after b-run	Cont. after c-run	Store No.	Data Read In	Cont. after a-run	Cont. after b-run	Cont. after c-run	Store No.	Data Read In	Cont. after a-run	Cont. after b-run	Cont. after c-run
	$a = 1$	1	x_{111}				17	x_{112}				33				
	2	2	x_{211}	$\underset{a}{\Delta}$	$\underset{b\,a}{\Delta\Delta}$	$\underset{c\,b\,a}{\Delta\Delta\Delta}$	18	x_{212}	$\underset{a}{\Delta}$	$\underset{b\,a}{\Delta\Delta}$	$\underset{c\,b\,a}{\Delta\Delta\Delta}$	34				$\underset{c\,b\,a}{\Sigma\Delta\Delta}$
$b = 1$	3	3	x_{311}				19	x_{312}				35				
	4	4		$\underset{a}{\Sigma}$	$\underset{b\,a}{\Delta\Sigma}$	$\underset{c\,b\,a}{\Delta\Delta\Sigma}$	20		$\underset{a}{\Sigma}$	$\underset{b\,a}{\Delta\Sigma}$	$\underset{c\,b\,a}{\Delta\Delta\Sigma}$	36				$\underset{c\,b\,a}{\Sigma\Delta\Sigma}$
	$a = 1$	5	x_{121}				21	x_{122}				37				
	2	6	x_{221}	$\underset{a}{\Delta}$	$\underset{b\,a}{\Delta\Delta}$	$\underset{c\,b\,a}{\Delta\Delta\Delta}$	22	x_{222}	$\underset{a}{\Delta}$	$\underset{b\,a}{\Delta\Delta}$	$\underset{c\,b\,a}{\Delta\Delta\Delta}$	38				$\underset{c\,b\,a}{\Sigma\Delta\Delta}$
$b = 2$	3	7	x_{321}				23	x_{322}				39				
	4	8		$\underset{a}{\Sigma}$	$\underset{b\,a}{\Delta\Sigma}$	$\underset{c\,b\,a}{\Delta\Delta\Sigma}$	24		$\underset{a}{\Sigma}$	$\underset{b\,a}{\Delta\Sigma}$	$\underset{c\,b\,a}{\Delta\Delta\Sigma}$	40				$\underset{c\,b\,a}{\Sigma\Delta\Sigma}$
	$a = 1$	9	x_{131}				25	x_{132}				41				
	2	10	x_{231}	$\underset{a}{\Delta}$	$\underset{b\,a}{\Delta\Delta}$	$\underset{c\,b\,a}{\Delta\Delta\Delta}$	26	x_{232}	$\underset{a}{\Delta}$	$\underset{b\,a}{\Delta\Delta}$	$\underset{c\,b\,a}{\Delta\Delta\Delta}$	42				$\underset{c\,b\,a}{\Sigma\Delta\Delta}$
$b = 3$	3	11	x_{331}				27	x_{332}				43				
	4	12		$\underset{a}{\Sigma}$	$\underset{b\,a}{\Delta\Sigma}$	$\underset{c\,b\,a}{\Delta\Delta\Sigma}$	28		$\underset{a}{\Sigma}$	$\underset{b\,a}{\Delta\Sigma}$	$\underset{c\,b\,a}{\Delta\Delta\Sigma}$	44				$\underset{c\,b\,a}{\Sigma\Delta\Sigma}$
	$a = 1$	13					29					45				
	2	14			$\underset{b\,a}{\Sigma\Delta}$	$\underset{c\,b\,a}{\Delta\Sigma\Delta}$	30			$\underset{b\,a}{\Sigma\Delta}$	$\underset{c\,b\,a}{\Delta\Sigma\Delta}$	46				$\underset{c\,b\,a}{\Sigma\Sigma\Delta}$
$b = 4$	3	15					31					47				
	4	16			$\underset{b\,a}{\Sigma\Sigma}$	$\underset{c\,b\,a}{\Delta\Sigma\Sigma}$	32			$\underset{b\,a}{\Sigma\Sigma}$	$\underset{c\,b\,a}{\Delta\Sigma\Sigma}$	48				$\underset{c\,b\,a}{\Sigma\Sigma\Sigma}$

reached. This means that during all but the last (kth) run we operate unnecessarily on some of the stores with zero content. This is convenient but somewhat wasteful of machine time.

Box 13: Control for going out of range of data block.

Boxes 14 and 15: Alter stores S and S' to initial values required for the operation on the next group of items.

The Δ operation is completed for a group of items.

Box 16: Step up of (factor) run count f.

Box 17: Control for entering square run.

Box 18: Initialization for next (factor) run.

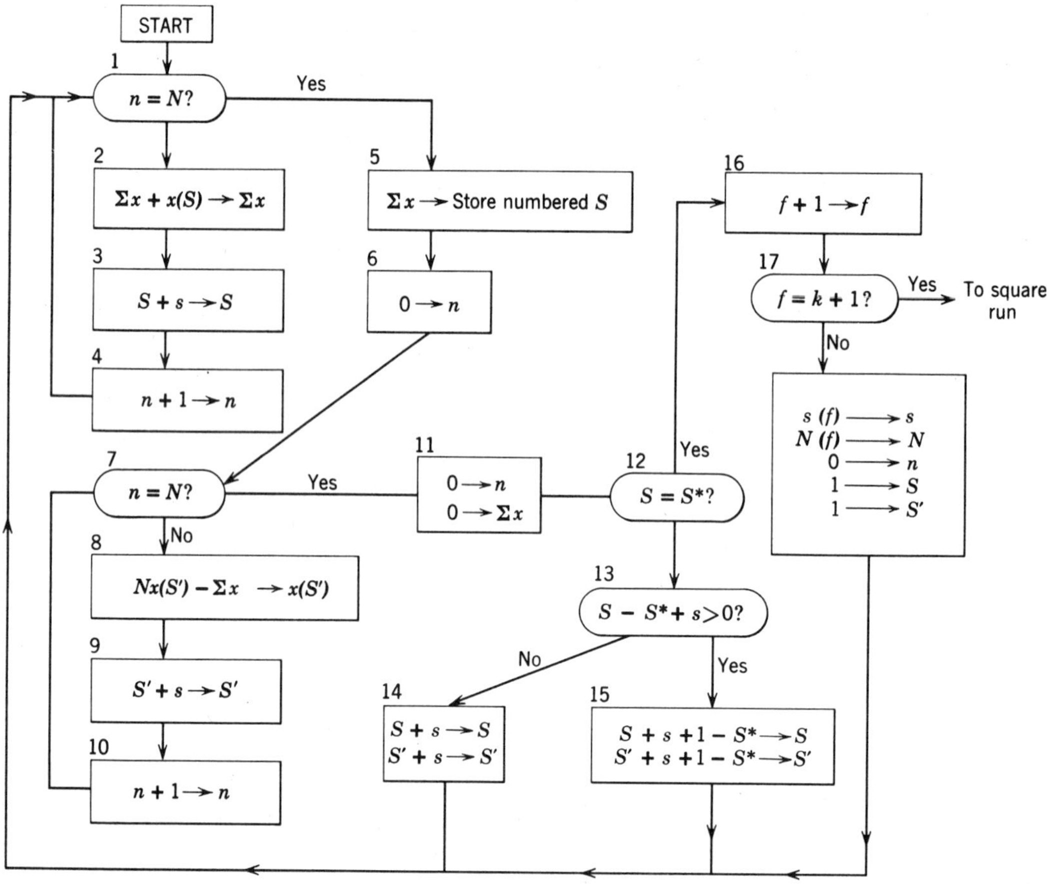

Flow chart 1. The Σ and Δ operations.

5. THE MEAN SQUARE OPERATION ( )², THE OUTPUT AND DESCRIPTION OF FLOW CHART 2

As mentioned in Section 2a, the $2^3 - 1 = 7$ sets of deviates formed at the end of the kth (3rd) factor run, and shown in lines 9 through 15 of Table 2, provide the input for the $(\ )^2$ operation which will form the sums of squares components of the analysis of variance and customarily are denoted by the symbols A, B, C; AB, AC, BC; ABC; we require therefore $2^3 - 1 = 7$ stores ($2^k - 1$ stores in the general case) to accumulate these squares. These cumulative sums of squares are denoted by $[A]^2$, $[B]^2$, $[C]^2$, $[AB]^2$, $[AC]^2$, $[BC]^2$, $[ABC]^2$, respectively. Simultaneously with the forming of these sums of squares the program makes provision for the optional output of certain of the deviates. Usually the sets $\sum_c \sum_b \Delta_a$, $\sum_c \Delta_b \sum_a$, $\Delta_c \sum_b \sum_a$ are required as they form the basis for the comparison of group means $\bar{x}_{a\cdot\cdot} - \bar{x}$, $\bar{x}_{\cdot b\cdot} - \bar{x}$, $\bar{x}_{\cdot\cdot c} - \bar{x}$. Also of particular interest to statisticians is the provision for the optional output of the deviates (residuals) for the 3-factor interaction $\Delta_c \Delta_b \Delta_a$ (i.e., the highest order interaction).

At the beginning of the square run the following stores must be initialized:

Counters a, b, c to one.

Progressive sums of squares $[ABC]^2$, $[AB]^2$, $\cdots$, $[C]^2$, to zero.

Current item store number $S'' = 0$.

EXPLANATION OF BOXES

Box 1: Step up of item store number S'' by one.

Box 2: Control for A-total; for $a = A + 1$ the current deviate or total denoted by x resulted from a $\sum_a$ operation, for $a < A + 1$ it resulted from a Δ_a operation.

Box 3: Step up of a-count.

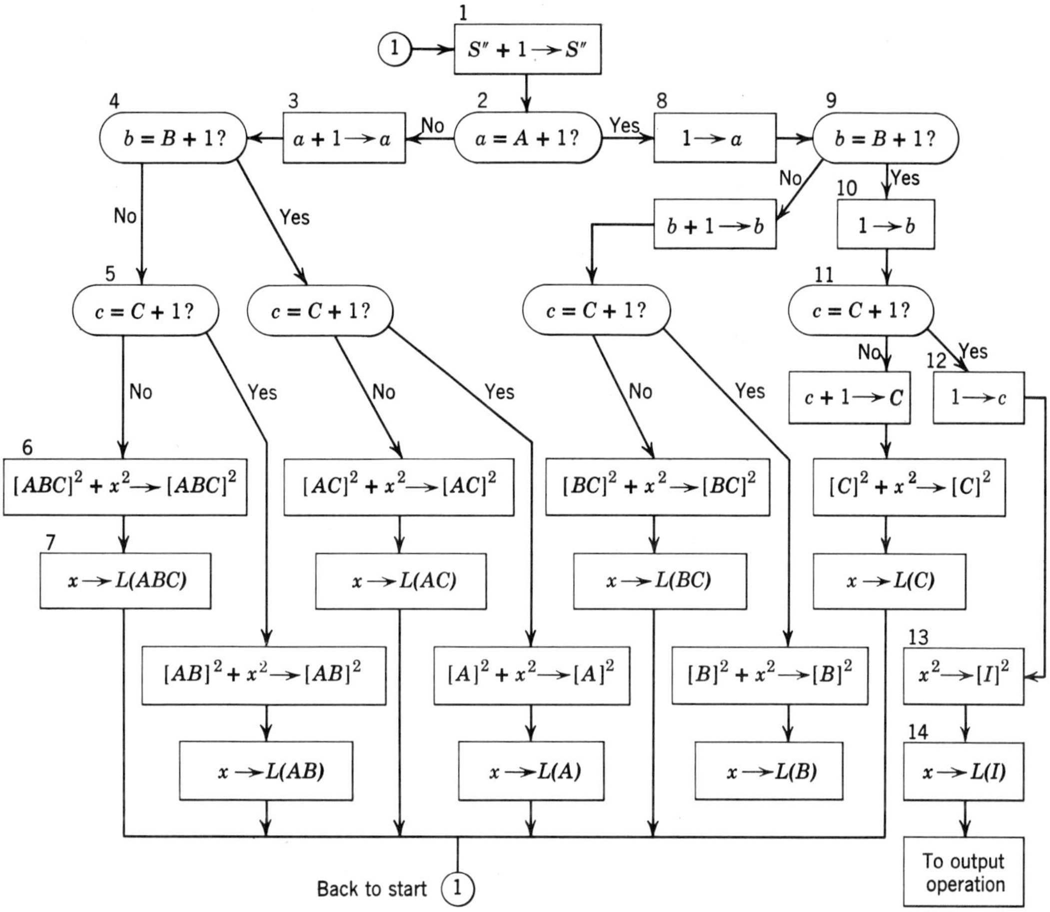

Flow chart 2. The mean square run and the output.

Box 4: Control for B-total.

Box 5: Control for C-total.

Box 6: Squaring of current item x and addition of x^2 to progressive sum of squares $[ABC]^2$.

Box 7: (Optional) disposal of current deviate x to appropriate output store $L(ABC)$.

Most of the remaining boxes are now self-explanatory. However, we will explain the branch of the loop on the right side leading to the end.

Box 8: Reset a-counter to unity.

Box 9: Control for B-total.

Box 10: Reset b-counter to unity.

Box 11: Control for C-total.

Box 12: Reset c-counter to unity.

Box 13: Square the grand total, X, and transfer to store I^2. The symbol I is customary for the grand total of the experimental readings and the quantity X^2 is what is known as the "correction for mean."

Box 14: (Optional) transfer of X to appropriate output store.

The flow chart has $2^3 = 8$ branches, the ends of which correspond respectively to the $2^3 - 1$ sums of squares $[ABC]^2, \cdots, [C]^2$, and to the store $[I]^2$ reserved for X^2.

In order to provide for a general program for k factors one has two choices: Either provide for a master assembly program which computes all the instructions involved in flow chart 2 from the given design characteristics k; $A, B, C, \cdots$; or make a decision as to a reasonable maximum number of factors and then provide in the flow chart for all the channels and stores which are neccessary to accumulate the analysis of variance components.

c. The Output Operation

At the end of the flow chart 2 operations, the stores $[ABC]^2, \cdots, [C]^2$ contain the (likewise

Table 6

Designation of Store and of Quantity Contained in It	1st Divisor Required to Form Sum of Squares of Analysis of Variance	2nd Divisor = ν (Degrees of Freedom) Required to Form Mean Squares
$(ABC)^2$	$ABC \cdot ABC$	$(A-1)(B-1)(C-1)$
$(AB)^2$	$ABC \cdot AB$	$(A-1)(B-1)$
$(AC)^2$	$ABC \cdot AC$	$(A-1)(C-1)$
$(BC)^2$	$ABC \cdot BC$	$(A-1)(C-1)$
$(A)^2$	$ABC \cdot A$	$(A-1)$
$(B)^2$	$ABC \cdot B$	$(B-1)$
$(C)^2$	$ABC \cdot C$	$(C-1)$

designated) sums of squares of the deviates. These must finally be divided by two sets of divisors as shown in Table 6 to form the quantities which are of interest to statisticians.

For our example the sums of squares $[ABC]^2/ABC \cdot ABC, \cdots, [C]^2/ABC \cdot C$, the number of degrees of freedom (as defined in Table 6) and mean squares = sum of squares/ν are shown in Table 7 in the usual order of tabulation:

Table 7. Analysis of variance table for $3 \times 3 \times 2$ experiment

Symbol for Analysis of Variance Component	Sum of Squares	Degrees of Freedom	Mean Square
A	0.33	2	0.17
B	4.33	2	2.17
C	0.22	1	0.11
AB	10.33	4	2.58
AC	10.11	2	5.06
BC	10.11	2	5.06
ABC	2.56	4	6.40

6. SUBROUTINES

None.

7. SAMPLE PROBLEM

The data of Tables 1, 4, and 7 constitute a sample problem.

8. MEMORY REQUIREMENTS

	Number of Stores ← k factors →
Input and deviates	$(A+1)(B+1)(C+1)\cdots$
Output stores	2^k
Approximate number of stores for program	$3(2^{k+1}) + 200$

9. ESTIMATION OF THE RUNNING TIME

An approximate formula for the arithmetic time T is

$$T = (19\nu + 4\mu)(A+1)(B+1)(C+1) + 2^{k+1}\mu$$

where μ and ν are the multiply and add times on the computer. This allows for neither logical operations nor for input and output time. The latter may be considerable and on machines with buffers should be interspersed with the arithmetic operations.

10. REFERENCES

Of the numerous textbooks giving an account of the statistical usage of analysis of variance we have only mentioned three [6], [7], [8]. Likewise, the papers cited represent a small selection of the growing number of papers dealing with analysis of variance on high-speed computers.

1. F. Yates, M. J. R. Healy, and S. Lipton, Routine Analysis of Replicated Experiments on an Electronic Computer, *J. Roy. Stat. Soc., B*, vol. 19, 1957, p. 234.
2. J. G. Rowell, The Analysis of a Factorial Experiment (With Confounding) on an Electronic Calculator, *J. Roy. Stat. Soc., B*, vol. 16, 1954, p. 242.
3. H. O. Hartley, A Plan for Programming Analysis of Variance for General Purpose Computers, *Biometrics*, vol. 12, 1956, p. 110.
4. M. J. R. Healy and M. H. Westmacott, Missing Values in Experiments Analyzed on Automatic Computers, *Appl. Stat.*, vol. 5, 1956, p. 203.
5. K. D. Tocher, The Design and Analysis of Block Experiments, *J. Roy. Stat. Soc., B*, vol. 14, 1952, p. 45.
6. W. T. Federer, *Experimental Design*, The Macmillan Co., New York, 1955.
7. O. Kempthorne, *The Design and Analysis of Experiments*, John Wiley & Sons, New York, 1952.
8. G. W. Snedecor, *Statistical Methods*, The Iowa State College Press, 5th ed., 1956.

PART VI MISCELLANEOUS METHODS

The numerical solution of polynomial equations

21

Herbert S. Wilf
The University of Illinois

1. FUNCTION

The purpose of this program is to find the roots of a polynomial equation with real coefficients.

2. MATHEMATICAL DISCUSSION

One of the most important talents that can be possessed by a digital computing specialist is the ability to recognize when his services are not needed. All too often, problems are run on computers, and then some very gross property of the solution is extracted, which could have been deduced by analysis without using the machine at all.

Problems in ordinary or partial differential equations, for example, are frequently processed in detail when only the asymptotic behavior of the solutions or bounds on their rate of growth are desired. The eigenvalues of matrices are all computed when knowledge that they all lie in, say, the left half-plane is the only information necessary, etc.

The numerical solution of polynomial equations is another example of an area where such unnecessary calculations are often done. Therefore, before diving into a detailed discussion of numerical procedures, we will spend some time with the theory of the location of roots in the complex plane, since this theory is in a very advanced state indeed, and can furnish precise and easily calculable information. The development will proceed in two stages. First we will give some sufficient conditions for the location of roots and then we will summarize some of the slightly more complicated, but very useful, necessary and sufficient conditions for localizing a root on an interval of the real axis, a circle in the complex plane, a half-plane, or, finally, in an arbitrary closed contour.

a. Sufficient Conditions

We will deal here with the equation

$$f(z) = z^N + a_1 z^{N-1} + \cdots + a_{N-1} z + a_N = 0 \tag{1}$$

where the a_j are real numbers.

Our first remark is that (1) is the characteristic equation of a real $N \times N$ matrix, for consider the matrix

$$A = \begin{pmatrix} -a_1 & -a_2 & -a_3 \cdots & -a_{N-1} & -a_N \\ 1 & 0 & 0 \cdots & 0 & 0 \\ 0 & 1 & 0 \cdots & 0 & 0 \\ \cdot & 0 & 1 \cdots & 0 & 0 \\ \cdot & \cdot & 0 \cdots & 0 & 0 \\ \cdot & \cdot & \cdot & \cdot & \cdot \\ \cdot & \cdot & \cdot & \cdot & \cdot \\ \cdot & \cdot & \cdot & \cdot & \cdot \\ 0 & 0 & 0 \cdots & 1 & 0 \end{pmatrix} \tag{2}$$

The characteristic equation of A

$$\det (zI - A) = 0 \tag{3}$$

is of the form

$$\begin{vmatrix} a_1 + z & a_2 & a_3 \cdots a_N \\ -1 & z & 0 \cdots 0 \\ 0 & -1 & z \cdots 0 \\ \cdot & \cdot & \cdot \quad \cdot \\ \cdot & \cdot & \cdot \quad \cdot \\ \cdot & \cdot & \cdot \quad \cdot \\ 0 & 0 & 0 \cdots z \end{vmatrix} = 0 \tag{4}$$

However, this determinant can be evaluated, and will be seen to have the value $f(z)$, if we multiply the first column by z and add it to the second column, multiply the second column of the result by z and add to the third column, etc. The end result is

$$\begin{vmatrix} * & * & * \cdots f(z) \\ -1 & 0 & 0 \cdots 0 \\ 0 & -1 & 0 \cdots 0 \\ 0 & 0 & -1 \cdots 0 \\ \cdot & \cdot & \cdot \quad \cdot \\ \cdot & \cdot & \cdot \quad \cdot \\ \cdot & \cdot & \cdot \quad \cdot \\ 0 & 0 & 0 \cdots 0 \end{vmatrix} = 0$$

which is obviously (1). We have shown

THEOREM 1: The characteristic equation of (2) is (1). This result is classical [1]; the matrix A is called the companion matrix of the polynomial (1).

The importance of this observation rests in the fact that there exist numerous methods for finding and computing the eigenvalues of matrices without using the characteristic equation per se. Any of these, when applied to the matrix A of (2), is a method for finding or computing the roots of (1). Some examples follow.

THEOREM 2 [2]: Let A be an $N \times N$ matrix. Then every eigenvalue of A lies in at least one of the circles

$$|z - A_{ii}| \leqq P_i \tag{5}$$

where

$$P_i = \sum_{j \neq i} |A_{ij}| \tag{6}$$

Furthermore, every eigenvalue of A lies in at least one of the circles

$$|z - A_{ii}| \leqq Q_i \tag{7}$$

where

$$Q_i = \sum_{j \neq i} |A_{ji}| \tag{8}$$

By direct application to (2), we find

THEOREM 3: Every root of (1) lies in one of the two circles

$$|z| \leqq 1 \tag{9}$$

$$|z + a_1| \leqq \sum_{2}^{N} |a_j| \tag{10}$$

Also, every root of (1) lies in one of the two circles

$$|z| \leqq 1 + \max_{j \geqq 2} |a_j| \tag{11}$$

$$|z + a_1| \leqq 1 \tag{12}$$

As another example, we have [3]

THEOREM 4: A sufficient condition that all the eigenvalues of an $N \times N$ matrix lie interior to the unit circle is

$$\max_i \sum_{j=1}^{N} |A_{ij}| \leqq 1$$

This translates immediately into

THEOREM 5: A sufficient condition that all the roots of (1) lie interior to the unit circle is

$$\sum_{j=1}^{N} |a_j| \leqq 1 \tag{13}$$

The mathematical literature is rife with eigenvalue location theorems of a much sharper type than the above. Most of these have not yet been translated into the equivalent statements about the roots of polynomial equations. This chapter will have served its purpose if, when the reader seeks bounds for the roots of polynomial equations, he turns first to the matrix literature.

As a further remark, the reader is probably familiar with the well-known power method for calculating the dominant eigenvalue of a matrix:

$$x_0; \qquad x_{n+1} = Ax_n; \qquad \frac{(x_{n+1}, x_n)}{(x_n, x_n)} \to \lambda \qquad (14)$$

where x_0 is an initial guess for an eigenvector of A. He will have no trouble verifying that the iteration (14) applied to the matrix (2) is precisely the Bernoulli iteration for the solution of (1) which will be discussed in greater detail below. Also, if the matrix (2) is first raised to a high power to separate the eigenvalues before iterating, the Graeffe process for solving (1) is the result, and, in fact, can conveniently be carried out in precisely the above manner. It follows also that any of the numerous devices for the acceleration of (14) is a device for accelerating the Graeffe and Bernoulli processes.

As a final conclusion, any computing installation which has a program for finding all the eigenvalues of a real matrix (no mean task) has no need for another routine to solve algebraic equations, at the expense of some inefficiency in computing time.

b. Necessary and Sufficient Conditions

Precise information about the location of the roots of an algebraic polynomial follows in a natural manner from the theory of continued fractions. For a thorough and systematic account of these developments the reader is referred to [4]. We give here only a summary of some salient results.

First, for the location of roots on an interval of the real axis, Sturm's theorem gives exact results. For the polynomials $f(x)$ given by (1), consider a sequence of $N+1$ polynomials $h_r(x)$, each of degree r, in general, defined by

$$h_N(x) = f(x); \qquad h_{N-1}(x) = f'(x) \qquad (15)$$

$$h_r(x) = q_r(x)h_{r-1}(x) - h_{r-2}(x) \qquad (r = N, N-1, \cdots, 2) \qquad (16)$$

where

$$\deg h_{r-2}(x) < \deg h_{r-1}(x)$$

For any real number x, let $V(x)$ denote the number of variations of sign in the sequence

$$h_0(x), h_1(x), \cdots, h_N(x) \qquad (17)$$

THEOREM 6 (Sturm): The number of roots of (1) that lie in the interval $[a, b]$ of the real axis is exactly $V(a) - V(b)$.

A proof can be found in [2]. If we now let

$$h_r(x) = \sum_{s=0}^{r} \lambda_{rs} x^s \qquad (18)$$

then the λ_{rs} can be calculated recursively by equating the coefficients of like powers of x on both sides of (16). The result is to take

$$\lambda_{Ns} = a_{N-s} \qquad (s = 0, 1, \cdots, N) \qquad (19)$$

and

$$\lambda_{N-1,s} = (s+1)a_{N-s-1} \qquad (s = 0, 1, \cdots, N-1) \qquad (20)$$

and then to calculate, recursively,

$$\alpha_r = \frac{\lambda_{rr}}{\lambda_{r-1,r-1}} \qquad (21)$$

$$\beta_r = \frac{1}{\lambda_{r-1,r-1}}[\lambda_{r-1,r-1} - \alpha_r \lambda_{r-1,r-2}] \qquad (22)$$

$$\lambda_{r-2,r-j} = \alpha_r \lambda_{r-1,r-j-1} + \beta_r \lambda_{r-1,r-j} - \lambda_{r,r-j}$$
$$(j = 2, 3, \cdots, r; \quad r = N, N-1, \cdots, 2) \qquad (23)$$

If we think of the λ_{rs} as the elements of an $(N+L) \times (N+1)$ lower triangular matrix Λ, then easy corollaries of Sturm's theorem are

THEOREM 7: All the roots of (1) are real and positive if and only if the first column of Λ alternates in sign and the diagonal elements of Λ are of constant sign.

THEOREM 8: All the roots of (1) are real if and only if the diagonal elements of Λ are of constant sign.

These theorems settle the question of the location of the roots of (1) on the real axis. Other results of interest are

THEOREM 9: All the roots of (1) lie in the left half-plane (i.e., have negative real parts) if and only if the matrix

$$Q = \begin{pmatrix} p_1 & p_3 & p_5 & \cdots & p_{2N-1} \\ 1 & p_2 & p_4 & \cdots & p_{2N-2} \\ 0 & p_1 & p_3 & \cdots & p_{2N-3} \\ 0 & 1 & p_2 & \cdots & p_{2N-4} \\ \cdot & \cdot & \cdot & & \\ \cdot & \cdot & \cdot & & \\ \cdot & \cdot & \cdot & & \\ 0 & 0 & 0 & \cdots & p_N \end{pmatrix} \qquad (24)$$

has positive principal minors, where

$$p_j = \begin{cases} a_j & 1 \le j \le N \\ 0 & j > N \end{cases}$$

A proof is in [4].

THEOREM 10: All the roots of (1) lie inside the unit circle in the complex plane if and only if the matrix

$$A_{rs} = \sum_{l=0}^{\min(r,s)} \{a_{N+l-r}a_{N+l-s} - a_{r-l}a_{r-s}\} \quad (a_0 = 1) \tag{25}$$

is positive definite.

The proof may be found in [5]. Finally, let C be a simple, closed contour in the complex plane. A classical result is [6]

THEOREM 11: The number of zeros of (1) inside C is equal to the variation of the amplitude of $f(z)$ along C, divided by 2π.

This theorem may be exploited numerically as follows: Choose a starting point z_0 on C, and a set of points z_i which march around C in the counterclockwise direction, as shown in Fig. 1. Draw the image of C under the

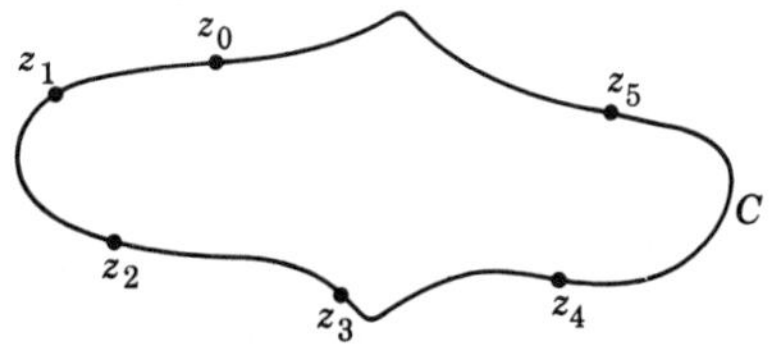

Fig. 1

mapping $z \to f(z)$ by calculating $f(z_i)$ for $i = 0, 1, \cdots$. The number of times that this image curve winds around the origin is equal to the number of zeros of $f(z)$ inside C. For a convenient algorithm for the evaluation of a polynomial at a complex value of the argument, see Chapter 24 on Fourier analysis.

We turn now to the question of calculating, rather than locating, the roots.

c. Calculation of the Roots

The methods which are presently available to solve (1) fall, roughly speaking, into two classes:

(a) Those which are undesirable for automatic computers because their convergence is not assured for every initial guess.

(b) Those which are undesirable for automatic computers because convergence, though certain, is excessively slow when a pair of nearly equal roots is present, and because the size of the numbers involved in the calculation grows exponentially from iteration to iteration.

In class (a) are the methods of Newton-Raphson, Birge-Vieta [7], Lin [8], and Bairstow [9]. In class (b) are those of Bernoulli [7], Graeffe [2], and Lehmer [2]. For these reasons the development of a machine program that will, with virtual certainty, produce all the roots of (1) is a matter of some difficulty. Since no single method appears to be wholly satisfactory, the procedure to be described below is a combination of a modification of the Bernoulli method, designed to prevent the growth of the numbers, for a good guess to the root, and an extension of the Birge-Vieta algorithm to the complex plane for refinement of the guess.

Let $x_1, x_2, \cdots, x_N$ denote the roots of (1). We have the

THEOREM (Newton):
Defining

$$S_p = \sum_{i=1}^{N} x_i^p \qquad (p = 1, 2, \cdots) \tag{26}$$

we have

$$S_m + a_1S_{m-1} + a_2S_{m-2} + \cdots + a_{m-1}S_1 + ma_m = 0 \qquad (m \leqq N) \tag{27}$$

$$S_{N+j} + a_1S_{N+j-1} + \cdots + a_NS_j = 0 \qquad (j = 1, 2, \cdots) \tag{28}$$

The proof can be found in [10], p. 84. These identities allow the recursive calculation of the S_p, starting with $S_1 = -a_1$.

Now suppose the roots x_i are ordered so that

$$|x_1| \geqq |x_2| \geqq \cdots \geqq |x_N| \tag{29}$$

Then, since

$$S_p = x_1^p + x_2^p + \cdots + x_N^p = x_1^p\left\{1 + \left(\frac{x_2}{x_1}\right)^p + \cdots + \left(\frac{x_N}{x_1}\right)^p\right\} \tag{30}$$

it follows that

$$\lim_{p\to\infty} \frac{S_{p+1}}{S_p} = x_1 \tag{31}$$

if x_1 is real and unrepeated in modulus.

On the other hand, if

$$x_1 = re^{i\theta} = r(\cos\theta + i\sin\theta) \tag{32}$$

then for large p

$$S_p \approx 2r^p \cos p\theta \tag{33}$$

so that, approximately,

$$S_{p+1} - 2r\cos\theta S_p + r^2S_{p-1} = 0 \tag{34}$$

$$S_p - 2r\cos\theta S_{p-1} + r^2S_{p-1} = 0 \tag{35}$$

From which it follows that

$$\lim_{p\to\infty} \frac{S_p^2 - S_{p-1}S_{p+1}}{S_{p-1}^2 - S_pS_{p-2}} = r^2 \tag{36}$$

$$\lim_{p\to\infty} \frac{S_pS_{p-1} - S_{p+1}S_{p-2}}{S_{p-1}^2 - S_pS_{p-2}} = 2r\cos\theta \tag{37}$$

The case of multiple roots can also be handled [7], but we will not concern ourselves with this here.

The method of Bernoulli for the solution of (1) consists in generating the S_p recursively by (27) and (28), noting which of the sequences (36)–(37) or (31) converges, and evaluating the root as the appropriate limit. One sees easily from (30) that the S_p will tend to vary exponentially in magnitude, and that convergence will be slow if $|x_2/x_1| \approx 1$.

Noting, however, that the limits involved in (31) and (36)–(37) depend only on the ratios of consecutive S_p, one is led naturally to the desirability of reframing the Bernoulli method so as to deal only with these ratios at every stage of the calculation, and thus avoid the rapid growth of the numbers.

Thus, consider (27), in the form

$$S_m = -a_1S_{m-1} - a_2S_{m-2} - \cdots - a_{m-1}S_1 - ma_m$$

so that

$$\frac{S_m}{S_{m-1}} = -a_1 - a_2\frac{S_{m-2}}{S_{m-1}} - a_3\frac{S_{m-3}}{S_{m-2}}\frac{S_{m-2}}{S_{m-1}} - \cdots - \left(\frac{S_1}{S_2}\cdots\frac{S_{m-2}}{S_{m-1}}\right)a_{m-1} - ma_m\left(\frac{1}{S_1}\frac{S_1}{S_2}\cdots\frac{S_{m-2}}{S_{m-1}}\right) \tag{38}$$

Substituting

$$\lambda_r = \frac{S_{r-1}}{S_r} \tag{39}$$

in (38), we get

$$-\frac{1}{\lambda_m} = a_1 + a_2\lambda_{m-1} + a_3\lambda_{m-2}\lambda_{m-1} + \cdots + a_{m-1}(\lambda_2\lambda_3\cdots\lambda_{m-1}) + \frac{ma_m}{a_1}(\lambda_2\lambda_3\cdots\lambda_{m-1}) \tag{40}$$

It is clear that the right-hand side of (40) can be evaluated conveniently by the recurrence

$$\xi_{m1} = a_{m-1} - \frac{ma_m}{a_1} \tag{41}$$

$$\xi_{m,r} = \xi_{m,r-1}\lambda_r + a_{m-r} \qquad (r = 2, 3, \cdots, m-1) \tag{42}$$

$$\lambda_m = -[\xi_{m,m-1}]^{-1} \tag{43}$$

In terms of the λ_r, the limiting expressions for the roots (31) and (36)–(37) become

$$x = \lim_{p\to\infty} \frac{1}{\lambda_p} \qquad (x \text{ real, unrepeated}) \tag{44}$$

$$r^2 = \lim_{p\to\infty} \frac{1}{\lambda_p\lambda_{p+1}}\left[\frac{\lambda_{p+1} - \lambda_p}{\lambda_p - \lambda_{p-1}}\right] \qquad (x \text{ complex, unrepeated}) \tag{45}$$

$$2r\cos\theta = \lim_{p\to\infty} \frac{1}{\lambda_{p+1}}\left[\frac{\lambda_{p+1} - \lambda_{p-1}}{\lambda_p - \lambda_{p-1}}\right] \tag{46}$$

Now suppose that we have arrived, by the process described above, at a reasonable guess at the real and imaginary parts, x and y, of a root, say, 3-figure agreement between successive iterates.

Consider the division of $f(z)$ by $z - z_0$, which gives, successively,

$$\begin{aligned} f(z) &= z^N + a_1z^{N-1} + \cdots + a_{N-1}z + a_N \\ &= (z - z_0)(z^{N-1} + b_1z^{N-2} + \cdots + b_{N-2}z + b_{N-1}) + f(z_0) \\ &= (z - z_0)^2(z^{N-2} + C_1z^{N-2} + \cdots + C_{N-2}) + (z - z_0)f'(z_0) + f(z_0) \end{aligned}$$

Equating coefficients

$$b_0 = 1 \tag{47}$$

$$b_r = a_r + z_0b_{r-1} \qquad (r = 1, 2, \cdots, N) \tag{48}$$

$$f(z_0) = b_N \tag{49}$$

$$C_0 = 1 \tag{50}$$

$$C_r = b_r + z_0b_{r-1} \qquad (r = 1, 2, \cdots, N-1) \tag{51}$$

$$f'(z_0) = C_{N-1} \tag{52}$$

Writing

$$z_0 = x + iy \tag{53}$$

$$b_r = \alpha_r + i\beta_r \tag{54}$$

$$C_r = \gamma_r + i\delta_r \tag{55}$$

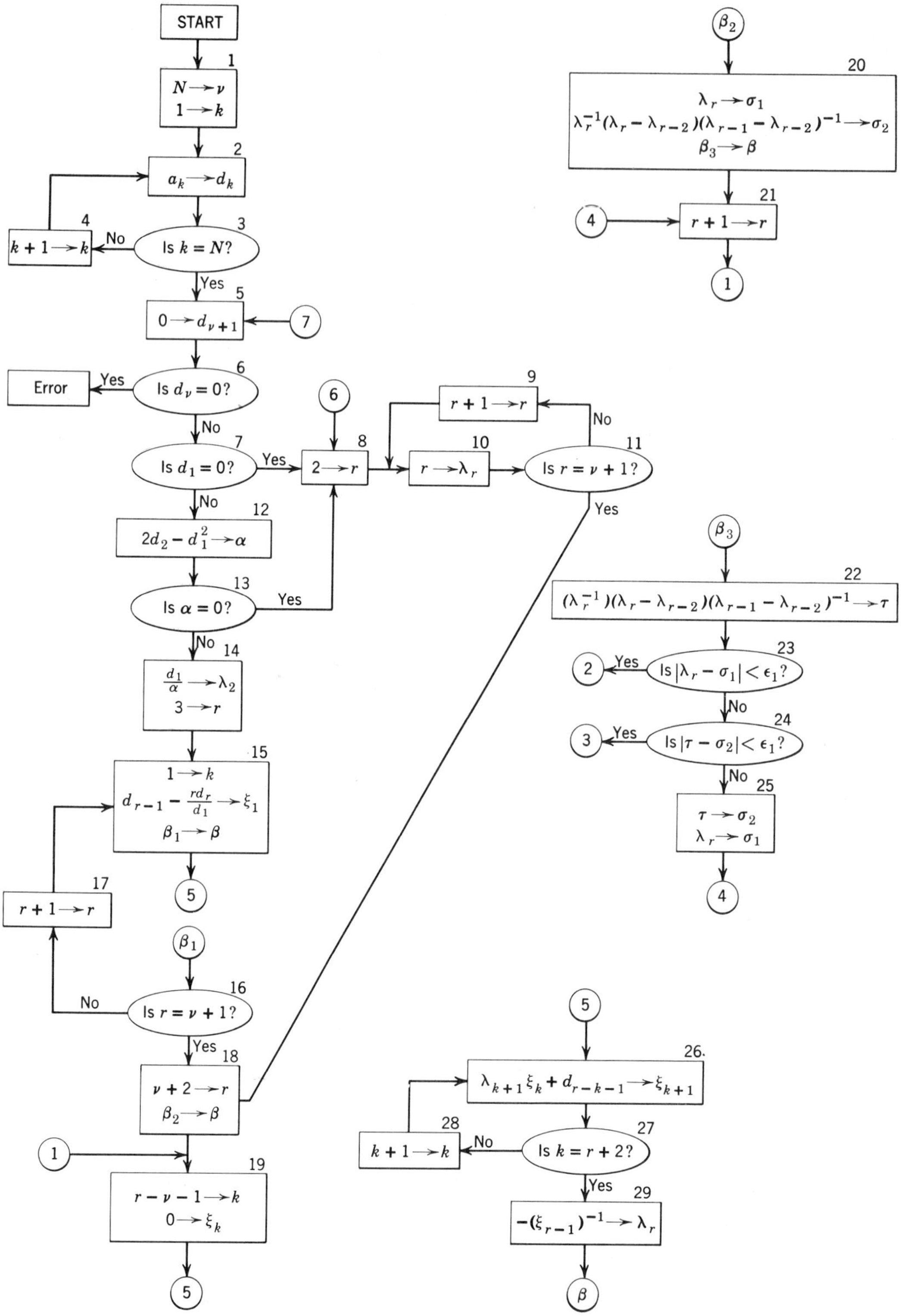
START
N → ν
1 → k
a_k → d_k
Is k = N?
k + 1 → k
No
Yes
0 → d_{ν+1}
Is d_ν = 0?
Error
Is d_1 = 0?
2 → r
r + 1 → r
r → λ_r
Is r = ν + 1?
2d_2 − d_1^2 → α
Is α = 0?
d_1/α → λ_2
3 → r
1 → k
d_{r−1} − r d_r/d_1 → ξ_1
β_1 → β
r + 1 → r
Is r = ν + 1?
ν + 2 → r
β_2 → β
r − ν − 1 → k
0 → ξ_k
λ_r → σ_1
λ_r^{−1}(λ_r − λ_{r−2})(λ_{r−1} − λ_{r−2})^{−1} → σ_2
β_3 → β
r + 1 → r
(λ_r^{−1})(λ_r − λ_{r−2})(λ_{r−1} − λ_{r−2})^{−1} → τ
Is |λ_r − σ_1| < ε_1?
Is |τ − σ_2| < ε_1?
τ → σ_2
λ_r → σ_1
λ_{k+1} ξ_k + d_{r−k−1} → ξ_{k+1}
Is k = r + 2?
k + 1 → k
−(ξ_{r−1})^{−1} → λ_r

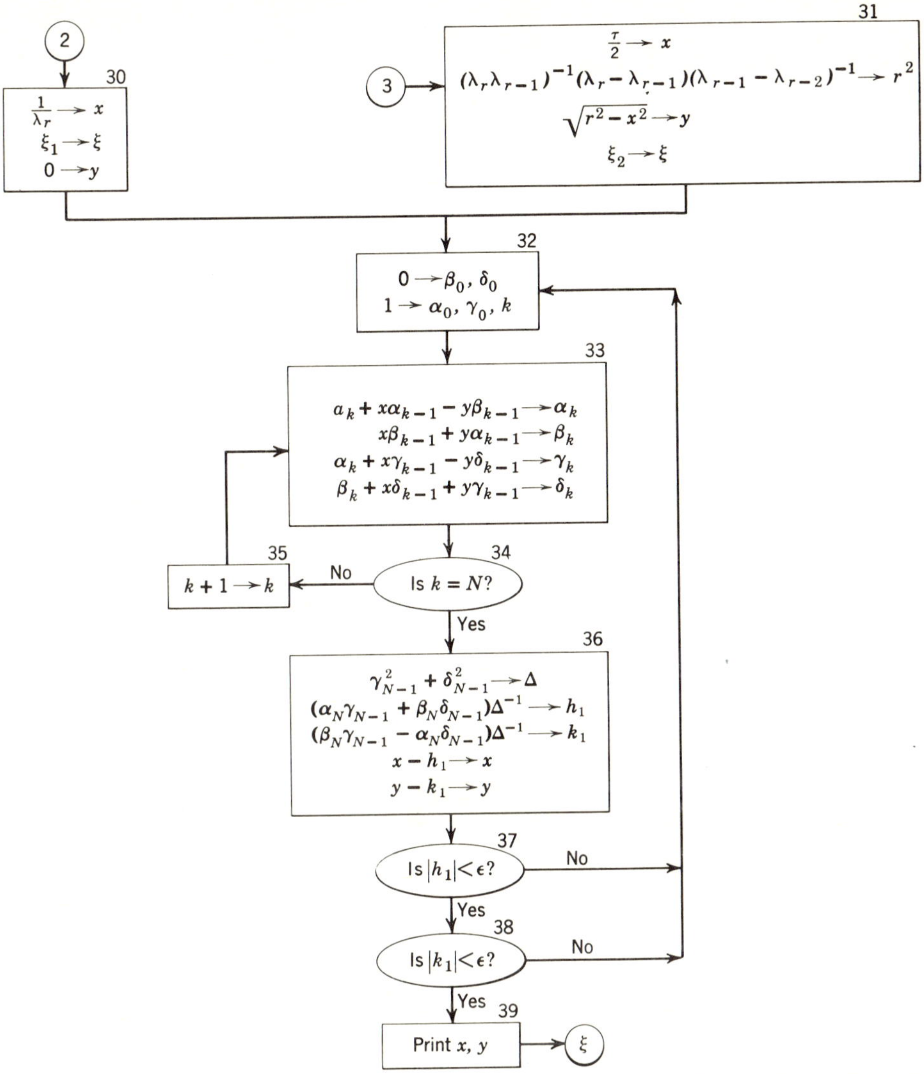
2
30
$\frac{1}{\lambda_r} \to x$
$\xi_1 \to \xi$
$0 \to y$
3
31
$\frac{\tau}{2} \to x$
$(\lambda_r\lambda_{r-1})^{-1}(\lambda_r - \lambda_{r-1})(\lambda_{r-1} - \lambda_{r-2})^{-1} \to r^2$
$\sqrt{r^2 - x^2} \to y$
$\xi_2 \to \xi$
32
$0 \to \beta_0, \delta_0$
$1 \to \alpha_0, \gamma_0, k$
33
$\alpha_k + x\alpha_{k-1} - y\beta_{k-1} \to \alpha_k$
$x\beta_{k-1} + y\alpha_{k-1} \to \beta_k$
$\alpha_k + x\gamma_{k-1} - y\delta_{k-1} \to \gamma_k$
$\beta_k + x\delta_{k-1} + y\gamma_{k-1} \to \delta_k$
35
$k + 1 \to k$
No
34
Is $k = N$?
Yes
36
$\gamma_{N-1}^2 + \delta_{N-1}^2 \to \Delta$
$(\alpha_N\gamma_{N-1} + \beta_N\delta_{N-1})\Delta^{-1} \to h_1$
$(\beta_N\gamma_{N-1} - \alpha_N\delta_{N-1})\Delta^{-1} \to k_1$
$x - h_1 \to x$
$y - k_1 \to y$
37
Is $|h_1| < \epsilon$?
No
Yes
38
Is $|k_1| < \epsilon$?
No
Yes
39
Print x, y
ξ

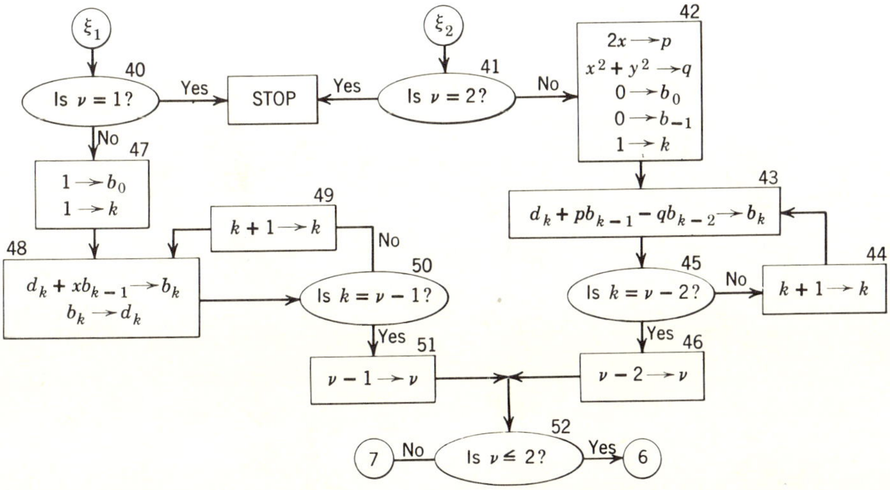
ξ_1
40
Is $\nu = 1$?
Yes
STOP
Yes
ξ_2
41
Is $\nu = 2$?
No
42
$2x \to p$
$x^2 + y^2 \to q$
$0 \to b_0$
$0 \to b_{-1}$
$1 \to k$
No
47
$1 \to b_0$
$1 \to k$
49
$k + 1 \to k$
No
48
$d_k + xb_{k-1} \to b_k$
$b_k \to d_k$
50
Is $k = \nu - 1$?
Yes
51
$\nu - 1 \to \nu$
43
$d_k + pb_{k-1} - qb_{k-2} \to b_k$
45
Is $k = \nu - 2$?
No
44
$k + 1 \to k$
Yes
46
$\nu - 2 \to \nu$
52
7
No
Is $\nu \leq 2$?
Yes
6

the recurrences (47)–(52) become

$$\alpha_0 = \gamma_0 = 1, \qquad \beta_0 = \delta_0 = 0 \tag{56}$$

$$\alpha_r = a_r + x\alpha_{r-1} - y\beta_{r-1} \qquad (r = 1, 2, \cdots, N) \tag{57}$$

$$\beta_r = x\beta_{r-1} + y\alpha_{r-1} \tag{58}$$

$$\gamma_r = \alpha_r + x\gamma_{r-1} - y\delta_{r-1} \tag{59}$$

$$\delta_r = \beta_r + x\delta_{r-1} + y\gamma_{r-1} \tag{60}$$

and the familiar Newton-Raphson iteration

$$z^* = z - \frac{f(z)}{f'(z)} = x^* + iy^* \tag{61}$$

becomes [see (49), (52)],

$$x^* = x - \frac{\alpha_N\gamma_{N-1} + \beta_N\delta_{N-1}}{\gamma_{N-1}^2 + \delta_{N-1}^2} \tag{62}$$

$$y^* = y - \frac{\beta_N\gamma_{N-1} - \alpha_N\delta_{N-1}}{\gamma_{N-1}^2 + \delta_{N-1}^2} \tag{63}$$

If $y = 0$, one readily verifies that (56)–(63) reduce to the familiar Birge-Vieta technique in the real case.

3. SUMMARY OF THE CALCULATION PROCEDURE

(a) Take $\lambda_2 = a_1/(2a_2 - a_1^2)$.

(b) For each $m = 3, 4, \cdots$ generate λ_m by (41)–(43).

(c) Continue until it is clear which of the sequences (44) or (45)–(46) is converging.

(d) At that time, calculate the first guess to the root, $z_0 = x_0 + iy_0$, by the appropriate one of (44) or (45)–(46).

(e) Refine this guess to convergence by repeated application of (56)–(60), (62)–(63).

(f) Depress the order of the equation by one if a real root was found, by synthetic division. If a complex root was found, depress the order of the equation by two through division by $x^2 - 2x_0x + (x_0^2 + y_0^2)$.

(g) Repeat steps (a)–(f) until all roots have been found, taking care to apply step (e) to the original equation, and all other steps to the depressed equation so as to avoid the build-up of round-off error.

4. FLOW CHART

The flow chart appears on pages 238–239.

5. DESCRIPTION OF THE FLOW CHART

Boxes 1–6: N is the degree of the equation to be solved, ν is the degree of the reduced equation. The a_k are the given coefficients and the d_k are the coefficients of the reduced equation. Initially ν is set to N and the d_k are set equal to the a_k.

Boxes 7–11: These boxes are executed only in the exceptional case that the tests of box 7 or 13 fail. In that case the starting value λ_2 for (41)–(43) cannot be computed by the formula of box 14, and therefore the value r is arbitrarily given to each of the first $(\nu - 2)$ λ_r's.

Boxes 12–15: In these boxes the initial values of the λ_r and ξ_{r1} for the recurrence (42) are calculated, and use is made of the subroutine in boxes 26–29 for calculating the λ_r's.

Boxes 16–17: Here we test to see if the first $(\nu - 1)$ λ_r's have been done. If not, r is increased and control returned to box 15 to do the next one.

Boxes 18–19: Counters and subroutine exits are set up here to do the λ_r with $r \geqq \nu$.

Box 20: We now try to discover whether the root is real or complex. In this box we store the latest λ_r and the latest value of $2r \cos \theta$. We also set the subroutine exit to bring us out in box 22.

Box 21: r is increased prior to the calculation of the next λ_r.

Box 22: We store here the next latest value of $2r \cos \theta$.

Boxes 23–24: The quantities λ_r and $2r \cos \theta$ which have now been calculated for two successive iterations are compared to see which, if either, is converging. For a real root we go to box 30, and for a complex root to box 31.

Box 25: In case neither sequence has converged sufficiently, we move the most recent values of λ_r and $2r \cos \theta$ into the comparison locations.

Boxes 26–29: This is a subroutine for the calculation of one of the λ_r using the recurrences (17)–(19).

Box 30: If a real root was discovered, the quantity $1/\lambda_r$ is stored as our best guess to the real part of the root to be found by Newton's method, and zero is stored in the imaginary part. A variable exit is set for later use.

Box 31: If a complex root was found, the

real and imaginary parts of the latest guess are computed from the polar forms (45)–(46) and stored, and the variable exit is set for later use.

Boxes 32–39: These boxes carry out, in a perfectly transparent manner, the modified Birge-Vieta algorithm for improvement of the Bernoulli estimates. Note that the procedure is quite indifferent to whether or not the root is real. After convergence is reached, the real and imaginary parts of the root are printed.

Box 40: If the root was real and the degree of the reduced equation was one, the calculation is finished.

Box 41: If the root was complex and the degree of the reduced equation was two, the calculation is finished.

Boxes 42–46: By synthetic division, the reduced equation is further reduced by dividing by the quadratic polynomial in x which vanishes at the complex root just found or its conjugate. The degree of the reduced equation is depressed by two.

Boxes 47–51: By synthetic division, the reduced equation has the real root just found divided out. Its degree is reduced by one.

Box 52: If the reduced equation is now linear or quadratic the logic of the starting procedure in boxes 12–14 fails and control is changed to the special starting procedure of boxes 8–11. We have assumed that the original equation is not linear or quadratic.

6. SUBROUTINES

Aside from the one built into the code (boxes 26–29) the only subroutine needed is a square root routine. By some ingenuity in logic, boxes 32–38 of the Birge-Vieta algorithm can be made to serve this function in addition to their present duties.

7. SAMPLE PROBLEM

$$f(x) = x^3 - \tfrac{14}{3}x^2 + \tfrac{109}{36}x - \tfrac{13}{9}$$

$$(x_j = 4, \tfrac{1}{3} \pm \tfrac{1}{2}i)$$

(a) $\left\{\dfrac{1}{\lambda_n}\right\} = \{3.3690,\ 4.0436,\ 4.0239,\ 4.0030,\ 3.99996, \cdots\}.$

(b) One cycle of Birge-Vieta now gives the root to ten figures.

(c) The reduced equation is

$$x^2 - \tfrac{2}{3}x + \tfrac{13}{36} = 0$$

(d) $\left\{\dfrac{1}{\lambda_n}\right\} = \{-.4167, 1.5333, 0.4312, -.1709, \cdots\}.$

The presence of complex roots is clearly indicated. Equations (45)–(46) converge immediately in this case to $r^2 = \frac{13}{36}$, $2r\cos\theta = \frac{2}{3}$.

8. MEMORY REQUIREMENTS

Program	500 words
a_k, d_k	$2N$ words
ξ_k, λ_k	$4N$ words
$\alpha_k, \beta_k, \gamma_k, \delta_k$	$4N$ words

Total: $500 + 10N$ words, approximately.

9. ESTIMATION OF THE RUNNING TIME

Assuming that on the average thirty Bernoulli iterations and three Birge-Vieta cycles will be needed to produce a root, the number of multiplications is approximately $30N^2$. The amount of bookkeeping required is rather extensive, and so we may estimate the time by

$$T \cong 120N^2(\mu + \nu)$$

where μ and ν are the multiply and add times of the computer.

10. REFERENCES

1. C. MacDuffie, *Theory of Matrices*, Chelsea Publishing Co., New York, 1946.
2. A. S. Householder, *Principles of Numerical Analysis*, McGraw-Hill Book Co., New York, 1953.
3. Y. K. Wong, Some Properties of the Proper Values of a Matrix *Proc. Amer. Math. Soc.*, vol. 6, 1955, pp. 891–899.
4. H. S. Wall, *Analytic Theory of Continued Fractions*, D. van Nostrand Co., New York, 1948.
5. H. S. Wilf, A Stability Criterion for Numerical Integration, *J. Assoc. Comp. Mach.*, vol. 6, 1959, p. 3.
6. J. F. Ritt, *Theory of Functions*, King's Crown Press, New York, 1947.
7. F. B. Hildebrand, *Introduction to Numerical Analysis*, McGraw-Hill Book Co., New York, 1955.
8. S. Lin, A Method for Finding Roots of Algebraic Equations, *J. Math. and Phys.*, vol. 22, 1943, pp. 60–77.
9. L. Bairstow, Investigations Relating to the Stability of the Aeroplane, NACA TRM 154, 1914.
10. F. Cajori, *An Introduction to the Modern Theory of Equations*, The Macmillan Co., New York, 1943.

Methods for numerical quadrature

22

Anthony Ralston
Bell Telephone Laboratories

1. FUNCTION

There are two types of problems which occur under the heading of numerical quadrature:

1. the evaluation of a definite integral, and

2. the evaluation of an indefinite integral as a function of the upper limit.

The second case occurs when we wish to tabulate the integral of a function. Since the problem of finding an indefinite integral can always be restated as the problem of finding a sequence of definite integrals, in this chapter we will consider explicitly only the first case above and will indicate, where pertinent, how the results we get can be used to solve the second problem.

2. MATHEMATICAL DISCUSSION

We wish, then, to find an approximation to the definite integral of a function $f(x)$ in the form*

$$\int_b^a f(x)\,dx \approx \sum_{j=0}^{n} H_j f(a_j) \tag{1}$$

* Until Section 2d we will consider only the case where a and b are finite.

where the problem is to specify the abscissas a_j and the weights H_j (where either or both may have restraints placed upon them; e.g., equal spacing of the abscissas) so that the right-hand side of (1) represents the left-hand side exactly for as many powers x^k ($k = 0, 1, \cdots, m - 1$) as possible. Associated with (1) there is an error term which, when added to the right-hand side of (1), makes (1) an equality.

When numerically integrating a function $f(x)$ over an interval $[a, b]$ there are two possible general approaches:

1. use a single quadrature formula over the whole interval $[a, b]$, or
2. break up the interval $[a, b]$ into a number of subintervals and use a single quadrature formula over each subinterval.

In both cases the resultant quadrature formula can be expressed in the form (1) but in the latter case, if the same quadrature formula is used in each subinterval, the right-hand side of (1) can be rewritten in the form

$$\sum_{j=0}^{n} H_j f(a_j) = \sum_{p=0}^{r-1} \sum_{q=0}^{s} H_q f(a_q + p\alpha) \tag{2}$$

where r is the number of subintervals, $s + 1$ is the number of points used in each subinterval, and $\alpha = (b - a)/r$. The general line of our discussion will be to derive a quadrature formula for the integration over a complete interval and then to consider the advantages of performing the integration by using this quadrature formula over a number of subintervals.

Quadrature formulas may be divided into two classes:

1. Newton-Cotes formulas in which the abscissas a_j are equally spaced in the interval $[a, b]$, and
2. Gaussian type formulas in which no restriction of equal spacing of the abscissas is made although there may be other specific restrictions on the abscissas and weights.

Newton-Cotes formulas can readily be used for the integration of functions given in the form of tabular data at equally spaced intervals. They are thus well suited for finding the indefinite integral of such a function. To integrate functions which are given analytically, either Newton-Cotes or Gaussian type formulas can be used but, as we shall see, the latter are generally to be preferred.

When (1) is exact for $f(x) = x^k$ $(k = 0, 1, \cdots, m - 1)$, it can often be shown that the error term is of the form $cf^{(m)}(\xi)$, where c is a constant, m is the highest power of x not integrated exactly by (1), and ξ is in (a, b). When this is so, c can be determined by considering $f(x) = x^m$. However, the proof that the error term can be expressed in this form is not trivial and is beyond the scope of this chapter, as are the derivations of the error terms when they cannot be expressed in the above form. For an excellent discussion of error terms in quadrature formulas we refer the reader to [1].

a. Newton-Cotes Quadrature Formulas

Since the abscissas are equally spaced, (1) can be rewritten as:

$$\int_a^b f(x)\,dx \approx \sum_{j=0}^{n} H_j f(a + jh) \qquad (3)$$

where $h = (b - a)/n$. If H_0 and H_n are postulated equal to zero, (3) is called a Newton-Cotes formula of the open type since the end points are not used. Conversely, if no restrictions are put on H_0 and H_n, then (3) is called a Newton-Cotes formula of the closed type.

For formulas of the closed type we have $n + 1$ parameters $H_0, \cdots, H_n$ at our disposal and thus we would expect to be able to make (3) an equality for $n + 1$ powers of x from 1 to x^n. If we let $f(x) = x^k$ $(k = 0, 1, \cdots, n)$ and substitute into (3), we get $n + 1$ equations for the $n + 1$ unknown weights

$$\alpha_k = \sum_{j=0}^{n} H_j (a + jh)^k, \qquad k = 0, 1, \cdots, n \qquad (4)$$

where

$$\alpha_k = \int_a^b x^k\,dx = \frac{b^{k+1} - a^{k+1}}{k + 1}$$

Since the determinant of the coefficients of the weights is a Vandermonde determinant, it is nonzero [2] and so (4) can be solved for the weights.* Thus for every n we can derive a closed type Newton-Cotes quadrature formula. Using the notation $f(a + jh) = f_j$, some examples of these formulas with their corresponding error terms are:

$$n = 1: \quad \int_a^b f(x)\,dx = \frac{h}{2}(f_0 + f_1) - \frac{h^3}{12} f''(\xi) \qquad (5a)$$

$$n = 2: \quad \int_a^b f(x)\,dx = \frac{h}{3}(f_0 + 4f_1 + f_2) - \frac{h^5}{90} f^{\text{iv}}(\xi) \qquad (5b)$$

$$n = 3: \quad \int_a^b f(x)\,dx = \frac{3h}{8}(f_0 + 3f_1 + 3f_2 + f_3) - \frac{3h^5}{80} f^{\text{iv}}(\xi) \qquad (5c)$$

where in all cases ξ is in (a, b).

* It can also be shown [1] that

$$H_j = \int_a^b l_j(x)\,dx$$

where $l_j(x)$ is the Lagrangian interpolation polynomial corresponding to the abscissas $a, a + h, \cdots, b$. For actual computation of the weights this formula is easier to use than solving the above system.

In a similar fashion we can derive Newton-Cotes open type formulas. Some examples of these are:

$$n = 3: \quad \int_a^b f(x)\,dx = \frac{3h}{2}(f_1 + f_2) + \frac{3h^3}{4} f''(\xi) \qquad (6a)$$

$$n = 4: \quad \int_a^b f(x)\,dx = \frac{4h}{3}(2f_1 - f_2 + 2f_3) + \frac{14h^5}{45} f^{\text{iv}}(\xi) \qquad (6b)$$

$$n = 5: \quad \int_a^b f(x)\,dx = \frac{5h}{24}(11f_1 + f_2 + f_3 + 11f_4) + \frac{95h^5}{144} f^{\text{iv}}(\xi) \qquad (6c)$$

where in all cases ξ is in (a, b).

A study of (5*b*) and (6*b*) reveals an interesting phenomenon. In both cases the weights were determined so that the quadrature formula was exact for powers of x through x^2. But an inspection of the error term shows that we have gotten a dividend in that x^3 is also integrated exactly. Indeed, it can be shown [1] that for any even n the Newton-Cotes formulas are exact for one higher power of x than expected. Thus, the formulas with $n = 2m$ and $n = 2m + 1$ both achieve the same order of accuracy (i.e., they both integrate exactly the same powers of x). Thus, it is seldom worthwhile to use the formula with $n = 2m + 1$ instead of the formula with $n = 2m$.

A comparison of (6*a*), (6*b*), and (6*c*) with (5*a*), (5*b*), and (5*c*), respectively, indicates that for formulas using the same number of computed ordinates, Newton-Cotes closed type formulas are significantly more accurate than open type formulas. For this reason open type formulas are virtually never used for numerical quadrature and we will consider them no further here. Their main application is in the numerical solution of differential equations (see Chapter 8).

It is clear from (5) and (6) that the greater the number of abscissas used in a quadrature formula, the higher is the power of h and the order of the derivative in the error term. Since $h < 1$ in virtually all practical cases, higher powers of h reduce the error. Conversely, however, it is well known that higher derivatives tend to grow rapidly, ultimately faster than powers of h decrease. From a practical point of view higher derivatives are generally very difficult to estimate so that the use of high order Newton-Cotes formulas is not recommended.

Another factor which works against the use of high order Newton-Cotes formulas is the matter of round-off error, i.e., the error caused by adding together a weighted sum of values of $f(x)$ each of which has been rounded. An estimate of the value of the round-off error from a quadrature formula can be measured by computing the sum of the squares of the weights. An inspection of (1) reveals immediately that, if the quadrature formula is exact for $f(x) = 1$, then the sum of the weights is $(b - a)$. Thus, the sum of the squares of the weights will be minimized if all the weights are equal in magnitude and positive. Equations (5) and (6) indicate that the higher the order of the quadrature formula, the more the magnitude of the weights fluctuates and, thus, the greater the average round-off error. The higher order formulas presented in [1] show that this tendency to greater fluctuation of the weights increases with order.

Thus, in order to avoid having high order derivatives in the error term and to minimize the round-off error we are led to the use of low order Newton-Cotes closed type formulas. However, if we use a single low order quadrature formula over the complete interval, the error term may be larger than desired. This leads us to break up the interval $[a, b]$ into subintervals and to apply the low order quadrature formula over each of the subintervals. The result of this is a so-called "composite formula." For example, if we divide $[a, b]$ into n intervals of length h and apply (5*a*) to each subinterval, we get

$$\int_a^b f(x)\,dx = \frac{h}{2}(f_0 + 2f_1 + 2f_2 + 2f_3 + \cdots + 2f_{n-1} + f_n) - \frac{nh^3}{12} f''(\xi) \qquad (7)$$

where ξ is in (a, b). Similarly, if we divide $[a, b]$ into $n/2$ subintervals (n even) of length $2h$ and apply (5*b*) over each subinterval, we get

$$\int_a^b f(x)\,dx = \frac{h}{3}(f_0 + 4f_1 + 2f_2 + 4f_3 + \cdots + 2f_{n-2} + 4f_{n-1} + f_n) - \frac{nh^5}{180} f^{\text{iv}}(\xi) \quad (8)$$

where again ξ is in (a, b). Equation (7) is the familiar trapezoidal rule and (8) is the also familiar parabolic rule. We note that, since all the abscissas, except the two at the ends of the complete interval, are common to the two subintervals, the trapezoidal rule requires $n + 1$ points for the n subintervals and the parabolic rule requires $n + 1$ points in $n/2$ subintervals.

In (7) and (8) $h = (b - a)/n$ so that as we increase the number of subintervals the error in (7) decreases by $1/n^2$ while the error in (8) decreases by $1/n^4$. For this reason and because the numerical coefficient in the error term of (8) is significantly less than that of (7), (8) is generally used in preference to (7) unless bad behavior of the fourth derivative is suspected. The sum of the squares of the coefficients in (7) is $h^2(n - \frac{1}{2})$ while in (8) it is $h^2(10n/9 - \frac{2}{9})$ so that the round-off properties of (8) are not very much worse than those of (7).

A useful method of improving the results of a numerical integration is to use a Richardson type extrapolative procedure [3]. Following [1], we illustrate this method for the case of the parabolic rule. Suppose that we use (8) to compute the integral of $f(x)$ over $[a, b]$ twice, first using $n_1/2$ subintervals and then using $n_2/2$ subintervals. Let the results be denoted respectively by I_1 and I_2. Then, using the error term in (8) we can express the true value of the integral I as

$$I = I_1 - \frac{(b - a)^4}{180 n_1^4} f^{\text{iv}}(\xi_1) \quad (9a)$$

$$I = I_2 - \frac{(b - a)^4}{180 n_2^4} f^{\text{iv}}(\xi_2) \quad (9b)$$

where both ξ_1 and ξ_2 are in (a, b). Now, if we assume that the fourth derivative does not vary much in (a, b), we can set $f^{\text{iv}}(\xi_1) = f^{\text{iv}}(\xi_2)$ and eliminate the fourth derivative between (9*a*) and (9*b*). If we do this the result is

$$I \approx I_2 + \frac{n_1^4}{n_2^4 - n_1^4}(I_2 - I_1) \quad (10)$$

As stated in [1], the approximation (10) can generally be used confidently if the previous assumption about the fourth derivative is valid or if successive approximations to I using greater values of n seem to be converging to I without oscillating around it and if the second term in (10) is only a small correction to I_2. Formulas analogous to (10) can clearly be derived for other quadrature formulas.*

b. Gaussian Type Quadrature Formulas

Until recently Gaussian type quadrature formulas were seldom used because in most cases the abscissas are irrational numbers, which, of course, makes them inconvenient for hand computation. However, on digital computers this argument has little weight and so in recent years the power of the Gaussian technique has caused these formulas to be used with increasing frequency.

If we put no restrictions on the weights and abscissas of (1), then we have $2n + 2$ parameters at our disposal and we would expect to be able to make (1) exact for polynomials of degree $2n + 1$ or less. We now proceed to determine the H_j's and a_j's so that this will be the case. First we consider the functions $f(x) = x^k$ $(k = 0, 1, \cdots, n)$. Since (1) is to be exact for these functions we have, as before,

$$\alpha_k = \sum_{j=0}^{n} H_j a_j^k \quad (11)$$

which can be solved for the H_j's if the a_j's are distinct.† To determine the abscissas we first note that any polynomial of degree $2n + 1$ can be written, without loss of generality, as

$$P_{2n+1}(x) = f_n(x) + \pi(x) g_n(x) \quad (12)$$

where $f_n(x)$ and $g_n(x)$ are both polynomials of degree n and

$$\pi(x) = (x - a_0) \cdots (x - a_n) \quad (13)$$

If (1) is to be exact for (12), then we have

$$\int_a^b [f_n(x) + \pi(x) g_n(x)]\,dx = \sum_{j=0}^{n} H_j f_n(a_j) \quad (14)$$

since the right-hand side of (1) vanishes for the function $\pi(x) g_n(x)$. We have shown that, if

* Note the similarity between this technique and that used to estimate the error in numerical integration techniques for the solution of ordinary differential equations in Chapter 8.

† Again it is true that

$$H_j = \int_a^b l_j(x)\,dx$$

the H_j's satisfy (11), then

$$\int_a^b f_n(x)\,dx = \sum_{j=0}^{n} H_j f_n(a_j)$$

since $f_n(x)$ is a linear combination of powers of $x \leq n$. Therefore, from (14), we get the result that (1) is exact for $P_{2n+1}(x)$ if the abscissas a_j are chosen so that $\pi(x)$ is orthogonal over $[a, b]$ to any polynomial of degree n. If, by a change of variable, we convert the interval $[a, b]$ to $[-1, 1]$, then it is a well-known fact [4] that the a_j's are roots of the Legendre polynomial of degree $n + 1$. Therefore, we have the result that if the H_j's are chosen to satisfy (11) and if the a_j's are roots of the Legendre polynomial of degree $n + 1$ (properly related to the interval $[a, b]$), then (1) is exact for all polynomials of degree $2n + 1$ or less. The particular class of quadrature formulas derived in this way is called Gaussian quadrature formulas.

The error of a Gaussian quadrature formula, when applied to a function $f(x)$, can be derived in a number of ways, perhaps most easily by using the Hermite interpolation formula to derive the Gaussian quadrature formula and then suitably operating on the error term of the Hermite formula [4]. Here we present only the result which is that the error in (1) is

$$E_n = \frac{f^{2n+2}(\xi)}{(2n+2)!}\int_a^b [\pi(x)]^2\,dx \cdot \qquad (15)$$

where ξ is somewhere in (a, b).

From (15) we see that Gaussian quadrature formulas of high order suffer even more than Newton-Cotes formulas from having high order derivatives in the error terms. For this reason we will be chiefly interested in low order Gaussian quadrature formulas. For $n = 0$, 1, and 2 we have, using the interval $[-1, 1]$,*

$$n = 0: \quad \int_{-1}^{1} f(x)\,dx = 2f(0) + \tfrac{1}{3}f''(\xi) \qquad (16a)$$

$$n = 1: \quad \int_{-1}^{1} f(x)\,dx = f(-1/\sqrt{3}) + f(1/\sqrt{3}) + \frac{1}{135}f^{\text{iv}}(\xi) \qquad (16b)$$

$$n = 2: \quad \int_{-1}^{1} f(x)\,dx = \tfrac{1}{9}[5f(-\sqrt{\tfrac{3}{5}}) + 8f(0) + 5f(\sqrt{\tfrac{3}{5}})] + \frac{1}{15{,}750}f^{\text{vi}}(\xi) \qquad (16c)$$

As in the case of Newton-Cotes formulas we can break up the interval $[a, b]$ into subintervals and use a Gaussian formula in each subinterval. Note, however, that since the end points of the subintervals are not abscissas of the quadrature formulas, we do not get the advantage of having some of the abscissas common to two subintervals. Dividing $[a, b]$ into n subintervals of length h and using (16a), we get the formula analogous to the trapezoidal rule

$$\int_a^b f(x)\,dx = h\sum_{m=0}^{n-1} f[a + h(m + \tfrac{1}{2})] + \frac{nh^3}{24}f''(\xi) \qquad (17)$$

where ξ is in (a, b). Similarly, using (16b) over $n/2$ subintervals (n even) of length $2h$, we get the formula analogous to the parabolic rule

$$\int_a^b f(x)\,dx = h\sum_{m=0}^{\frac{1}{2}n-1} (f\{a + h[(1 - 1/\sqrt{3}) + 2m]\} + f\{a + h[(1 + 1/\sqrt{3}) + 2m]\}) + \frac{nh^5}{270}f^{\text{iv}}(\xi) \qquad (18)$$

where again ξ is in (a, b). Equations (17) and (18) each require the computation of one less point than their analogs, (7) and (8), and they have, respectively, smaller numerical coefficients in the error term. Note also that since the weights in (17) and (18) are equal, both these formulas have ideal round-off properties. The extrapolation techniques of the previous section can also be applied to (17) and (18).† It is also worthwhile to note that since (17) uses equally spaced abscissas, it can be used to tabulate the integral of a tabulated function.

It is clearly possible to derive quadrature formulas in which some of the abscissas are assigned and some are not. The most interesting of these are the ones in which the end points of the interval are assigned abscissas. In this case we again reap the advantages of Newton-Cotes formulas in that we have abscissas common to two subintervals when

* To convert $\int_a^b g(y)\,dy$ to an integral over the interval $[-1, 1]$ let

$$x = \frac{2y - (b + a)}{b - a}$$

† Note, however, that with (17) or (18) we cannot choose n_1 and n_2 so that the abscissas for $n = n_1$ are a subset of those for $n = n_2$. With (7) or (8) this can be done if we choose $n_2 = 2n_1$.

the formula is used as part of a composite rule.

If we assign the end points of the interval as abscissas and *also* force the weights associated with these end points to be of equal magnitude and opposite sign, we get a class of formulas [5] with the interesting property that when used in composite rules the end points of each subinterval drop out, except for the ends of the complete interval. Compared to their analogous Newton-Cotes or Gaussian formulas, these formulas achieve higher accuracy at the cost of computing one or two more ordinates, respectively. Furthermore, the round-off properties of this class of formulas are somewhat better than those of Newton-Cotes quadrature formulas and nearly as good as Gaussian formulas.

One further quadrature formula which should be mentioned is that of Chebyshev in which all the weights are postulated equal. The resulting formula then has $n + 2$ parameters (the $n + 1$ abscissas and common weight) and so an accuracy of order $n + 1$ would be expected. For n even this is what we obtain but for n odd an accuracy of order $n + 2$ is obtained. The advantage of Chebyshev's formula is that, of course, the round-off properties are ideal since all the weights are equal. As we would expect since (17) and (18) both have equal weights, the Gaussian formulas with $n = 0$ and 1 are identical with the Chebyshev formulas for $n = 0$ and 1. Discussions of Chebyshev's quadrature formula can be found in [1] or [4].

c. Weight Functions

Instead of (1) we can consider deriving a formula of the form

$$\int_a^b w(x) f(x)\, dx \approx \sum_{j=0}^{n} H_j f(a_j) \qquad (19)$$

where $w(x)$, the weight function, is known analytically. A large number of quadrature formulas corresponding to different weight functions have been developed but it is beyond the scope of this chapter to go into these in detail. Most of the weight functions treated in the literature are of the form $(1 - x)^{\alpha}(1 - x)^{\beta}$. Discussions of these can be found in both [1] and [4]. An interesting treatment of the weight function $\sin mx$, where m is an integer, has been given by Filon [6].

d. Infinite Intervals

If the interval of integration is infinite there are two possible approaches:

1. if that part of the integral for $|x| > N$ can be assumed small or can be estimated analytically, then use one of the quadrature formulas previously presented for $|x| < N$, or
2. employ a quadrature formula derived for use over the complete infinite interval.

Consider the case where $[a, b] = [0, \infty]$. In this case we commonly consider (19) with the weight function e^{-x} since this assures convergence of the integral when $f(x)$ is of less than exponential order. The formula we wish to develop is, then,

$$\int_0^{\infty} e^{-x} f(x)\, dx = \sum_{j=0}^{n} H_j f(a_j) + E_1 \qquad (20)$$

Following the technique used to develop Gaussian quadratures, we find [4] that the a_j's are the roots of $L_{n+1}(x)$, the Laguerre polynomial of order $n + 1$, the weights are given by

$$H_j = \frac{(n+1)!}{L'_{n+1}(a_j) L_{n+2}(a_j)}, \quad j = 0, 1, \cdots, n \qquad (21)$$

and the error is given by

$$E_1 = \frac{[(n+1)!]^2}{(2n+2)!} f^{2n+2}(\xi) \qquad (22)$$

where $0 < \xi < \infty$.

When $[a, b] = [-\infty, \infty]$ we consider the weight function e^{-x^2} so that the quadrature formula is

$$\int_{-\infty}^{\infty} e^{-x^2} f(x)\, dx = \sum_{j=0}^{n} H_j f(a_j) + E_2 \qquad (23)$$

In this case we find [4] that the abscissas are the roots of $H_{n+1}(x)$, the Hermite polynomial of order $n + 1$, the weights are given by

$$H_j = \frac{2^{n+1} n! \sqrt{\pi}}{H'_{n+1}(a_j) H_n(a_j)}, \quad j = 0, 1, \cdots, n \qquad (24)$$

and the error is given by

$$E_2 = \frac{(n+1)! \sqrt{\pi}}{2^{n+1}(2n+2)!} f^{2n+2}(\xi) \qquad (25)$$

where $-\infty < \xi < \infty$.

Tables of the abscissas and weights for these formulas as well as the Gaussian and Chebyshev formulas of Section 2b can be found in [4].

3. SUMMARY OF THE CALCULATION PROCEDURE

We consider the procedure when either (7), (8), (17), or (18) is to be used. The input quantities are h, a, n (or b), and the constants required by the formula in question.

The steps in the calculation are then just a series of evaluations of the function which are multiplied by the appropriate weights and summed. The output is of course just the value of the integral.

The flow chart that follows is for equation (18).

4. FLOW CHART

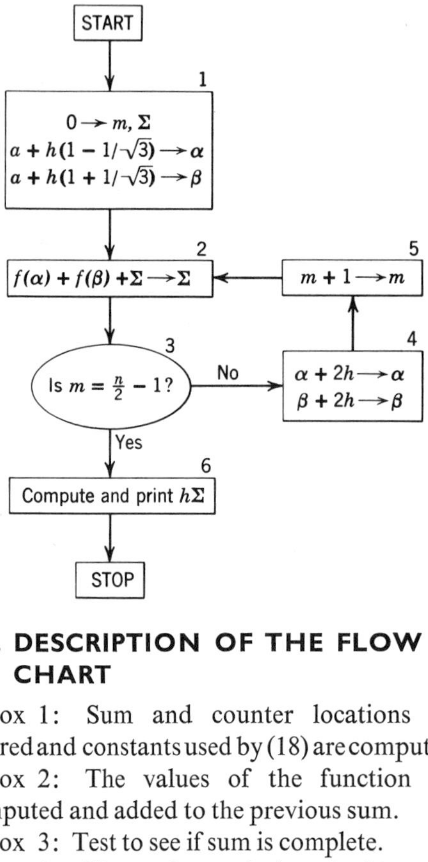

5. DESCRIPTION OF THE FLOW CHART

Box 1: Sum and counter locations are cleared and constants used by (18) are computed.

Box 2: The values of the function are computed and added to the previous sum.

Box 3: Test to see if sum is complete.

Box 4: New values of the abscissas are computed.

Box 5: Value of the counter is increased.

Box 6: Value of the integral is computed and printed. At this point provision can be made for changing parameters and computing another integral.

6. SUBROUTINES

Only those subroutines, if any, to evaluate $f(x)$ are required.

7. SAMPLE PROBLEM

Consider the following integral.

$$\int_0^1 e^x\,dx \approx 1.71828$$

Using (18) with $h = \frac{1}{4}$ we must compute

$$e^{\frac{1}{4}(1-1/\sqrt{3})} = 1.11144$$

$$e^{\frac{1}{4}(1+1/\sqrt{3})} = 1.48340$$

$$e^{\frac{1}{4}(1-1/\sqrt{3})+\frac{1}{2}} = 1.83246$$

$$e^{\frac{1}{4}(1+1/\sqrt{3})+\frac{1}{2}} = 2.44572$$

Summing these and multiplying by h, we get

$$\int_0^1 e^x\,dx = 1.71826$$

Corresponding results using (7), (8), and (17) with $h = \frac{1}{4}$ are, respectively, 1.72722, 1.71831, and 1.71382.

8. MEMORY REQUIREMENTS

The requirement for the constants and the program except for that part used to evaluate $f(x)$ is about thirty registers. Therefore, if M is the number of words required for the program to evaluate $f(x)$, the total storage required is about $30 + M$ words.

9. ESTIMATION OF THE RUNNING TIME

If T is the time required to evaluate $f(x)$ and v is the addition time of the computer, then the time required to evaluate (18) is approximately

$$n(T + v)$$

There are only three multiplications required and the bookkeeping time is negligible.

10. REFERENCES

1. F. B. Hildebrand, *Introduction to Numerical Analysis*, McGraw-Hill Book Co., New York, 1956.
2. J. V. Uspensky, *Theory of Equations*, McGraw-Hill Book Co., New York, 1948.
3. L. F. Richardson and J. A. Gaunt, The Deferred Approach to the Limit, *Trans. Roy. Soc. London*, vol. 226A, 1927, p. 300.
4. Z. Kopal, *Numerical Analysis*, John Wiley & Sons, New York, 1955.
5. A. Ralston, A Family of Quadrature Formulas Which Achieve High Accuracy in Composite Rules, *J. Assoc. Comp. Mach.* vol. 6, 1959, pp. 384–394.
6. L. N. G. Filon, On a Quadrature Formula for Trigonometric Integrals, *Proc. Roy. Soc. Edinburgh*, vol. 49, 1928, pp. 38–47.

Multiple quadrature by Monte Carlo methods

23

Herman Kahn
The RAND Corporation

I. INTRODUCTION AND FUNCTION

A short description of the Monte Carlo method can be given as follows: The expected score of a player in any reasonable game of chance, however complicated, can in principle be estimated by averaging the results of a large number of plays of the game. Such estimation can often be rendered more efficient by various devices which replace the original game with another known to have the same expected score. The new game may lead to a more efficient estimate by being less erratic, i.e., having a score of lower variance, or by being cheaper to play with the equipment on hand. There are obviously many problems about probability that can be viewed as problems of calculating the expected score of a game. Still more, there are situations that do not concern probability but are nonetheless equivalent for some purposes to the calculation of an expected score. The Monte Carlo method refers simply to the exploitation of these remarks.

As an example of a very simple game which leads to the evaluation of an integral

$$p = \int_0^1 g(x)\, dx, \qquad 0 \leq g(x) \leq 1$$

consider the following:

1. A point (x, y) is picked at random with uniform probability over the unit square (later in the chapter a method of doing this will be described).
2. If this point is underneath the curve $y = g(x)$, a success is scored. If the point is above the curve, i.e., between the curve $y = g(x)$ and the line $y = 1$, it is called a failure.
3. The above process is repeated many times and the proportion of successes is calculated.

It is clear that the experimentally estimated probability of winning this very simple game will give the area underneath the curve $y = g(x)$, since the probability of landing in that area is numerically equal to the area. It should also be clear that it would be very easy to generalize the above process to a multidimensional integral. Lastly, it should be plain that the efficiency of the process does not depend at all upon the detailed microscopic character of the function $g(x)$ but only on a very gross characteristic—the total area, or in the case of a multidimensional integral, the total multidimensional volume.

The above example illustrates that one of the

simplest and most direct applications of the Monte Carlo method is to the evaluation of integrals. However, if we wish to use Monte Carlo on a more general class of integrals, we must rely on the following two theorems (which will be given without proof) to justify this wider application.

THEOREM I (THE STRONG LAW OF LARGE NUMBERS) [1]: If a sequence of N random variables x_1 to x_N are picked from a population with the probability density function* $f(x)$, and a random variable $\hat{z}_N$ defined by the equation

$$\hat{z}_N = \frac{1}{N}\sum_1^N z(x_i) \tag{1}$$

where $z(x)$ is a given integrable function, and if the integral

$$\bar{z} = \int_{-\infty}^{\infty} z(x) f(x)\, dx \tag{2}$$

exists in the ordinary sense, then $\hat{z}_N$ will, with probability 1, approach $\bar{z}$ as a limit as N approaches ∞.

The integral (2) is called the expected value† of the function $z(x)$, and $\hat{z}_N$ is called an estimate of $\bar{z}$. If $\overline{z^2}$, the expected value of $z^2(x)$, also exists, an estimate can be made about the amount that $\hat{z}_N$ deviates from $\bar{z}$ for large N. Denote the variance of $z(x)$ by either σ^2 or V; define it by the equation

$$\begin{aligned}\sigma^2 = V &= \overline{(z - \bar{z})^2} \\ &= \int (z - \bar{z})^2 f(x)\, dx \\ &= \int z^2 f(x)\, dx - 2\bar{z}\int z f(x)\, dx + \bar{z}^2 \int f(x)\, dx \\ &= \overline{z^2} - 2\bar{z}^2 + \bar{z}^2 \\ &= \overline{z^2} - \bar{z}^2 \end{aligned} \tag{3}$$

and then apply Theorem II.

* $f(x)\,\Delta x$ gives the approximate probability that the random variable x lies in the "small" region $(x, x + \Delta x)$. The only properties that a function $f(x)$ needs in order to be a probability density are that it be nonnegative and that its integral be 1.

† Most readers will be familiar with the fact that the "expected value" may be very unexpected. For example, if a random variable has a probability of 1/2 of taking on the value 0, and the same probability of taking on the value 1, then the expected value of this random variable is 1/2 though the random variable itself never takes on this value.

THEOREM II (THE CENTRAL LIMIT THEOREM):‡ For large N the probability that the event $\bar{z} - \delta \le \hat{z} \le \bar{z} + \delta$ occurs is asymptotically independent of the exact nature of $z(x)$ or $f(x)$ but depends only on N and σ^2. In fact,

$$\Pr\{\bar{z} - \delta \le \hat{z} \le \bar{z} + \delta\} = \sqrt{\frac{1}{2\pi}} \int_{-\frac{\delta}{\sigma}\sqrt{N}}^{\frac{\delta}{\sigma}\sqrt{N}} e^{-x^2/2}\, dx + \text{terms of order } \frac{1}{\sqrt{N}} \tag{4}$$

The probability that the deviation of $\hat{z}$ from $\bar{z}$ will exceed $\pm\lambda\sigma/\sqrt{N}$ is given by substituting $\delta = \lambda\sigma/\sqrt{N}$ in (4) and then taking

$$1 - \Pr\{\bar{z} - \delta \le \hat{z} \le \bar{z} + \delta\}$$

which turns out to be

$$\sqrt{\frac{2}{\pi}} \int_{\lambda}^{\infty} e^{-x^2/2}\, dx$$

An abbreviated table of this integral is given in Table 1. It can be seen from Table 1 that deviations greater than $\pm\sigma/\sqrt{N}$ will be frequent, deviations greater than $\pm 2\sigma/\sqrt{N}$ not uncommon, and deviations greater than $\pm 3\sigma/\sqrt{N}$ so uncommon that if the table applies the possibility that this last event may occur can usually be ignored. $\sigma/\sqrt{N}$ is called the standard deviation (s.d.) and is conventionally used as a measure of the sampling error of the estimate $\hat{z}_N$. It decreases as the sample size N gets larger, but because it goes down as $\sqrt{N}$ one must increase the sample size by a factor of 4 in order to halve the error.

Table I

λ	Probability
0.6745	.5000
1.0000	.3173
2.0000	.0455
3.0000	.0027
4.0000	.0001

There are two problems which arise in the application of Table 1. The first is that it holds only for "sufficiently large N" and in many applied problems there is no obvious prescription for calculating how large N should be. However, the sweet bye and bye character of the

‡ Almost any book on statistics discusses this theorem. Cramér [2] is especially full and interesting on this theorem and its variants.

theorem does not give rise to much trouble in most applications. The second problem, the estimate of σ, is also routine, but sometimes give trouble. Normally the most convenient way to evaluate σ^2 in (3) is to substitute the estimates $\widehat{z}^2$ and $\hat{z}$ for $\overline{z^2}$ and $\bar{z}$. While this too is all right for "sufficiently large N," now the bye and bye character is sometimes serious in practice. In any case where the problem may be "pathological" it is necessary to do subsidiary calculations—either by Monte Carlo or by the use of some approximation—to bound the error reliably. This problem is alleviated in most cases because only the roughest bound is needed, since one can normally live with very high errors in the estimate of the error. If the estimate of the error is 5% and if this estimate was low by a factor of 2, it is likely to be all right; in most problems in which a 5% error is acceptable, a 10% error would also be considered acceptable.

The reason that sampling is useful in evaluating multiple integrals of a high order is that neither of the two theorems depends on the dimensionality of the integral. The number of points required to evaluate a multidimensional integral to a fixed level of accuracy depends only on σ (or $\sigma/\bar{z}$ if a fixed per cent accuracy is desired). While it is true that in this perverse world σ, or more likely $\sigma/\bar{z}$, often increases with the dimensionality of the integral, there is no reason in principle why this should be so. By contrast, in almost all standard techniques the number of points required to evaluate an integral goes up in geometrical progression with its dimensionality. In part this undoubtedly is due to a defect in the theory of integration in many dimensions,* but it often seems to be unavoidable. This exponential increase almost never occurs if the integration is done by Monte Carlo. The other occasional advantage of Monte Carlo is that it may be cheaper to compute points by Monte Carlo than in the standard ways. This shows up most sharply in trying to solve certain Green's function types of problems (Curtiss in [3], pp. 191–233)—or when only a very small amount of information is desired; for example, the solution of the potential equation at a single point rather than everywhere.

On the whole, though, it must be admitted that Monte Carlo has not shown up very well in competition with standard techniques when these techniques have been at all reasonable. It has been most useful where the standard numerical techniques have been inadequate. In this sense it is a method of last resort.

2. A BRIEF DESCRIPTION OF THE MAJOR IDEAS

When doing a Monte Carlo problem one focuses attention on three main topics. They are:

1. Choosing or analogizing the probability process.
2. Designing and using variance reduction techniques.
3. Generating sample values of the random variables on the computing machine to be used.

In the application to integration the first topic usually corresponds to the way one breaks the integrand into the product $z(x) f(x)$. The $f(x)$ chosen, of course, determines the

* One way to calculate $\int \cdots \int z(x_1, \cdots, x_n)\, dx_1 \cdots dx_n$ is to use a variant of the trapezoidal rule and evaluate z at the M^n points obtained by dividing each x_i space into M intervals and taking the mid-points of these intervals. Even if more powerful methods of integration are used, as long as they are straightforward generalizations of ordinary one-dimensional formulas, the number of points required goes up exponentially with the number of dimensions.

However, there are other possibilities. One that has been suggested to the author by George Brown goes like this: The function $z(x_1, \ldots, x_n)$ would be expanded in the form $\sum_{k=1}^{K} \prod_{i=1}^{n} \phi_{ik}(x_i)$. The corresponding trapezoidal integration formula would then only require MnK points. Routine methods exist for making such expansions, but if the function $z(x_1, \cdots, x_n)$ is in any way unsuitable, a vary large K may be needed to make the expansion accurate enough.

In general, in any definite problem special techniques can be used to reduce the number of points to less than M^n but in a large class of problems not enough less to make the classical numerical integration competitive with the number of points required by Monte Carlo. However, it also seems to be true, in the past, that Monte Carlo itself has been of specialized usefulness, mostly in evaluating integrals that have arisen out of probabilistic situations. The author knows of relatively few serious nonprobabilistic integrals evaluated by Monte Carlo. It is hard to say whether this is coincidental or symptomatic.

probability process. In complicated integrals it is often advantageous to work directly with the model which generated the integral that one is trying to compute rather than to just look at the final result.

The second topic, variance reduction, is often of great importance. In many Monte Carlo problems, doing the sampling in a straightforward way would require an extremely large number of samples to obtain a satisfactory level of accuracy. It is always possible, sometimes by simple means and sometimes only by subtle ones, to decrease the sample variances, often by many orders of magnitude. Therefore, on any problems that will take more than a few minutes of machine time, the analyst should consider using variance-reducing techniques. Of course, on very small problems it may not be advisable, since the additional complexity in both the analysis and coding can be more costly than the saving in running time. However, if the applied mathematician wishes to increase his skill he should probably try to use variance reduction techniques on even these very simple problems, because it is easier to learn on them. Six of the most important variance-reducing techniques that are available are listed below:

1. Importance Sampling.
2. Russian Roulette and Splitting.
3. Use of Expected Values.
4. Correlation and Regression.
5. Systematic Sampling.
6. Stratified Sampling (Quota Sampling).

The idea of Importance Sampling is that the probability process can be modified so that most of the samples are drawn from "interesting" regions to the neglect of "uninteresting" regions. The words "interesting" and "uninteresting" here refer to the error in the final answers that is attributable to these regions. It can be shown that for almost any single definite question there always exists an Importance Sampling scheme with zero variance. While in practice one cannot hope to approach this limit, startling reduction in variance can often be made if a suitable scheme is chosen.

Russian Roulette and Splitting is related to the sequential sampling used in industry. As opposed to Importance Sampling, in which the sampling process is modified, the sampling is done in a straightforward fashion, but samples that go into "interesting" regions are split into many independent branches while those that go into "uninteresting" regions are often terminated. The net effect is the same as in Importance Sampling in that the "interesting" regions receive proportionately more attention than the "uninteresting" ones.

The Use of Expected Values refers to the possibility of combining analytical and probabilistic methods. There is no necessity for the Monte Carlo calculator to use a pure probabilistic approach on all of the probabilistic elements of a problem. The analytical methods can be used on that portion of the problem where it is easy to do so, and the sampling done only where the analytical methods are inconvenient or impossible. It often turns out that those parts of the problem which are simple to do analytically would have contributed a great deal to the variance had they been done by sampling.

Correlation and Regression refers to the possibility that one can exploit structural similarities between different problems or questions in order to reduce either the error or the amount of work that is done. It is particularly valuable when one is doing a large parametric study or when one has a simple, analytically tractable model or approximation of a situation and is using the Monte Carlo only to find the difference between the simple model or approximation and a more complicated and realistic but analytically untractable model.

Systematic Sampling refers to the possibility of combining, in a quite natural fashion, what is in effect a numerical integration with the sampling.

Stratified Sampling can be looked on as a combination of Importance Sampling and Systematic Sampling.

The six techniques are discussed at greater length in an article by the author in [3], pp. 146–190. In this same reference are articles by G. E. Albert, H. F. Trotter and J. W. Tukey, M. J. Berger, L. A. Beach and R. B. Theus, and A. W. Marshall, which discuss various aspects of the problem of reduction. Finally, [4] contains a systematic and detailed account of the application of such methods to integrals, random walks, and integral equations.

The third topic concerns the generation of sample values of the random variables. In order to do a Monte Carlo problem a gambling device, such as a roulette wheel, dice, or coins must be available. From the viewpoint of the computer, the simplest and most practical gambling device is a table of uniform Random numbers.* Such a table usually consists of a long sequence of the digits 0 to 9 arranged so that the probability that the nth entry is x is 0.1 and so that every entry is statistically independent of the other entries. By using this table it is possible to select a random equidistributed point on the line 0 to 1 to any desired degree of accuracy by combining the digits in the form $0 \cdot xyz \cdots$.

Such numbers are available from many sources. The largest supply is in a book called *A Million Random Digits with a Hundred Thousand Normal Deviates*, published by The RAND Corporation. These same numbers are available on punched cards from The RAND Corporation.

In the modern high-speed electronic computer, these Random numbers are usually generated internally. While it would be possible for the machine to have a built-in random process which could be exploited to generate Random numbers, there seems to be no particular advantage to this, and some very serious disadvantages. Most problems are done with what is known as "pseudo" Random numbers where an exact deterministic process is used for generating the next number in the series from its predecessors. Many deterministic processes exist which generate series that, for all practical purposes, look like they were generated by random processes.

Probably the simplest method of generating pseudo Random numbers in a binary digital machine is by the equation

$$R_{n+1} = KR_n \bmod 2^N \tag{5}$$

where R_n is the nth random number

R_{n+1} is the $(n + 1)$st random number

K is a constant multiplier—any odd power of 5 will do, but it is desirable to use as large a power as the machine's word length will hold

N is usually the number of binary digits in each word of the machine

* Whenever we refer to *uniform* Random numbers, we will capitalize the word Random.

In most machines it is trivially simple to do the mod 2^N operation by just doing an exact K times R_n operation and then taking R_{n+1} equal to the least significant half of the result. It can be shown that starting with an odd R_0, one will run through 2^{N-2} numbers before repeating a number, and that for all practical purposes those numbers can be treated as being independent and uniformly distributed between 0 and 1. If we desired to play the very simple game described at the beginning of this paper, which required picking a point (x, y) uniformly in the unit square, it would only be necessary to generate two random numbers, R_n and R_{n+1}, and take

$$x = R_n \qquad \text{and} \qquad y = R_{n+1} \tag{6}$$

This method of generating random numbers and some other methods are discussed in an article by Taussky and Todd in [4], pp. 15–28.

Once we have a method of getting Random numbers, we must examine methods of using these Random numbers to get values of random variables with nonuniform probability density functions. It is easy to show that if we have a probability density function $f(x)$ and solve the equation

$$\int_{-\infty}^{x_n} f(x)\, dx = R_n \tag{7}$$

for x_n, that x_n has been picked out of a population distributed according to $f(x)$. The above method can be generalized by the use of iterated integrals to multidimensional random variables. In practice, though, other methods of generating random variables are often more convenient for both one-dimensional and multidimensional random variables.

Two examples of picking values of random variables by other methods will be given; we will need them in treating the example discussed at the end of this chapter. First, let us assume that we wish to select a point uniformly and randomly within the sphere $x^2 + y^2 + z^2 \leq 1$. This can be done by first picking three Random numbers R_1, R_2, R_3, and then provisionally setting

$$\begin{aligned} x &= 2R_1 - 1 \\ y &= 2R_2 - 1 \\ z &= 2R_3 - 1 \end{aligned} \tag{8}$$

The provisional point (x, y, z) has clearly been picked uniformly from the cube bounded by the planes $x = \pm 1, y = \pm 1, z = \pm 1$. Now, if we pick many such points, but save only the ones which satisfy the unequality

$$x^2 + y^2 + z^2 \leq 1 \tag{9}$$

the resulting collection of points will have been picked uniformly at random from the proper unit sphere.

The above is an example of what is called a rejection technique. Other useful types of manipulations are available. For example, suppose we wished to select values of x out of the probability density function

$$f(x) = 3x^2, \qquad 0 \leq x \leq 1 \tag{10}$$

If we tried to use (7), we would have to solve the equation

$$\int_0^x 3x^2\, dx = R, \qquad 0 \leq x \leq 1 \tag{11a}$$

or

$$x = \sqrt[3]{R} \tag{11b}$$

Taking a cube root on an electronic computer is a relatively lengthy procedure. It is much simpler to generate three random numbers (R_1, R_2, and R_3) and let x be equal to the largest one of these numbers. The reader can verify for himself that this process leads to the proper probability density function for x.

A systematic development of various methods of generating random variables with many examples can be found in Chapter 1 of [4] or the article by Butler in [3], pp. 249–264.

3. A SAMPLE MONTE CARLO PROBLEM

The example we will consider is the evaluation of a multiple integral that arises in a problem in electron physics [5].*

$$I = \int dV_1 \int_{\substack{\beta \leq 1 \\ |\vec{a}+\vec{b}|>1}} dV_2 \int_{\substack{\gamma \leq 1 \\ |\vec{a}+\vec{c}|>1}} dV_3 g(a_1, a_2, \cdots, c_2, c_3) \tag{12}$$

* The evaluation in [5] was done by J. Marcum and the author.

where

$$g(a_1, a_2, \cdots, c_2, c_3) = \frac{1}{(\vec{a} + \vec{b} + \vec{c})^2} \frac{1}{\alpha^2 + \vec{a} \cdot (\vec{b} + \vec{c})} \tag{13}$$

$$\begin{aligned}
\vec{a} &= a_1 i + a_2 j + a_3 k \\
\vec{b} &= b_1 i + b_2 j + b_3 k \\
\vec{c} &= c_1 i + c_2 j + c_3 k \\
dV_1 &= da_1 da_2 da_3 \\
dV_2 &= db_1 db_2 db_3 \\
dV_3 &= dc_1 dc_2 dc_3 \\
\alpha &= |a| = \sqrt{a_1^2 + a_2^2 + a_3^2} \\
\beta &= |b| = \sqrt{b_1^2 + b_2^2 + b_3^2} \\
\gamma &= |c| = \sqrt{c_1^2 + c_2^2 + c_3^2}
\end{aligned}$$

We will now discuss the evaluation of the integral of (12).

First Method

This was the method that was actually used when the problem came up. It is by no means the best, but it was the first one that came to mind and because it seems to be a good pedagogical example we will discuss it here. Because the problem was so simple and short, no thought was put into using more refined methods.

The integral (12) can be written (because of symmetry we need only consider $\vec{a} = \alpha i$) in the form

$$I = 4\pi \int K(\alpha)\, d\alpha \tag{14a}$$

where

$$K(\alpha) = \int_{\beta \leq 1} dV_2 \int_{\gamma \leq 1} dV_3 g(\alpha, b_1, \cdots, c_2, c_3) \tag{14b}$$

$$[(\alpha + b_1)^2 + b_2^2 + b_3^2] > 1$$

$$[(\alpha + c_1)^2 + c_2^2 + c_3^2] > 1$$

$$g(\alpha, b_1, \cdots, c_2, c_3) = \frac{1}{(\alpha + b_1 + c_1)^2 + (b_2 + c_2)^2 + (b_3 + c_3)^2} \times \frac{1}{\alpha^2 + \alpha b_1 + \alpha c_1} \tag{14c}$$

Instead of trying to evaluate I directly by Monte Carlo, we will evaluate $K(\alpha)$ by Monte

Carlo for many values of α and then evaluate I by a numerical integration on α.

It turns out that for $\alpha > 2$, $K(\alpha)$ can be reduced to a numerical integration in one dimension, so we will not use Monte Carlo for $\alpha > 2$. We also note by examination that $K(0) = 0$. This means that when we evaluate $K(\alpha)$ for $0 < \alpha < 2$ by Monte Carlo, we can check the end points of the curve we calculate by independent methods.

The probability density that we will pick will be:

$$f(b_1, b_2, b_3, c_1, c_2, c_3) = \frac{1}{(\frac{4}{3}\pi)^2} \quad \text{if } \beta \leq 1 \text{ and } \gamma \leq 1 \qquad (15a)$$

$$f(b_1, b_2, b_3, c_1, c_2, c_3) = 0 \quad \text{if } \beta > 1 \text{ or } \gamma > 1 \qquad (15b)$$

That is, we are picking vectors $\vec{b}$ and $\vec{c}$ uniformly in their respective unit spheres. This means that the z function whose expected value we will be calculating is

$$z(\alpha, b_1, \cdots, c_2, c_3) = (\tfrac{4}{3}\pi)^2 g(\alpha, b_1, \cdots, c_2, c_3) \quad \text{if } [(\alpha + b_1)^2 + b_2^2 + b]_3^2 \geq 1 \text{ and } [(\alpha + c_1)^2 + c_2^2 + c_3^2] \geq 1 \qquad (16a)$$

$$z(\alpha, b_1, \cdots, c_2, c_3) = 0 \quad \text{otherwise} \qquad (16b)$$

and the integration will now be performed over the entire space $\beta \leq 1$, $\gamma \leq 1$; the other restriction is taken account of by the definition of z in (16*b*). The extra factor of $(\frac{4}{3}\pi)^2$ in (16*a*) is needed because we must pick f and z so that

$$zf = g \qquad (17)$$

4. THE FLOW CHART FOR THE EXAMPLE

The flow chart given on page 256 is almost identical with the one we actually used in calculating the example. While it may seem complicated, particularly to an inexperienced person, it is possible for an experienced person to draw up a flow chart of this type in an hour or so. While we do not recall how long it took to code the problem we would not be surprised if it took less than a day, probably significantly less.

On this flow chart α denotes the independent variable of the function $K(\alpha)$ which we are trying to evaluate. Every time we draw a sample value we refer to it as being a trial. For bookkeeping purposes we give every separate printout a serial number which we denote by n. While not strictly necessary it is desirable to have more than one printout for all the trials associated with an α value. That is, we like to break a problem into subportions both for error control and to enable certain statistical checks to be carried on outside the machine. We denote by n_0 the serial number of an initial subportion and by n_f the serial number of the final subportion. In other words, n_0 denotes the first printout of a new α value and n_f the last printout with such an α value.

In order to be able to reproduce the results, in case there is any trouble, we will record the initial Random number, R_0, that is used to start the problem. In fact, we print the initial Random number which starts every printout and we also temporarily save in the machine the initial Random number which starts a trial. Doing this is very helpful in trouble-shooting.

i denotes the serial number of an individual trial. N is the number of individual trials per printout. Therefore, $N(n_f - n_0 + 1)$ is the total number of independent samples or trials for a particular α value. k is the index number denoting a particular value of the parameter α, and K is the total number of α values to be run; i.e., k runs from 1 to K. τ is an auxiliary index which tells us whether we are sampling in the b space or the c space. T.S. denotes a temporary storage.

5. DESCRIPTION OF THE FLOW CHART

Box 1: Inputs all of the information needed to determine the problem to be run and sets the k tally to zero. It should be noted that the number of printouts per α value vary according to the α value. It is desirable to do this because we may need a varying number of samples for different values of α.

Box 2: Advances k tally; arranges for changing the α value.

Box 3: Sets things up for running a new printout.

Box 4: Sets the τ index to zero so that we will start sampling on the b space and stores

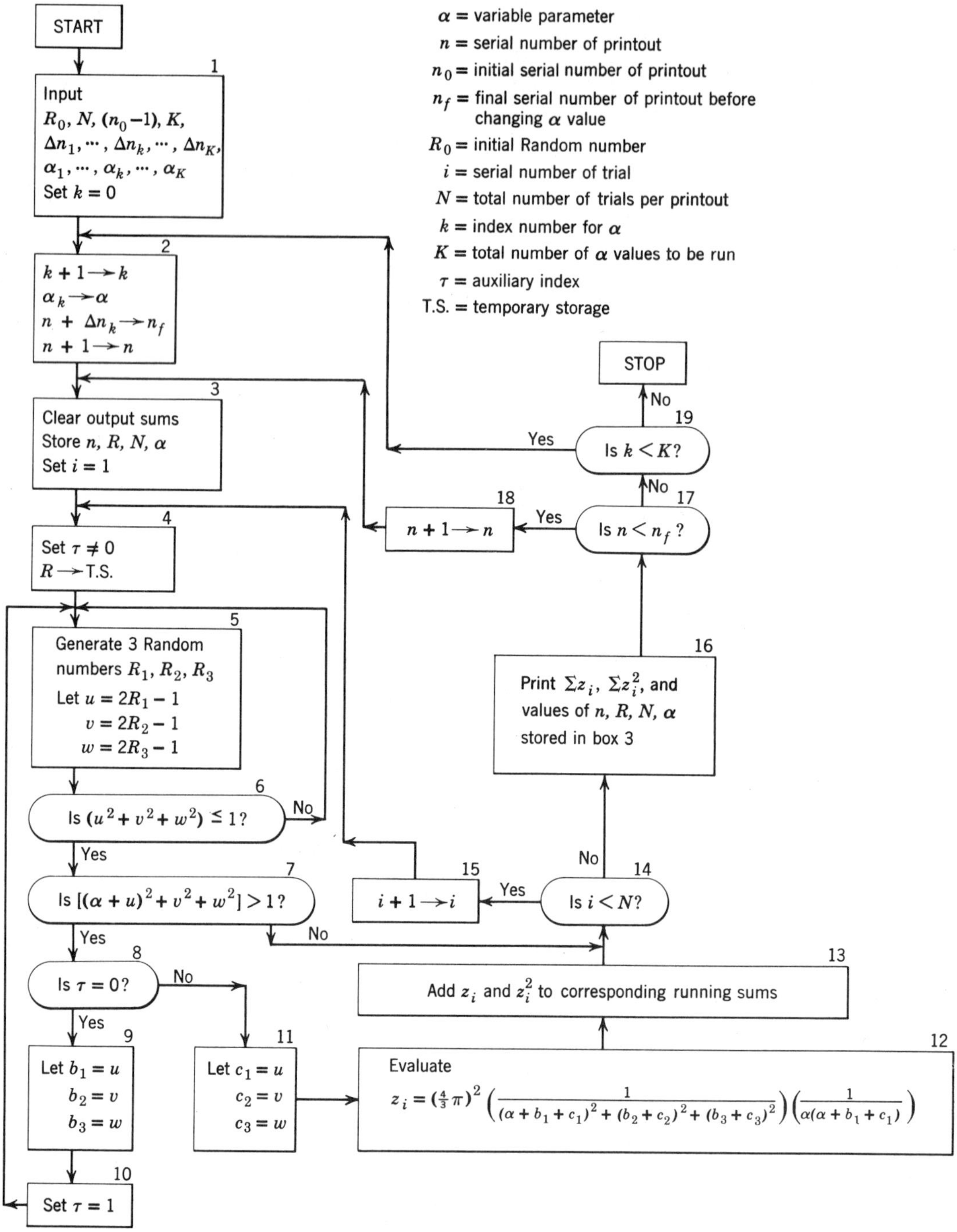

the initial Random number which starts that trial in a temporary storage. If the problem runs into any trouble, which results in a machine stop, we do not need to go back to the beginning of the run in order to trouble-shoot but can start with this Random number which begins the trial which ran into trouble.

Boxes 5 and 6: Pick a three-dimensional point uniformly within the unit sphere.

Box 7: Checks if this point is outside a unit sphere circumscribed around the point $x = \alpha$. If the point is not outside, the value of z is taken to be zero [see (16*b*)] so we simply skip the rest of the calculation (in effect recording a zero for z_i and z_i^2) and go on to the next trial. If the point is inside the unit sphere we go on to box 8.

Box 8: Checks the tally to see if it is the b or c space which is being sampled.

Boxes 9 and 11: Assign values to either the b or c vector, whichever is appropriate.

Box 10: Resets the auxiliary τ index if we happened to be working on the b space so that the next time box 8 will send us to the c space.

Box 12: Evaluates the function $z(x)$ when it is not zero. (See box 7.) Some readers will be offended by the fact that we multiply the constant $(\frac{4}{3}\pi)^2$ every time rather than do this operation outside the machine but this simplifies the bookkeeping slightly.

Box 13: Sums z_i and z_i^2. The reason we collect z_i^2 is to be able to estimate the error, as explained on page 250.

Box 14: Checks if we have finished running enough trials for that printout.

Box 15: Varies the index if we have not finished running enough trials for that printout.

Box 16: Prints. We not only print the serial number that identifies the printout but we print the value of R that started this printout so that if any question arises later we will be able to rerun this particular portion of the problem without rerunning the entire problem. Printing N reminds us how many individual trials we ran; likewise we print the value of α to remind us what the parameter value is. In theory, of course, both of these last two quantities are determined by the serial number n but experience has shown us that it is valuable to overlabel each printout in this fashion.

Box 17: Checks if we have finished the assigned number (Δn_k) of printouts for that α value.

Box 18: Advances n if we have not run Δn_k printouts.

Box 19: Checks if we have run all k of our α values or not.

6. SOME ADDITIONAL COMMENTS ON THE EXAMPLE

It need not have been necessary, of course, to evaluate $K(\alpha)$ at a number of points. We could have also sampled the α space by, for example, drawing α values from the simple probability density function

$$f(\alpha) = 1/2, \qquad 0 \leq \alpha \leq 2 \tag{18}$$

The reason we did not was because we knew the value of $K(\alpha)$ for $\alpha = 0$ and 2 (and also the slope of the curve at $\alpha = 0$ and 2) so we could check the end points of the $K(\alpha)$ curve very accurately. If we had not been able to evaluate $K(\alpha)$ for $\alpha \geq 2$ rather easily and had to do Monte Carlos on the entire space from zero to infinity, we probably would have done sampling on α too. An appropriate $f(\alpha)$ to be used now might be

$$f(\alpha) = \frac{3\delta^3}{(\delta + \alpha)^4} \tag{19}$$

where δ is a constant that should probably be chosen equal to about one. The z function to be evaluated must be multiplied by the reciprocal factor $(\delta + \alpha)^4/3\delta^3$. The reason why the above $f(\alpha)$ is appropriate is that it has the same asymptotic character as the integrand making the new z function essentially equal to a constant. It was shown in the previously mentioned references on Importance Sampling that this is a good thing to do.

To select α's according to the probability density function of (19), we might solve the equation

$$\int_0^{\alpha_i} f(\alpha)\, d\alpha = R_i \tag{20}$$

for α_i. This gives

$$\alpha_i = \frac{\delta}{\sqrt[3]{R_i}} - \delta \tag{21}$$

We pointed out previously on page 254 how to avoid taking the cube root.

7. REFERENCES

1. J. L. Doob, *Stochastic Processes*, John Wiley & Sons, New York, 1953.
2. H. Cramér, *Mathematical Methods of Statistics*, Princeton University Press, 1946.
3. Herbert A. Meyer (ed.), *Symposium on Monte Carlo Methods*, John Wiley & Sons, New York, 1956.
4. Herman Kahn, Applications of Monte Carlo, Research Memorandum RM-1237-AEC, The RAND Corporation, Apr. 19, 1954.
5. M. Gell-Mann and K. A. Brueckner, Correlation Energy of an Electron Gas at High Density, *Phys. Rev.*, vol. 106, Apr. 15, 1957, eq. 9.

24 Fourier analysis

G. Goertzel
Nuclear Development Corporation of America

1. FUNCTION

The purpose of this program is to determine $2N + 1$ constants α_p $(p = 0, 1, \cdots, N)$, b_p $(p = 1, 2, \cdots, N)$ in such a way that the equations

$$f_n = \tfrac{1}{2}\alpha_0 + \sum_{p=1}^{N}\left(\alpha_p \cos\frac{2\pi np}{2N+1} + b_p \sin\frac{2\pi np}{2N+1}\right) \quad (n = 0, 1, \cdots, 2N) \qquad (1)$$

are satisfied, where the f_n are given numbers. The f_n may be thought of as the values of a function $f(x)$ at the points

$$x_n = \frac{2\pi n}{2N+1} \qquad (n = 0, 1, \cdots, 2N) \qquad (2)$$

2. MATHEMATICAL DISCUSSION

The solution of the system (1) is well known [1]:

$$\alpha_p = \frac{2}{2N+1}\sum_{n=0}^{2N} f_n \cos\frac{2\pi np}{2N+1} \qquad (3)$$

$$b_p = \frac{2}{2N+1}\sum_{n=1}^{2N} f_n \sin\frac{2\pi np}{2N+1} \qquad (4)$$

The following discussion will be concerned with the efficient calculation of α_p and b_p from (3) and (4). In particular, it will be shown how this can be accomplished with only two trigonometric function look-ups, namely for the numbers $\cos 2\pi/(2N + 1)$ and $\sin 2\pi/(2N + 1)$.

For this purpose consider, for each $p = 0, 1, 2, \cdots, N$, the numbers U_{np} defined recursively by

$$U_{2N+2,p} = U_{2N+1,p} = 0 \qquad (5)$$

$$U_{np} = f_n + 2\cos\frac{2\pi p}{2N+1}U_{n+1,p} - U_{n+2,p} \quad (n = 2N, 2N-1, \cdots, 1) \qquad (6)$$

It is now shown, following [2], that

$$\alpha_p = \frac{2}{2N+1}\left(f_0 + U_{1p}\cos\frac{2\pi p}{2N+1} - U_{2p}\right) \qquad (7)$$

$$b_p = \frac{2}{2N+1}U_{1p}\sin\frac{2\pi p}{2N+1} \qquad (8)$$

To see this, suppose that

$$U_{n+1}\sin x = \sum_{j=n+1}^{2N} f_j \sin(j-n)x = \sum_{j=n}^{2N} f_j \sin(j-n)x \qquad (9)$$

Then

$$(f_n + 2\cos x U_{n+1} - U_{n+2})\sin x$$
$$= f_n \sin x + \sum_{j=n+1}^{2N} f_j[2\cos x \sin(j-n)x - \sin(j-n-1)x]$$
$$= f_n \sin x + \sum_{j=n+1}^{2N} f_j \sin(j-n+1)x$$
$$= U_n \sin x$$

which yields (5) and (6).

To obtain (7), one considers, for $n = 0$,

$$(f_n + U_{n+1}\cos x - U_{n+2})\sin x$$
$$= f_n \sin x + \sum_{j=n+1}^{2N} f_j[\cos x \sin(j-n)x - \sin(j-n-1)x]$$
$$= \left[f_n + \sum_{j=n+1}^{2N} f_j \cos(j-n)x\right]\sin x$$
$$= \left[\sum_{j=n}^{2N} f_j \cos(j-n)x\right]\sin x$$

Equation (9) with $n = 0$ gives (8).

Concerning the remark about only two trigonometric look-ups, observe that if

$$\rho_p = \cos p\theta \tag{10}$$

and

$$\tau_p = \sin p\theta \tag{11}$$

then

$$\begin{bmatrix}\rho_{p+1}\\ \tau_{p+1}\end{bmatrix} = \begin{bmatrix}\rho_1 & -\tau_1\\ \tau_1 & \rho_1\end{bmatrix}\begin{bmatrix}\rho_p\\ \tau_p\end{bmatrix}. \tag{12}$$

so that one need only know ρ_1 and τ_1.

The algorithm noted above is applicable to two other interesting problems aside from those of conventional Fourier analysis.

First, suppose one is given a polynomial

$$f(Z) = \sum_{n=0}^{N} \alpha_n Z^n \tag{13}$$

It can easily be verified that the evaluation

$$\varphi + i\psi = f(re^{i\theta}) \tag{14}$$

is conveniently carried out by the algorithm

$$V_{N+1} = V_{N+2} = 0 \tag{15}$$

$$V_n = \alpha_n + 2r\cos\theta V_{n+1} - r^2 V_{n+2} \qquad (n = N, N-1, \cdots, 1) \tag{16}$$

$$\varphi = \alpha_0 + r\cos\theta V_1 - r^2 V_2 \tag{17}$$

$$\psi = r\sin\theta V_1 \tag{18}$$

Finally, if it is desired to evaluate a Chebyshev expansion

$$f(\omega) = \sum_{n=0}^{N} \alpha_n T_n(\omega) \tag{19}$$

without looking in a table, where

$$T_n(\omega) = \cos(n\cos^{-1}\omega) \tag{20}$$

one may write, as usual,

$$U_{N+1} = U_{N+2} = 0 \tag{21}$$

$$U_n = \alpha_n + 2\omega U_{n+1} - U_{n+2} \qquad (n = N, N-1, \cdots, 1) \tag{22}$$

to obtain

$$f(\omega) = \alpha_0 + U_1\omega - U_2 \tag{23}$$

The reader will recognize the generalization to trigonometric polynomials of classical synthetic division. This observation has immediate consequences, such as a generalization of Lin's method [1] for solving (13) in the complex plane based on (14)–(18). This will not be explored further here.

3. SUMMARY OF THE CALCULATION PROCEDURE

Input quantities are

$$N, \quad C_1 = \cos\frac{2\pi}{2N+1}, \quad S_1 = \sin\frac{2\pi}{2N+1}$$

$$f_n \text{ for } n = 0, 1, \cdots, 2N$$

(1) To start, set

$$C_p = 1, \quad S_0 = 0 \qquad (p = 0)$$

(2) Then calculate (for each p)

$$U_{2N+2} = U_{2N+1} = 0$$

$$U_n = f_n + 2C_p U_{n+1}, - U_{n+2} \qquad (n = 1, 2, \cdots, 2N)$$

$$\alpha_p = \frac{2}{2N+1}(f_0 + C_p U_1 - U_2)$$

$$b_p = \frac{2}{2N+1} S_p U_1$$

α_p and b_p are output.

(3) If $p = N$, stop.
If $p < N$, calculate

$$C_{p+1} = C_1 C_p - S_1 S_p$$

$$S_{p+1} = C_1 S_p + S_1 C_p$$

Replace $p + 1$ by p and repeat step (2).

Table I

Box	p	n	C	S	u_0	u_1	u_2	Remarks
1			1					$1 \to C$
				0				$0 \to S$
	0							$0 \to p$
2							0	$0 \to u_2$
						0		$0 \to u_1$
		2						$2N \to n$
3					.250			$f_2 = .250; f_n + 2Cu_1 - u_2 \to u_0$
							0	$u_1 \to u_2$
						.250		$u_0 \to u_1$
		1						$n - 1 \to n$
4								Is $n = 0$? No!
3					1.000			$f_1 = .500$
							.250	
						1.000		
		0						
4								Yes!
5								Print $p = 0$
								Print $\alpha_p = \frac{2}{3}(1 + 1 - .25) = 1.167$
								Print $b_p = \frac{2}{3} \cdot 0 \cdot 1 = 0$
6								Is $p = N$? No!
7								$C_1C - S_1S = -.500 \to Q$
				866				$C_1S + S_1C \to S$
			−.500					$Q \to C$
	1							$p + 1 \to p$
2							0	
						0		
		2						
3					.250			
							0	
						.250		
		1						
4								No!
	1	1	−.500	.866	.250	.250	0	
3					.250			$f_1 = .500$
						.250	.250	
		0						
4								Yes!
5								Print $p = 1$
								Print $\alpha_p = 0.417$
								Print $b_p = 0.144$
6								Yes!
STOP								

4. FLOW CHART

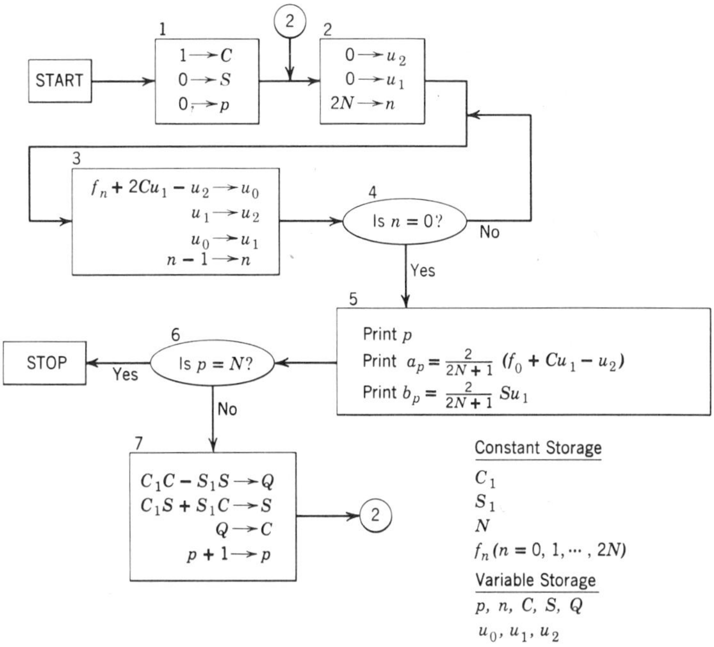

5. DESCRIPTION OF THE FLOW CHART

Box 1: The counter p is set to zero. The initial values of S and C are set.

Box 2: $U_{2N+2} = U_{2N+1} = 0$ is set in u_2 and u_1. The counter n is set to $2N$.

Box 3: U_n is calculated from U_{n+1}, U_{n+2}, and f_n. The counter n is decreased by one.

Box 4: If $n = 0$, $u_1 = U_1$ and $u_2 = U_2$ so that α_p and b_p may be calculated. If not, box 3 must be repeated.

Box 5: α_p and b_p are calculated and printed, along with p.

Box 6: If $p = N$, the problem is done. If not, must calculate for next p.

Box 7: Calculate S and C for next value of p. Also step counter p up by one. Repeat with box 2.

6. SUBROUTINES

No subroutines are needed. If sine and cosine subroutines are available, one might eliminate C_1 and S_1 from input and calculate them in box 1.

7. SAMPLE PROBLEM

Input

$$N = 1$$

$$C_1 = \cos\frac{2\pi}{3} = -0.500$$

$$S_1 = \sin\frac{2\pi}{3} = 0.866$$

$$f_0 = 1.000$$

$$f_1 = 0.500$$

$$f_2 = 0.250$$

The calculations are as indicated in Table 1.

Output

p	α_p	b_p
0	1.167	.000
1	.417	.144

8 MEMORY REQUIREMENTS

Storage	No. of Locations
Program	75
Intermediate results	9
Constants	1
Input	$4 + 2N$
Total	$89 + 2N$

The input may not be overwritten.

9. ESTIMATION OF THE RUNNING TIME

This program may be readily coded on a fixed point computer. In so doing, it has been assumed that the rescaling after multiplications is accomplished by a division. With this assumption, the running time will be

$$[12 + 29(N + 1) + 14(N + 1)2N]\mu + [1 + 14(N + 1) + 2(N + 1)2N]\nu$$

where μ is the average time for all internal operations but multiplication and division (add, store, transfer, shift, etc.) and ν is the average time for multiplication and division.

10. REFERENCES

1. F. B. Hildebrand, *Numerical Analysis*, McGraw-Hill Book Co., New York, 1955.
2. G. Goertzel, An Algorithm for the Evaluation of Finite Trigonometric Series, *Amer. Math. Monthly*, vol. 65, 1958, pp. 34–35.

The solution of linear programming problems

25

Dean N. Arden
Massachusetts Institute of Technology

1. FUNCTION

This chapter discusses briefly the major properties of linear programming problems and the algorithms in most common use for the solution of these problems, the simplex method and the dual simplex method.

The proofs and descriptions of various algorithms for the solution of the linear programming problem are given with attention to the details which arise in computational work. In particular, the question of keeping track of permutations of variables is handled explicitly with the aid of matrices which play a role similar to permutation matrices. The formal mechanism is consequently somewhat more involved than would be necessary if these problems were ignored. However, in computational practice a formal method of handling these problems is useful and is therefore employed in the proofs, which are computationally oriented. However, an appreciation of the problem is certainly not complete without a knowledge of the elegant geometrical and algebraic aspects of it. A description of a geometrical interpretation of the simplex method can be found in [1] and of the dual simplex method in [2]. The elegant algebraic proofs of the fundamental theorems due to Gale, Kuhn, Tucker, and others can be found in [3].

Recently, a new algorithm has been discussed by Dantzig, Ford, and Fulkerson [4]. This method, called the primal-dual algorithm, shows much promise but has not, to the author's knowledge, been applied computationally to the general linear programming problem to any great degree at the present time.

It should be emphasized that large special classes of programming problems can be solved by simpler techniques. In particular, problems which can be stated as the optimization of linear functionals of the flow in the branches of capacitated networks have an especially attractive solution algorithm closely related to the primal-dual algorithm mentioned above. For a description of these methods see [5]–[8].

2. MATHEMATICAL DISCUSSION

a. Fundamental Theorems

The problem of linear programming can be expressed as the calculation of the n-dimensional column vector X which maximizes an objective form

$$CX \tag{1}$$

subject to the restrictions

$$AX \leq B \tag{2}$$

$$X \geq 0 \tag{3}$$

where C and B are the given $1 \times n$ and $m \times 1$ matrices respectively, A is a given $m \times n$ matrix, and X is a variable $n \times 1$ matrix. A solution to (2) and (3) is called a feasible point. A maximizing or optimal vector X^* may not exist if there is no solution to (2) and (3) or if the values of CX for solutions of (2) and (3) have no upper bound.

It is well known that linear programming problems occur in dual pairs. The problem defined above will be called the primal problem. The problem dual to it, which will be called the dual problem, is the determination of a $1 \times m$ matrix Y which minimizes the objective form

$$YB \tag{4}$$

subject to the restrictions

$$YA \geq C \tag{5}$$

$$Y \geq 0 \tag{6}$$

Again, minimizing or optimal solutions may not exist for reasons analogous to those mentioned above in connection with the primal problem. The situation is clarified by Theorem 1.

THEOREM 1: If there exist solutions to the primal and dual restrictions, the values of the forms CX and YB for all solutions of their respective restrictions are bounded above and below respectively.

Proof: Let X' and Y' be any solutions of $AX \leq B$, $YA \geq C$, $X \geq 0$, and $Y \geq 0$ respectively. Then $B \geq AX'$ and, since $Y' \geq 0$, it follows that $Y'B \geq Y'AX'$. Similarly, since $Y'A \geq C$ and $X' \geq 0$, it follows that $Y'AX' \geq CX'$. These results yield $CX' \leq Y'B$. Due to the arbitrariness of Y' and X' it is apparent that the value of CX for any *particular* solution to the primal restrictions (2) and (3) is a lower bound to the values of YB for *all* solutions to the dual restrictions (5) and (6). Similarly, the values of CX for all solutions of (2) and (3) are bounded above by the value of YB for any solution of (5) and (6).

COROLLARY: If X' and Y' satisfy the primal and dual restrictions respectively and $CX' = Y'B$, then X' and Y' are optimal.

THEOREM 2: If X^* and Y^* are optimal, then $CX^* = Y^*B$.

The proof of this result is not trivial and is postponed until Section c where the result will appear as a consequence of the discussion of the simplex method for the solution of linear programming problems.

THEOREM 3: Let $U = B - A\bar{X}$ and $V = \bar{Y}A - C$, where $\bar{X}$ and $\bar{Y}$ are feasible points. The $\bar{X}$ and $\bar{Y}$ are optimal if and only if $\bar{Y}U = V\bar{X} = 0$.

Proof:

$$\bar{Y}U = \bar{Y}B - \bar{Y}A\bar{X}$$

$$V\bar{X} = \bar{Y}A\bar{X} - C\bar{X}$$

Hence

$$\bar{Y}U + V\bar{X} = \bar{Y}B - \bar{X}C$$

and

$$\bar{Y}U + V\bar{X} = 0$$

if and only if $\bar{X}$ and $\bar{Y}$ are optimal by Theorem 2 and Corollary 1. Since U, V, $\bar{X}$, $\bar{Y} \geq 0$, then $\bar{Y}U \geq 0$ and $V\bar{X} \geq 0$ so that $\bar{Y}U + V\bar{X} = 0$ implies that $\bar{Y}U = V\bar{X} = 0$.

COROLLARY: At least one of the two corresponding components of U and $\bar{Y}$ is zero and similarly for V and $\bar{X}$.

Proof: By an extension of the argument above it follows that since all components of $\bar{X}$, $\bar{Y}$, U, V are nonnegative, each term of $\bar{Y}U$ and $V\bar{X}$ is zero which gives the result.

This will be designated "the principle of complementary slackness" since U and V are often called slack variables. (The term is due to Tucker, who used it in a stronger sense: see [3], where it is shown that if optimal solutions exist, at least one can be found which satisfies $\bar{Y}U = V\bar{X}$ *and* $U^T + \bar{Y} > 0$, $V^T + \bar{X} > 0$.)

b. Systematic Elimination

In computational work the restrictions (2) are generally transformed to equations by the addition of m new variables called slack variables. This procedure gives

$$AX + I_m U = B$$

$$X, U \geq 0$$

or in matrix form

$$(A, I_m)\begin{pmatrix} X \\ U \end{pmatrix} = B \tag{7}$$

$$\begin{pmatrix} X \\ U \end{pmatrix} \geq 0 \tag{8}$$

where the m components of U are the slack variables and I_m is the unit $m \times m$ matrix.

The primary computational difficulty in the solution of the linear programming problem by presently available techniques is the solution of (7) and the evaluation of CX. The simplex method, which will be discussed in the next section, is essentially a rule for proceeding from one solution of (7) and (8) to another which yields a larger value of CX.

The problem of the evaluation of CX is usually handled by adding yet another variable W and considering the system

$$\begin{aligned} -CX + 1 \cdot W + O_m^T U &= 0 \\ AX + O_m W + I_m U &= B \end{aligned} \tag{9}$$

of $m + 1$ equations in $n + m + 1$ variables, where O_m is a column vector of m zero components. Any solution $\bar{X}, \bar{U}, \bar{W}$ of this system satisfies $A\bar{X} + I\bar{U} = B$, $\bar{W} = C\bar{X}$ so that the value of the linear form CX is given by the value of W.

Solutions to this set of equations can be found by the process of elimination, and the remainder of this section will be devoted to a discussion of the elimination process for systems without inequality constraints.

For this discussion, C will denote an arbitrary $p \times q$ matrix with elements c_{ij} and columns c_j ($i = 1, 2, \cdots, p$; $j = 1, 2, \cdots, q$). The elements and columns of the unit $p \times p$ matrix I_p will be denoted by e_{ij} and e_j respectively ($i, j = 1, \cdots, p$).

The notation $E(i, j)$ will denote the *operation* of dividing the ith row of an array by the element at the intersection of its ith row and jth column and subtracting from every other row that multiple of this result which causes the entry in the jth column to be zero. The operation will be assumed to apply to the array denoted by the symbols occurring to the right of $E(i, j)$ unless the contrary is indicated by appropriate use of parentheses. For example, the result of elimination of the jth variable from all equations except the ith of the set

$$CX = D$$

or equivalently

$$(C, D)\begin{pmatrix} X \\ -1 \end{pmatrix} = 0$$

can be denoted by

$$[E(i, j)(C, D)]\begin{pmatrix} X \\ -1 \end{pmatrix} = 0$$

The following are not correct

$$[E(i, j)C]X - E(i, j)D = 0$$

$$[E(i, j)C]X - D = 0$$

since in the first case $E(i, j)D$ is not defined if $j > 1$ and in the second the *operation* $E(i, j)$ is not applied to D.

The ambiguity which results from the use of $E(i, j)C$ to denote both an operation and its result will be tolerable here. We will call (C, D) the array corresponding to the system $CX = D$.

The elimination operation is not defined by the above statements if $c_{ij} = 0$.

For a particular choice of C and if $c_{ij} \neq 0$, the operation $E(i, j)C$ yields exactly the same result as multiplication by a matrix formed by replacing the ith column of I_p by the column

$$\begin{matrix} \dfrac{-c_{1j}}{c_{ij}} \\ \dfrac{-c_{2j}}{c_{ij}} \\ \cdot \\ \cdot \\ \cdot \\ \dfrac{-c_{i-1,j}}{c_{ij}} \\ \dfrac{1}{c_{ij}} \\ \dfrac{-c_{i+1,j}}{c_{ij}} \\ \cdot \\ \cdot \\ \cdot \\ \dfrac{-c_{pj}}{c_{ij}} \end{matrix}$$

whose determinant is $1/c_{ij}$ and hence is nonsingular. This matrix will be denoted by $E_C(i, j)$ and will be said to correspond to the operation $E(i,j)$ on C, with the reference to C omitted if it is clear from context. The product of the matrices corresponding to each of a sequence of elimination steps will be said to correspond to the sequence,

If $c_{ij} \neq 0$,

$$E_C(i,j)e_k = e_k \quad \text{if } k \neq i \tag{10}$$

$$E_C(i,j)c_j = e_i \tag{11}$$

Consequently, any column vector whose ith component is zero is left unchanged after multiplication by $E_C(i,j)$ since it is a linear combination of the e_k, $k \neq i$. From (10) and (11) it also follows that the inverse of $E_C(i,j)$ can be formed from I_p by replacing its ith column by c_j. With these results the following theorem can be immediately verified.

THEOREM 4: The inverse of the matrix corresponding to a sequence

$$E(i_k, j_k) \cdots E(i_2, j_2)E(i_1, j_1)C$$

of k elimination operations can be formed from the unit matrix by successively replacing the i_1st column by c_{j_1}, replacing the i_2nd column of the result by c_{j_2}, and in general replacing the i_rth column of the result at the end of the $(r-1)$st step by c_{j_r}, $r \leq k$.

Proof: By induction on k. If $k = 1$, the result has already been noted as an immediate consequence of (10) and (11). If $k = r$, and B^{-1} is the matrix corresponding to the first $k - 1$ elimination steps, let B with columns b_j, $j = 1, 2, \cdots, p$, denote the inverse of B^{-1}. Then

$$\begin{aligned}
[E_{B^{-1}C}(i_r, j_r)B^{-1}]^{-1} &= BE_{B^{-1}C}^{-1}(i_r, j_r) \\
&= B(e_1, e_2, \cdots, e_{i_{r-1}}, B^{-1}c_{j_r}, e_{i_{r+1}}, e_p) \\
&= (B_1, B_2, \cdots, B_{i_{r-1}}, c_{j_r}, B_{i_{r+1}}, \cdots, B_p)
\end{aligned}$$

which is formed from B by replacing its i_rth column by c_{j_r} as required.

It should be noted that $B^{-1}C_{j_r}$ are the coefficients in the expansion of c_{j_r} as a linear combination of the columns of B since, if

$$\begin{aligned}
c_{j_r} &= a_1B_1 + a_2B_2 + \cdots + a_pB_p \\
&= BA
\end{aligned}$$

where A is a column vector with components $a_1, a_2, \cdots, a_p$, then

$$A = B^{-1}c_{j_r}$$

The columns of B are called basis vectors. Consequently, after a sequence of elimination operations with corresponding matrix B^{-1} each column of the array contains the coefficients in the expansion of the corresponding column in the original array in terms of the basis vectors. The determinant of B^{-1} is nonzero since each elimination was definable and hence had a nonsingular corresponding matrix. The matrix corresponding to the sequence of eliminations is the product of these and hence is nonsingular.

A solution to the set

$$CX = D \tag{12}$$

of p equations in q variables, $q \geq p$, with no restrictions on X, can be found by performing a sequence of p eliminations

$$E(i_p, j_p) \cdots E(i_2, j_2)E(i_1, j_1)(C, D)$$

on the $p \times (q+1)$ array (C, D) with $i_p \neq i_{p-1} \neq \cdots \neq i_2 \neq i_1$ and $j_1, j_2, \cdots, j_p \leq q$. The result in the $(q+1)$st column is then the expression of D as a linear combination of the basis vectors and hence can be taken as the values of the corresponding variables giving a solution to (12) with the values of all other variables taken to be zero. This corresponds to combining equations of (12) until each of a set of p variables occurs in only one equation and setting all other variables equal to zero.

If at any state $k < p$ an elimination $E(i_k, j_k)$ cannot be made, i.e., there is no c_j such that $c_{ij} \neq 0$, $i \neq i_{k-1}, i_{k-2}, \cdots, i_1$, then C has rank $k - 1$ and there is a solution if and only if the only nonzero components of the $(q+1)$st column of (C, D) occur in rows $i_1, i_2, \cdots, i_{k-1}$, which are then the values of $x_{j_1}, x_{j_2}, \cdots, x_{j_{k-1}}$.

The nonzero components can be identified computationally by storing i_k in the j_kth of a sequence of q registers which otherwise hold zero and by printing the final solutions in order by scanning the table sequentially and by printing the contents of the register if zero or the contents of the jth number in the $(q+1)$st column if the contents are j. An alternative is to store j_k in the i_kth of p registers and print the $(q+1)$st column in order with the variable subscript taken from the corresponding register of the table.

It is not unusual in computational practice, after a suitable normalization of the columns (or rows) of an array, for the elimination operations $E(i_k, j_k)$ to be chosen so that i_k is the row in which the element of largest magnitude in the j_kth column occurs. This technique tends to reduce rounding error and is commonly called "positioning for size." The elements in the right-hand column of the array must be appropriately permuted to get the solution components in their original order by a method of the type described above.

c. The Simplex Method

We will now apply the method of elimination described in the previous section to the solution of the linear programming problem in the form

$$\max CX \tag{13}$$

subject to

$$AX \leq B \qquad \text{and} \qquad X \geq 0$$

The method can also be applied to the dual of this problem since it can be put in the above form by manipulation of signs and inequalities.

Successive elimination operations will be performed on the $(m + 1) \times (n + m + 2)$ array

$$\begin{pmatrix} -C & 1 & O_m^T & O \\ A & O_m & I_m & B \end{pmatrix} \tag{14}$$

corresponding to the system

$$\begin{aligned} -CX + W &= 0 \\ AX + I_m U &= B \end{aligned} \tag{15}$$

For brevity the arrays

$$\begin{pmatrix} -C & 1 & O_m^T \\ \\ A & O_m & I_m \end{pmatrix}, \quad \begin{pmatrix} O \\ \\ B \end{pmatrix}, \quad \begin{pmatrix} X \\ W \\ U \end{pmatrix}$$

will be denoted by

$$C_0, \qquad D_0, \qquad Z$$

respectively, so that the system (15) becomes

$$C_0 Z = D_0$$

A solution of (15) consisting of at most $m + 1$ nonzero variables $z_{i_1}, z_{i_2}, \cdots, z_{i_{m+1}}$ will be called basic and the ordered set of positive integers $(i_1, i_2, i_3, \cdots, i_{m+1})$ will be called a basis description and the corresponding set of columns of C_0 will be called the basis.

If $i_1 = n + 1$ and $z_{i_2}, z_{i_3}, \cdots, z_{i_{m+1}} \geq 0$, the solution will be called a basic feasible solution. Since $W = z_{n+1}$, this implies that a value of W is always the first component of a basic feasible solution.

It will be convenient in the discussion of the simplex method to make use of a matrix $P(i_1, i_2, \cdots, i_{m+1})$ corresponding to a basis description $(i_1, i_2, \cdots, i_{m+1})$ defined as an $(n + m + 1) \times (m + 1)$ matrix whose only nonzero component in the kth column is a one in the i_kth position. The dependence on the basis description may not be explicitly indicated if it is clear from the context.

This matrix has the property

$$P^T P = I_{m+1}$$

and PP^T is an $(n + m + 1) \times (n + m + 1)$ matrix which is identical to I_{n+m+1} in columns $(i_1, i_2, \cdots, i_{m+1})$ and zero elsewhere. Multiplication of a row vector on the right by P gives a row vector formed from the $i_1, i_2, \cdots, i_{m+1}$ components of the original vector with all except the $i_1, i_2, \cdots, i_{m+1}$ components replaced by zero. Corresponding results obtain for the multiplication of column vectors on the left. These matrices behave somewhat like permutation matrices since they permute a subset of the components of the vector on which they operate and enable a precise description of the simplex and similar algorithms to be made quite easily.

The basis corresponding to the basis description $(i_1, i_2, \cdots, i_{m+1})$ is then

$$C_0 P$$

and the corresponding components of the basic solution $z_{i_1}, z_{i_2}, \cdots, z_{i_{m+1}}$ are denoted succinctly by

$$Z^T P$$

According to the results of Theorem 4, the inverse of the matrix corresponding to a sequence of l elimination steps is of the form $C_0 P_l$. Hence a sequence of l elimination steps is equivalent to multiplication on the left by $(C_0 P_l)^{-1}$, where P_l corresponds to a basis description formed from $(n + 1, n + 2, \cdots, n + m + 1)$, which corresponds to the basis I_{m+1}, by replacing the i_kth integer of the basis description by j_k at the kth elimination step $E(i_k, j_k)$ for k equal successively to $1, 2, \cdots, l$.

If $D_0 \geq 0$, then

$$X = 0, \qquad W = 0, \qquad U = B \tag{16}$$

is a basic feasible solution corresponding to the basis description $(n+1, n+2, \cdots, n+m+1)$. For the present it will be assumed that this is the case and we denote $P(n+1, n+2, \cdots, n+m+1)$ by P_0 and the corresponding solution (16) by Z_0. Hence

$$Z_0^T P_0 = D_0$$

The following lemma stipulates the conditions under which an elimination operation applied to an array corresponding to a basic feasible solution will yield a new array corresponding to a basic feasible solution with an undiminished value of CX. Subscripted lower case letters will be used to denote elements of arrays denoted by the corresponding upper case letter.

LEMMA 1: Given an array

$$(C, D) = (C_0P)^{-1}(C_0, D_0),$$

$D \geq 0$, where P corresponds to the basis description $(i_1, i_2, \cdots, i_{m+1})$ with $i_1 = n+1$, the elimination operation $E(p, q)$ $(2 \leq p \leq m+1,\ 1 \leq q \leq n+m+1)$ is defined and results in a new array

$$(C', D') = (C_0P')^{-1}(C_0, D_0),$$

with $d_1' \geq d_1$ and $D' \geq 0$, where P' corresponds to the basis description obtained by replacing i_p by q in the sequence $(i_1, i_2, \cdots, i_{m+1})$, if

(a) $c_{1q} < 0$.

(b) $\dfrac{d_p}{c_{pq}} = \min\limits_k \dfrac{d_k}{c_{kq}}$ for all k for which $c_{kq} > 0$.

Proof: Applying $E(p, q)$ to the array (C, D) gives

$$d_k' = d_k - \frac{c_{kq}}{c_{pq}} d_p \quad \text{for } k \neq p$$

$$d_p' = \frac{d_p}{c_{pq}}$$

Condition (b) requires $c_{pq} > 0$ and since $d_p \geq 0$, $d_p' = \dfrac{d_p}{c_{pq}} \geq 0$. From (b) it follows that

$$\frac{d_k}{c_{kq}} \geq \frac{d_p}{c_{pq}} \quad \text{for } c_{kq} > 0$$

or $d_k - \dfrac{c_{kq}}{c_{pq}} d_p = d_k' \geq 0$ if $c_{kq} > 0$. But if $c_{kq} \leq 0$, $d_k - \dfrac{c_{kq}}{c_{pq}} d_p = d_k' \geq 0$ since $d_k \geq 0$. Hence $D' \geq 0$. Since

$$d_1' - d_1 = -\frac{c_{1q}}{c_{pq}} d_p \tag{17}$$

and $d_p \geq 0$, $c_{pq} > 0$, it follows that $d_1' - d_1 \geq 0$ if $c_{1q} < 0$ and therefore $d_1' \geq d_1$. But $d_1 = z_{i_1} = z_{n+1} = W$ so that the value of CX is undiminished.

The simplex algorithm consists of repeated elimination steps satisfying (a) and (b) and is said to terminate if either condition (a) cannot be met or if (a) is satisfied but no p can be found for which $c_{pq} > 0$ and hence condition (b) cannot be satisfied. In computational practice q is chosen so that $c_{1q} \leq c_{1j}$, $1 \leq j \leq n+m+1$. This gives the following alternative.

THEOREM 5: If the simplex algorithm terminates, it yields either

(a) a basic feasible solution X to the primal problem which is optimal and a basic feasible solution Y to the dual problem which is optimal and for which $CX = YB$, or

(b) a family of solutions to the primal restrictions for which CX has no upper bound.

Proof: Case (a) occurs if condition (a) of Lemma 1 cannot be satisfied. In this case

$$c_{1l} \geq 0, \qquad 1 \leq l \leq n+m+1$$

Since $p \geq 2$, the vector $c_{n+1} = e_1$ cannot be eliminated as its only nonvanishing component is the first and consequently $E(p, n+1)$ is not defined for $p \geq 1$. Hence the first basis vector and therefore the first column of C_0P will be e_1. Under these conditions it is easily verified that $(C_0P)^{-1}$ can be written in the form

$$\begin{pmatrix} 1 & Y \\ O_m & B^{-1} \end{pmatrix}$$

Recalling that

$$C_0 = \begin{pmatrix} -C & 1 & O_m^T \\ A & O_m & I_m \end{pmatrix}$$

and $D_0 = \begin{pmatrix} O \\ B \end{pmatrix}$, $(C_0P)^{-1}(C_0, D_0)$ can be expanded to

$$\begin{pmatrix} -C + YA & 1 & Y & YB \\ B^{-1}A & O_m^T & B^{-1} & B^{-1}B \end{pmatrix}$$

$c_{1l} \geq 0$, $l = 1, 2, \cdots, n+m+1$, implies that the first row is nonnegative except for the last entry. Therefore we have

$$-C + YA \geq 0 \quad \text{and} \quad Y \geq 0$$

or

$$YA \geq C \quad \text{and} \quad Y \geq 0$$

so that Y is a basic feasible solution to the dual. The values of the basic variables in the order determined by the basis description are found in the right-hand column. But since the basis description always begins with $n + 1$, the value of $z_{n+1} = W$ which is the value of CX is found in the first entry of the last column. The above result shows that this is also the value of YB, and by Corollary 1 to Theorem 1 the solution is optimal.

Case (b) occurs if for some q, $1 \leq q \leq n + m$,

$$c_{1q} < 0 \quad \text{and} \quad c_{kq} \leq 0, \quad 2 \leq k \leq m + 1$$

In this event the solution whose only nonzero components are

$$z_q = a \quad \text{and} \quad Z^T P = D - c_q a, \quad a > 0$$

satisfies the system $CZ = D$ or $(C_0P)^{-1}C_0Z = (C_0P)^{-1}D_0$ and hence the system $C_0Z = D_0$ for arbitrarily large a. But since $c_{1q} < 0$, $d_1 - ac_{1q}$, which is the value of W, is unbounded for these solutions.

If at every elimination $E(p, q)$, $d_p > 0$, it is a consequence of (17) that the value of CX increases and identical bases cannot occur since the solution vector for a given basis is unique. There are at most $\binom{n+m}{m}$ possible bases, however, so the procedure must converge in a finite number of steps. However, if $d_p = 0$ it is possible that identical bases may occur and the simplex algorithm may not terminate since a sequence of bases may recur repetitively. Such situations are said to be degenerate. For a discussion of a method for avoiding degeneracy or cycling, see [1], [9]. Computationally, the method described is quite time-consuming on high-speed computers, especially if magnetic tapes are used for secondary storage, and since no instance of cycling seems to have arisen in practical calculations, no provision for this eventuality is usually made in practical routines.

The simplex method provides an algorithm for proceeding from a given basic feasible solution to a new basic feasible solution with an increase in the form to be maximized. The method must therefore start with a basic feasible solution. As has been noted above, if $B \geq 0$, then $U = B$ is such a solution. If it is not the case that $B \geq 0$, then the equations of the system (9) corresponding to negative components of B can be multiplied by -1 so that a system of equations with nonnegative nonhomogeneous part is obtained. If the resulting system is denoted by

$$\bar{C}Z = D, \quad Z \geq 0$$

then $D \geq 0$ and the system

$$\bar{C}Z + I_m R = D, \quad Z, R \geq 0$$

has the obvious basic feasible solution

$$R = D$$

The components of the vector R are called artificial variables. If the problem

$$\min HR$$
$$\bar{C}Z + I_m R = D$$

where H is any row vector with m positive components, is solved by the simplex method, the result will be a basic feasible solution to $\bar{C}Z = D, Z \geq 0$, if $\min HR = 0$. If $\min HR > 0$ the restrictions

$$\bar{C}Z = D, \quad Z \geq 0$$

are incompatible or infeasible, since $\min HR > 0$ implies that some component of R is positive which contradicts the existence of a solution.

Since

$$H\bar{C}Z + HR = HD$$

the minimum of HR on the restriction set will correspond to the maximum of $H\bar{C}Z$ so that the feasibility problem can be expressed in the form

$$\max H\bar{C}Z$$
$$\bar{C}Z + I_m R = D$$
$$Z, R \geq 0$$

The solution of the feasibility problem is often required in practice and is called phase I, and the solution of the optimality problem phase II of the complete solution of the linear programming problem by the simplex method. In most applications H is taken to be the vector with unit components. In practice, artificial variables are necessary only for those equations with negative nonhomogeneous part after the inclusion of the slack variables.

An alternative procedure is to add slack variables and a *single* artificial variable corresponding to a column of negative ones and occurring in the objective form with a large negative coefficient. A single elimination step with pivot element in this column and in the row corresponding to the negative component of B with largest magnitude will then yield a basic feasible solution. If the artificial variable is present in the maximal solution, infeasibility is indicated, since the fact that a large increase in the form would be possible if it were eliminated indicates that such elimination is not possible and hence the original problem without the added variable is infeasible.

Theorem 5 demonstrates that if solutions to the primal restrictions exist and have bounded values of CX, then the simplex method gives optimal solutions to the primal and dual problems with equal values of CX and YB. By Theorem 1, if optimal solutions exist, then feasible points for both problems have bounded values of CX and YB. These results yield a proof of Theorem 2.

d. The Dual Simplex Algorithm

Elimination methods can always be applied in two ways: first, as described in the previous sections, by the formation of new arrays from linear combinations of rows (or equations) of the old, which corresponds to multiplication on the left by elimination matrices; second, by the formation of new arrays from linear combinations of columns of the original array. The process seems to be more difficult conceptually but only amounts to a transformation of coordinates and corresponds to multiplication on the right by the transpose of an elimination matrix. Before discussing the dual simplex method the simplex method will be described from this point of view. For this purpose the system of linear forms

$$\begin{matrix} +BW - AX \\ 1\cdot W + O_n^T X \\ 0\cdot W + I_n X \\ 0\cdot W + CX \end{matrix} = \begin{pmatrix} B & -A \\ 1 & O_n^T \\ O_n & I_n \\ 0 & C \end{pmatrix}\begin{pmatrix} W \\ X \end{pmatrix} \qquad (18)$$

will be denoted by

$$\begin{pmatrix} C_0 \\ D_0 \end{pmatrix} Z$$

where

$$D_0 = (0 \;\; C), \quad C_0 = \begin{pmatrix} B & -A \\ 1 & O_n^T \\ O_n & I_n \end{pmatrix}, \quad Z = \begin{pmatrix} W \\ X \end{pmatrix}$$

The primal problem can now be stated by the requirement that with $W = 1$, $D_0 Z$ should be a maximum for

$$C_0 Z \geq 0$$

The simplex algorithm provides a rule for the generation of a sequence of arrays derived from the array $\begin{pmatrix} C_0 \\ D_0 \end{pmatrix}$ by transformations of coordinates of the form

$$\begin{pmatrix} W' \\ X' \end{pmatrix} = Z' = \tilde{E}^{-1}(i,j)Z$$

applied to an array $\begin{pmatrix} C \\ D \end{pmatrix}$, initially taken to be $\begin{pmatrix} C_0 \\ D_0 \end{pmatrix}$, where $\tilde{E}^{-1}(i,j)$ is a matrix formed from the unit $(n+1)\times(n+1)$ matrix by replacing its jth row by the ith row of $\begin{pmatrix} C \\ D \end{pmatrix}$.

This change of coordinates applied to the system $\begin{pmatrix} C \\ D \end{pmatrix} Z$ gives a new system $\begin{pmatrix} C' \\ D' \end{pmatrix} Z'$ in which the ith form of $\begin{pmatrix} C \\ D \end{pmatrix} Z$ is transformed into the jth coordinate variable of Z'. It is apparent that

$$\begin{pmatrix} C \\ D \end{pmatrix} Z = \begin{pmatrix} C \\ D \end{pmatrix} \tilde{E}(i,j) Z' \qquad (19)$$

so that the coefficient array of the forms (18) in the new variables Z' is $\begin{pmatrix} C \\ D \end{pmatrix} \tilde{E}(i,j) = \begin{pmatrix} C' \\ D' \end{pmatrix}$. The index i is restricted to the range $1 \leq i \leq n + m + 1$ so that the form $0 \cdot W + CX$ in general does not become a coordinate variable, and j is restricted to $2 \leq j \leq n + 1$ so that the coordinate W is never transformed and is always a coordinate variable.

The values of the forms $\begin{pmatrix} C' \\ D' \end{pmatrix} Z'$ for the substitution

$$X' = 0, \qquad W' = W = 1$$

can be found in the left column of $\begin{pmatrix} C' \\ D' \end{pmatrix}$, and since the first $n + m + 1$ forms of (18) are required to be positive for solutions of the linear programming problem the values of the *original variables* given by the $(m + 1)$st to $(m + n)$th forms of $\begin{pmatrix} C' \\ D' \end{pmatrix} Z'$ are said to define a feasible point or solution if this is the case. If at most m of the first $n + m$ forms are nonzero [or at most $m + 1$ of the first $n + m + 1$ forms since the $(n + m + 1)$st form, W, is always one for this substitution], the solution so defined is said to be basic. The value of CX for $X' = 0$, $W' = W = 1$ is given by the last form and should not decrease.

The lemma below contains the criterion of the simplex algorithm for proceeding from an array $\begin{pmatrix} C \\ D \end{pmatrix}$ corresponding to a basic feasible solution $Z = 0$, $W = 1$ to an array $\begin{pmatrix} C' \\ D' \end{pmatrix}$ and basic feasible solution $Z' = 0$, $W' = W = 1$ by a transformation of variables of the form $Z' = \tilde{E}^{-1}(i, j)Z$.

The operations described in this section can be reduced to the case of Section c by replacing the arrays $\begin{pmatrix} C \\ D \end{pmatrix}$ by $(C^T D^T)$. The results of the multiplications on the right by the $\tilde{E}(i, j)$ then become multiplications on the left by the transpose of these matrices which correspond to the elimination operations $E(j, i)$ carried out in the order of the subscript. Hence the effect of the first k elimination steps can be summarized by the use of Theorem 4 where the kth basis is then the ordered set of $n + 1$ *rows* determined by a *basis description* $(i_1, i_2, \cdots, i_{n+1})$ with $i_1 = m + 1$ and defined in terms of the original array $\begin{pmatrix} C_0 \\ D_0 \end{pmatrix}$ by the use of an $(n + 1) \times (n + m + 1)$ matrix $P(i_1, i_2, \cdots, i_{n+1})$ whose only nonzero elements occur in the i_kth column of the kth row ($i_k = 1, 2, \cdots, n + m + 2$; $k = 1, 2, \cdots, n + 1$). The kth basis is then PC_0 and the effect of the first k steps is to transform the array C_0 into

$$\begin{pmatrix} C \\ D \end{pmatrix} = \begin{pmatrix} C_0 \\ D_0 \end{pmatrix} (PC_0)^{-1}$$

Lemma 2 contains the criterion of the simplex method for proceeding from a basic feasible solution to another basic feasible solution with an increased value of CX.

LEMMA 2: Given a system of linear forms $\begin{pmatrix} C \\ D \end{pmatrix} Z$ with $c_{k1} \geq 0$ and $\begin{pmatrix} C \\ D \end{pmatrix} = \begin{pmatrix} C_0 \\ D_0 \end{pmatrix} (PC_0)^{-1}$, where P corresponds to the basis description $(m + 1, i_1, i_2, \cdots, i_{m+n+1})$, then the change of variable defined by

$$Z = \tilde{E}(q, p)Z', \qquad 2 \leq q \leq n + 1, \quad 1 \leq p \leq m + n + 1$$

yields a new system $\begin{pmatrix} C' \\ D' \end{pmatrix} Z'$ with

$$\begin{pmatrix} C' \\ D' \end{pmatrix} = \begin{pmatrix} C \\ D \end{pmatrix} \tilde{E}(q, p) = \begin{pmatrix} C_0 \\ D_0 \end{pmatrix} (P'C_0)^{-1}$$

where P' corresponds to the basis description formed from $(m + 1, i_2, i_3, \cdots, i_{m+n+1})$ by replacing i_q by p. The new system satisfies

$$d_1' \geq d_1 \qquad \text{and} \qquad c_{k1}' \geq 0$$

if p and q are chosen so that

(a) $d_q > 0$.

(b) $\dfrac{c_{p1}}{c_{pq}} = \max\limits_k \dfrac{c_{k1}}{c_{kq}}$ for all k for which $c_{kq} < 0$.

Proof: The general form of the new system follows from the results of Section b. A calculation of the first column of $\begin{pmatrix} C \\ D \end{pmatrix} \tilde{E}(q, p)$ gives

$$c_{k1}' = c_{k1} - \frac{c_{kq}}{c_{pq}} c_{p1}$$

$$d_1' - d_1 = -\frac{d_q}{c_{pq}} c_{p1}$$

Condition (b) requires that $c_{pq} < 0$ and since $d_q \geq 0$, it follows that $d_1' - d_1 \geq 0$ and hence $d_1' \geq d_1$. From (a)

$$\frac{c_{p1}}{c_{pq}} \geq \frac{c_{k1}}{c_{kq}}, \qquad c_{kq} < 0$$

and therefore

$$\frac{c_{p1}}{c_{pq}} c_{kq} \leq c_{k1}$$

and

$$c_{k1} - \frac{c_{p1}}{c_{pq}} c_{kq} = c_{k1}' > 0$$

THEOREM 6: If the algorithm terminates, then it yields either:

(a) optimal solutions for both the primal and the dual problems, or

(b) a family of solutions to the primal problem for which the objective form CX has no upper bound.

Proof: If condition (a) of Lemma 2 cannot be met, then for some basis and basis description $(m + 1, i_2, i_3, \cdots, i_n)$

$$\begin{pmatrix} C \\ D \end{pmatrix} = \begin{pmatrix} C_0 \\ D_0 \end{pmatrix} (PC_0)^{-1}$$

and $d_q \leq 0$ for $2 \leq q \leq n + 1$. Since the first row of the basis is $(1, O_m^T)$, its inverse $(PC_0)^{-1}$ can be written in the form

$$\begin{pmatrix} 1 & O_m^T \\ X & B^{-1} \end{pmatrix}$$

Recalling the meaning of C_0 and D_0, this gives

$$\begin{pmatrix} C \\ D \end{pmatrix} = \begin{pmatrix} B & -A \\ 1 & O_n^T \\ O_n & I_n \\ 0 & C \end{pmatrix} \begin{pmatrix} 1 & O_n^T \\ X & B^{-1} \end{pmatrix}.$$

$$= \begin{pmatrix} B - AX & -AB^{-1} \\ 1 & O_n^T \\ X & B^{-1} \\ CX & CB^{-1} \end{pmatrix}$$

Since the left-hand column of this result is positive, X is a feasible point of the primal problem with value CX for the primal objective form. The dual problem can be written

$$YC_0 + D_0 = 0$$

where Y is a variable $(n + m + 1)$ component row vector and it is apparent that

$$Y = -D_0(PC_0)^{-1}P = -DP \geq 0$$

is a solution with d_1 as the value of YB. Therefore from Theorem 1 both X and Y are optimal with $CX = YB$ which is case (a).

If condition (a) of Lemma 2 but not (b) of Lemma 2 can be met, then for some q

$$d_q > 0, \qquad c_{kq} \geq 0, \qquad 1 \leq k \leq n + m + 1$$

In this case $z_q = M$ satisfies the requirements of the primal problem for arbitrarily large $M \geq 0$ since all the forms of $\begin{pmatrix} C \\ D \end{pmatrix} Z$ with positive coefficient of z_q are unbounded. In particular, DZ is not bounded above, which gives case (b).

All of the results of Section c can be extended to this variation of the simplex method in an entirely analogous fashion. Again $B \geq 0$ is a requirement for the initiation of the algorithm and, if this is not the case, the devices of Section c can be applied in phases I and II.

It is important to note that this method applied to the *dual* problem operates on the transpose of the array of coefficients of Section c. The resulting algorithm requires $C \geq 0$ for its initiation and is called the dual simplex algorithm [2]. Most recent routines provide for the application of either algorithm.

The condition $C \geq 0$ can be met by the use of a device due to Beale [10]. The dual problem can be written as

$$\begin{aligned} YC + 1 \cdot W + O_m^T \cdot V &= 0 \\ -YA + 0 \cdot W + I_m V &= -B \\ Y, V &\geq 0 \end{aligned}$$

after the addition of slack variables. Let the restriction

$$EY + s = M$$

be added to this system, where E is a row vector with n unit components, s is a single slack variable, and M is sufficiently large so that the solutions of

$$YA \geq B$$

if bounded satisfy

$$EY \leq M$$

If $EY + s = M$ is solved for the component of Y corresponding to the largest negative component of C and the result substituted in the dual restrictions and form, the resulting form will have nonnegative coefficients.

3. COMPUTATIONAL CONSIDERATIONS AND CALCULATION PROCEDURE

a. Variations of the Simplex Method

The computational variations of the simplex method differ primarily in the extent to which the complete eliminations are carried out. It is clearly only necessary that the basis inverse $(C_0P)^{-1}$ be available. In the product-inverse method [11], the matrices of the successive elimination steps are made available for recall at each iteration by storing the column in which they differ from the unit matrix and its number. Then the first row of $(C_0P)^{-1}$,

$(C_0P)^{-1}e_1$, can be computed by multiplying e_1 on the left by the transposes of the elimination matrices in the reverse of the order of the elimination steps themselves (since the transpose of a product is the product of the individual transposes in the reverse order). From this result the first row of the transformed array

$$C_0^T(C_0P)^{-1}e_1 = [e_1^T(C_0P)^{-1}C_0]^T$$

can be computed and the subscript, q, its negative component of largest magnitude, determined. Then the qth column of $(C_0P)^{-1}C_0$ and $(C_0P)^{-1}D_0$ can be computed by multiplying both D_0 and the qth column of C_0 on the left by the product of the elimination matrices in the order of the corresponding elimination steps. Ratios of corresponding components of the qth column of $(C_0P)^{-1}C_0$ and $(C_0P)^{-1}D_0$ can be calculated and the integer p for the elimination step $E(p, q)$ can be computed by the rule of Lemma 1, and the new elimination matrix (or practically, the column in which it differs from the unit matrix) can be stored. This algorithm is tailored to a storage medium for which sequential access *in both directions* is possible, e.g., magnetic tape.

Schematically the data can be arranged as follows

$$C_0D_0E_1E_2, \cdots, E_N$$

where the E_k denote matrices of elimination steps with subscripts indicating their order of application. The data is used from right to left forming $(C_0P)^{-1}e_1$, as $E_1^T, \cdots, E_{N-1}^TE_N^Te_1$, and then $e_1^T(C_0P)^{-1}C_0$. Then moving to the right $E_N \cdots E_2E_1D_0 = (C_0P)^{-1}D_0$ and $E_N \cdots E_2E_1$ is applied to the qth column of C_0 to give the qth column of $(C_0P)^{-1}C_0$ and from these an E_{N+1} is determined which is recorded at the right and the procedure repeated. Rapid-access storage for at most two $(m + 1)$ vectors $(C_0P)^{-1}D_0$ and the qth column of $(C_0P)^{-1}C_0$ is required. Each iteration increases the required storage by at least $m + 1$ numbers so that after $m + 1$ iterations the amount of storage required to generate $(C_0P)^{-1}$ compares unfavorably with the amount required to store it directly. In principle, however, the entire inverse $(C_0P)^{-1}$ can be recovered at any stage and some other method used. The storage medium in question is usually magnetic tape and since considerable difficulties arise if access in both directions is not possible, the failure of many tape systems to provide such access will probably doom this ingenious technique.

Another useful method of utilization of magnetic tape is to store the data

$$(C_0P)^{-1}, \qquad C_0, \qquad D_0$$

by columns on magnetic tape with the rewind point on the left. Moving to the right of the rewind point, the first row of the basis inverse, $e_1^T(C_0P)^{-1}$, is stored in rapid-access memory and the first row $e_1^T(C_0P)^{-1}C_0$ of $(C_0P)^{-1}C_0$ calculated, keeping in rapid-access memory D_0 and the column of C_0 corresponding to the algebraically smallest component of $e_1^T(C_0P)^{-1}C_0$ and its column number q. Tape is then rewound and the qth column of $(C_0P)^{-1}C_0$ and $(C_0P)^{-1}D$ are formed. Since $(C_0P)^{-1}$ is stored by columns, this requires storage space for four rather than two columns in high-speed memory, since the components of $(C_0P)^{-1}D_0$ and the product of $(C_0P)^{-1}$ and the qth column of C_0 are developed in parallel from the columns of $(C_0P)^{-1}D_0$ and the qth column of C_0 and hence require two columns of temporary storage. This could be avoided by storing $(C_0P)^{-1}$ by rows. The elimination $E(p, q)$ is then determined by the rule of Lemma 1, the tape is rewound, and the elimination is performed on $(C_0P)^{-1}$, forming the new inverse and storing the last row of the inverse in rapid-access memory. The inverse and original data can be stored on different tapes and the two passes over the inverse can be made in different directions if it is possible to read tape backwards.

The array (I_{m+1}, C_0, D_0) can be initially stored on tape, for this method. If the array C_0 contains I_{m+1}, which will be the case if slack variables have been added and $D_0 \geq 0$, it need not be recorded twice, since the elimination matrix $(C_0P)^{-1}$ can be built up in the I_{m+1} array in C_0. But in this case the qth column of C_0 may be one of the columns of I_{m+1}. If this is the case, then the product of $(C_0P)^{-1}$ and the qth column of C_0 will be already recorded in place of the qth column of C_0 and need not be calculated either to determine the algebraic minimum of the components of $e_1^T(C_0P)^{-1}C_0$ or to calculate the qth column of $(C_0P)^{-1}C_0$. The rest of the data would be left unaltered, however, and these calculations would have to be made at each iteration.

This method is due to Dantzig [12] and is called the revised or modified simplex method.

A final alternative is to perform the successive eliminations on the entire array

$$[(C_0P)^{-1}C_0, (C_0P)^{-1}D]$$

at each step. This variation is described by Charnes, Cooper, and Henderson in [1].

b. Calculation Procedure

The following calculation procedure makes use of the modified simplex method. The input data consists of the problem parameters $m + 1$ and $n + m + 1$, the $(m + 1) \times (n + m + 2)$ array (C_0, D_0), the basis description $(n + 1, n + 2, \cdots, n + m + 1)$, and I_{m+1}, which is the inverse of the matrix formed by the $m + 1$ columns of C_0 numbered $n + 1, n + 2, \cdots, n + m + 1$ in accordance with the basis description.

The array (C_0, D_0) is stored by columns, i.e., with elements adjacent in a column being adjacent in memory, and the array I_{m+1} is stored by rows. The columns of C_0 are denoted by c_j $(j = 1, \cdots, n + m + 1)$ and the elements of C_0 by c_{ij} $(i = 1, 2, \cdots, m + 1; j = 1, 2, \cdots, n + m + 1)$. The elements of D_0 are denoted by d_j. These quantities are not altered in the course of the calculation.

The space originally occupied by I_{m+1} will be occupied at the kth iteration by the current basis inverse $(C_0P_k)^{-1}$. The notation b_i $(i = 1, 2, \cdots, m + 1)$ will be used to denote the sequence of registers containing the ith row of this array and its contents.

The registers containing the basis description will be denoted by $l_1, l_2, \cdots, l_{m+1}$ and their contents will be altered in the course of the algorithm.

Two columns of temporary storage are required whose registers and contents will be denoted by s_i and t_i $(i = 1, 2, \cdots, m + 1)$ respectively.

A single temporary register r, two counters i and j, and two registers p and q, whose contents at the kth iteration, p_k and q_k, determine the elimination matrix to be applied to the current $(C_0P_{k-1})^{-1}$, are also required.

In principle the simplex method requires at the kth iteration the determination of a row index p and a column index q according to the requirements of Lemma 1 and the application of an elimination operation $E(p_k, q_k)$ to an $(m + 1) \times (n + m + 2)$ array which is initially (C_0, D_0). The modified simplex method differs in that the original data (C_0, D_0) is left unaltered and only the basis inverse $(C_0P_k)^{-1}$ is maintained by left multiplication of $(C_0P_{k-1})^{-1}$ by the matrix corresponding to $E(p_k, q_k)$. It has been shown in Section 2b that the modified data could be recovered by the multiplication $(C_0P_k)^{-1}(C_0, D_0)$ since the multiplication by $(C_0P_k)^{-1}$ summarizes the first k elimination operations. However, only the first row of this result is required to determine q_{k+1} and then only the corresponding column and the $n + m + 2$ column are required to determine p_{k+1} so that the entire multiplication $(C_0P_k)^{-1}(C_0, D_0)$ is not performed.

The quantities required in the calculation are the following:

1. The first row of $(C_0P_k)^{-1}C_0$. (This calculation requires only the first row of $(C_0P_k)^{-1}$.)
2. The value of q_{k+1} $(1 \leq q \leq n + m + 1)$ which is the column number of the minimal negative element of the first row of $(C_0P_k)^{-1}C_0$.
3. The value of p_{k+1} $(1 \leq p \leq m + 1)$ which is the number of the row for which the positive ratios of corresponding elements of $(C_0P_k)^{-1}D_0$ and $(C_0P_k)^{-1}C_0$ assume their minimum.
4. The new basis inverse $(C_0P_{k+1})^{-1}$ which is obtained by multiplying $(C_0P_k)^{-1}$ by the matrix corresponding to $E(p_{k+1}, q_{k+1})$.
5. The new basis description obtained by replacing $l_{p_{k+1}}$ by q_{k+1}.

It is important to note that in this and the following sections the notations c_{ij} and c_j refer to elements and columns of the original data in contradistinction to the usage of Section 2b.

The nonzero components of the solution will be located in the registers s_i in the order determined by the basis description located in registers l_i $(i = 1, 2, \cdots, m + 1)$.

4. FLOW CHART

The flow chart appears on pages 275, 276, and 277.

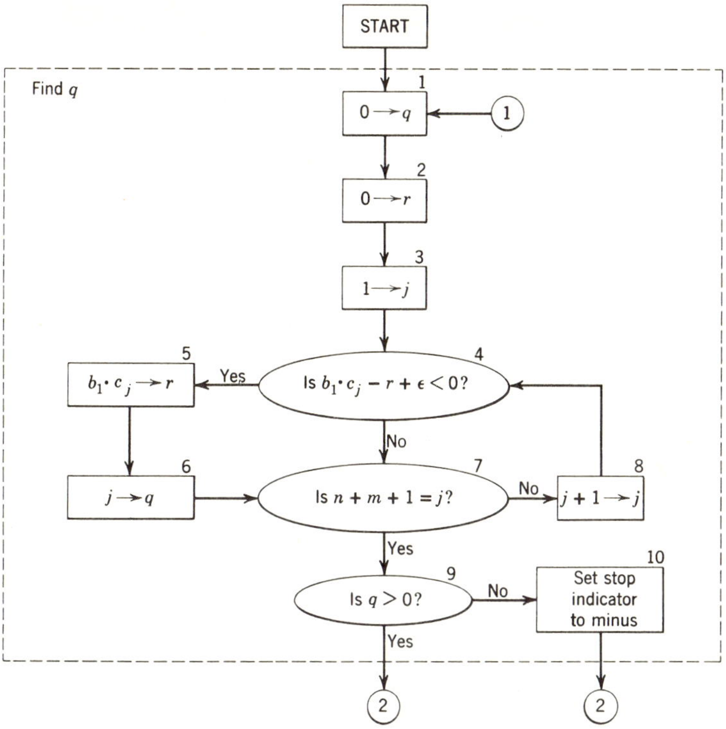

5. DESCRIPTION OF THE FLOW CHART

Box 1: Set q to zero. q will contain the index of the column of $(C_0P)^{-1}C_0$ with minimum negative element in the first row: If no negative element is found, q remains zero and the algorithm will terminate.

Box 2: Clear r. r will hold the smallest negative element thus far discovered.

Box 3: Initialize the loop counter.

Box 4: Compute the jth element of the first row of $(C_0P)^{-1}C_0$ and compare with smallest negative element found thus far. The maximum magnitude of the round-off eiror in $b_1 \cdot c_j - r$ is denoted by ϵ.

Box 5: If $b_1 \cdot c_j$ is less than the contents of r, replace r by $b_1 \cdot c_j$.

Box 6: Place current value of j in q.

Box 7: Has every element in the first row been tested?

Box 8: Test next element in the row.

Box 9: Has a negative element been found? This could also be determined by testing for $r > 0$ in which case box 2 could be omitted.

Box 10: The current basis is optimal but at this point the results in the registers s_i have not been calculated and therefore the stop must be delayed.

Box 11: M denotes the largest number in the machine arithmetic.

Box 12: Initialize the loop counter. This will not allow $p = 1$ since t_1 will always be negative and so the Yes branch at box 16 will not be taken when $i = 1$.

Box 13: Set p to zero.

Box 14: Calculate the first component of $(C_0P)^{-1}D_0$.

Box 15: Calculate the first component of $(C_0P)^{-1}c_q$.

Box 16: Is t_i positive? Again ϵ is the magnitude of the maximum rounding error.

Box 17: Is s_i/t_i the minimum positive ratio? The ratio is positive since s_i is positive or zero.

Box 18: Replace r by the new minimum ratio.

Box 19: Store index of row corresponding to minimum ratio in p.

Box 20: Has every ratio been tested?

Box 21: Increase the counter.

Box 22: Test stop indicator.

Box 23: The final solution is in the registers s_i.

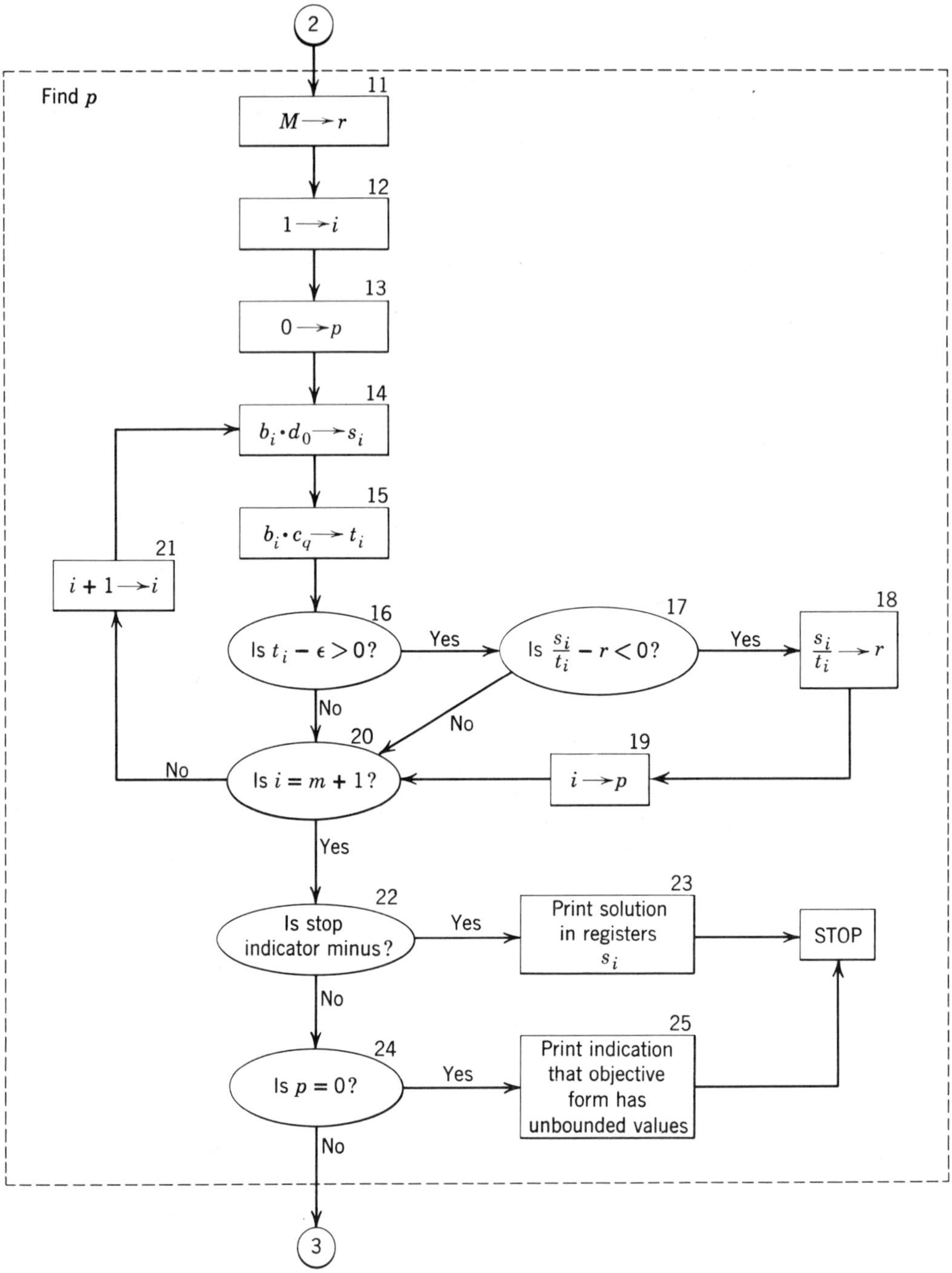

Box 24: Was there a positive ratio?

Box 25: No. In this event $s_i - at_i$ is feasible for arbitrary $a > 0$ and yields unbounded values for the form to be maximized.

Box 26: Set column index j to one.

Box 27: Divide each element by t_p.

Box 28: Is the row complete?

Box 29: Advance to the next element.

Box 30: Set i to one. The index i determines the row being calculated.

Box 31: Is this the pth row?

Box 32: Begin the row calculation.

Box 33: Calculate the new b_{ij}.

Box 34: Is the row complete?

Box 35: Advance to the next element.

Box 36: Are all the rows computed?

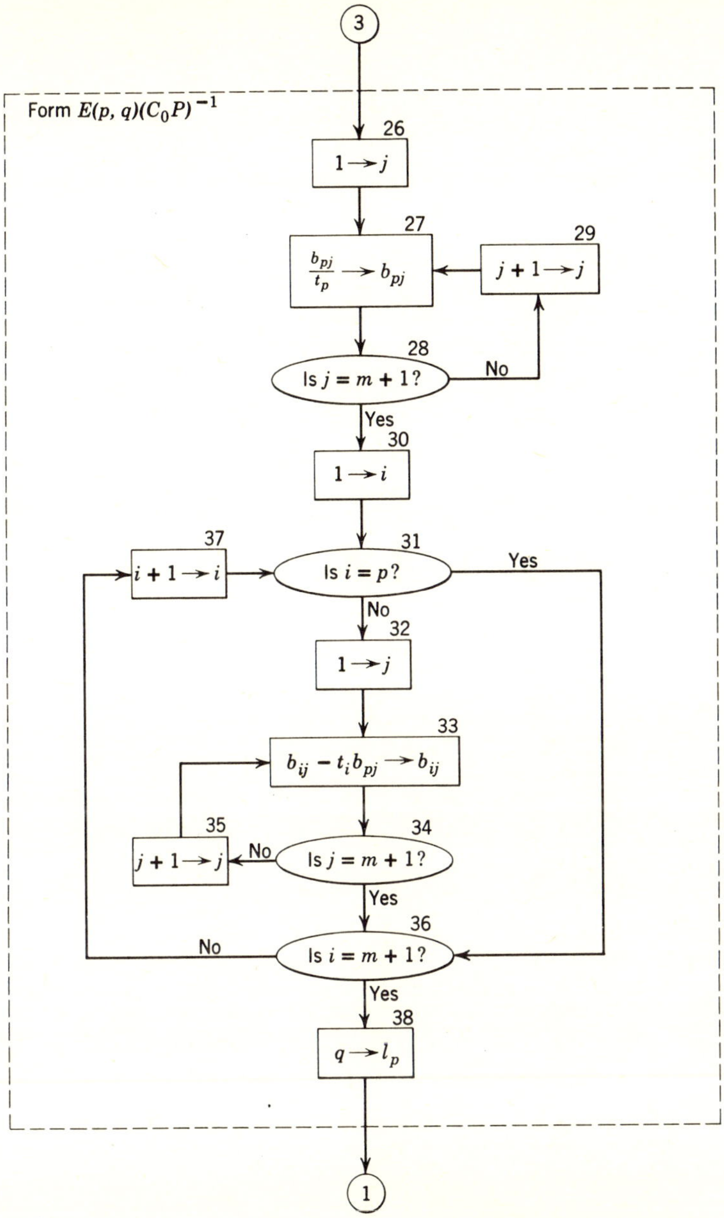

Box 37: Do the next row.
Box 38: Form the new basis description.

6. SUBROUTINES

A subroutine which forms the scalar product of two vectors is necessary and is used in boxes 4, 14, and 15. The dimension of the operand vectors should be preset at $m + 1$ since this parameter remains unchanged in the course of the routine. In tape operations this subroutine is usually combined with a tape-reading routine.

7. SAMPLE PROBLEM

Following equations (1), (2), and (3), consider the following problem with $m = 3$ and $n = 2$:

$$\text{Maximize } 2x_1 + x_2$$

subject to the restrictions

$$\begin{aligned} x_1 \quad\;\; &\leq 4 \\ x_2 &\leq 3 \\ x_1 + x_2 &\leq 8 \\ x_1, x_2 &\geq 0 \end{aligned}$$

The alternative form (9) then is

$$\text{Maximize } w$$

$$\begin{aligned}
-2x_1 - x_2 + w \qquad\qquad &= 0\\
x_1 \qquad + u_1 \qquad\qquad &= 4\\
x_2 \qquad + u_2 \qquad &= 3\\
x_1 + x_2 \qquad\qquad + u_3 &= 8\\
x_1, x_2, u_1, u_2, u_3 &\geq 0
\end{aligned}$$

Then initially we have the following array:

$(C_0P)^{-1}$	C_0	D_0	s_i	Basis Description
1 0 0 0	−2 −1 1 0 0 0	0	0	3
0 1 0 0	1 0 0 1 0 0	4	4	4
0 0 1 0	0 1 0 0 1 0	3	3	5
0 0 0 1	1 1 0 0 0 1	8	8	6

Note that in the computer we need not store the last four columns of C_0. Since -2 is the smallest element in the first row, $q = 1$ and by Lemma 1 $p = 2$. Therefore c_1 enters the basis and we have:

$(C_0P)^{-1}$	s_i	Basis Description
1 2 0 0	8	3
0 1 0 0	4	1
0 0 1 0	3	5
0 0 0 1	4	6

$$b_1 \cdot c_j = (0, -1, 1, 2, 0, 0)$$

$$t_i = b_i \cdot c_2 = (-1, 0, 1, 1)$$

This time we find $q = 2$ and $p = 3$ and we get:

$(C_0P)^{-1}$	s_i	Basis Description
1 2 1 0	11	3
0 1 0 0	4	1
0 0 1 0	3	2
0 −1 −1 1	1	6

$$b_1 \cdot c_j = (0, 0, 1, 2, 1, 0)$$

Since all the components of $b_1 \cdot c_j$ are nonnegative, we have found the optimum solution. Using the basis description this solution is:

$$\begin{aligned}
z_3 &= w = 11\\
z_1 &= x_1 = 4\\
z_2 &= x_2 = 3\\
z_6 &= u_3 = 1
\end{aligned}$$

8. MEMORY REQUIREMENTS

About 75 instructions are required for the algorithm described by the flow chart. The inclusion of phase I and phase II options and the dual simplex algorithm would add somewhat to this total.

The array (C_0, D_0) requires $(m + 1) \times (n + m + 2)$ registers. If the basis inverse is stored separately, it requires $(m + 1)^2$ registers. Two registers are required for the parameters $m + 1$ and $n + m + 1$ and about $2m + 3$ temporary storage registers are needed.

The output appears in the temporary storage registers. The input data cannot be overwritten except when, as noted before, an I_{m+1} array in C_0 is used to build up the basis inverse.

9. ESTIMATION OF THE RUNNING TIME

About $3m + 2$ divisions, $(m + 1)(n + 4m + 3)$ multiplications, and $(m + 1)(n + 7m + 6)$ additions are needed for each iteration of the routine diagrammed in the flow chart. Thus the running time per iteration is about

$$[(m + 1)(n + 4m + 3) + 3m + 2]\mu + (m + 1)(n + 7m + 6)\nu$$

where μ is the multiplication/division time of the computer and ν is the addition time.

No good bound for the number of iterations is known but at most $\binom{n + m + 1}{m + 1}$ bases are possible and computational experience seems to predict that the number of iterations is a linear function of the problem dimension, e.g., $2m$ or $3m$.

Many routines suppress zero data in the input or intermediate results or both. This can result in great efficiencies for certain classes of problems.

10. REFERENCES

1. A. Charnes, W. W. Cooper, and A. Henderson, *An Introduction to Linear Programming*, John Wiley & Sons, New York, 1953.
2. C. E. Lemke, The Dual Method of Solving the Linear Programming Problem, *Nav. Res. Logistics Quart.*, Vol. 1, 1954, pp. 36–47.
3. A. W. Tucker, Dual Systems of Homogeneous Linear Relations, in *Linear Inequalities and Related Systems*, H. W. Kuhn and A. W. Tucker (eds.), Princeton University Press, 1956, pp. 3–18.
4. G. B. Dantzig, I. R. Ford, Jr., and D. R. Fulkerson, A Primal-Dual Algorithm for Linear Programs, in *Linear Inequalities and Related Systems*, H. W. Kuhn and A. W. Tucker (eds.), Princeton University Press, 1956, pp. 171–181.
5. Jack B. Dennis, *Mathematical Programming and Electrical Networks*, Technology Press, Cambridge, and John Wiley & Sons, New York, 1959.
6. L. R. Ford, Jr., and D. R. Fulkerson, Solving the Transportation Problem, Paper P-895, The RAND Corporation, June 20, 1956.
7. H. W. Kuhn, The Hungarian Method for the Assignment Problem, *Nav. Res. Logistics Quart.*, vol. 2, nos. 1 and 2, March–June 1955.
8. L. R. Ford, Jr., and D. R. Fulkerson, A Simple Algorithm for Finding Maximal Network Flows and an Application to the Hitchcock Problem, Paper P-743, The RAND Corporation, Sept. 1955.
9. S. Vajda, *The Theory of Games and Linear Programming*, Methuen and Co., London, 1956.
10. E. M. L. Beale, An Alternative Method for Linear Programming *Proc. Camb. Phil. Soc.*, vol. 50, 1954, pp. 513–523.
11. G. B. Dantzig and W. Orchard-Hays, The Product Form for the Inverse in the Simplex Method, *MTAC*, vol. 8, no. 46, Apr. 1954, pp. 64–67.
12. G. B. Dantzig, Maximization of a Linear Function of Variables Subject to Linear Inequalities, in *Activity Analysis of Production and Allocation*, T. C. Koopmans (ed.), Cowles Commission Monograph 13, John Wiley & Sons, New York, Chap. 21, pp. 339–349, 1951.

26 Network analysis

T. R. Bashkow*
Columbia University†

1. FUNCTION

The problem to be considered is that of a computer technique for the steady-state analysis of linear electric networks which may or may not be passive. Ideally, a computer solution should require for input only that information from a circuit diagram which can be obtained by someone almost completely ignorant of circuit analysis. The commonly used method of mesh analysis requires application of a set of rather complicated rules. Nodal analysis is perhaps simpler to formulate but both techniques lead to high order determinant evaluations for large networks. This gives rise to inaccuracy not only because of the size but also of the nature of such determinants. They contain many terms which exactly cancel one another algebraically but not numerically because of the usual computational errors.‡

* A note of acknowledgement should be made to C. L. Semmelman, who programmed the method of the A sections, and to Miss D. J. Bierman, who coded both methods. Both are at the Bell Telephone Laboratories, Murray Hill, N.J.

† Formerly of Bell Telephone Laboratories.

‡ There are other approaches to this general problem which are topological. However, discussion of these would require considerably more space than is available. See [1], [2], [3].

Consequently, discussion will be restricted to the special class of ladder networks (Fig. 1). The required input data can easily be obtained from a circuit diagram and the following problems can be solved in a simple iterative manner.§

Given: Circuit diagram and element values.

(A) Find steady-state magnitude and phase curves of input-output ratios as a function of frequency. (In this case, a set of desired frequencies is also required as part of the input data.)

(B) Find any network rational function, e.g., E_{in}/I_{in}, I_{out}/E_{in}, etc. That is, determine the coefficients of the numerator and denominator polynomials.

The remainder of the discussion will be divided into two parts. The sections marked A refer to problem A above and those marked B to problem B.

2A. MATHEMATICAL DISCUSSION

In Fig. 1 (and discussion) Z and Y are the 2-terminal impedance and admittance, respectively, of the indicated "black box"; I and V

§ Much work on solutions of large network problems has been done by G. Kron, who uses a very different approach. See [4], [5].

are the current and voltage, respectively, of the branch in which it is located. E is an ideal voltage generator. The variables will be complex numbers at any single frequency. They will be rational functions of the complex frequency $p = \sigma + j\omega$, where ω is 2π times the real frequency f, if general analysis is required as in the B sections.

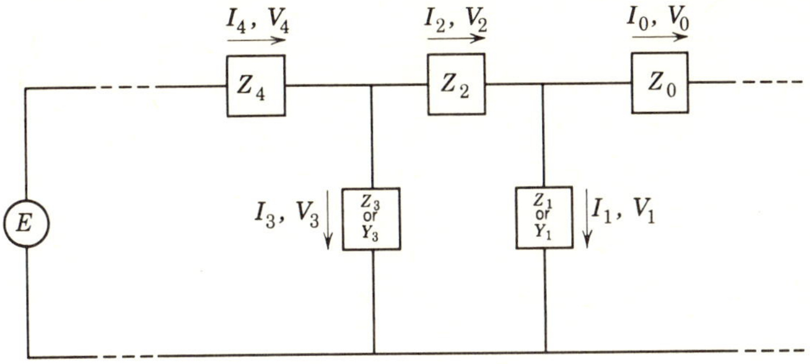

Fig. 1

An analysis of the network of Fig. 1 can be made utilizing the indicated branch currents and voltages. Starting at the right-hand shunt branch, for example, one writes:

$V_1 = I_1 Z_1$	shunt branch	
$I_2 = I_1 + I_0$	series branch	
$V_2 = I_2 Z_2$	series branch	(1)
$V_3 = V_2 + V_1$	shunt branch	
$I_3 = V_3 Y_3$	shunt branch	

.
.
.

Although the analysis is seemingly restricted to the particular network shown, it should be mentioned that more general networks can often be analyzed by one of the following schemes:

(a) It may be possible to use known techniques (see [6], [7]) to convert other types of passive networks into equivalent networks of this form, e.g., a lattice to unbalanced T transformation. Then computations are performed on the equivalent circuit. Since negative elements are as computable as positive ones, the equivalent circuit need not be physically realizable.

(b) Active networks or transformers may be included either by the equivalent network technique or by considering the 4-terminal equations for Fig. 2,

$$\begin{aligned} V_i &= AV_0 + BI_0 \\ I_i &= CV_0 + DI_0 \end{aligned} \qquad (2)$$

Such 4-terminal networks may be inserted at any point in the ladder structure. I_0 and V_0 can be found by (1) and then (2) can be used to compute the quantities necessary to continue the iteration. The use of (2) implies using special subroutines and these will not be considered in what follows.

Equations (1) and (2) should be considered as single frequency equations in which all variables are complex numbers. Thus, at each

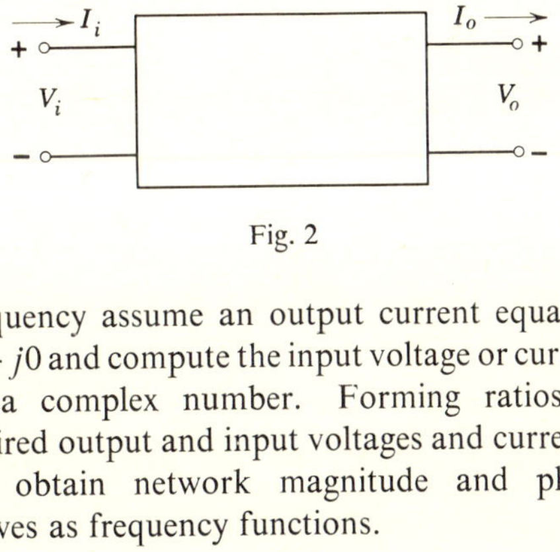

Fig. 2

frequency assume an output current equal to $1 + j0$ and compute the input voltage or current as a complex number. Forming ratios of desired output and input voltages and currents, we obtain network magnitude and phase curves as frequency functions.

3A. SUMMARY OF THE CALCULATION PROCEDURE

(a) The input quantities are a list of identification numbers specifying the particular element configuration comprising the circuit in each series and shunt "black box" of Fig. 1. Each such number is followed by a set of values for the particular resistances, inductances and capacitances found in this configuration. See Section 7A, for example. The order of the configuration numbers in the list is a "map" of the ladder itself starting with the output branch and proceeding to the input. Special numbers may be used to indicate that a shunt branch or series branch normally expected at this point is missing. A list of frequencies must be given and information as to which network functions are desired as output.

(b) The order of the calculation is determined in part by the particular network being analyzed, but in general will be as follows. (The particular flow chart which will be considered in detail in the next section assumes the rightmost branch is a series branch and the leftmost branch is a shunt branch. The flow chart does not include the subroutines or program control needed for computation of the circuit of Fig. 2. Any active networks which have been converted to the form of Fig. 1 can, of course, be computed as shown.)

(1) Get the next series configuration in the list and compute Z_k at this frequency. (This will be a complex number.) Compute

$$V_k = I_k Z_k \quad \text{(voltage across this series branch)}$$
$$V_{k+1} = V_k + V_{k-1} \quad \text{(voltage across next shunt branch)}$$

For the first series branch $I_k = 1 + j0$, $V_{k-1} = 0$.

(2) Get the next shunt configuration in the list and compute Y_{k+1} at this frequency. Compute

$$I_{k+1} = Y_{k+1} V_{k+1} \quad \text{(current through this shunt branch)}$$
$$I_{k+2} = I_{k+1} + I_k \quad \text{(current through next series branch)}$$

(3) Repeat steps (1) and (2) after replacing I_k by I_{k+2} and V_{k-1} by V_{k+1}.

(4) Make final computations as desired, e.g.,

$$I_{\text{in}}/I_{\text{out}} = I_{k+2}/(1 + j0)$$

(5) Repeat steps (1) through (4) at the next frequency.

(c) The output quantities will be a value of magnitude and phase for each input frequency. These may represent any network function which has been specified by the input data.

4A. FLOW CHART: MAGNITUDE AND PHASE

The flow chart appears on page 283.

5A. DESCRIPTION OF THE FLOW CHART

Box 1: An input frequency is stored. The value $1 + j0$ is put into the I_k locations and the value $0 + j0$ is put into the V_{k-1} locations. The counter k is set to one.

Box 2: The kth configuration number is selected from the input list. The first of these is assumed to specify a series branch.

Box 3: The configuration number specifies a particular 2-terminal network, e.g., a resistance, capacitance, and inductance in parallel. This number may thus be considered a "transfer to subroutine" instruction for the particular branch which has been selected.

Box 4: These are actually subroutines which compute a complex number for the impedance of a particular 2-terminal network. Some typical networks may be: a single resistance, capacitance, or inductance; a series combination of resistance, capacitance, and inductance; etc. As input these subroutines require the element values for the network and a frequency and give the real and imaginary parts of the impedance Z as output.

Box 5: ZI_k is computed and put into the V_k locations.

Box 6: $V_k + V_{k-1}$ is computed and put into the V_{k+1} location.

Box 7: The $(k + 1)$st configuration number is selected from the input list. It and all even-numbered branches on the list are assumed to specify a shunt branch.

Box 8: A transfer to subroutine is made. See box 3 details.

Box 9: These compute the real and imaginary parts of the admittance Y for a particular network. These subroutines may actually be the same as those specified by box 4 and require the same input information. If they are the same a computation of $1/Z$ must be made before going to the next step.

Box 10: Compute YV_{k+1} and put into I_{k+1} location.

Box 11: Compute $I_{k+1} + I_k$ and put into I_{k+2} location.

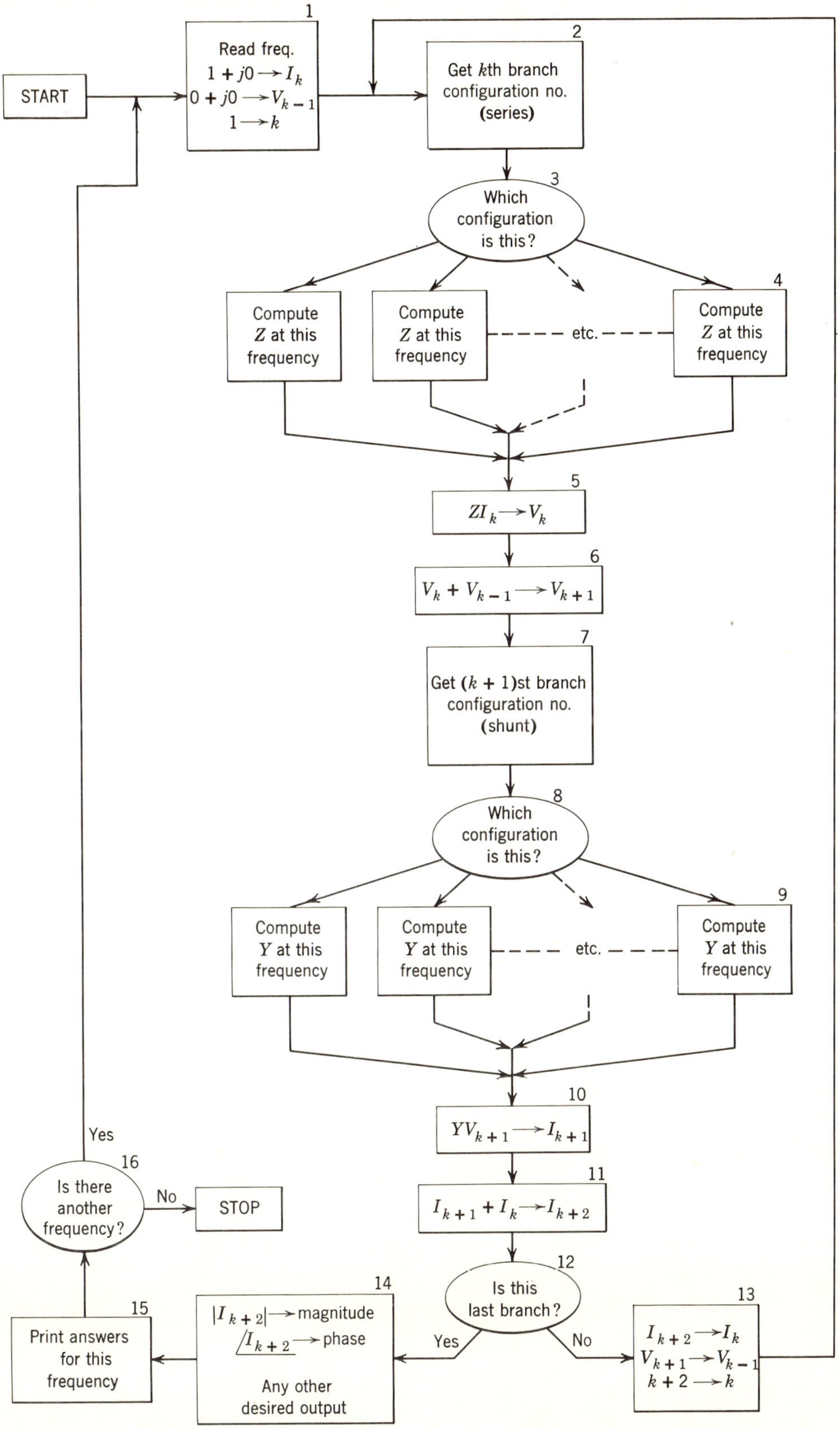
START
1
Read freq.
$1 + j0 \rightarrow I_k$
$0 + j0 \rightarrow V_{k-1}$
$1 \rightarrow k$
2
Get kth branch configuration no. (series)
3
Which configuration is this?
Compute Z at this frequency
Compute Z at this frequency
etc.
4
Compute Z at this frequency
5
$ZI_k \rightarrow V_k$
6
$V_k + V_{k-1} \rightarrow V_{k+1}$
7
Get $(k + 1)$st branch configuration no. (shunt)
8
Which configuration is this?
Compute Y at this frequency
Compute Y at this frequency
etc.
9
Compute Y at this frequency
10
$YV_{k+1} \rightarrow I_{k+1}$
11
$I_{k+1} + I_k \rightarrow I_{k+2}$
12
Is this last branch?
Yes
No
13
$I_{k+2} \rightarrow I_k$
$V_{k+1} \rightarrow V_{k-1}$
$k + 2 \rightarrow k$
14
$|I_{k+2}| \rightarrow$ magnitude
$\underline{/I_{k+2}} \rightarrow$ phase
Any other desired output
15
Print answers for this frequency
16
Is there another frequency?
Yes
No
STOP

Box 12: This test for the last branch may consist of a special number on the input list. If it is present, proceed to box 14. If it is not present, proceed to box 13.

Box 13: I_{k+2} is moved to the I_k location, V_{k+1} to the V_{k-1} location. The index k is increased to pick up the next series branch configuration number. Control is transferred to box 2.

Box 14: In the flow chart is it assumed that the desired network function is I_{in}/I_0, where I_{in} is a current source and the leftmost branch in the network is a shunt branch. Since

$$I_0 = 1 + j0, \qquad \frac{I_{in}}{I_0} = I_{k+2} = A + jB$$

The magnitude, $\sqrt{A^2 + B^2}$, and the phase, $\tan^{-1} B/A$, are now computed. Other possible network functions may also be computed at this point. It is assumed that the user of the program has been told how to exercise these options although for simplicity these details have not been shown.

Box 15: The answers for this frequency are now printed (or punched for later sorting).

Box 16: If additional input frequencies are given, control is returned to the start of the program at box 1; otherwise stop.

6A. SUBROUTINES

(a) SR1—Multiplication of complex numbers. Used in boxes 5 and 10.

(b) SR2—Addition of complex numbers. Used in boxes 6 and 11.

(c) SR3—The computations of Z and Y may be considered as subroutines. Details given in box 4 and box 9 explanations.

(d) SR4—Square root. Used in box 14.

(e) SR5—Arc tangent. Used in box 14.

(f) SR6—Logarithm. May be used in box 14 if answers are needed in decibels.

7A. SAMPLE PROBLEM

Input list. See Fig. 3. (Rather than give configuration numbers we specify by type.)

k

1. Single resistance 1.0 ohm
2. Single capacitance 2.0 farads
3. Single inductance 1.0 henry
4. Single capacitance 2.0 farads

   Find I_{in}/I_0 for $2\pi f_1 = 1$

   End

$$\begin{aligned}
& I_0 = 1 + j0 \\
& V_{-1} = 0 + j0 \\
k = 1 \quad & Z = 1 + j0 \\
& V_0 = (1 + j0)(1 + j0) = 1 + j0 \\
& V_1 = (1 + j0) + 0 + j0 = 1 + j0 \\
k = 2 \quad & Y = (j2\pi f_1)2.0 = 0 + j2.0 \\
& I_1 = (0 + j2)(1 + j0) = 0 + j2 \\
& I_2 = (0 + j2) + (1 + j0) = 1 + j2 \\
k = 3 \quad & Z = (j2\pi f_1)1 = 0 + j1 \\
& V_2 = j1(1 + j2) = -2 + j1 \\
& V_3 = (-2 + j1) + 1 = -1 + j1 \\
k = 4 \quad & Y = j2 \\
& I_3 = j2(-1 + j1) = -2 - j1 \\
I_{in} = {} & I_4 = (-2 - j1) + (1 + j2) \\
& = -1 + j1 \\
& I_{in}/I_0 = -1 + j1 \\
& \text{Mag.} = \sqrt{2} \\
& \phi = \tan^{-1}(1/-1) = 135^\circ
\end{aligned}$$

One might also compute

$$Z_{in} = V_3/I_{in} = \frac{-1 + j1}{-1 + j1} = 1 + j0$$

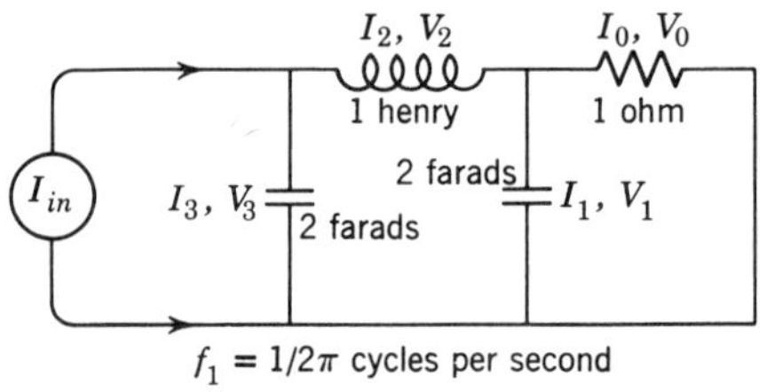

Fig. 3

8A. MEMORY REQUIREMENTS

The memory requirements for intermediate calculations are trivial:

V_k	2 words
V_{k-1}	2 words
V_{k+1}	2 words
I_{k+2}	2 words

For a network of N branches with m elements per branch to be evaluated at F frequencies:

Element values	mN words
Configuration numbers	N words
Frequencies	F words

The size of the program depends on the number of types of 2-terminal networks allowed and their complexity. If one assumes 6 different network configurations and 25 words for each, then 150 words are needed. Since the square root, arc tangent, and logarithm subroutines are essential for meaningful answers, they should probably be included in the count. The total will probably be of the order of 300 words of storage with an additional 300 words for bookkeeping in the main routine.

9A. ESTIMATION OF THE RUNNING TIME

There will be about 5 multiplication and 5 additions per element plus 8 multiplications and 4 additions per branch in the subroutines calculating Z or Y. Four multiplications and 5 additions are required per branch to compute V and I. As an estimate then, for N branches and m elements, a time of $(12N + 5m)(\mu + \nu)$ is required at each frequency, where μ is the multiplication time and ν the addition time.

In addition, a time S is required to compute the various special functions such as arc tangent, etc., at each frequency. Therefore, for F frequencies,

$$T \approx F[S + (12N + 5m)(\mu + \nu)]$$

is the total calculation time which should probably be at least doubled to account for "bookkeeping." It should be noted that this formula is fairly accurate for runs on a medium-speed machine using an interpretive system.

2B. MATHEMATICAL DISCUSSION

The following scheme is also applicable only to ladder networks. To simplify the analysis, assume that Fig. 1 has only the shunt and series branches Z_1 to Z_4. The extension to larger networks will be obvious.

One can now write the following equations of which the first is a mesh equation, the second a nodal equation, etc., in alternation. For a current source, the sequence would start with a nodal equation.

$$\begin{aligned} E &= Z_4I_4 + V_3 \\ 0 &= -I_4 + Y_3V_3 + I_2 \\ 0 &= 0 - V_3 + Z_2I_2 + V_1 \\ 0 &= 0 + 0 - I_2 + Y_1V_1 \end{aligned} \tag{3}$$

In matrix form this becomes

$$\begin{bmatrix} E \\ 0 \\ 0 \\ 0 \end{bmatrix} = \begin{bmatrix} Z_4 & +1 & 0 & 0 \\ -1 & Y_3 & +1 & 0 \\ 0 & -1 & Z_2 & +1 \\ 0 & 0 & -1 & Y_1 \end{bmatrix} \begin{bmatrix} I_4 \\ V_3 \\ I_2 \\ V_1 \end{bmatrix} \tag{4}$$

Since network functions may be expressed as ratios of determinants, assume that Cramer's rule is to be used for the solution of this system. Consider evaluation of the determinant by evaluating a sequence of determinants d_i, starting at the lower right:*

$$\begin{aligned} d_1 &= Y_1 \\ d_2 &= Z_2Y_1 + 1 = Z_2d_1 + 1 \\ d_3 &= Y_3d_2 + Y_1 = Y_3d_2 + d_1 \\ d_4 &= Z_4d_3 + d_2 \end{aligned}$$

If $d_0 = 1$, $d_{-1} = 0$ by definition, the iteration scheme is

$$d_n = (Z_n \text{ or } Y_n)d_{n-1} + d_{n-2} \tag{5}$$

The quantity in parentheses is Z or Y depending on whether a series or shunt branch is involved at a given stage.

Input admittances (e.g., I_4/E in Fig. 1) are seen to be d_{n-1}/d_n by Cramer's rule (e.g., d_3/d_4 for I_4/E of Fig. 1). Since d_{n-1} must be saved for (5), no extra computation is required. For a current source d_{n-1}/d_n is an input impedance. A transfer function, such as V_1/E, can be shown to be $1/d_n$ and I_1/E is thus Y_1/d_n, etc.

Equation (5) must not be used blindly, however, or the resulting rational function will contain polynomials of higher degree than necessary because of common factors in

* Suggested by a method of Lanczos [8] for a similar determinant.

numerator and denominator. This can be shown as follows:

Assume each Z_n or Y_n on the main diagonal of (4) is equal to N_n/D_n, where N and D are polynomials in the complex frequency p. Then:

$$d_1 = \frac{N_1}{D_1}$$

$$d_2 = \left(\frac{N_2}{D_2}\right)\left(\frac{N_1}{D_1}\right) + 1 = \frac{N_2N_1 + D_2D_1}{D_2D_1}$$

$$d_3 = \frac{N_3}{D_3}\left(\frac{N_2N_1 + D_2D_1}{D_2D_1}\right) + \frac{N_1}{D_1}$$

A mechanical cross multiplication of numerator and denominator at this point will give the common factor D_1. This can be avoided by noting the pattern of formation of these fractions and cross multiplying only by the proper factors. This requires saving certain information at each step to apply to succeeding steps. It can be seen that $d_n = A_n/B_n$, where $B_n = D_nD_{n-1} \cdots D_1$.

Once again, the method is not restricted to passive circuits if the actual network to be computed has an equivalent circuit of the ladder form.

3B. SUMMARY OF THE CALCULATION PROCEDURE

(a) The input quantities are essentially the same as in Section 3A, namely, a list of configuration numbers identifying the form of each Z or Y and following each number a set of element values. No frequencies need be given, however.

(b) This program assumes a network as in Fig. 1, except that the leftmost branch of the ladder will be a shunt branch if a current source is assumed. The rightmost branch is always assumed to be a shunt branch.

(1) Get the next shunt (series) configuration in the list and compute the coefficients of numerator N_n and denominator D_n polynomials of Y_n (Z_n). Save the degree of numerator and denominator of this rational fraction for subsequent multiplication operations which require this information.

Using (5), one computes

$$d_n = \frac{A_n}{B_n} = \frac{N_n}{D_n}\left(\frac{A_{n-1}}{B_{n-1}}\right) + \frac{A_{n-2}}{B_{n-2}}$$

After elimination of common factors this becomes

$$\frac{N_nA_{n-1} + D_nD_{n-1}A_{n-2}}{D_nB_{n-1}}$$

However, for transfer functions B_n/D_1 may be required and for insertion loss functions B_n/D_nD_1 is needed. Rather than attempt this division, it is simpler to compute $B_n' = D_{n-1}D_{n-2} \cdots D_2$ and obtain $B_n = D_nB_n'D_1$ at the end of the calculation [in step (3)] if it is needed. Thus, one computes $d_n' = A_n/B_n'$ instead of d_n. It is assumed that the first and all odd-numbered branches on the list are shunt branches and that

$$A_{n-2} = 0, \qquad A_{n-1} = B_{n-1}' = D_{n-1} = 1$$

(2) Replace B_{n-1}' by B_n', A_{n-1} by A_n, A_{n-2} by A_{n-1}. Replace D_{n-1} by D_n. Save degree of all these polynomials. Repeat step (1).

(3) Make final computations as desired; for example, for Fig. 1:

$$\frac{V_1}{E} = \frac{1}{d_n} = \frac{D_nB_n'D_1}{A_n}$$

$$\frac{I_1}{E} = \frac{D_nB_n'N_1}{A_n}$$

$$Z_{\text{in}} = \frac{E}{I_4} = \frac{d_n}{d_{n-1}} = \frac{A_n/D_nB_n'D_1}{A_{n-1}/B_n'D_1} = \frac{A_n}{D_nA_{n-1}}$$

(c) The output quantities which can be made available from these programs are limited only by the patience of the programmer and the demands of the users. Similarly, the input information can be given in other forms which may be more useful; for example, an arbitrary Z or Y can be specified by a list of real and imaginary values at the input frequencies for problem A or a list of numerator and denominator coefficients for problem B.

4B. FLOW CHART: RATIONAL FUNCTIONS

The flow chart appears on page 287.

5B. DESCRIPTION OF THE FLOW CHART

Box 1: The value zero is put into the location of the constant terms of the polynomial A_{n-2}, the value one put into the

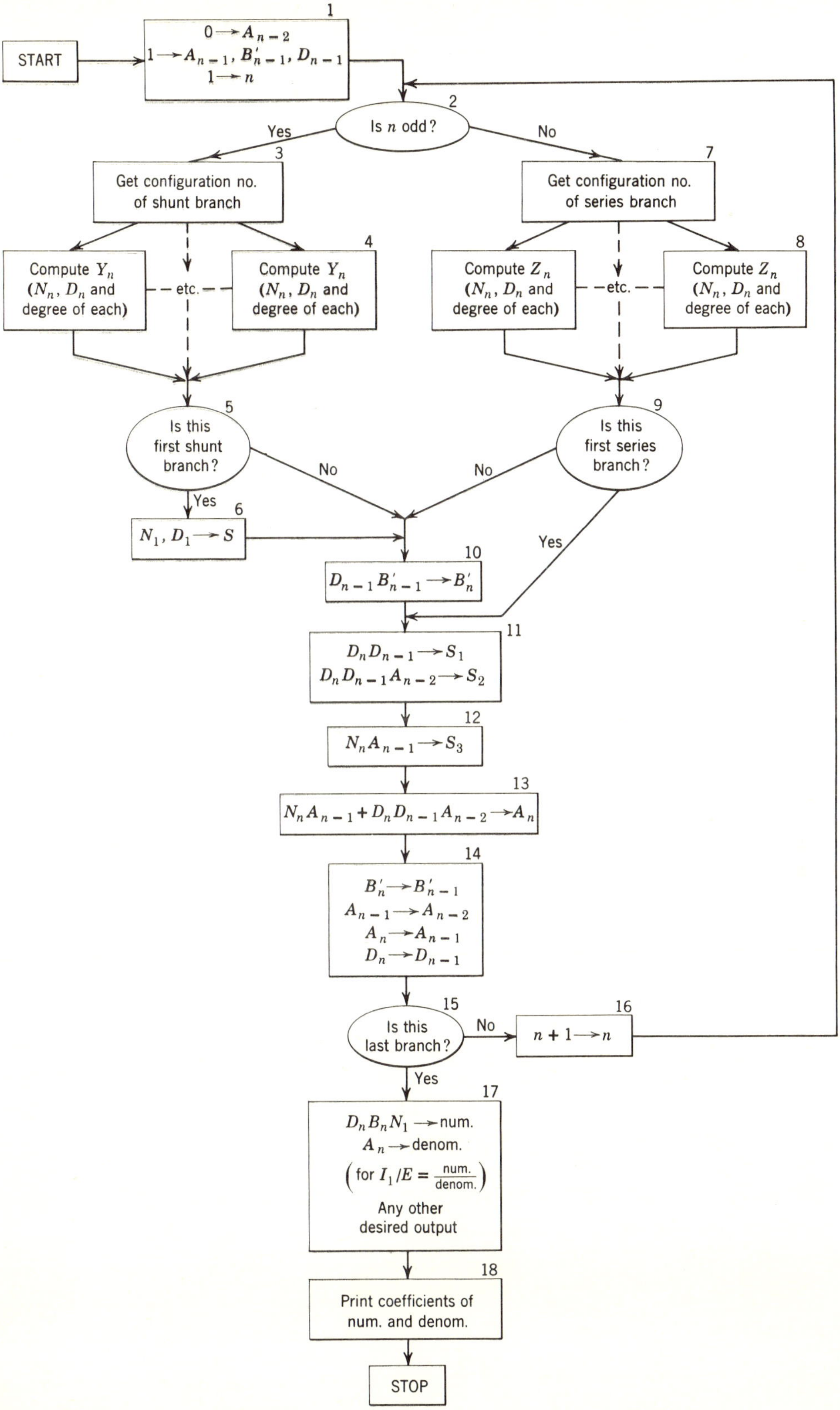
START
1
$0 \rightarrow A_{n-2}$
$1 \rightarrow A_{n-1}, B'_{n-1}, D_{n-1}$
$1 \rightarrow n$
2
Is n odd?
Yes
No
3
Get configuration no. of shunt branch
7
Get configuration no. of series branch
4
Compute Y_n (N_n, D_n and degree of each)
etc.
8
Compute Z_n (N_n, D_n and degree of each)
5
Is this first shunt branch?
9
Is this first series branch?
No
Yes
6
$N_1, D_1 \rightarrow S$
10
$D_{n-1}B'_{n-1} \rightarrow B'_n$
11
$D_nD_{n-1} \rightarrow S_1$
$D_nD_{n-1}A_{n-2} \rightarrow S_2$
12
$N_nA_{n-1} \rightarrow S_3$
13
$N_nA_{n-1} + D_nD_{n-1}A_{n-2} \rightarrow A_n$
14
$B'_n \rightarrow B'_{n-1}$
$A_{n-1} \rightarrow A_{n-2}$
$A_n \rightarrow A_{n-1}$
$D_n \rightarrow D_{n-1}$
15
Is this last branch?
16
$n + 1 \rightarrow n$
17
$D_nB_nN_1 \rightarrow$ num.
$A_n \rightarrow$ denom.
(for $I_1/E = \frac{\text{num.}}{\text{denom.}}$)
Any other desired output
18
Print coefficients of num. and denom.
STOP

location of the constant terms of the polynomials A_{n-1}, B'_{n-1}, D_{n-1}. They are all assigned degree zero. The counter n is set to one.

Box 2: The counter is tested for oddness. This may not be an actual test, but may simply be a transfer order altered every other time control comes here. If n is odd, control goes to box 3; otherwise to box 7.

Box 3: The nth configuration number is selected from the input list. The first of these is assumed to specify a shunt branch.

Box 4: The configuration number specifies a particular 2-terminal network. It may be considered a transfer to subroutine which computes a numerator and denominator polynomial for the admittance Y_n from the element values which have been listed. Since $Y_n = N_n/D_n$, we compute and store in fixed locations the coefficients and degrees of the polynomials N_n and D_n.

Box 5: If this is the first shunt branch, control goes to box 6; for all subsequent branches, control goes to box 10.

Box 6: N_1 and D_1 are stored in special locations for later use and control goes to box 10.

Box 7: This is the alternative path if n is even. The nth configuration number is selected from the input list.

Box 8: As in box 4, these may be considered subroutines which compute the coefficients and degrees of the polynomials N_n and D_n, where $Z_n = N_n/D_n$.

Box 9: If this is the first series branch, control goes to box 11; if not, control proceeds to box 10. This is to avoid computing $D_{n-1}B'_{n-1}$ for the first series branch since this would give $D_1 = B'_2$, which contradicts the definition of B'_n.

Box 10: The polynomial $B'_n = D_{n-1}B'_{n-1}$ is computed. After the first shunt branch has been selected, $B'_1 = 1$. B'_2 must also equal 1, which is the reason for the test in box 9.

Box 11: The polynomial D_nD_{n-1} is computed and stored in temporary locations S_1. This result is multiplied by A_{n-2} and stored in temporary locations S_2.

Box 12: The polynomial N_nA_{n-1} is computed and stored in S_3.

Box 13: $N_nA_{n-1} + D_nD_{n-1}A_{n-2}$ is computed and stored in the A_n locations.

Box 14: The various polynomials which have been computed are moved to new locations as indicated: B'_n to B'_{n-1} locations, A_{n-1} to A_{n-2} locations; A_n to the now available A_{n-1} locations; D_n to the D_{n-1} locations.

Box 15: This test for the last branch may be a special number on the input list (possibly specifying what network functions are desired as output). If it is present, control goes to box 17. If not, proceed to box 16.

Box 16: The index n is increased by one to pick up the next branch configuration number and control goes to box 2.

Box 17: The last computations are performed as desired. In the chart it is assumed that I_1/E is specified. Therefore, $D_nB'_nN_1$ is computed and put into the numerator printout locations. A_n is put into the denominator printout locations.

Box 18: The answers are printed and the program stopped.

6B. SUBROUTINES

(a) SR1—Multiplication of two polynomials of arbitrary degree. The input coefficients must have the degree associated with them and the subroutine must provide this information as output. Used in boxes 10, 11, 12, 17.

(b) SR2—Addition of two polynomials of arbitrary degree. Same input and output information as in SR1. Used in box 13 and possibly box 17 for certain functions.

(c) SR3—The computations of Z and Y may be considered as subroutines. Details given in box 4 and box 8 explanations.

7B. SAMPLE PROBLEM

Input list. See Fig. 4.

n

1. Single resistance 1.0 ohm

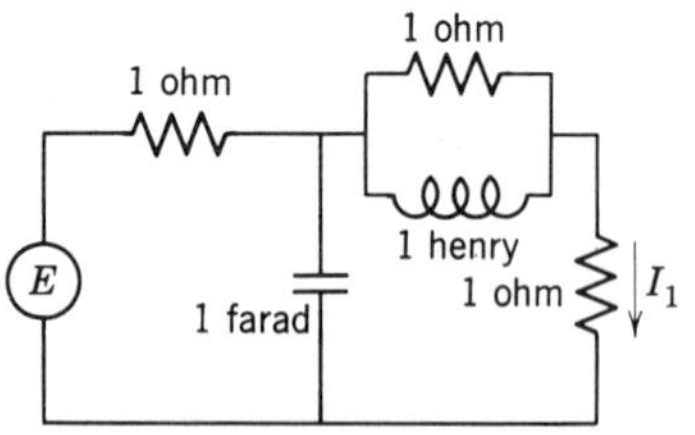

Fig. 4

2. Inductance and resistance in parallel 1.0 henry, 1.0 ohm
3. Single capacitance 1.0 farad
4. Single resistance 1.0 ohm

   Find I_1/E

   End

$A_{-1} = 0$

$A_0 = B'_0 = D_0 = 1$

$n = 1$ $Y_1 = 1/1 = 1, N_1 = D_1 = 1$

$B'_1 = 1 \times 1 = 1$

$D_1 D_0 = 1$

$D_1 D_0 A_{-1} = 0$

$N_1 A_0 = 1$

$A_1 = 1$

$n = 2$ $Z_2 = \dfrac{p}{p+1}, N_2 = p + 0, D_2 = p + 1$

$D_2 D_1 = p + 1$

$D_2 D_1 A_0 = p + 1$

$N_2 A_1 = p + 0$

$A_2 = 2p + 1$

$B'_2 = B'_1 = 1$ (no computation was made so this is the old value)

$n = 3$ $Y_3 = p, N_3 = p + 0, D_3 = 1$

$B'_3 = p + 1$

$D_3 D_2 = p + 1$

$D_3 D_2 A_1 = p + 1$

$N_3 A_2 = 2p^2 + p + 0$

$A_3 = 2p^2 + 2p + 1$

$n = 4$ $Z_4 = 1, N_4 = 1, D_4 = 1$

$B'_4 = p + 1$

$D_4 D_3 = 1$

$D_4 D_3 A_2 = 2p + 1$

$N_4 A_3 = 2p^2 + 2p + 1$

$A_4 = 2p^2 + 4p + 2$

$$\text{num.} = D_4 B'_4 N_1 = p + 1$$

$$\text{denom.} = A_4 = 2p^2 + 4p + 2$$

$$I_1/E = \frac{p+1}{2p^2 + 4p + 2}$$

8B. MEMORY REQUIREMENTS

The memory requirements for intermediate calculations are as follows, assuming that each of N branches contains m elements (if some elements are not reactive, then this estimate can be reduced):

N_n	$m + 2$ words	(m reactive elements produce an mth-degree polynomial plus one extra word to specify the degree)
D_n	$m + 2$ words	
D_{n-1}	$m + 2$ words	
B'_{n-1}	$mN + 2$ words	
S_1	$mN + 2$ words	
S_2	$mN + 2$ words	
S_3	$mN + 2$ words	
A_n	$mN + 2$ words	
A_{n-1}	$mN + 2$ words	
A_{n-2}	$mN + 2$ words	

No space is actually required for the B'_n computation in box 10 since the multiplication and addition subroutines must hold $mN + 2$ words in answer locations. After the computation of B'_n in box 10, the results can be immediately placed in the B'_{n-1} locations. Similarly, the temporary storage locations S_2 may actually be S_1 in box 11. However, an estimate of $7(mN + 2) + 2(m + 2)$ as given by the above list is probably good if it is assumed to include the space required for answers in the subroutines SR1 and SR2. An additional $2(mN + 2)$ locations will be needed for input data locations in these subroutines, giving an over-all memory requirement of

$$9(mN + 2) + 2(m + 2) = 9mN + 18 + 2m + 4$$
$$= m(9N + 2) + 22$$

The size of the program depends on the number of types of 2-terminal networks allowed as branches and their complexity. Six different network configurations averaging 15 words apiece requires about 90 words. An additional 150 words for the control instructions and polynomial arithmetic brings the total to about 240 words.

9B. ESTIMATION OF THE RUNNING TIME

This program is comprised essentially of the SR1, SR2, SR3 subroutines, which contain all the arithmetic, plus bookkeeping orders. Assume that most of the time for arithmetic operations is consumed in SR1 and SR2. The polynomials in B'_n and those in S_2 and S_3 are built up essentially in the following fashion. Take an mth-degree polynomial, multiply it by another mth-degree polynomial. Take this result and multiply by another mth-degree polynomial. Do this N times. The time required, where μ and ν are the multiply and add times, is approximately

$$\frac{m^2N(N-1)(\mu+\nu)}{2}$$

which must be done for each of the three polynomials indicated.

The polynomial A_n is built up as follows: Take two mth-degree polynomials and add. Take two $2m$th-degree polynomials and add. Do this N times. The time required is

$$\frac{N(N+1)m\nu}{2}$$

Aside from incidental bookkeeping in SR1 and SR2, the major time-consuming operation is moving the polynomial coefficients from their normal locations to the input locations required by SR1 and SR2.

To simplify programming, assume that the largest possible polynomial, of degree mN, is moved each time. There are 2 such polynomials involved for each addition or multiplication and they must be moved N times, giving $2mN^2$ numbers. In this program there are 4 multiplications or additions involving this size polynomial. Assuming it takes 2 addition times to move a number, we obtain $4(4mN^2\nu)$. In addition, in box 14, we must move 3 such polynomials from one location to another, which gives $3(2mN^2\nu)$ for a total of $22mN^2\nu$. The time required is thus

$$T \approx \frac{m^2N(N-1)(\mu+\nu)+N(N+1)m\nu}{2} + 22mN^2\nu$$

10. REFERENCES

1. W. Mayeda and M. E. Van Valkenburg, Network Analysis and Synthesis by Digital Computer, *WESCON Convention Record*, 1957, pp. 137–149.
2. W. S. Percival, The Solution of Passive Electrical Networks by Means of Mathematical Trees, *J. Inst. Elec. Engrs.* (*London*), vol. 100, part III, May 1953, pp. 143–150.
3. S. Mason, Feedback Theory—Some Properties of Signal Flow Graphs, *Proc. I.R.E.*, vol. 41, no. 9, Sept. 1953, pp. 1144–1156.
4. G. Kron, A Set of Principles to Interconnect the Solutions of Physical Systems, *J. Appl. Phys.*, vol 24, no. 8, Aug. 1953, pp. 965–980.
5. G. Kron, *Tensor Analysis of Networks*, John Wiley & Sons, New York, 1939.
6. K. S. Johnson, *Transmission Circuits for Telephonic Communication*, D. Van Nostrand Co., New York, 1929, pp. 280–282.
7. E. A Guillemin, *Synthesis of Passive Networks*, John Wiley & Sons, New York, 1957, pp 196–210.
8. C. Lanczos, An Iteration Method for the Solution of the Eigenvalue Problem of Linear Differential and Integral Operators, *J. Research Natl. Bur. Standards*, vol. 45, no. 4, Oct. 1950, pp. 255 ff.

INDEX